GRAVITY, DEFORMATION AND THE EARTH'S CRUST

GRAVITY, DEFORMATION AND THE EARTH'S CRUST

Second Edition

In theory, experiments and geological application

HANS RAMBERG

Division of Mineralogy and Petrology
Institute of Geology
University of Uppsala
Sweden

1981

ACADEMIC PRESS
A Subsidiary of Harcourt Brace Jovanovich, Publishers
London New York Toronto Sydney San Francisco

ACADEMIC PRESS INC. (LONDON) LTD.
24/28 Oval Road
London NW1

United States Edition published by
ACADEMIC PRESS INC.
111 Fifth Avenue
New York, New York 10003

First published 1967

Second edition 1981

British Library Cataloguing in Publication Data

Ramberg, Hans
Gravity, deformation and the earth's crust. –
2nd ed.
1. Rock deformation – Simulation methods
2. Gravity – Simulation methods
3. Earth – Internal structure – Simulation methods
I. Title
551.1'3 QE604 80-41317

ISBN 0-12-576860-5

Printed in Great Britain by
Page Bros (Norwich) Ltd

Preface to the Second Edition

The new edition has been extensively modified and enlarged. Much new material has been added concerning theoretical analysis, numerical calculations, experimental models, as well as the application of the theoretical and experimental results on geological-geophysical problems. Thus Chapters 7 and 8 contain essentially new theoretical analysis in support of which a large number of new numerical models are presented. Chapter 7 contains the theoretical basis for the rise of buoyant layers and the spacing of, for example, salt diapirs, basement uplifts and the regular distance between volcanic islands in some island arcs. In Chapter 8 the structural response to lateral compression of selected crustal layered models is treated theoretically and is supported by some experiments. The material in Chapter 5 is thoroughly revised. In Chapter 9 the theoretical treatment of thrust-sheet motion regarded as a consequence of gravitational collapse and spreading is new, as are the experimental simulations of the phenomenon. In Chapter 11 numerous new experimental models are presented and discussed; these partly emphasize significant details of buoyancy-generated structures, such as the interference between adjacent diapirs, and partly simulate new structures caused by buoyancy, such as the rise of rigid basement blocks into a plastic cover or the elevation of blocks of a brittle superstructure by a buoyant plastic substratum. Most of the material in Chapter 13 is also new: models are presented which show the result of the buoyant model material being much less viscous than the model overburden. Such models simulate the rise of magmatic melts through the Earth's solid crust. The experimental results deviate quite drastically from the structures produced when a crystalline though plastic diapir rises into an overburden with comparable competency. The experimental models presented in Chapter 15 are believed to simulate the evolution of orogens. They give a more detailed and comprehensive picture of the role of gravity in orogenesis than in the first edition.

Applications to the Appalachians and the Alps have been added. The experimental study of the evolution of submarine ridges, of the process of sea-floor spreading and of the rifting of continents has been considerably expanded, as documented in Chapter 16, as has the modelling of convective processes in the Earth's mantle, in Chapter 18. Throughout the book experimental and theoretical models have been compared with geological structures.

Not formally trained in mathematics and physics the author wishes to apoligize to readers for perhaps not always using standard forms. A case in point is the use of gravity-respective, spatial-potential as synonyms for potential gravitational energy, respective potential spatial energy.

In addition to the individuals and institutions whose help and support were acknowledged in the first edition, I am glad to have the opportunity here to express my gratitude to the following persons: F. K. Rolf Häll and F. K. Håkan Sjöström for preparing and running most of the centrifuged models, F. K. Harald Berner, who translated the analytical theory treated in Chapters 7 and 8 into a practical computer programme, photographers Assar Lindberg and K.-E. Alnavik for making almost all prints for the halftones in the book, Mr. Hans Filing, who made most line drawings, and Miss Kersti Glöersen, who patiently typed and retyped a sometimes illegible manuscript. My thanks go also to Christopher Adamson at Academic Press, who edited the book and offered many valuable suggestions during the proof reading.

Permission to reproduce illustrations from other books has been granted by Professor Asger Berthelsen (Figs 11.97 and 11.98), Professor Villard S. Griffin, Jr (Fig. 15.24), Thomas Nelson and Sons Limited (Fig. 11.8), Gulf Publishing Company (Figs 11.20, 11.23 and 11.60), and John Wiley and Sons (Figs 15.18 to 15.21). Much of the theoretical and experimental work contained in the book has been supported by grants from the Natural Science Council of Sweden, whose economic support has been necessary in order to obtain most of the experimental results presented.

Uppsala, March, 1981 *Hans Ramberg*

Preface to the First Edition

Few books deal with the role of gravity in tectonic deformation, fewer still deal with this subject from the point of view of dynamic analysis, and there are none, I believe, which treat tectonics in the light of experiments with centrifuged dynamic models. For these reasons, the present work should fill a real need, and I hope it will prove to be of some use to the seasoned earth scientists as well as to those who are struggling to reach that stage. For the latter group the book may perhaps best serve as a complementary text in courses in tectonics and structural geology.

The book could not have been written without the experience gained during the many summers spent in the Scandinavian Caledonides for the Geological Survey of Norway. For this I wish to express my gratitude to the directors, Dr. C. Bugge, his successor Dr. S. Føyn and the present head Dr. H. Bjørlykke. Discussions concerning Caledonian structures with Dr. H. Carstens and Dr. F. C. Wolff at the survey are also greatly appreciated.

The chief part of the book is based upon experimental studies in our tectonic laboratory in Uppsala. I am obliged to a number of persons for their helpful assistance. Fil.kand. O. Stephansson, Mr. G. Haglund, Mr. J. C. Lindström and Mrs. A. Eriksson prepared models and constructed instruments. The manuscript was typed by Mrs. A.-M. Ekström. The drawings were made by Mrs. A. Kaljusaar and Miss K. Andersson and some of the photographic work was done by Mr. G. Andersson.

My thanks are also directed toward the publishers for the excellent reproduction of all the halftones which play such a crucial role in experimental works of this kind.

Uppsala, March, 1967 *Hans Ramberg*

Contents

1 Scale Models and Similarity

INTRODUCTION

One of the chief objectives of endogenic geology is the study of the past and present motion of matter in the Earth. Such motion is of different types, some of a chemical nature, others of a mechanical. In this book we shall deal with mechanical motion—including strain—of rock and magma bodies, especially the control which gravity exercises over such processes. The atomistic and crystal-physical aspects of the phenomenon of rock flowage are outside the scope of the present work, in which rocks will be treated as continua exhibiting macroscopic mechanical properties such as yield point and ultimate strength, effective viscosity (partly due to recrystallization and transient strain recovery), rigidity, etc. These properties are both isotropic and anisotropic, anistropy being the rule rather than the exception with crystalline rocks.

By treating rocks as continua in a mechanical sense, the theories of fluid dynamics and strength of material may be applied in the study of the evolution of the deformation structure of rock complexes.

Unfortunately, the geometric pattern—architecture—of rock complexes often defies theoretical analysis even if the various rocks in the structure happen to exhibit simple rheological properties. Rocks, however, are far from simple in their mechanical behaviour, a fact which renders a rigorous theoretical dynamic treatment hopelessly complicated for any but the most simple structures.

Greatly intensified experimentation with dynamic scale models and increased application of numerical methods are therefore called for, if a better understanding of the evolution of the structure of the Earth is to be achieved.

SCALE MODELS

The significance of scale-model work in tectonic studies lie in the fact that a correctly constructed dynamic scale model passes through an evolution which simulates exactly that of the original (the prototype), though on a more convenient geometric scale (usually smaller) and with a conveniently changed rate (usually faster).

In the study of geologic structures the situation is generally such that we are in a position to inspect and observe by various means a given final structure without having had a chance to study the initial situation or to witness the actual evolution. This is not exactly a favourable basis on which to build a sound dynamic theory of the evolution of the structure in question. The situation is not improved by the fact that even the final picture is only fragmentary: significant evidence has been removed by erosion or is inaccessible owing to depth of burial. The final picture presented is, therefore, apt to be strongly coloured by the particular experience of the field geologist, and not least by his imagination.

Though it may sometimes be possible to make a reasonable guess as to the pattern of earlier evolutionary stages of a given structure based upon general geological knowledge (we are, for example, reasonably confident that sedimentary strata were deposited as more or less horizontal sheets, the younger above the older), it is certainly no surprise in the light of the above situation that the hypotheses offered as "explanations" of geologic structures are legion. Witness the several contrasted, and often mutually contradictory, hypotheses of orogenesis; of evolution of certain types of folds and of some domal structures; and even of such a universally occurring phenomenon as schistosity.

It seems inevitable that model experiments coupled with theoretical analysis of the dynamics of tectonic processes will contribute greatly to a sound, coherent theory of structural geology and tectonics. By running scale models of tectonic events, one may ultimately hope to separate the physically possible from the physically impossible hypotheses, and the former may be studied in detail to illustrate tectonic processes to an extent not otherwise possible.

SIMILARITY

A model is said to be *geometrically similar* to an original structure if it is a reduced (or enlarged) geometric replica of the original. One may speak of corresponding points in model and original, of corresponding lines or curves, of corresponding surfaces and corresponding volumes. The ratio between the distances between any two corresponding points in model and in original is constant for a given model–original pair. This ratio is called the model ratio of length,

$$\frac{l_m}{l_o} = l_r \tag{1.1}$$

(The model ratios employed in the following are chiefly from the classic work of Hubbert, 1937. See also the paper by Koeniqsberger and Morath, 1913).

The model ratios of area and volume, respectively, are consequently

$$\frac{A_m}{A_o} = l_r^2 \tag{1.2}$$

and

$$\frac{V_m}{V_o} = l_r^3 \tag{1.3}$$

A model that imitates movements in an original structures is said to be *kinematically similar* to the original if corresponding particles are found at

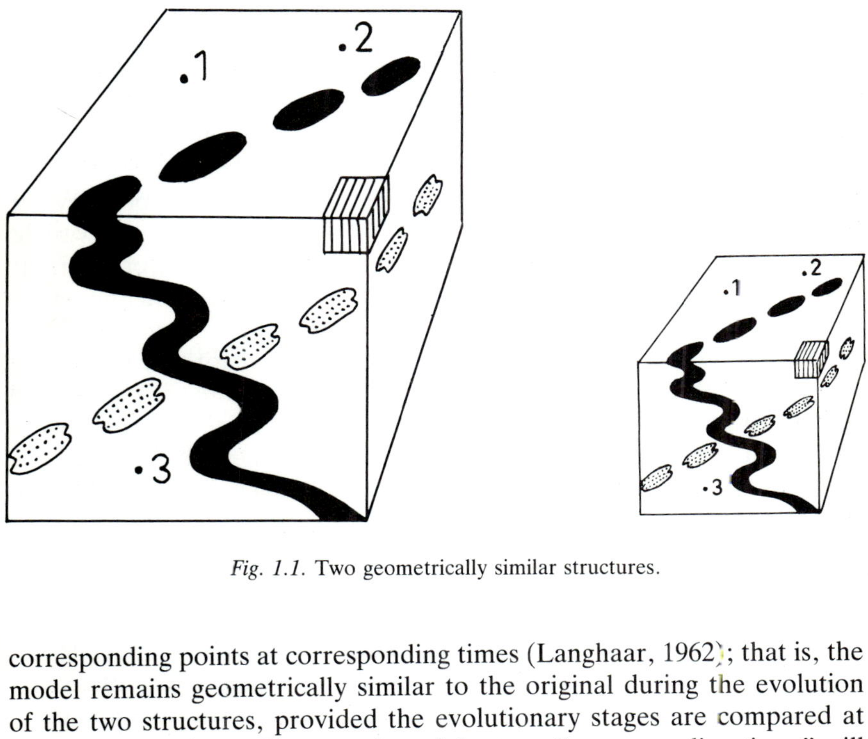

Fig. 1.1. Two geometrically similar structures.

corresponding points at corresponding times (Langhaar, 1962); that is, the model remains geometrically similar to the original during the evolution of the two structures, provided the evolutionary stages are compared at corresponding times. The meaning of the term "corresponding times" will become clear when we introduce the model ratio of time, t_r, which is the ratio between the lengths of time needed for the completion of corresponding movements in model and original. Thus

$$\frac{t_m}{t_o} = t_r \tag{1.4}$$

where t_o is the time needed for a given movement in the original and t_m the time needed for the corresponding motion in the model.

The model ratio of time is constant for motions anywhere throughout the two kinematically similar structures, otherwise the model becomes distorted relative to the original and the two structures cease to be geometrically similar.

Geometrically and kinematically similar structures are *dynamically similar* if the ratio is constant between the various kinds of mechanical forces, compared kind for kind, that act on any two corresponding particles in original and model. Assigning the symbol F_r to the model ratio of force, the above condition is written

$$F_r = \frac{F_{mg}}{F_{og}} = \frac{F_{mi}}{F_{oi}} = \frac{F_{mv}}{F_{ov}} = \frac{F_{me}}{F_{oe}} = \frac{F_{mf}}{F_{of}} \tag{1.5}$$

where F is strength of force on corresponding particles (volumes). Subscript o and m refer to original and model, while g, i, v, e and f refer to gravity, inertia, viscous, elastic and frictional force, respectively.

F_{og}, for example, equals the acceleration due to gravity multiplied by the mass contained in the volume l_o^3 of the original structure, while F_{mg} equals the acceleration to which the model is exposed multiplied by the mass contained in the volume l_m^3. F_{ov} is the force caused by the viscous stress acting on the surface area of a defined region within the original, say the area l_o^2, and F_{mv} is the force caused by the viscous stress acting on the area l_m^2 of the corresponding region in the model.

If one constructs a model which is geometrically similar to the inital stage of a spontaneously evolving natural structure, and moreover makes sure that the mechanical forces in model and original are related as shown in expression (1.5), then the model will develop in a manner exactly similar to that of the original. A study of the evolution of such models is as valuable as the study of the evolution of the original structure. Quite obviously then, model testing is an ideal tool for the study of the slow geotectonic processes.

The above statement is made under the assumption that non-mechanical processes such as chemical reactions, diffusion, heat flow and nuclear processes do not occur. Hence the chemical processes which often are extremely significant during rock deformation (e.g. pressure-solution schistosity and the continuous filling of expanding fissures by growing minerals) are unfortunately not imitated in the scale-model experiments discussed in this book.

From the condition of constant model ratio of force in a dynamically similar model–original pair follows the condition of constant model ratio of mass per corresponding volume throughout the two structures. It is clear that the model ratio of the body force of inertia or of gravity or other body-force field is not constant throughout the two structures unless the ratio of mass is constant. (The ratio of mass refers to mass in corresponding volumes, just as the ratio of force refers to force acting on corresponding volumes of matter in the structures.)

The requirement of constant model ratio of mechanical forces when compared kind for kind in model and in prototype leads to the condition that the ratio between any two dissimilar forces in the model must be equal to the ratio between the same two forces in the prototype. Thus, for example, in models of fluid flow in which both inertia and viscosity are

Table 1.1 Model ratios of some mechanical quantities

Quantity	Dimensional formula	Model ratio
Length	L	$\frac{l_m}{l_o}=l_r$
Mass	M	$\frac{m_m}{m_o}=m_r$
Time	T	$\frac{t_m}{t_o}=t_r$
Area	L^2	$\frac{l_m^2}{l_o^2}=l_r^2$
Volume	L^3	$\frac{l_m^3}{l_o^3}=l_r^3$
Velocity	LT^{-1}	$\frac{l_m t_m^{-1}}{l_o t_o^{-1}}=l_r t_r^{-1}$
Acceleration	LT^{-2}	$\frac{l_m t_m^{-2}}{l_o t_o^{-2}}=l_r t_r^{-2}$
Density	ML^{-3}	$\frac{m_m l_m^{-3}}{m_o l_o^{-3}}=m_r l_r^{-3}$
Force	MLT^{-2}	$\frac{m_m l_m t_m^{-2}}{m_o l_o t_o^{-2}}=m_r l_r t_r^{-2}$
Stress	$ML^{-1}T^{-2}$	$\frac{m_m l_m^{-1} t_m^{-2}}{m_o l_o^{-1} t_o^{-2}}=m_r l_r^{-1} t_r^{-2}$
Strain	L^o	$\frac{\Delta l_m l_m^{-1}}{\Delta l_o l_\sigma^{-1}}=1$
Viscosity	$ML^{-1}T^{-1}$	$\frac{m_m l_m^{-1} t_m^{-1}}{m_o l_o^{-1} t_o^{-1}}=m_r l_r^{-1} t_r^{-1}$

Subscript m refers to model, o to original and r to model ratio.

important, the Reynolds number must be the same in both model and prototype if dynamic similarity is sought. The Reynolds number is defined as the ratio between the inertial and the viscous forces (Table 6.1, p. 41).

Length, time and mass are basic concepts in the conventional theory of mechanics. All other mechanical concepts, such as velocity, acceleration, strength, stress, viscosity, etc., may be defined in terms of length, time and mass. Similarly, the model ratios of all mechanical concepts are determined for a model–original pair by the model ratios of the three basic parameters, length, time and mass. Table 1.1 shows some pertinent relationships. For a comprehensive study of model theory, the works by Hubbert (1937), Porter (1958), Langhaar (1962) and Kline (1965) should be consulted. The symposium on model works in geophysics edited by Long (1953) is also important.

For a study of tectonic problems by means of model experiments, it is of course desirable that the models be dynamically and kinematically as similar to the original as possible, to ensure that the tectonic evolution in question is really imitated by the model. Only then will an analysis of the mechanics of the model throw light on the tectonic event under study.

For several reasons, which we shall discuss subsequently, we are generally obliged to work with models that are but partially similar to the geologic structures. Fortunately, it is often possible to select model materials and conditions such that the lack of strict dynamic similarity does not seriously affect the particular processes one wishes to study.

Before discussing the practical limitations of model investigation of tectonic phenomena, let us consider a theoretical model of tectonic evolution.

2 Theoretical Model of a Rock Complex in Tectonic Evolution

FORCES ACTIVE IN TECTONIC PROCESSES

A dynamic treatment of tectonic evolution is conveniently introduced by considering the mechanical forces that act on the various parts of an arbitrarily limited three-dimensional region of the Earth that contains the structure in question. Such a region will be called a tectonic system—or simply a system—in the following discussion.

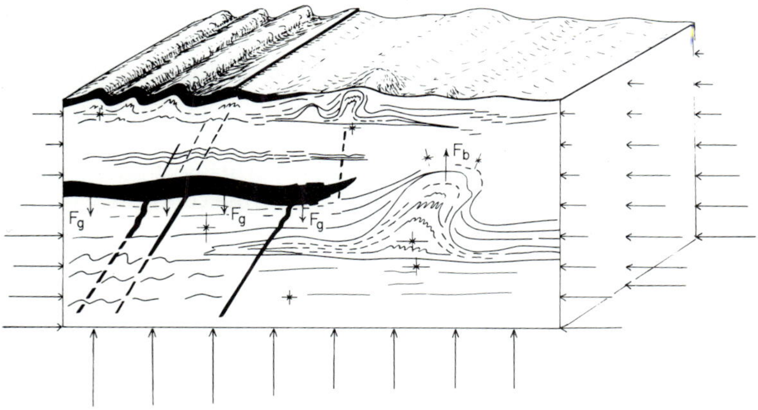

Fig. 2.1. A tectonic system as defined in the text. Arrows at boundary indicate a stress system not corresponding to static stability in the field of gravity. (Possible shear stress at boundary not indicated.) F_g is gravitational force acting on a basic sill; F_b is buoyant force acting on a granite dome. Local stresses are indicated at selected points.

A tectonic system is generally affected by the following types of mechanical forces that control its dynamic evolution (Fig. 2.1).

(1) The stresses that act at the boundary of the system. These stresses are transmitted through the interior of the system in a manner controlled by the geometric pattern of the system, by the mechanical

properties of the constituent rock and magma bodies, and by the pull of gravity or other body forces.

(2) The force of gravity, which acts on every material particle in the system.

(3) The inertial force of masses in accelerating or retarding motion.

(4) The internal stresses due to viscous drag and elastic strain within the system.

(5) The internal stresses not produced by any of the agencies listed above. For example, stress generated by a change in volume of a part of the system due to phase change (e.g. melting) and stress generated around suddenly developing fractures.

Complete information on these forces implies in fact, through the theories of fluid dynamics, elasticity and rheology, information on the deformation of the region as a whole and of its individual bodies as well as motions of the latter in relation to one another. (The motion of the system as a unit in relation to the adjacent parts of the Earth, however, is partly controlled by the dynamic conditions in the surroundings.)

Some of the forces listed are less important than others in tectonic systems. Because of the exceedingly slow movements in most tectonic processes, inertial forces are negligible relative to other mechanical forces. (This is shown by very small Reynolds number.) No essential error is therefore introduced by disregarding inertia altogether. Unless otherwise stated, inertial forces are accordingly disregarded in the analyses throughout this book. An important exception is, of course, the centrifugal force which is used to imitate the force of gravity in our model study.

Though the stresses listed under (5) are influential in geodynamics, the mechanism by which these stresses are produced is generally neglected in this book unless specifically considered in occasional analyses.

The experimental models to be presented comprise only purely mechanical processes thus excluding phase-change reactions, such as melting and crystallization or, e.g., the change from gabbro to eclogite. Moreover, the development of fractures in rocks seems as yet an erratic affair controlled by conditions not easily reproduced. Fractures, therefore, have not been systematically studied in our models, though fracturing is by no means uncommon in the experimental runs.

Expansion or contraction due to phase change means a change in density and a consequent buoyant or subsiding tendency in the field of gravity. In this respect, volume changes are not neglected in the experimental models, whose chief importance lies in their ability to imitate global readjustment of unstable mass distribution in the field of gravity.

In most tectonic deformations the permanent viscous or plastic deformation and movements overshadow any elastic strain. The elastic stresses and strains and not therefore considered.

Of the three remaining classes of force, the body force of gravity and the stresses transmitted through the surroundings to the boundary of the system are to be regarded as the potential driving forces in a mechanically unstable tectonic system. The viscous and plastic stresses are passive, resistive forces operating only when motion occurs within the system.

STABLE–UNSTABLE STATES, MECHANICAL POTENTIAL

In an isolated body, such as the solid earth, which is kept coherent by its own field of gravity, the relative motion of its constituent parts is generated by local imbalance between the stresses that act at the boundary of the moving and/or deforming region and by the body force of gravity that acts on the mass of the region.

At static mechanical equilibrium the static stress at the boundary exactly balances the gravitational pull, and there is no tendency to movement parallel to the force lines of the gravitational field. At imbalance between the static boundary stress and the body force movements occurs, and the sum of the static and dynamic boundary stresses matches the body force of gravity. True mechanical stability also requires that the stress field is hydrostatic or lithostatic. That is, deviatoric stress is not consistent with the conditions of complete mechanical stability because such stress causes movement and changing strain even in materials that ordinarily exhibit mechanical strength. Given enough time, micro-creep and similar phenomena make crystalline and solid materials relax their elastic shear stress even at stresses below the ordinary yield point.

It is a trivial consequence of the conditions for mechanical equilibrium that a stable globe containing several immiscible chemical phases must exhibit a concentric configuration of homogeneous single-phase shells whose density increases towards the centre. Only if several immiscible phases happen to show equal density and equal compressibility should stable multiphase shells occur.

Within each shell at equilibrium the lithostatic pressure increases downward with a gradient equal to the specific gravity of the shell: $\delta P/\delta h = -\rho g$, where h is taken as positive upward, ρ is density and g acceleration due to gravity. Fortunately for the human race and other air-breathing creatures, this mechanically impotent situation has not yet materialized on our globe, and probably never will. The ocean therefore can never cover the entire globe. Instead, the pull of gravity acting on rocks and magma masses is generally not statically balanced by the stress transmitted to their boundaries. It is this lack of static balance which is the immediate driving force behind tectonic processes. The difference between the static boundary stresses and the pull of gravity is instead balanced by the resistive forces of viscous flow and other dynamic strain.

There is a certain advantage in invoking potential mechanical energy rather than utilizing only the concepts of force and stress when analysing the dynamics of tectonic phenomena. One recognizes the concept of potential gravitational energy–or simply *gravity potential*—of a mass of matter, and one recognizes its potential spatial energy—or *spatial potential*—of a volume of matter in a pressure field.

The gravity potential of a mass m in the field of gravity is plainly

$$\phi_{gr} = mgh \tag{2.1}$$

where h is the vertical distance from a datum level in the field and g the acceleration due to gravity, assumed constant through the distance h.

The concept of spatial potential is perhaps less obvious. By virtue of the space which a body of matter occupies in a pressure field, it possesses a potential ability to do work by moving to regions of lower pressure, or to regions of greater tensile stress.

An incompressible body of volume V existing in a lithostatic pressure field of intensity P may be regarded as "possessing" a spatial potential of magnitude

$$\phi_{sp} = PV \tag{2.2}$$

where P is taken as positive for confining pressure and negative for confining tension.

PV corresponds to the drop in potential energy associated with a movement of the body from a region of pressure P to one of vanishing pressure (provided that other forces, e.g. gravity, do not act at the same time). The datum level in the pressure field here assumed is one of vanishing pressure, though any other level may be defined to suit the special case in question.

The difference in spatial potential of an incompressible body with volume V existing at regions with pressure P and P_0, respectively, is then

$$\phi_{sp} = (P - P_0)V \tag{2.3}$$

where pressure need not assume positive values only. For compressible solids or liquids the formula for the variation of the potential with pressure is a little more complicated, but for the present argument the compressibility of rocks and magmas may be disregarded.

The sum of the gravity potential and the spatial potential of a given mass of matter is conveniently termed its *mechanical potential.* When reckoned per unit mass, say 1 g, we speak of the *specific* mechanical potential. The *bulk* mechanical potential refers to the total body we happen to consider. Tectonic movements and deformations, being of a mechanical nature and generally too slow for inertial forces to count, are always so directed that the mechanical potential decreases, the mechanical energy being dissipated to heat in the process. Non-mechanical processes, e.g. chemical, thermal or nuclear processes, do not generally lead to decreasing mechanical potential. In fact, non-mechanical processes often increase locally the mechanical potential, thus starting and maintaining geodynamic and tectonic processes. (The growth of a garnet porphyroblast is a case in point. Here the low chemical potential of the garnet nucleus causes elements to diffuse toward the nucleus which therefore builds a high pressure—and hence a high mechanical potential—in its immediate surroundings. An osmotic system is another example on how a mechanical-potential difference can be created by a chemically instable situation and associated diffusion from low to high pressure.)

Tectonic movements are thus propelled by the net drop in the mechanical potential of a system comprising all the bodies that participate in the process. Locally the potential of some bodies may increase while the potential of other bodies decreases. The significant condition is that the sum of the potential changes is negative, i.e. that the net variation is a decrement of mechanical potential.

Mathematically, the bulk mechanical potential of incompressible media can be expressed in terms of given volume or given mass, respectively:

$$\phi_{\text{mech}} = V\rho gh + PV \tag{2.4}$$

$$\phi_{\text{mech}} = mgh + P\frac{m}{\rho} \tag{2.5}$$

where the mass m of the body is its volume times its density, $m = V\rho$. Pressure and gravity are assumed constant throughout the body to avoid using differential forms.

For our purpose rocks and magmas may be treated as incompressible media without causing significant errors. This is assumed throughout this book unless otherwise stated.

Let us briefly consider vertical motion within a uniform medium in which lithostatic equilibrium exists, e.g. within a uniform concentric shell in the earth. (Since hydrostatic—or lithostatic—equilibrium exists, the vertical movements are only imaginary or must be thought of as having been brought about by an auxiliary force not included in the definition of the mechanical potential.)

Lithostatic equilibrium requires that the vertical pressure gradient be

$$\frac{\partial P}{\partial h} = -\rho g \tag{2.6}$$

Combining this with equations (2.4) and (2.5) shows that the change in gravity potential experienced by a portion of the medium which is moved up or down is exactly equal in magnitude, but opposite in sign, to the change in the spatial potential of the same portion. Consequently, the net mechanical potential remains constant during the process. In short, there is no driving force, as must be expected from a system in static equilibrium.

On the other hand, consider a body of density ρ and volume V, immersed in a uniform medium of density ρ' and which itself is in a state of lithostatic equilibrium.

A vertical pressure gradient of magnitude

$$\frac{\partial P}{\partial h} = -\rho' g \tag{2.7}$$

exists in the enclosing medium. While moving vertically through the enclosing medium the immersed body is therefore subject to a change in spatial potential given by

$$\mathrm{d}\phi_{sp} = V\,\mathrm{d}P = -V\rho' g\,\mathrm{d}h. \tag{2.8}$$

But the change is in gravitational potential is

$$\mathrm{d}\phi_{gr} = mg\,\mathrm{d}h = V\rho g\,\mathrm{d}h. \tag{2.9}$$

As $\rho' \gtrless \rho$, the variation in spatial potential of the vertically moving immersed body is not balanced by an opposite variation in its gravity potential. In fortunate agreement with Archimedes, the net mechanical potential of the body drops when the body ascends if $\rho < \rho'$, and the potential drops when the body descends if $\rho > \rho'$.

It is not sufficient, however, to limit the consideration to the movement of the immersed body, because simultaneously with its rise or descent an equal volume of the enclosing medium moves in the opposite direction to leave space for the immersed body and to fill the space left by it. A possible change in mechanical potential associated with this motion must also be included in the total energy balance sheet.

Under vertical movement of a small portion, of volume V, of the enclosing medium the change in gravity potential is

$$\mathrm{d}\phi_{gr} = \rho' gV\,\mathrm{d}h \tag{2.10}$$

and the change in spatial potential is

$$d\phi_{sp} = -\rho' gV\,dh. \tag{2.11}$$

Since, thus, the variations of the two potentials cancel one another, the net mechanical potential of the enclosing medium remains constant in this example, a condition which does not always prevail, however.

POTENTIAL ENERGY OF HORIZONTAL LAYERS IN THE FIELD OF GRAVITY

As noted above, it is a consequence of the force of gravity that the potential mechanical energy of a polyphase assemblage is at a minimum when arranged in concentric shells of increasing density downward. This is obvious enough with respect to the gravity potential, which is the product of mass and height from a datum level. It is perhaps less obvious that the stable state characterized by layers arranged according to their density also represents minimized spatial potential by virtue of the fact that such an arrangement generates less average lithostatic pressure throughout the system. Since this is of significance in an analysis of gravity tectonics, we shall illustrate the principle by a schematic example.

Assume two incompressible materials A and B (say sial and sima), of density ρ_A and ρ_B, respectively; $\rho_A < \rho_B$. Let the two substances constitute two horizontal (concentric) equally thick layers on the Earth, the one above the other, as in Fig. 2.2. As indicated in the figure, the pressure in the interior of the layers is higher if B rests on A than if A is on top. Since the spatial potential is the produce of pressure and volume it follows that

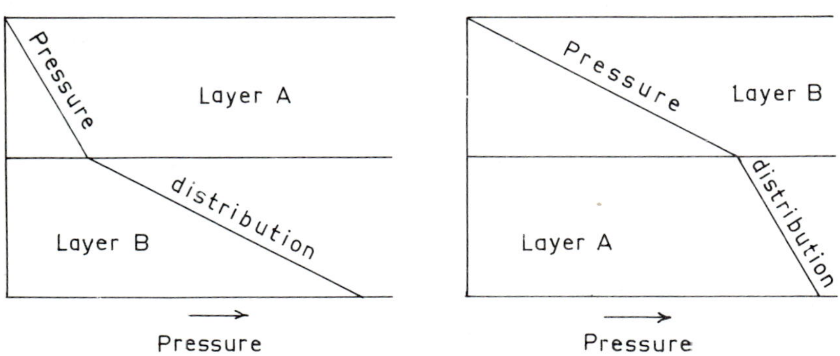

Fig. 2.2. Two layers of contrasted densities ($\rho_B > \rho_A$) in different arrangements in the field of gravity. See text.

an arrangement with the more dense substance resting above the less dense one represents increased bulk spatial potential. We shall demonstrate this mathematically.

Assume first that A rests on B, referred to as arrangement 1. The spatial potential of layer A, as calculated per given horizontal area, is then

$$\phi'_{A(sp)} = ah\rho_A g\tfrac{1}{2}h = \tfrac{1}{2}a\rho_A gh^2 \tag{2.12}$$

where a is the horizontal area considered and ah, consequently, the volume for which the potential is calculated. $\rho_A g\frac{1}{2}h$ is the average pressure in the top layer. The density is considered constant.

Similarly, the spatial potential of the bottom layer B in arrangement 1 is

$$\phi'_{B(sp)} = ah(\rho_A gh + \rho_B g\tfrac{1}{2}h) = ah^2 g(\rho_A + \tfrac{1}{2}\rho_B) \tag{2.13}$$

where, again, ah is the volume and $(\rho_A + \frac{1}{2}\rho_B)\, gh$ the average pressure in the bottom layer. The total spatial potential of the complex, as calculated per horizontal area a, is consequently

$$\phi'_{sp} = ah^2 g(\tfrac{3}{2}\rho_A + \tfrac{1}{2}\rho_B) = Vhg(\tfrac{3}{2}\rho_A + \tfrac{1}{2}\rho_B) \tag{2.14}$$

in which the volume $V = ah$ is inserted.

In arrangement 2, with B on the top and A at the bottom, the total spatial potential per area a can be shown to be

$$\phi''_{sp} = ah^2 g(\tfrac{1}{2}\rho_A + \tfrac{3}{2}\rho_B) = Vgh(\tfrac{1}{2}\rho_A + \tfrac{3}{2}\rho_B) \tag{2.15}$$

The spatial potential drop following a change from the unstable arrangement 2 to the stable arrangement 1 is

$$\Delta\phi_{sp} = \phi''_{sp} - \phi'_{sp} = ah^2 g(\rho_B - \rho_A) = Vgh(\rho_B - \rho_A) \tag{2.16}$$

This should be compared with the difference in gravity potential between the two arrangements.

In arrangement 1 the gravity potential of the top layer, as reckoned per horizontal area a, is

$$\phi'_{A(gr)} = ah\rho_A g\tfrac{3}{2}h = \tfrac{3}{2}\, ah^2 \rho_A g = \tfrac{3}{2}\, Vh\, \rho_A g \tag{2.17}$$

For layer B the potential is

$$\phi'_{B(gr)} = ah\rho_B g\tfrac{1}{2}h = \tfrac{1}{2}ah^2\rho_B g = \tfrac{1}{2}Vh\rho_B g \tag{2.18}$$

Similarily, arrangement 2 yields the following potentials for the two layers:

$$\phi''_{A(gr)} = \tfrac{1}{2} V\rho_A hg \tag{2.19}$$

$$\phi''_{B(gr)} = \tfrac{3}{2} V\rho_B hg, \tag{2.20}$$

the difference between arrangement 2 and arrangement 1 being

$$\Delta\phi_{gr} = \phi''_{gr} - \phi'_{gr} = V(\rho_B - \rho_A)hg \tag{2.21}$$

In other words, the rearrangement from the unstable state to the stable state in the field of gravity means a drop in both the gravity potential and the spatial potential. In the above example both potentials decrease by the same amount. For an analysis of the dynamics of the rearrangement process see Chapter 7.

It is worth noting that the spatial-potential change in this example is not related to a displacement of the boundary of the total system in a spatially variable stress- or pressure-field, as is usually the situation (see below), but that the potential change is due to an *in situ* change of the internal pressure as produced by the internal rearrangement in the system. The pressure at the top and bottom faces is the same before and after the rearrangement, but the pressure within the layers has changed.

GRAVITY AND SPATIAL POTENTIALS OF A DEFORMABLE BODY IN A COMBINED GRAVITY AND PRESSURE FIELD

For the analysis of tectonic phenomena it is useful to consider the behaviour of a fairly large deformable body with arbitrary shape exposed simultaneously to the pull of gravity and to the force of a pressure gradient created by the surrounding materials. The body may be any plastico-viscous rock or magma mass with negligible yield point. We assume that the motion is so slow that not only is inertia negligible, but so is also the magnitude of the deviatoric stress relative to the confining pressure.

Inasmuch as the pressure varies along the boundary a change of the shape of the body and its movement through the surroundings generally means that the bulk spatial potentical changes when the body moves. This change of the bulk spatial potential of the body associated with the boundary displacement should not be confused with the fact that even if the body is immobile and in a state of static equilibrium, its specific spatial potential varies continuously in the vertical direction. This is simply because the lithostatic pressure varies in the vertical direction.

(a)

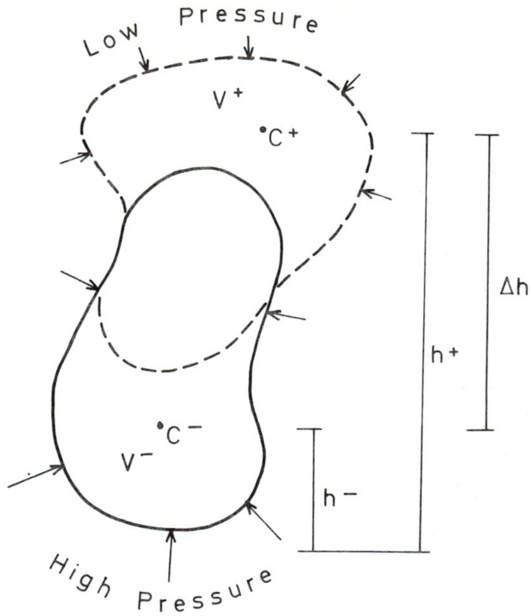
Low Pressure
V⁺
C⁺
Δh
h⁺
C⁻
V⁻
h⁻
High Pressure

(b)

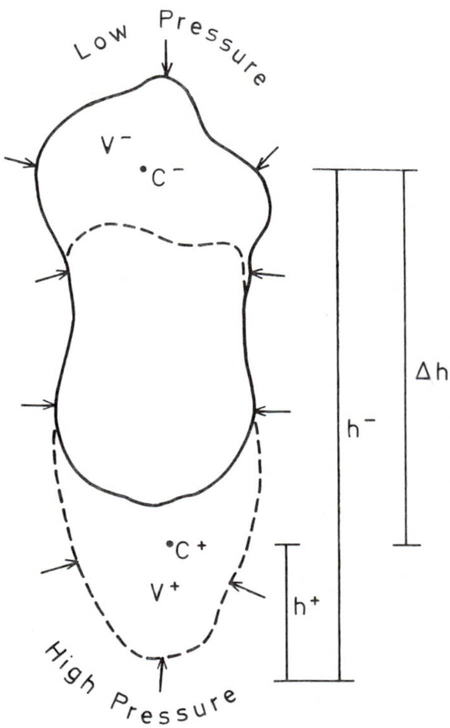
Low Pressure
V⁻
C⁻
Δh
h⁻
C⁺
V⁺
h⁺
High Pressure

(c)

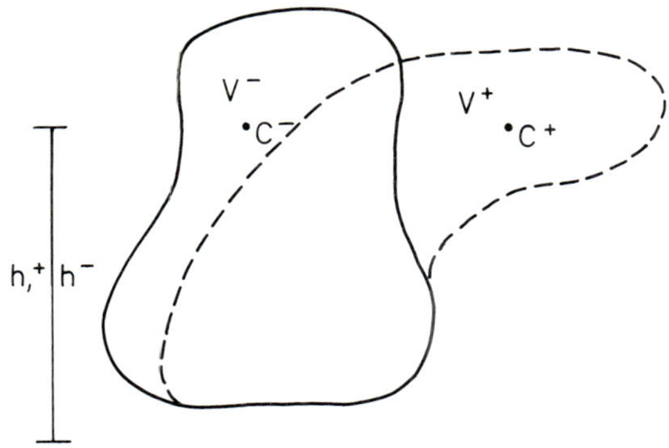

Fig. 2.3 Bodies whose shape and position in the gravitational field have changed as indicated by solid-line boundary (initial state) and stippled-line boundary (final state). (*a*) indicates increased gravity potential but decreased spatial potential, (*b*) shows a case of decreased gravity potential but increased spatial potential. In (*c*) the fact that the gravity potential has not varied during the change of shape necessarily means that the spatial potential has decreased, provided the deformation is not frictionless. C^+ and C^- indicate the centres of gravity of the gained volume V^+ and the lost volume V^- respectively.

We shall consider the change of shape of the body and its movement through the surroundings in terms of the displacement of its boundary. Some segments of the boundary are displaced toward the interior of the body; this we define as negative displacement of the boundary. Other boundary segments are displaced away from the interior of the body; this we define as positive displacement of the boundary. The condition of incompressibility requires that the volume swept through by the inward-moving boundary segments exactly equals the volume swept through by the outward-moving boundary segments, see Fig. 2.3. The former volume is defined as negative and indicated by the notation V^-. It may be regarded as the volume "lost" by the body as it moves. The latter volume, which is indicated by the notation V^+, may be regarded as the volume "gained" by the moving body and hence defined as positive. By virtue of incompressibility the net volume change is zero: $V^- + V^+ = 0$.

The "gained" volume V^+ generally covers a region in which the pressure is different from that within the region covered by the "lost" volume V^-. Accordingly, there will in general be a difference between the spatial potential which is "lost" by the body, say V^-P^-, and the spatial potential which is "gained" by the body, say V^+P^+. Here P^- and P^+ are the average pressures within the two regions.

The movement and change of shape of the body are defined by the geometric character of the regions V^+ and V^- and are accordingly associated with a change in bulk spatial potential equal to

$$\Delta\phi_{sp} = (P^+ - P^-)V^* \tag{2.22}$$

provided no change of pressure has occurred in the body outside the two regions V^- and V^+. Here $|V^*| = |V^-| = |V^+|$. (The vertical lines indicate numerical values.)

Simultaneously with the change of bulk spatial potential a change of the bulk gravity potential generally takes place since the centres of gravity of the "lost" and "gained" volumes are in general not located at the same level in the field of gravity. Let the centres of gravity of the two volumes be located at heights h^+ and h^+, respectively, from a given datum level. Then the change in bulk gravity potential of the body, associated with the movement and change of shape as defined above, is

$$\Delta\phi_{gr} = \rho g(h^+ - h^-)V^* \tag{2.23}$$

The bulk mechanical-potential change experienced by the body when it moves and changes shape as defined is then

$$\Delta\phi_{mech} = \Delta\phi_{sp} + \Delta\phi_{gr} = [P^+ - P^- + \rho g(h^+ - h^-)]V^* \tag{2.27}$$

If $\Delta\phi_{mech} < 0$ the movement and change of shape in question will proceed spontaneously.

The potential-energy change associated with the change of shape and the motion of the body as calculated in terms of the "lost" and "gained" volume can also be found by integration over the whole body before and after the motion. The change of bulk spatial potential is then given by the triple integral over the volume:

$$\Delta\phi_{sp} = \iiint_V P_2 \mathrm{d}V - \iiint_V P_1 \mathrm{d}V \tag{2.25}$$

where the pressures before and after the deformation, P_1 and P_2 respectively, vary with position within the volume, V, of the body.

Similarly the change in bulk gravity potential is

$$\Delta\phi_{gr} = \rho g \iiint_V h_2\, \mathrm{d}V - \rho g \iiint_V h_1\, \mathrm{d}V \tag{2.26}$$

where h_1 and h_2 respectively, are the varying heights from a reference level to the differential volume dV before and after the deformation.

The bulk mechanical potential change thus calculated is identical to the result obtained when the process is considered in terms of the "lost" and "gained" volumes.

$\Delta\phi_{sp}$ is also the same as the volume of the body multiplied by the difference in *average* pressure within the body before and after the deformation. Likewise $\Delta\phi_{gr}$ can be regarded as the weight of the body multiplied by the difference in height of its centre of gravity before and after the deformation.

In plastico-viscous flow of rocks and magmas it is the drop in mechanical potential which drives the movements and deformations, the potential energy being dissipated to heat by the internal resistance.

A part of the mechanical-potential drop may also be bound as internal energy due to strain-hardening and elastic strain. For simplicity, however, we assume complete internal relaxation within the body after the deformation. That means the atomistic-molecular structure is the same before and after the deformation. Only the external shape and the pattern of internal macrostructure and supramolecular structure (e.g. as visualized by markers such as sedimentary layering and orientation of mineral grains) have changed during the deformation. Under such circumstances the following condition is valid

$$\Delta\phi_{gr} + \Delta\phi_{sp} + Q = 0 \tag{2.27}$$

where Q is the energy dissipated to heat by the resistive forces as put in action by the movements.

If only one of the two mechanical potentials decreases during deformation and motion of the body, then this decreasing potential is regarded as the driving energy of the process. A drop in the spatial potential, for example—which is equivalent to saying that an energy of magnitude $\Delta\phi_{sp}$ is introduced into the body by virtue of the pressure acting on the moving boundary segments—may be regarded as producing a deformation which dissipates an energy Q and at the same time possibly gives rise to an increase in the gravity potential of the body: $-\Delta\phi_{sp} = \Delta\phi_{gr} + Q$. By the same token, a drop in the gravity potential alone is regarded as the power which propels the dissipative deformation and simultaneously may cause a rise in the spatial potential. A few schematic examples of movement and deformation of uniform bodies driven by gravity- and spatial-potentials, respectively, are shown in Fig. 2.3.

3 Comments on the Definition of Gravity Tectonics

The terms "gravity tectonics" and "gravity-collapse tectonics" usually refer to down-slope gliding of rock bodies, e.g. sedimentary complexes—or the collapse of rock structures under the pull of gravity (see, for example, Gillet-Lamont, 1799; Kühn, 1819; Reyer, 1888; Schardt, 1893; Haarmann, 1930; van Bemmelen, 1933 b, 1960 and 1973; Dennis, 1967; Gary *et al.*, 1972 and DeJong and Scholten, 1973). Thus, gravity tectonics comprises tectonic evolutions propelled by the force of gravity. The evolution of many dome and diapir structures, then, is also a phenomenon of gravity tectonics despite the fact that the conspicuous part of the movement goes againt the pull of gravity.

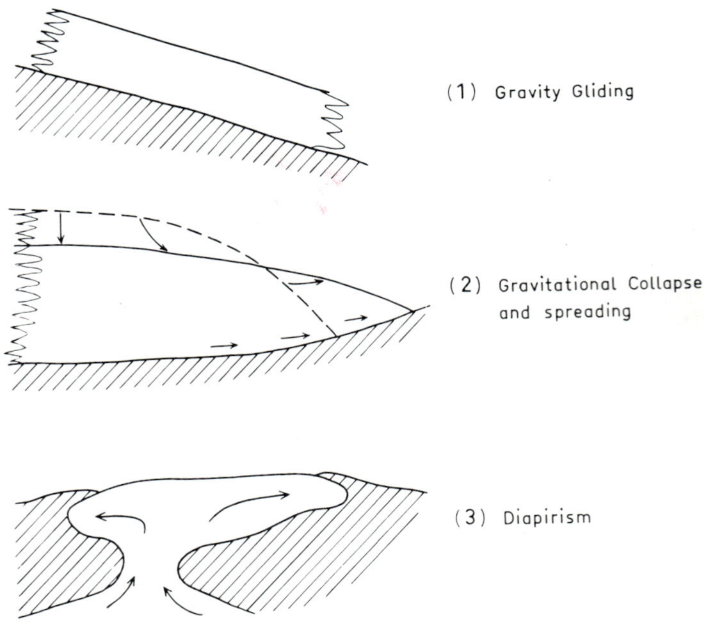

Fig. 3.1. The three chief types of gravity tectonics (from Ramberg, 1980*a*).

In a broad scheme of classification of gravity tectonics one may distinguish between structures formed by (1) gravity gliding, (2) gravity collapse-and-spreading and (3) diapirism. The three classes are indicated in Fig. 3.1 (from Ramberg, 1980).

It would seem natural to apply the basic concepts of gravity potential and spatial potential in an attempt to distinguish formally between "gravity tectonics" and other kinds of tectonic phenomena.

Obviously the ever-present force of gravity exercises some control on nearly all types of tectonic process, and plays a major part in moulding the final pattern of many deformation structures. For example, the pull of gravity modifies markedly the wavelength of certain buckle folds generated by stresses not necessarily themselves of gravitational origin, and this is discussed in Chapter 8.

In fact, gravity sets an upper limit to the wavelength of physically possible crustal buckles, as discussed by Smoluchowski (1909, 1910), Biot (1959), Ramberg (1963a, 1964b, 1970) and Ramberg and Stephansson (1964). Moreover, lithostatic pressure in the Earth is plainly a direct result of the gravitational field. Pressure in its turn controls, in part, the rheological character of rocks and magmas, and the structural pattern is affected by these properties of the building materials of the Earth.

Besides being a mere modifying factor in structure production gravity is also the generating power of many deformation structures.

In using the mechanical potentials as means of defining gravity tectonics it seems reasonable to suggest that tectonic features generated by a drop in the gravity potential be classified as gravity tectonics. It is not sufficient to show that the gravity potential of the tectonic system in question decreases during the movements and deformation: it must also be demonstrated that the potential drop is of a magnitude corresponding to the energy dissipated during the viscous and plastic processes. Alternatively, it may perhaps be possible to show that no energy is available for the process other than the gravity potential, in which case the causal relationship between gravity and tectonic processes follows.

Tectonic phenomena produced solely by decreasing spatial potential of the system in question—chiefly in consequence to the displacement of the boundary in the stress field—should then belong to a class distinct from gravity tectonics.

Just as structures produced by a drop in the gravity potential may represent an increment of the spatial potential, so structures produced by a drop in the spatial potential may represent increased gravity potential.

A third class of tectonic phenomena would comprise structures produced by both the pull of gravity and the stress at the boundary as transmitted through the surroundings. In this class gravity potential and the special potential decrease simultaneously during the structural evolution. A case of this nature is discussed on p. 13.

To this seemingly reasonable scheme of dynamic classification there is the valid objection that, when all comes about, virtually all tectonic features and deformation structures are produced by adjustment of masses in the field of gravity, resulting in a reduced gravity potential. This may be shown if the system one chooses to consider is made so large that it encompasses all mechanical processes and all exchanges of mechanical energy which are coupled with the tectonic phenomenon in question.

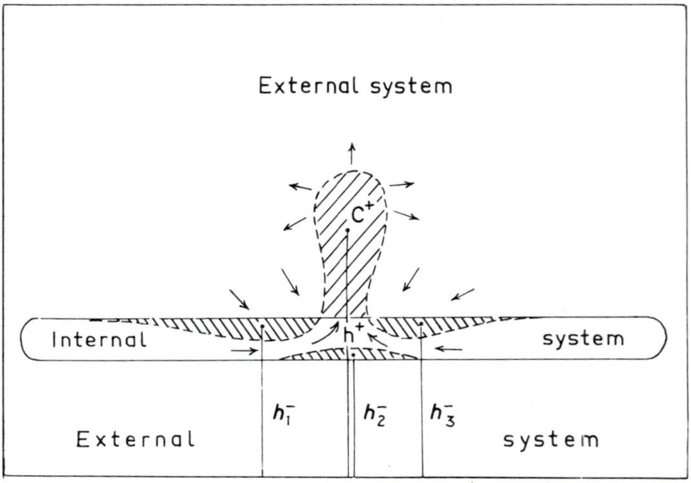

Fig. 3.2. Spontaneous movement of buoyant sheet (internal system) embedded in more dense material (external system). For spatial-potential change and gravity-potential change during rise, (see pp. 15 ff.) //// gross volume gained, V_i^+, by the internal system (= gross volume lost, V_e^-, by the external system), \\\\ gross volume lost, V_i^-, by the internal system (= gross gained, V_e^+, by the external system); h_i indicates position of the centres of gravity above the basal reference level.

To demonstrate the principles involved it will suffice to treat a simple model which may be obtained by expanding the system of deforming uniform bodies discussed above (p. 16). We shall in particular consider the body whose movement and change of shape represents increased gravity potential but decreased spatial potential. That is, the body whose deep part shrinks under the high pressure and whose upper part expands into regions where pressure is low. An ascending diapir is a natural example (Fig. 3.2).

Treating the diapir as an isolated system, it seems at first glance that the growth of the structure does not belong to gravity tectonics as defined above, because the gravity potential of the body actually increases while its spatial potential decreases. Since the boundary of the body (diapir and source layer) is differentially displaced in the pressure field, the deforming body is open with reference to energy, and its internal processes are necessarily coupled with processes simultaneously taking place outside the body. This means that the processes taking place within the body cannot be understood unless seen in relation to the surroundings.

In the following discussion we shall refer to the original system—in this case the deforming body—as the *internal* system, and to the region around it as the *external* system, whose outer limits in turn border on the *environment*. The external system is assumed to be so large that it encompasses all processes connected with the deformation and movement of the internal system. In short, the external system is closed towards the environment but not towards the internal system, the internal and external systems together constituting a self-contained whole.

Under these circumstances the causes of the deformation of the internal system are limited to two: (1) a spontaneous volume expansion of the material of the external system immediately outside the inward-moving portion of the internal system and/or a volume contraction of the external material outside the outward-moving parts of the boundary; (2) spontaneous flow of the external material away from the outward-moving portions of the boundary and towards the inward-moving boundary segments.

The expansion and/or contractions in the external system may be due to phase changes (e.g. melting, change from gabbro to eclogite), to local thermal expansion connected with flow of heat, or to, e.g. release of local remanent elastic stress accumulated during the lifetime of the system. In sediments compaction may be a reasonable cause.

Deformations in the internal system produced by such volume changes in the surroundings are not to be classed as gravity tectonics. Deformations of this type are generally quite limited unless the size of the internal system is small relative to the part of the external system that undergoes volume changes. This is a consequence of the percentage-wise small volume change associated with phase transitions, thermal expansion or elastic compression of condensed materials.

Tectonic processes of this nature do occur, but they are probably of quite limited significance.

The process described under (2) above may well be propelled by gravity—i.e. the spontaneous flow may correspond to a drop in the gravity potential of the external system. This process is capable of causing extensive displacement of matter and quite drastic rearrangement of materials in the systems. For gravity to work in the manner indicated, let the external system consist of a material of density ρ_{ex} somewhat greater than the density ρ_{in} of the internal system. The hydrostatic pressure gradient is then steeper in the external system than in the internal. This creates an unstable situation along any part of the common boundary of the two systems. The material of the outer system exerts excess pressure on the lower part of the boundary, which consequently is pushed inward and upward. Similarly, the material in the internal system exerts excess pressure on the upper boundary, which consequently is displaced upward and outward (Fig. 3.2).

Since the materials are incompressible, the volume of the external material pushed away from the upper part of the boundary corresponds exactly to the volume of external material moved towards the rising and contracting lower part of the boundary. This means a displacement of the material of the external system to lower levels in the field of gravity, and a consequent reduction of its gravity potential amounting to $V\rho_{ex}\, gh$ per unit time, where V is the volume displaced down a distance h in unit time. At the same time an equal volume of the internal system is moved up a distance h, the resultant rise in the gravity potential of the internal system being $V\rho_{in}\, gh$. However, the significant point is that, looked upon from the combined system point of view, the process is associated with a drop in the gravity potential of $V(\rho_{ex}-\rho_{in})\, gh$. This drop in gravity potential is then to be regarded as the driving power of the deformation of both systems and the movement of the internal system relative to the external.

The rearrangement of the structure may also mean a change in the spatial potential. It is easy to see that this change is zero provided the pressure does not change within the "lost" and "gained" volumes as marked by hatching in Fig. 3.2. The internal system "loses" a net spatial potential $V_i(P_2 - P_1)$ and the external system "gains" a net spatial potential $V_e(P_1 - P_2)$ where $V_i = V_e$. P_1 is the average pressure within the region covering the "gained" volume of the internal system (= "lost" volume of the external system), and P_2 is the average pressure within the region covering the "lost" volume of the internal system (= the gained volume of the external system). In all likelihood, however, P_1 and P_2 change somewhat during the rearrangement of the internal and external systems, causing a small change in the spatial potential.

It is evident that in general the movements and deformations thus caused by gravitational instability are less limited than those caused by the stress associated with phase changes.

The gravitational instability does not disappear until the internal system has moved on top of the external system and spread out as a horizontal sheet. The unstable stresses caused by volume changes in the materials, on the other hand, are stabilized after only very limited movement within the surrounding region.

From the relations between the external and internal systems, which together constitute an energetically closed supersystem, we conclude that for closed global systems gravitational instability is the chief cause of mechanical movement and deformation. The value of the gravity potential of the real system, minus the value of the gravity potential of an ideally stable system, is a measure of the mechanical work—in geological systems, usually meaning tectonic work—which the system may ideally produce.

Only purely mechanical processes are considered in the above analysis. The important coupling of mechanical processes—i.e. bodily motion of matter—and mechanical energy with processes and energies of chemical, thermal or nuclear nature are not considered. Gravitational or other mechanical energy may well be produced in a closed system by some chemical, thermal or nuclear processes that alter the density of materials. Thus, caused by chemical phase changes or nuclear reactions in the Earth, a mechanically stable arrangement may well become gravitationally unstable with a consequent tendency towards bodily rearrangement of matter.

The Earth as a whole is a dynamic system which is practically closed with respect to direct communication of mechanical energy—the "leak" (tidal forces) is negligible as compared with the energy involved in tectonic processes—because the stress vanishes at the boundary which coincides with the outer limit of the atmosphere. Even the consolidated Earth, including the hydrosphere, may be regarded as practically closed with respect to mechanical energy, because the energy change caused by a possible volume change of the globe under the pressure of the atmosphere is negligible compared with the energy involved in large-scale crustal and subcrustal motions. On the other hand, the solid globe exclusive of the hydrosphere and its crystalline counterpart, the ice-caps, is not a closed system with respect to spatial energy. This fact is well illustrated by the isostatic depression of the crust under a growing ice-cap. During the downward movement of the surface of the crust below the ice-cap the solid Earth receives energy from the external system, namely the ice-cap. This spatial energy acts partly to overcome the viscous resistance of the crust and mantle, partly to raise the gravity potential of the somewhat elevated crustal region around the ice-cap.

In the light of the above discussion, the term "gravity tectonics" has little meaning if defined as tectonic features produced by the force of gravity or by a drop in the gravity potential of closed tectonic systems, inasmuch as such a definition includes virtually all tectonic features except a few types of peripheral significance.

Global scale processes such as thermal convection in the mantle and mantle diapirism are gravity-driven phenomena whose motion may be transmitted to the crust in the form of continental drift and sea-floor spreading. Thus, when all comes about even the modern plate tectonics with its large horizontal displacements may be driven by gravity.

There are, however, more meaningful ways of defining gravity tectonics, which distinguish a certain class of tectonic phenomena from other classes.

With alternative definitions in mind, consider a structure produced by the gravitational pull on an unstable mass distribution such as a dome of granite or salt raised by the buoyant force through a layered complex. A second example would be a complex of surface rocks sagged down to the bottom of a basin, so-called "full-trough" gliding (Haarmann, 1930). A third example could be the slow plastic collapse and spreading of an orogenic nappe such as discussed on p. 198 ff.

We may say that all these structures were generated by the force of gravity acting on the entire system whose bulk gravity potential decreased during the tectonic evolution. The large-scale architecture of the domes with their rim synclines represents a lower gravity potential than the initial arrangement of a light layer of rock below a heavy overburden, but the local small-scale pattern of buckled or fragmented strata does not generally represent lower gravity potential than the initial undeformed local structure. Buckles on an originally horizontal sheet, for example, do not represent a drop in gravity potential relative to the straight sheet, unless the rock below the sheet is less dense than the rock above (see p. 56 ff.). If the rocks below the sheet are more dense than the sheet and the superincumbent rocks then the buckled state represents higher gravity potential than the straight layering. Even if the rock below is less dense, the wavelength of the folds must exceed several hundred metres if the drop in gravity potential should be sufficent to produce the folds.

In connection with the ascent of the dome or "full-trough" gliding in the basin, a variety of deformation structures generally develop. Owing to local concentration of compression and extension buckle folds, drag folds, boudins, pinch-and-swell and similar structural patterns evolve where layers with the proper competency occur (see Figs 11.13, 11.16, 11.19, and many other model discussed later).

It is also noteworthy that the geometric pattern of the gravity-produced folds will differ from that of buckle folds (cf. difference between buckle folds and supertenous folds: Hills, 1963; Ramberg, 1964a). With equal density of rocks above and below, the folds on the sheet do not represent a change in gravity potential at all, whereas a situation of a buckled layer with low-density rock above and high-density rock below means increased gravity potential. Likewise, the presence of waves on the free surface of the Earth means increased gravity potential compared with a smooth surface parellel to an equipotential surface.

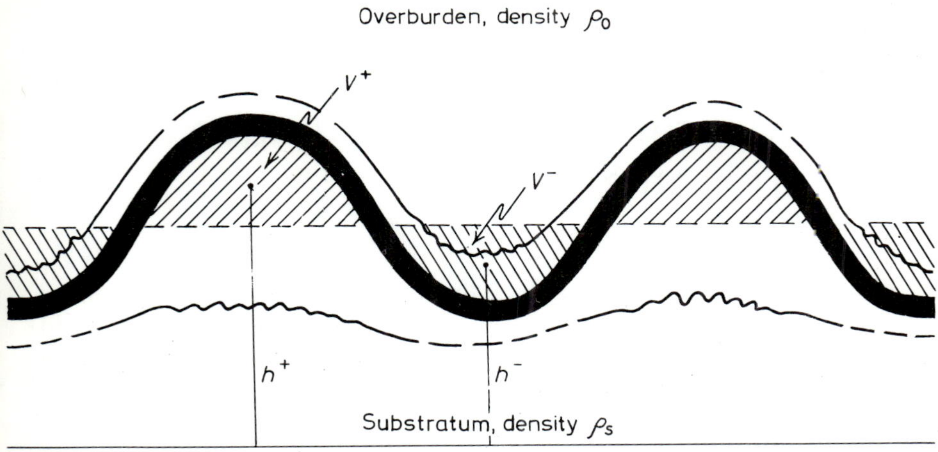

Fig. 3.3. Spontaneously formed waves on the boundary between a low-density substratum and a more dense overburden. The large folds represent decreased gravity potential relative to the original flat layering, the change of potential being $(\rho_s - \rho_0)V^+h^+ + (\rho_c - \rho_s)V^-h^- = (\rho_s - \rho_0)(h^+ - h^-)V$ which is negative. However, the small buckles on the thin competent sheets and their boundinage structure do not represent lower gravity potential than the initial smooth non-buckled state of these sheets, as there is no change in density on either side of each of the thin sheets. V^+ is numerically equal to V^-

As a possible alternative definition of gravity tectonics, the term should be applied only to structures whose pattern represents a decreased gravity potential relative to the initial pattern. Figure 3.3. shows folds which represent a lower gravity potential than the unfolded state, together with folds which represent unchanged gravity potential relative to the unfolded state.

In consequence of the relationship between the geometric scale of structures and the efficiency of gravity as the driving power (see Chapter 7), gravity tectonics as defined above can only involve rather large structural patterns. The more obvious classes of gravity-tectonic processes are:

(1) The down-slope sliding of rock masses: *gravity sliding.*
(2) The slow plastic vertical collapse and the complementary lateral spreading of rock masses: *gravity collapse* or *gravity spreading.*
(3) Gravitational readjustment of masses with unstable density-distribution. The readjustment often occurs in the form of diapirism and the rise of domes and diapirs. We may therefore refer to this type of gravity-generated tectonics as *gravitational diapirism.*

Dennis (1967, p. 85) defines gravity tectonics as "The process and result of rock deformation primarily under the influence of gravity".

4 Mental Model of an Active Tectonic System

A realistic tectonic system is generally immensely complex, as it contains a number of rock and magma bodies that exhibit a variety of shapes and sizes and possess complicated non-linear rheological properties and densities varying over a considerable range.

In our mental model the bodies are tightly packed, allowing no voids within the bulk of the system. A part of its boundary is the free surface of the Earth. The system is closed with respect to matter—there is no diffusion across the boundary, so metasomatism is excluded from our model, whose processes are solely of a mechanical nature.

Initially there is no relative movement among the bodies or among separate parts of single bodies, nor is the system in motion relative to its immediate surroundings. In short, the system contains no kinetic energy (except, of course, on a molecular scale). This is to prevent inertia from becoming a source of subsequent deformations. For most tectonic deformations the rotational energy of the Earth and the resulting Coriolis force may be disregarded.

We do not want elastic deviatoric stress to interfere with our analyses either, therefore the initial state of the system is relaxed with reference to elastic strain.

It follows that the system contains only two kinds of mechanical energy; potential gravitational energy by virtue of the distribution of its mass in the field of gravity, and potential spatial energy by virtue of the volume and shape it occupies in the field of stress transmitted to its boundary through the surrounding matter.

Now, it would be a rare situation if a system like this happened to be in a state of stable mechanical equilibrium.

In general, a realistic system of above nature would be mechanically unstable in the sense that its parts are affected by forces tending to change spontaneously the position and the shape of its constituent bodies of rocks and magmas. In other words, the mechanical potential of the tectonic system is higher than that of other possible arrangements and shapes of its individual bodies, the state representing the lowest mechanical potential being the mechanically stable state. Processes therefore tend to occur, controlled by a decreasing mechanical potential, in attempts to reach a more stable mechanical state.

Generally, the stress at the boundary is not lithostatic in the sense that it varies with depth as the specific gravity, therefore in general:

$$\sigma_x \gtreqless \sigma_y \gtreqless \sigma_z; \qquad \frac{\partial P}{\partial h} \gtreqless -\rho g; \qquad P = \frac{\sigma_x + \sigma_y + \sigma_z}{3}$$

In consequence of the unbalanced boundary stress, various movements occur in the plastico-viscous materials, possibly altering the shape of the system and its internal configuration and displacing its bulk in relation to the surroundings.

Regarding rocks and magmas as incompressible and the system as closed with respect to communication of matter—though not energy—there is no net volume change during the distortion of shape. The change of shape of the system and its movement in bulk relative to the surrounding framework are associated with changing mechanical potential of the delimited system.

Displacement of the boundary toward regions where pressure is low and away from regions where pressure is high means decreasing spatial potential, the decrement balancing the energy dissipated in the plastico-viscous strains and a possible rise in the gravity potential of the system. Conversely, boundary displacement away from regions where pressure is low and volume-compensative displacement into regions where pressure is high mean increased spatial potential of the region. This must be balanced by a drop in the gravity potential, for example by subsidence or spreading. The gravity-potential drop must also be sufficient to cover the dissipated strain energy associated with the viscous and plastic deformations.

Besides the instability manifested by the spatial and gravity-potential changes of the whole system, the internal distribution of bodies with unlike densities is also in general unstable. In most real tectonic systems the true distribution of rock and magma bodies of dissimilar densities represents a higher gravity potential than do other arrangements. Thus, in addition to a possible change of outer shape of the system in bulk, internal rearrangements work towards the stable configuration in which the various types of material are arranged in horizontal (concentric) shells, increasingly dense towards the bottom. Vertical motion (rising of light material, sinking of dense) and lateral spreading are the chief trends of this internal rearrangement, which minimizes not only the gravity potential but also the spatial potential, because the arrangement minimizes the average pressure in the system, as discussed on p. 13. It is interesting that in simple tectonic systems consisting of but one homogeneous rock or magma body—e.g. a salt diapir, a granite magma, a thick plastic sedimentary stratum, etc.—the movement of the boundary in fact determines the strain-rate distribution within the body. The strain in the bulk of a homogeneous system is controlled by general fluid dynamic differential equations whose solution depends on the boundary condition (e.g. the shape and movement of the boundary) and on the rheological properties of the materials. In this connection see the application of the *stream-function* on pp. 207ff.

Only if the arbitrarily delimited tectonic system is very small can the strain within be homogeneous. Generally, the movement of the boundary is of non-affine type, in which case the geometric character of the strain changes with distance from the boundary; that is, the internal strain is heterogeneous.

A heterogeneous movement at the boundary—for some reason or other the boundary may be thrown into sinuous buckles, for example—penetrates into the system with rapidly diminishing amplitude, the sinuous strain practically vanishing at distances less than a wavelength from the boundary (Ramberg, 1963b, 1964a, 1968, 1970). See also Chapters 7 and 8.

The penetration of heterogeneous strain into a uniform single-body system may be made visible by the pattern of passive markers. These are markers with rheological properties and density identical to those of the bulk of the rock. Examples are lithologically and structurally indistinguishable strata or other shapes of contrasting colours. Such markers do not influence the strain in their neighbourhood.

In contrast to the passive markers, most strain detectors in tectonic system are rheologically unlike the adjacent rock, thus causing disturbance of the strain pattern. These markers are surrounded by a contact zone of a strain pattern arising from an interference between the "general strain" within the region and the perturbed strain caused by the active marker. "General strain" is simply the strain that would have occurred at the location had it not been perturbed by active markers.

The active markers are generally more competent than the adjacent or enclosing rock or magma. For this reason sheet-shaped active markers become buckled if oriented parallel to a component of compressive strain, the resulting wavy strain pattern penetrating through a contact zone less than a wavelength wide. Layers parallel to a component of extensive strain become stretched into pinch-and-swell structures or fragmented to boudins.

The larger the active markers and the more competent they are relative to the main part of the system, the more they disturb the strain pattern controlled by the motion of the boundary of the system.

The tectonic system now considered is heterogeneous and realistic, the strain distribution in the bulk being in part controlled by the movement at the boundary, in part by the rheological properties, and the shape and the size of the constituent bodies. As gravity is a significant factor, the mean density of the system relative to the surroundings and the density contrast between the various constituent bodies are also of prime consequence for the dynamic evolution of the system and the strain pattern within it.

Summarizing, the mental model we have of a realistic tectonic system is a heterogeneous complex of dissimilar rock and magma bodies, the complex being arbitrarily delimited against the surroundings through which stress is transferred to the system. The system is mechanically unstable in the sense that its boundary may become displaced in the course of time, though in such a manner that the volume remains constant; the whole system is being strained or moved relative to the surroundings, and/or the individual bodies within the system rise or sink in relation to one another and change shape in various ways.

The driving power behind these processes is a declining mechanical potential; either the spatial potential may drop or the gravity potential, or both may decrease simultaneously. Often when one potential drops, the other rises.

The local pattern of the changing strain is controlled by a certain economy principle: the movements that take place dissipate least energy or release stress most rapidly.

5 Energy Associated with the Change of Shape of Bodies Exposed to Non-hydrostatic Stress

As shown in Chapter 2, a body "possesses" a certain potential energy simply by virtue of the volume which it occupies in space when present in a hydrostatic stress field. This spatial potential energy of the body equals in magnitude the energy that would be released if its volume were to shrink to zero under the given constant pressure or if the body were displaced from the pressure field to a site with vanishing pressure.

As discussed in various connections above, the concept of spatial potential is a useful one in tectonic analysis. When rocks deform, however, the stress field is generally non-lithostatic, and we may ask whether a potential energy analogous to the spatial potential may be assigned to bodies exposed to a non-hydrostatic stress field. Useful as such a quantity would seem to be as a driving energy of strain in stressed rocks, it appears impossible to assign an uniquely defined potential to a viscous or plastic body in a field of non-hydrostatic stress. Here we are in the same confusing situation as the thermodynamicists find themselves when they seek a quantity to apply to solids under deviatoric stress analogous to the Gibbs free energy of solids under hydrostatic pressure. (See McDonald, 1957; Verhoogen, 1951; Ramberg, 1960; Kamb, 1961; McLellan, 1968 and 1970; Paterson, 1973; and of course Gibbs classic treatment, republished 1948).

For solids or liquids exposed to hydrostatic pressure the energy associated with the movement of a given mass from one pressure to another is independent of the path. The energy change defines the potential change and is uniquely determined by the two pressures and the volume of the incompressible material. (For compressible materials the potential change is a little more complicated, but the change is still uniquely defined by the two end situations.) That is, the spatial potential is a *state function*. This is not so for bodies under non-hydrostatic stress. When the stress field varies with direction in space the energy associated with the movement of a given mass of a material between two different states of stress is not uniquely determined by the two states of stress. The energy change also depends on the path followed during the movement from the one state to the other. Indeed, energy may be absorbed or released even if the two states of stress are identical but the paths are different.

Although difficulties may be encountered when it comes to the question of a rigorous definition of a potential for bodies under non-hydrostatic stress, it is still useful to discuss the energy changes which take place when a deformable rock or a magma mass is exposed to a non-hydrostatic but homogeneous field of stress.

In a brief quantitative analysis of the change of energy associated with homogeneous deformation we shall consider a case with simple symmetry: a rectangular body whose edges are parallel to the principal stress axes in a non-hydrostatic stress field (Fig. 5.1).

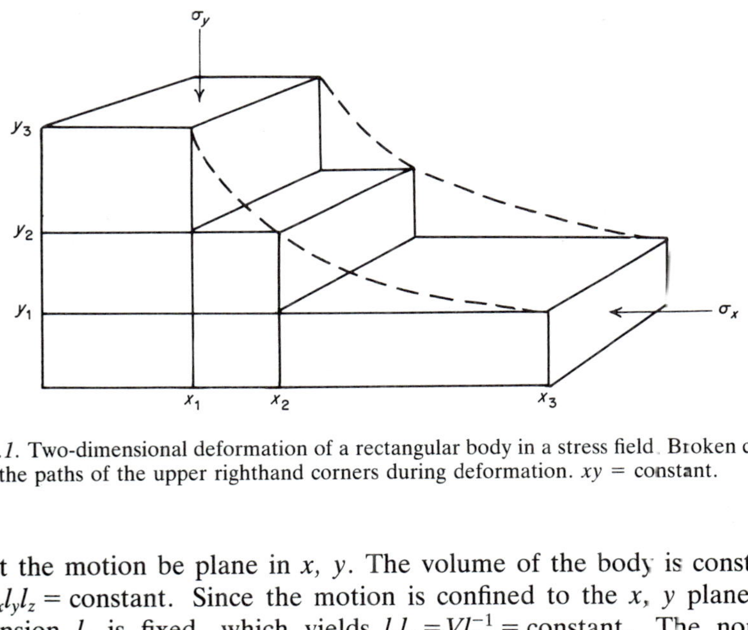

Fig. 5.1. Two-dimensional deformation of a rectangular body in a stress field. Broken curves show the paths of the upper righthand corners during deformation. xy = constant.

Let the motion be plane in x, y. The volume of the body is constant: $V = l_x l_y l_z$ = constant. Since the motion is confined to the x, y plane the dimension l_z is fixed, which yields $l_x l_y = V l_z^{-1}$ = constant. The normal stresses parallel to x and y are σ_x and σ_y. The stresses are kept constant during the defomations. As the body changes shape in the field of compressive stress, it receives energy from the surroundings when the surface moves inwards, while the body loses energy when the surface is displaced outward. If the stress is tensile, outward movement of the surface means gain of energy and inward movement means loss of energy. In the above example σ_y is compressive, whereas σ_x may be compressive or tensile. σ

is reckoned positive for tension, negative for compression. Now, a compression of magnitude $\mathrm{d}l_y$ corresponds to a gross energy change of magnitude

$$\mathrm{d}E_y = \sigma_y l_x l_z \, \mathrm{d}l_y. \tag{5.1}$$

Positive E means that the body receives energy, negative E signifies loss of energy by the body. In the above equation we introduce $l_x = Vl_z^{-1}l_y^{-1}$, hence

$$\mathrm{d}E_y = \sigma_y V l_y^{-1} \, \mathrm{d}l_y. \tag{5.2}$$

Integration at constant σ_y and V yields

$$\Delta E_y = \sigma_y V \ln \frac{l'_y}{l^\circ_y} \tag{5.3}$$

which is the gross change of energy associated with a compression along y from l°_y to l'_y.

Simultaneously with the compression along y, a volume-preserving extension occurs parallel to x. The associated change of energy is

$$\mathrm{d}E_x = \sigma_x l_y l_z \, \mathrm{d}l_x, \tag{5.4}$$

and

$$\Delta E_x = \sigma_x V \ln \frac{l'_x}{l^\circ_x} \tag{5.5}$$

or, since $l^\circ_x l^\circ_y = l'_x l'_y$

$$\Delta E_x = \sigma_x V \ln \frac{l^\circ_y}{l'_y} = -\sigma_x V \ln \frac{l'_y}{l^\circ_y}. \tag{5.6}$$

During this pure shear-type deformation of the body with volume V, the *net* change in energy is accordingly

$$\Delta E = \Delta E_y + \Delta E_x = (\sigma_y - \sigma_x) V \ln \frac{l'_y}{l^\circ_y} \tag{5.7}$$

or, in terms of dimension parallel to x,

$$\Delta E = (\sigma_x - \sigma_y) V \ln \frac{l'_x}{l^\circ_x}. \tag{5.8}$$

Generally $\ln(l'_x/l^\circ_x)$ is negative when $\sigma_x - \sigma_y$ is negative and positive when $\sigma_x - \sigma_y$ is positive such that ΔE is always positive. This shows that the body always receives energy from the surroundings by the displacement of the boundaries in a non-hydrostatic stress field, provided that energies of other nature are not involved, such as introduction of thermal energy or a change of its gravity potential.

Assuming that the stresses are kept constant during the deformation, we obtain through differentiation and division by time the rate of energy input, thus

$$\frac{\mathrm{d}\Delta E}{\mathrm{d}t} = (\sigma_x - \sigma_y)V\frac{\mathrm{d}l_x}{l_x\mathrm{d}t}, \text{ or} \tag{5.9}$$

$$\Delta\dot{E} = (\sigma_x - \sigma_y)V\,\dot{\varepsilon}_x \tag{5.10}$$

where $\dot{\varepsilon}_x$ is the strain rate and V the volume of the body.

For pure shear the general relationship between stress and strain rate of Newtonian materials is

$$(\sigma_x - \sigma_y) = 4\mu\dot{\varepsilon}_x. \tag{5.11}$$

This equation, in combination with eqn (5.10) yields

$$\Delta\dot{E} = 4\mu\dot{\varepsilon}_x^2 \tag{5.12}$$

per unit volume. But this is the general equation for the rate of change of longitudinal strain energy. That is, the rate of energy input due to the displacement of the boundaries in a non-hydrostatic stress field balances exactly the rate of strain-energy change.

We have in the above analysis assumed that the internal energy does not change during the deformation, i.e. that no chemical mineral-reaction or energy-changing textural alteration takes place and that the gravity potential before and after the deformation is the same. Under these conditions the input rate of energy due to the displacement of the boundaries in the stress field is completely dissipated to heat because the strain-energy rate term $4\mu\dot{\varepsilon}_x^2$ is also a measure of the amount of heat produced by the viscous resistance per unit of time.

In the gravity field of the Earth, however, the displacement of the boundaries of a tectonic system, because of statically unbalanced forces, means not only change of shape and strain but often also a change of the gravity potential of the system as illustrated by a vertical displacement of the centre of gravity.

As an example consider a homogeneous mass of a Newtonian material exposed to a horizontal compression (Fig. 5.2). To make the kinematics of the system simple without losing the essence of its dynamic behaviour let the rate of strain be uniform throughout. This requires that the stress

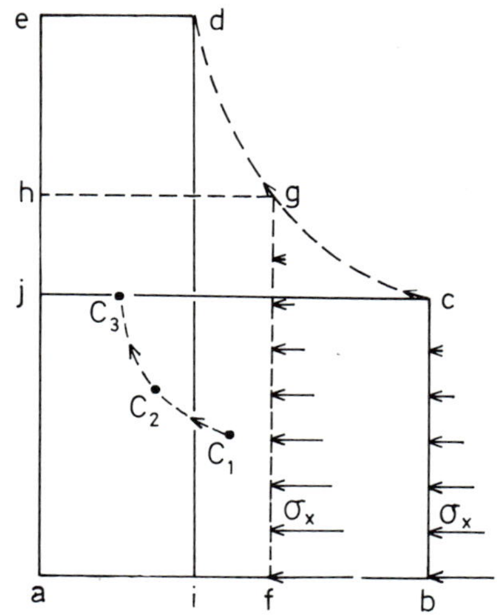

Fig. 5.2. Cross-section of a parallelepiped of viscous material being deformed homogeneously while resting in the gravitational field. See text.

contrast $\sigma_x - \sigma_y$ is the same throughout the system as shown by the general condition for plane, non-rotational strain

$$\dot{\varepsilon}_x = \frac{\sigma_x - \sigma_y}{4\mu}. \tag{5.13}$$

Due to the weight of the superincumbent mass, the stress σ_y at any level h below the surface is

$$\sigma_y = \rho g h \tag{5.14}$$

where h is negative in agreement with the fact that the vertical stress is compressive below the Earth's surface. Consequently

$$\sigma_x - \rho g h = 4\mu\dot{\varepsilon}_x, \tag{5.15}$$

and the horizontal normal stress at any level h below the surface becomes

$$\sigma_x = 4\mu\dot{\varepsilon}_x + \rho g h \tag{5.16}$$

where $\dot{\varepsilon}_x$ is constant.

Let the basal boundary of the system remain fixed such that the rise of the free surface represents the entire vertical extension of the system. The rate of energy input related to the horizontal displacement of the vertical sides is

$$d\dot{E}_x = \sigma_x \dot{\varepsilon}_x l \, dh \tag{5.17}$$

reckoned per strip with thickness dh, length l and unit width in the direction normal to the plane hl. Introducing the expression for σ_x as a function of h, and integration between the limit o and h_0 yield

$$\dot{E}_x = (4\mu\dot{\varepsilon}_x^2 - \tfrac{1}{2}\rho g h_0 \dot{\varepsilon}_x)V. \tag{5.18}$$

Here $h_0 l$ has been replaced by the volume of the system per unit length normal to the section considered and the height h_0 is reckoned positive.

Since the normal stress at the free surface vanishes, there is no energy change due to surface forces associated with the rise of the free surface. Hence the time rate of the *net* energy input is as given by eqn (5.18). Positive $\dot{E}$ means that the system has received energy from the surroundings, negative $\dot{E}$ means that the system has imparted energy to the surroundings.

In the model under consideration $\dot{\varepsilon}_x$ is negative such that both terms in eqn (5.18) are positive. The term $4\mu\dot{\varepsilon}_x^2$ is the rate at which energy is dissipated to heat because of viscous friction, and $-\frac{1}{2}\rho g h_0 \dot{\varepsilon}_x$ is the rate at which the gravity potential changes during the horizontal compression and vertical extension, all per unit volume. In the rectangular cross-section of the system the centre of gravity is at depth $-\frac{1}{2}h_0$ below the surface and the velocity of rise of the centre of gravity is $-\frac{1}{2}h_0\dot{\varepsilon}_x$, or, since for incompressible materials $-\dot{\varepsilon}_x = \dot{\varepsilon}_y$, the velocity of rise is $\frac{1}{2}h_0\dot{\varepsilon}_y$. The gravity potential is accordingly augmented at a rate $\frac{1}{2}h_0\dot{\varepsilon}_x\rho g$ per unit volume. We conclude then that the energy introduced into the system by a horizontal compression in response to a horizontal compressive stress which exceeds the vertical stress, partly changes to heat by dissipation, partly goes to raise the gravity potential of the system.

It is also possible to reverse the process described above in such a manner that the rate of decreasing gravity potential goes partly to expand the system horizontally against a stress σ_x, partly to support the strain rate. This occurs when in the system described above $\dot{\varepsilon}_x$ is extensive rather than compressive, a condition which requires that σ_x is either positive or, if negative, numerically less than $\sigma_y = h\rho g$. This is the situation characteristic of tectonic systems in the process of gravity collapse and spreading. Considerations of this nature are useful when the mechanism of nappe movement is treated, see Chapter 15.

6 Difficulties Encountered in Experiments with Tectonic Models

For a number of reasons it is not possible in practice to construct experimental dynamic models which in every respect are true-to-scale replicas of tectonic phenomena. Models that only partly satisfy the requirements of the theory of dynamic similarity have to suffice.

The difficulties encountered in experimental model investigation of tectonic processes are chiefly of the following kinds.

(1) Incomplete knowledge of rheological data of the rocks in the prototype. Very little is known about the strength, effective viscosity, yield point, etc. of rocks under the conditions at which they yield in active tectonic systems. Data of this nature are unfortunately produced only slowly in contemporary experimental research.

(2) Difficulties in finding model materials with rheological properties suited to simulating rocks and other terrestrial materials in ordinary dynamic scale models. Particularly scarce are materials that mimic reasonably well the characteristic visco-elastic behaviour of rocks, i.e. their ability to yield coherently or plastically under gentle long-lasting stresses while being almost perfectly elastic at sudden impacts. (Stitching wax and silicone putty are among the very few solid substances exhibiting such behaviour suited for model work.)

In models intended to imitate global-scale processes (e.g. convection currents in the mantle) impossible demands are made on the compressibility of the model materials. From top to bottom of the mantle the density changes, presumably owing to compression of a homogeneous (?) solid substance, from 3·3 to 5·6 g cm^{-3}. No known solid would be correspondingly compressed by its own weight through a model mantle of manageable thickness, say a few decimetres at most.

(3) When materials with acceptable imitation properties are found, they are often soft and mechanically weak to a degree that makes model construction and model study—e.g. by sectioning and microscopic

inspection—quite impracticable or at best a most messy operation.

However, the softness and weakness required of model materials are limited to models of tectonic phenomena over which gravity exercises a significant control. In tectonic systems in which the gravitational pull is negligible relative to the stresses at the boundary, the strength and viscosity of the model materials are limited only by the maximum stresses which the experimenter is able to transmit to the system. In the study of buckle folds, for example, of a kind not affected by gravity (e.g. folds of rather short wavelength) the hardest of materials may well be used as model material if one has access to a press strong enough to produce buckling. The evolution of huge undulations, such as geosynclines and geanticlines, on the other hand, is closely governed by the force of gravity; models of such structures do not reproduce the phenomena realistically unless the materials are soft and weak enough to sag, and even fracture, under their own weight. This is even more true of models of such features as the rise and evolution of salt domes, batholiths and the like; structures of this kind do not form at all in models unless the materials are soft and weak.

(4) In models whose outer frame is at rest (or in non-accelerated motion) in the Earth's field of gravity, the model ratios of time and length are interrelated, $l_r t_r^{-2} = 1$. As model and tectonic prototype are both exposed to the same gravitational field, acceleration in free fall is the same in the two systems. The model ratio of acceleration, a_r, is therefore unity, which for strictly correct scale models must hold for acceleration in general throughout the structures. Consideration of dimensions, however, leads to $a_r = l_r t_r^{-2}$ and consequently $l_r t_r^{-2} = 1$ (see Table 1.1, p. 5).

The requirement $l_r t_r^{-2} = 1$ would make model study of tectonics quite impossible by virtue of the great contrast in time and geometric dimensions of tectonic systems and manageable models. For example, for a tectonic study a model ratio of length of, say, $l_r = 10^{-5}$ is a common figure (10 km in nature corresponds to 10 cm in model). A time span of, say, 10^6 years may be a reasonable figure for the evolution of the structure in question.

Now, from the model ratio of length follows the ratio of time

$$t_r = \sqrt{10^{-5}} \approx 3{\cdot}16 \times 10^{-3}, \tag{6.1}$$

which means that a strictly dynamically similar model requires $t_m = 3{\cdot}16 \times 10^{-3} \times 10^6 = 3160$ years to undergo the evolution we want to study—not exactly a practical time span for the experimenter.

However, the acceleration—in terms of rate of change of velocity, but certainly not in terms of force per unit mass in a body–force field—of most tectonic processes is negligible except, for example, in earthquakes and the flow of magmas. This is equivalent to saying that the inertial terms are negligible in the fluid-dynamic equations when applied to tectonic processes. For this reason no significant error is introduced by disregarding the condition $l_r t_r^{-2} = 1$ and treating l_r and t_r as independent variable in most tectonic problems.

The reasoning implicit in the above comments is not limited to negligible inertial forces but is generally applicable in model work. Although strict dynamic similarity does require that the model ratio of force be constant throughout for a model–prototype pair, if some types of force can be shown to be negligible for the phenomenon under study, then no important error is introduced by relaxing the prototype–model ratio requirement for just these types of force. Such relaxations of the strict theoretical requirements with respect to insignificant types of force in experimental models not only simplify the construction and handling of models but often open the only avenue to experimental approach. The relaxation of the restriction $l_r\, t_r^{-2} = 1$ due to negligible inertial forces is a case in point.

Other examples are provided by tectonic structures not significantly controlled by the force of gravity. For models designed for the study of such structures (e.g. practically all small-scale folds, boudins, and pinch-and-swell and similar structures in rocks), the ratio of the gravitational force acting on corresponding masses in model and natural system need not equal the ratios of the forces applied on the boundaries of model and natural system. This we can show by model theory.

The model ratio of the forces due to gravity equals the model ratio of mass times the model ratio of acceleration due to gravity, which is unity in ordinary model work. But the ratio of mass is identical with the ratio of density times the ratio of volume, hence

$$\frac{F_{mg}}{F_{og}} = F_r = m_r a_r = \rho_r l_r^3\, a_r = \rho_r l_r^3 \tag{6.2}$$

since $a_r = 1$. The ratio of the stress distribution caused by gravity in the two structures is correspondingly

$$\sigma_r = F_r l_r^{-2} = \rho_r l_r \tag{6.3}$$

showing that at constant ρ_r, the stress due to gravity decreases with decreasing linear dimension of the model of a given natural structure. In order to reproduce flowage and collapse correctly, the strength of the model material must correspondingly decrease with diminishing model size, hence the requirement of weak and soft material in gravity-controlled tectonic models. See Hubbert (1937, 1945) and Ramberg (1968, 1970).

However, if in nature the force of gravity is negligible relative to the viscous or elastic forces and therefore also relative to the boundary forces needed to produce the desired deformation, then the condition $\sigma_r = \rho_r l_r$ needs not be satisfied, and one may apply without restriction in our models any stresses necessary to produce the wanted deformation quite irrespective of the model ratio of force to which these stresses correspond.

The buckling and boudinage experiments performed by Ramberg (1955, 1963b), Huddleston and Stephansson (1973), Strömgård (1973), Ramberg and Strömgård (1971) and Ghosh and Ramberg (1968) are examples of models in which gravity exercises no appreciable control over the structural pattern. The relative significance of the compressive surface force and the body force of gravity in layered systems are discussed in detail in Chapter 8.

In fluid dynamics some standard dimensionless expressions for the ratios between various forces are in use, some of which are given in Table 6.1,

Table 6.1 Force ratios useful in model work

$F_v \mathrel{\widehat{=}} \mu v l$	$F_p \mathrel{\widehat{=}} \Delta p l^2$	$F_g \mathrel{\widehat{=}} \rho l^3 g$	
$\frac{F_i}{F_v} = \frac{v l \rho}{\mu}$ Reynolds number	$\frac{F_p}{F_i} = \frac{\Delta p}{\frac{1}{2}\rho v^2}$ Pressure coefficient	$\frac{F_i}{F_g} = \frac{v^2}{g l}$ Froude number	$F_i \mathrel{\widehat{=}} \rho v^2 l^2$
	$\frac{F_p}{F_v} = \frac{\Delta \mathrm{pl}}{\mu \mathrm{v}}$ Stokes number	$\frac{F_g}{F_v} = \frac{\rho l^2 g}{\mu v}$ Nameless ratio	F_v
		$\frac{F_g}{F_p} = \frac{\rho l g}{\Delta p}$ Suggested: Smoluchowski number	F_p

F_i = inertia force; ρ = density;
F_v = viscous force; l = length;
F_p = pressure force; v = velocity;
F_g = gravity force; μ = viscosity;
p = pressure or stress;
$\widehat{=}$. "dimensionally equal to"; g = acceleration of gravity.

reproduced from Kline (1965). Six of these ratios are so often used that they have acquired specific names, Reynolds number probably being the best-known.

The values which these dimensionless numbers achieve in a system show the relative importance of the forces involved. For example, a small Reynolds number means small inertial force relative to the viscous force, a high Froude number signifies strong inertial force as compared with the force of gravity, etc. For our purposes the Stokes number and especially the ratios F_g/F_v and F_g/F_p, which so far have not been honoured with a person's name (probably because the ratios between gravity and viscosity and gravity and pressure force are of little significance in ordinary fluid dynamic problems), are the most significant force ratios.

In general, the situation in tectonic systems is such that the value of both the Reynolds number and the Froude number is less than the limit above which inertial forces are influential. (Note, however, that this limit depends upon the geometry of the system; see also Chapters 7 and 8.)

At strict dynamic similarity, which requires that the ratio between forces in model and in prototype be constant if compared kind for kind, the ratio between two dissimilar forces, such as those given in Table 6.1, in the model must assume the same value as the ratio between the same forces in the prototype.

From

$$\frac{F_{mg}}{F_{og}} = \frac{F_{mi}}{F_{oi}} = \frac{F_{mv}}{F_{ov}} \tag{6.4}$$

follows that

$$\frac{F_{mg}}{F_{mi}} = \frac{F_{og}}{F_{oi}}; \frac{F_{mv}}{F_{mi}} = \frac{F_{ov}}{F_{oi}}; \frac{F_{mg}}{F_{mv}} = \frac{F_{og}}{F_{ov}} \tag{6.5}$$

As we have seen, it is not necessary to meet the last requirement if one (or both) of the two forces involved in the ratio is insignificant in both model and prototype. There is generally no need, therefore, to attempt to obtain the same Reynolds number or the same Froude number for a model as those which existed in the tectonic prototype during its evolution. One has only to ensure that the inertial forces are negligible also in the models.

On the other hand, the ratio F_g/F_v in models of most large-scale tectonic structures must approximate very closely the ratio in the natural structures because both gravitational and viscous forces are of paramount significance in tectonic evolution. This applies also to the ratio F_g/F_p and the Stokes number.

Tectonic deformations have been mentioned in which neither gravity nor inertia was a controlling force—e.g. the evolution of small buckle folds and boudins. In models of such structures, however, it is necessary that the Stokes number, which gives the ratio between pressure, or stress, force and viscous force in a system, be the same in model and in prototype.

The effect of surface tension forces on structural pattern in tectonics can be ignored, since the viscous and gravitational forces are so much stronger. In small models, however, whose fold pattern, for example, may be less than, say, 1 cm across, surface tension may play an important role relative to viscosity and gravity. In such cases the ratios between the surface tension force and the viscous and gravity forces should be watched lest the models deviate noticeably from dynamic similarity.

7 Instability in the Field of Gravity of Layered Systems not Horizontally Compressed

INTRODUCTORY EXPERIMENTAL TESTS OF DIAPIRISM†

It is now generally accepted that the diapirs and ridges of salt that cut through sedimentary strata in many regions throughout the globe have been emplaced by the buoyant force due to density contrast between salt and overburden. Simple and obvious as this view may seem to us it did not become the prevailing one until after 1930 though proposed by Arrhenius as early as 1912 for the German domes. In the comprehensive report of 1926 on salt-dome oil fields, for example, the motivating force was considered an open question (Barton, 1926; DeGolyer, 1926). In 1934, however, Nettleton showed how salt-dome structures would develop in models of two viscous fluids of dissimilar densities. These were the first dynamic models of domes driven only by the buoyant force in the field of gravity. Thus Nettleton's models were quite different from previous ones, in which the salt-imitating materials were squeezed through openings by stresses grossly out of proportion to the gravity force.

Until 1967, when the first edition of this book appeared, the most extensive study, involving especially such features as the control of dome initiation, structural deformation in the overburden, geometry of domes and marginal sinks, rate of ascent, etc. was that of Parker and McDowell (1955). In their study of some 800 models, of which about 20% were examined in detail, Parker and McDowell mainly employed asphalt as salt imitation, with wet mud and various loose powders as overburden. Many natural-looking structures developed spontaneously from the gravitationally unstable arrangement of light asphalt below heavier mud. Of special interest was the production of fracture and fault systems in the overburden, a feature not generated in models with only viscous fluids as rock-simulating materials.

ROLE OF GRAV...

† For notations used in this chapter, see list on p. 45.

SYMBOLS USED IN CHAPTERS 7 AND 8

A, B, C, D: coefficients in the velocity equations and stress equations valid for a single layer, Table 7.1

$a_i = \exp(-2\pi h_i/\lambda)$

$a, b, c, d, e, f, g, h, i, j, k, l$ elements in matrix to set of velocity equations valid for multilayers, Table 7.3

c_{ij}: coefficients in the amplitude growth equation $\bar{y}_i = f(\phi, t)$

g: acceleration due to gravity

G: ratio between stress force and gravity force:

$$\frac{2\sigma}{(\rho_1 - \rho_2)gh}$$

h_i: thickness of layer i

$h_{(i+1):i} \equiv h_{(i+1)}/h_i$.

k_{ji}: specific growth factors or coupling coefficients for growth of amplitudes in multilayers, eqn (8.55) and Table 8.2)

$P^{vi}_{i(i-1)}, P^{ui}_{ii}$, etc.: coefficients in equations for normal stress (eqns 7.24 and 7.25), referring to interface $(i - 1)$ or i on layer i, and operating on the velocity components v_i or u_i, respectively, Table 7.2

p: pressure.

$$q_i \equiv \frac{(\rho_i - \rho_{(i+1)})h_{(i+1)}g}{2\mu_{(i+1)}}$$

$q_{(i+1):i} \equiv q_{(i+1)}/q_i$

q_1^{-1}: model unit of time

$S^{vi}_{i(i-1)}, S^{ui}_{ii}$, etc.: coefficients in equations for shear stress (eqns 7.22 and 7.23), referring to interface $(i - 1)$ or i on layer i, and operating on the velocity components v_i or u_i respectively, Table 7.2

t: time

u: horizontal velocity component

$\bar{u}_{ii}$: rate of change of horitzontal tangential amplitude (at $x = 0$ and $x = n\lambda$) at lower boundary of layer i

$\bar{u}_{i(i-1)}$: same, at upper boundary of layer i

v: vertical velocity component

$\bar{v}_{ii}$: rate of change of vertical (normal) amplitude (at $x = \frac{1}{4}\lambda$ and $x = (\frac{1}{4} + n)\lambda$) at lower boundary of layer i

$\bar{v}_{i(i-1)}$: same, at upper boundary of layer i

$\bar{y}_{ii}, \bar{y}_{i(i-1)}$: normal amplitude on lower, respective upper boundary of layer i

$\bar{y}_{0i}$: initial amplitude on interface i at $t = 0$

$\dot{\bar{y}}_i \equiv \partial\bar{y}_i/\partial t \equiv \bar{v}_i$

$\bar{y}_{i:0i} \equiv \bar{y}_i/\bar{y}_{0i}$

α: thermal expansion coefficient

γ: shear strain

$\dot{\gamma} \equiv \partial\gamma/\partial t$: rate of change of shear strain

ε: longitudinal strain

$\dot{\varepsilon} \equiv \partial\varepsilon/\partial t$: rate of change of longitudinal strain

κ_i: exponent in equation $\bar{y}_i = f_i(\phi, t)$ eqns (7.64), (8.57)

λ: wavelength

λ_{max}: wavelength of dominant perturbation, i.e. of perturbation growing with maximum rate

ρ: density

$\sigma_{i(i-1)}, \sigma_{ii}$: normal stress at boundary $(i - 1)$, respective i, of layer i

$\tau_{i(i-1)}, \tau_{ii}$: tangential stress at boundary $(i - 1)$, respective i, of layer i

$\phi_i \equiv 2\pi h_i/\lambda$

$\phi_{(i+1):i} \equiv \phi_{(i+1)}/\phi_i$

ψ: stream function (solution of the biharmonic

$$\frac{\partial^4\psi}{\partial x^4} + 2\frac{\partial^4\psi}{\partial x^2\partial y^2} + \frac{\partial^4\psi}{\partial y^4} = 0)$$

$\omega \equiv 2\pi/\lambda$

μ_i: viscosity of layer i

$\mu_{(i+1):i} = \mu_{(i+1)}/\mu_i$

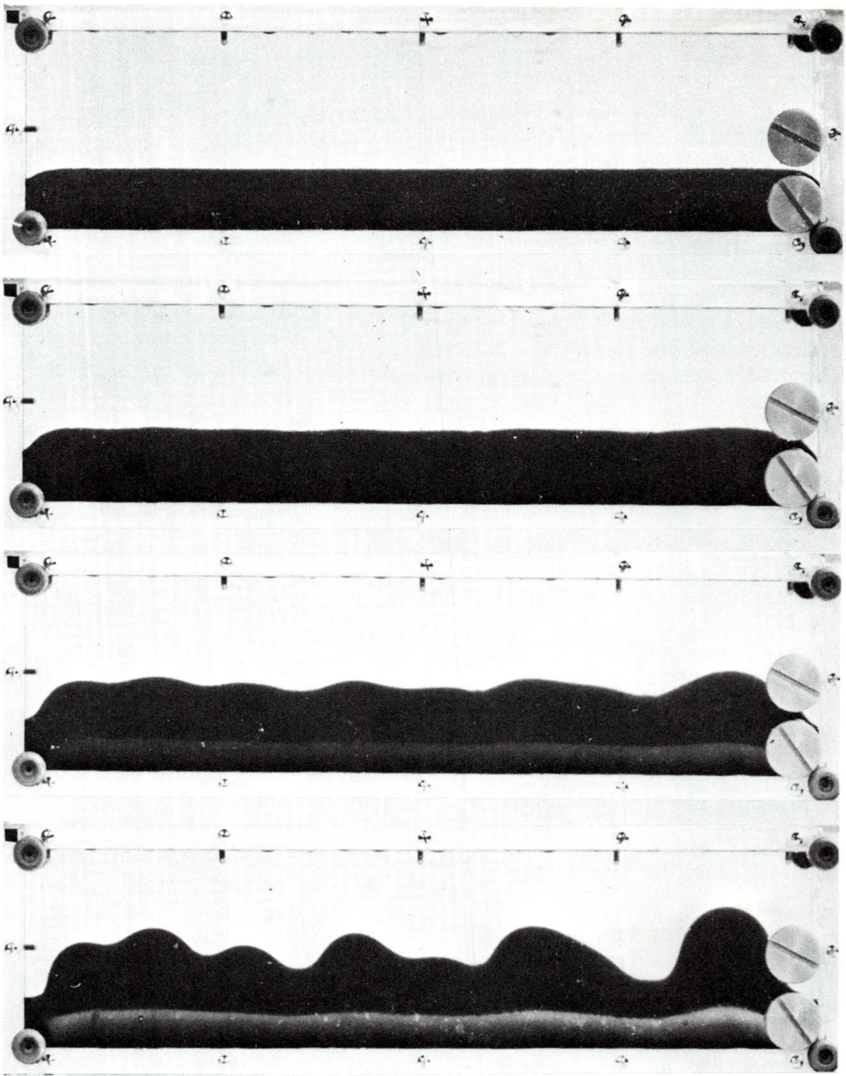

Since 1967 a large number of dome structures have been simulated by the centrifuge method, and some will be described in Chapter 11.

For the study of liquid models not run in a centrifuge, the author has adopted Nettleton's technique by using heavy oil, syrup and other liquids in closed transparent boxes of Plexiglass chiefly for demonstration in lectures. The method is also well suited to the study of certain still not fully solved problems relating to dome formation and growth, for example,

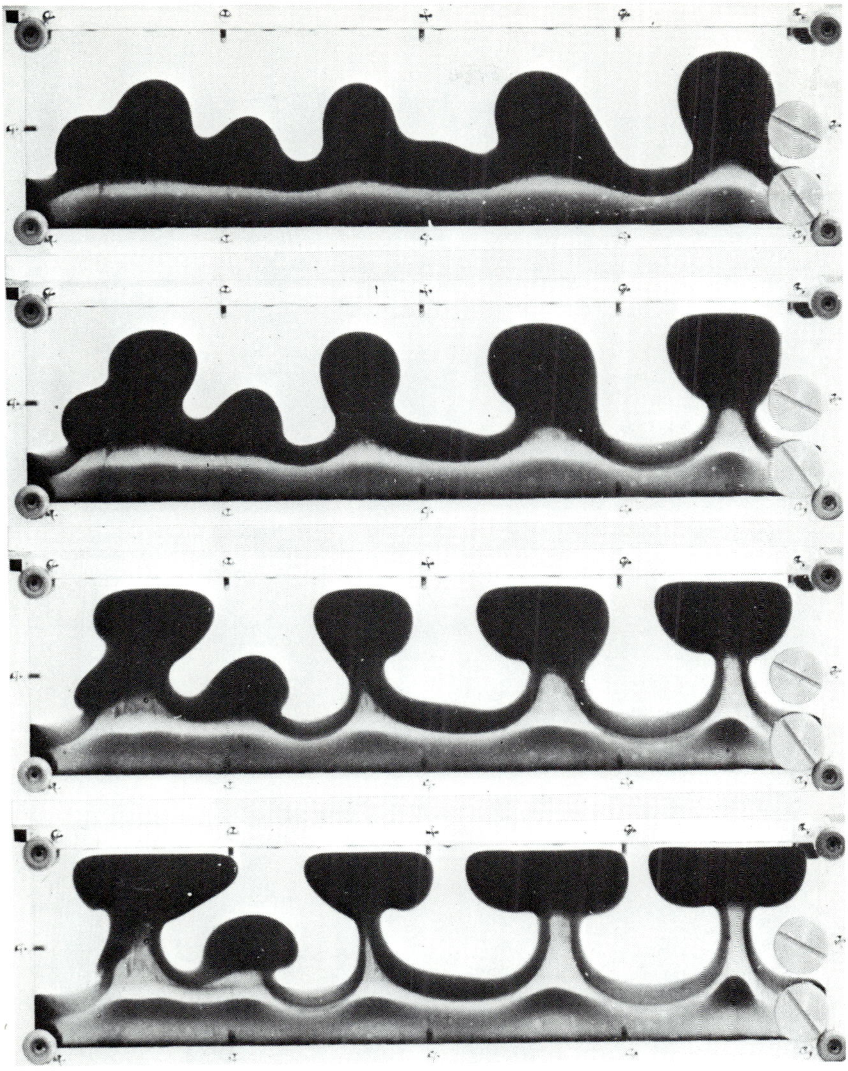

Fig. 7. 1. Successive states of diapir evolution of buoyant oil in overburden of syrup.

the control of initiation of domes from a source stratum, the relation between dome dimensions and thickness, and the rate of rise of domes relative to such parameters as the density and the effective viscosity of the materials, etc.

In the author's tests the tight rectangular Plexiglass boxes were filled

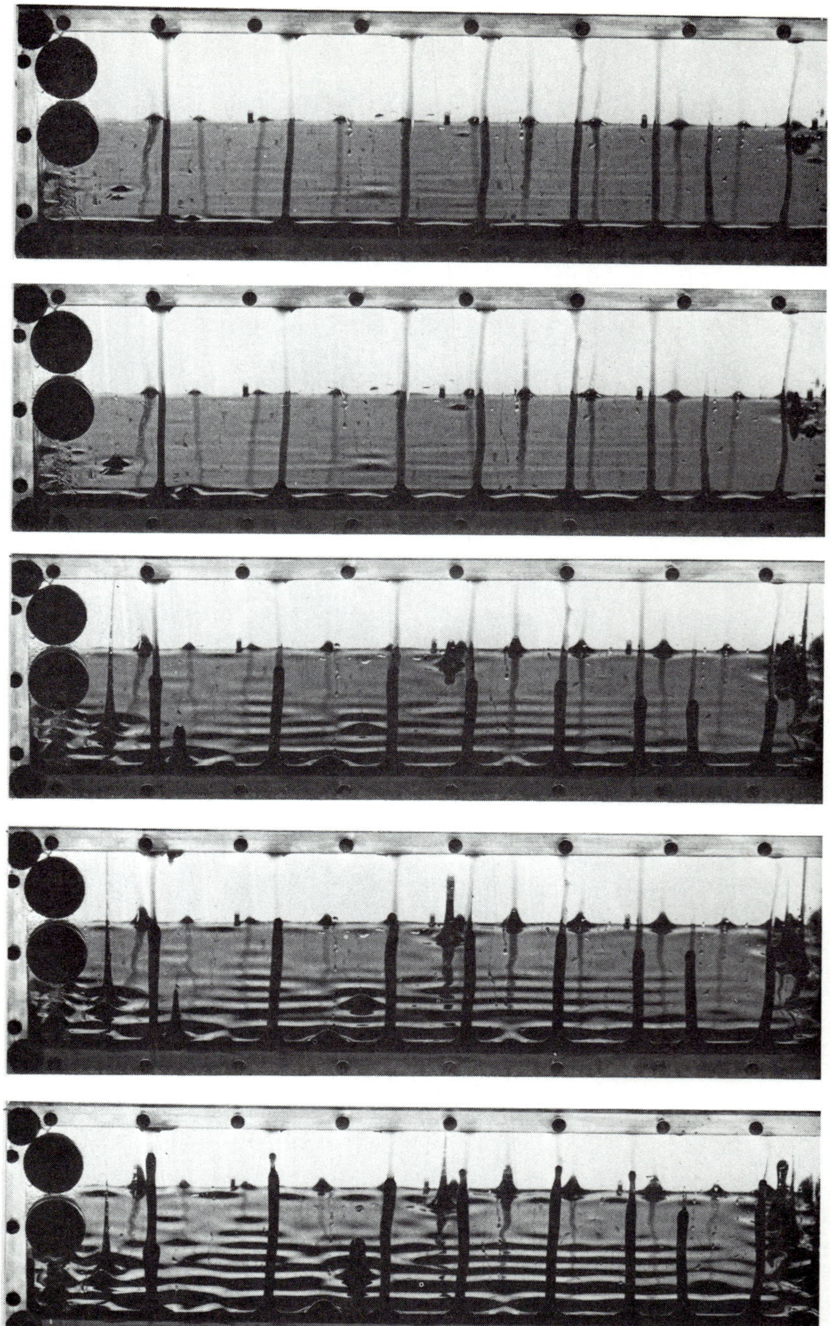

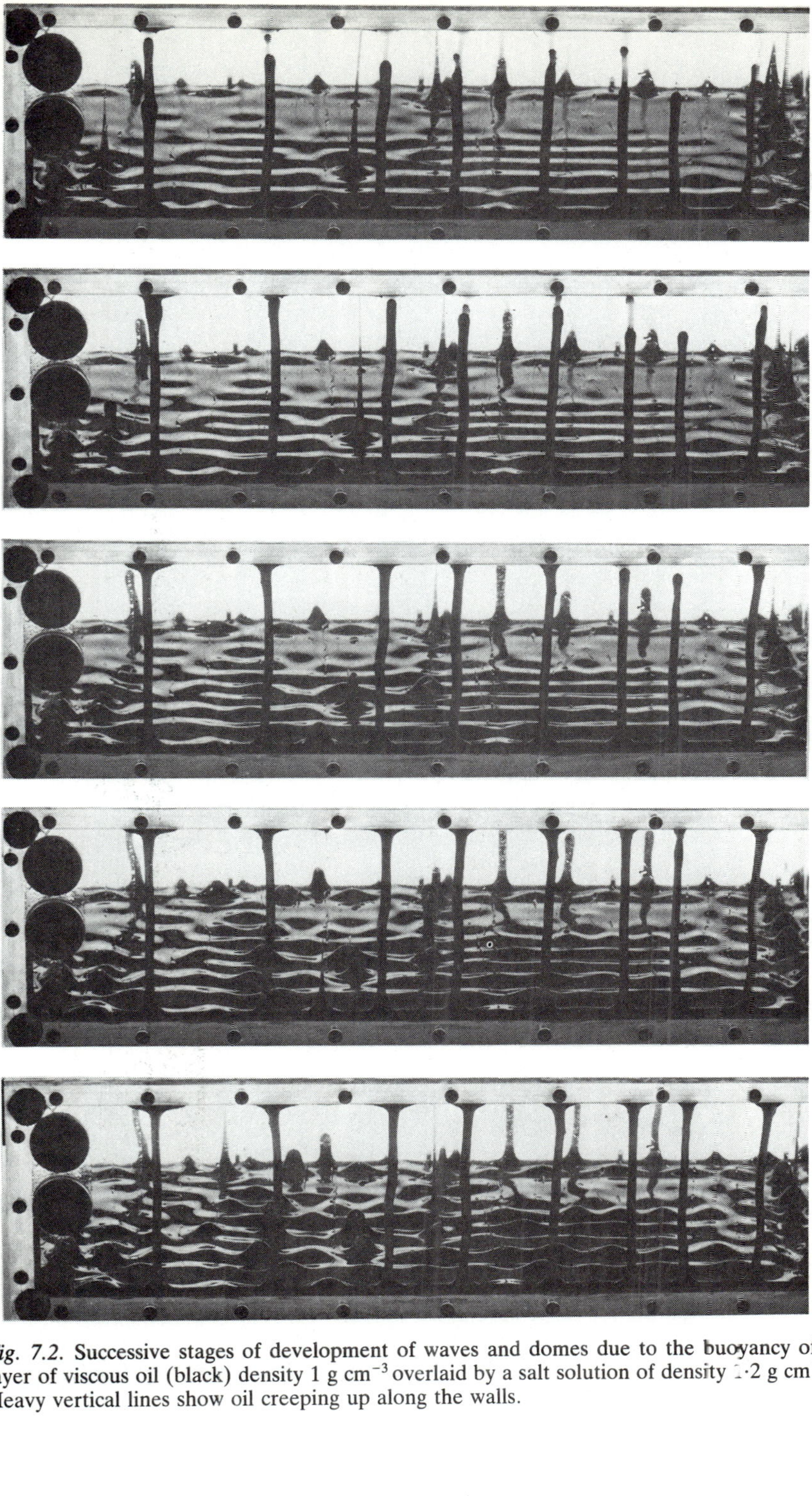

Fig. 7.2. Successive stages of development of waves and domes due to the buoyancy of a layer of viscous oil (black) density 1 g cm^{-3} overlaid by a salt solution of density 1·2 g cm^{-3}. Heavy vertical lines show oil creeping up along the walls.

(a)

(b)

(c)

(d)

(e)

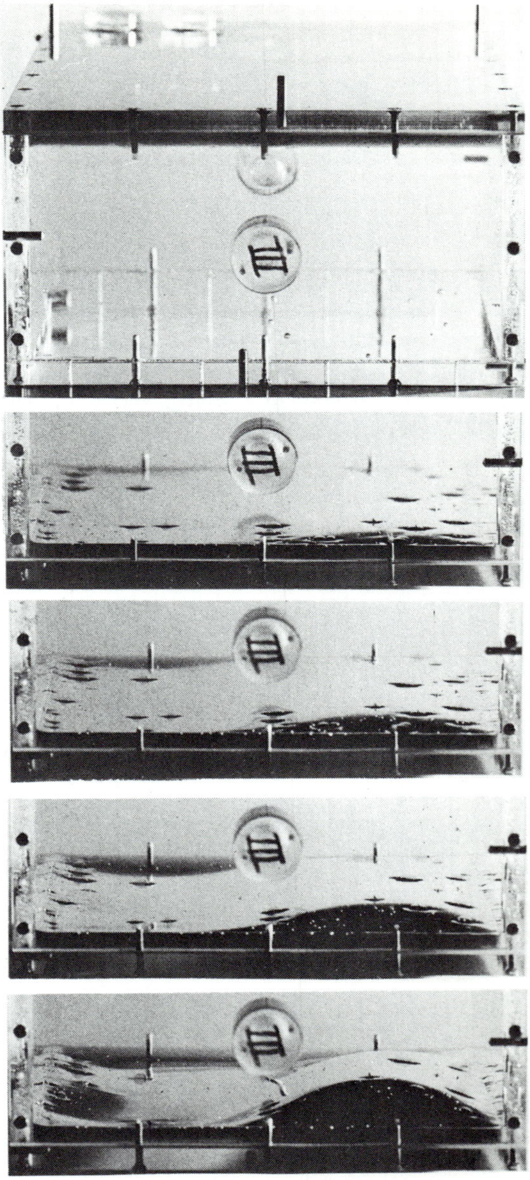

Fig. 7.3 Development of a single wave at the boundary between a buoyant oil layer 0.55 cm thick with viscosity 20.3 poise, and an overburden of stiff syrup with viscosity 5.2×10^4 poise: (*a*) stable arrangement with oil layer on top; (*b–e*) progressive stages of evolution after the container was turned upside down (from Ramberg, 1968a).

completely with the source layer of oil and the overburden of syrup or other fluid, which was somewhat more dense than the oil and did not mix with the latter. When the oil had stabilized itself as a uniform layer on top of the overburden fluid, the box was turned upside down and the evolution from this unstable situation was observed and photographed. Some results are presented in Figs 7.1 to 7.8.

If heterogeneities such as air bubbles and irregularly varying thickness of source layer or overburden occurred, dome initiation would of course be controlled by these irregularities. As might be expected, any small air bubble in the oil would initiate the rise of a dome, as would a local excess thickness of an irregular source stratum.

However, the effect of heterogeneities or irregularities in geometry or physical properties on the initiation and evolution of diapirs is best studied in the centrifuged models of solids and high-viscosity fluids to be described later. The low-viscosity liquid models now under discussion are best suited to a study of the doming process under homogeneous and regular conditions with homogeneous source and overburden materials which occur as uniformly thick layers, or strata whose thickness varies in a simple manner with distance across the model.

With a completely bubble-free source layer of oil whose uniform thickness was small relative to its lateral extension, the following conditions were observed.

On the originally plane but gravitationally unstable source-layer surface, waves with horizontal axes started to form whose amplitude increased, at first very slowly, then with increasing rate. Soon rather evenly spaced culminations developed on the anticlinal ridges, the culmination changing into circular diapirs and rising rapidly to the surface of the overburden.

The initially generated anticlines arranged themselves parallel to the outline of the container and were evenly spaced at a distance which depends upon the thickness of the source layer and the viscosity contrast. In the rectangular boxes the anticlines are parallel to the longer side of the box (Fig. 7.2); in a circular container the initial anticlines would curve parallel to the circumference.

The "edge effect" in the control of dome initiation—i.e. the condition that initial anticlines with subsequent rows of domes parallel the outline of the source layer— is very well exhibited if sheets of silicone putty (see below) are placed below an overburden of syrup. Figure 7.6 shows how in an oblong container an anticline rises along the median line, and Figs 7.7 and 7.8 show how a buoyant ridge rises parallel to the irregular outline of the source layer.

When the thickness of the source layer was changed from test to test, it was noted that the rate of dome initiation and the speed of ascent

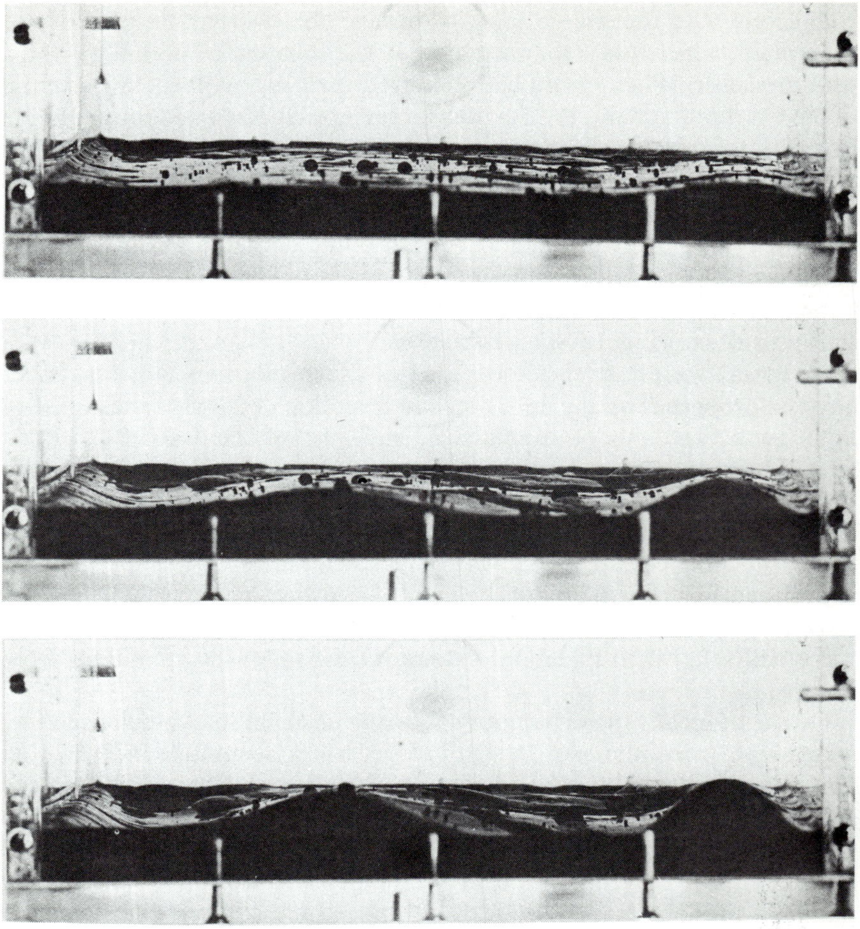

increased rapidly with increasing thickness of source layer, provided the same substances were used as imitation materials. This, incidentally, is in full agreement with model laws which claim that for similar structures with identical substances the velocity of gravity-generated movements increases in proportion to the square of the linear dimension (see p. 107). Of course, the less viscous the model materials employed, the more rapid the rise. For low-viscosity oils as source and water as overburden the rise would be so fast that turbulence occurred; usually, however, the flow within the domes was of laminar type, because the oil used was rather viscous (350–600 poise).

Increasing source-layer thickness caused enlarged spacing between

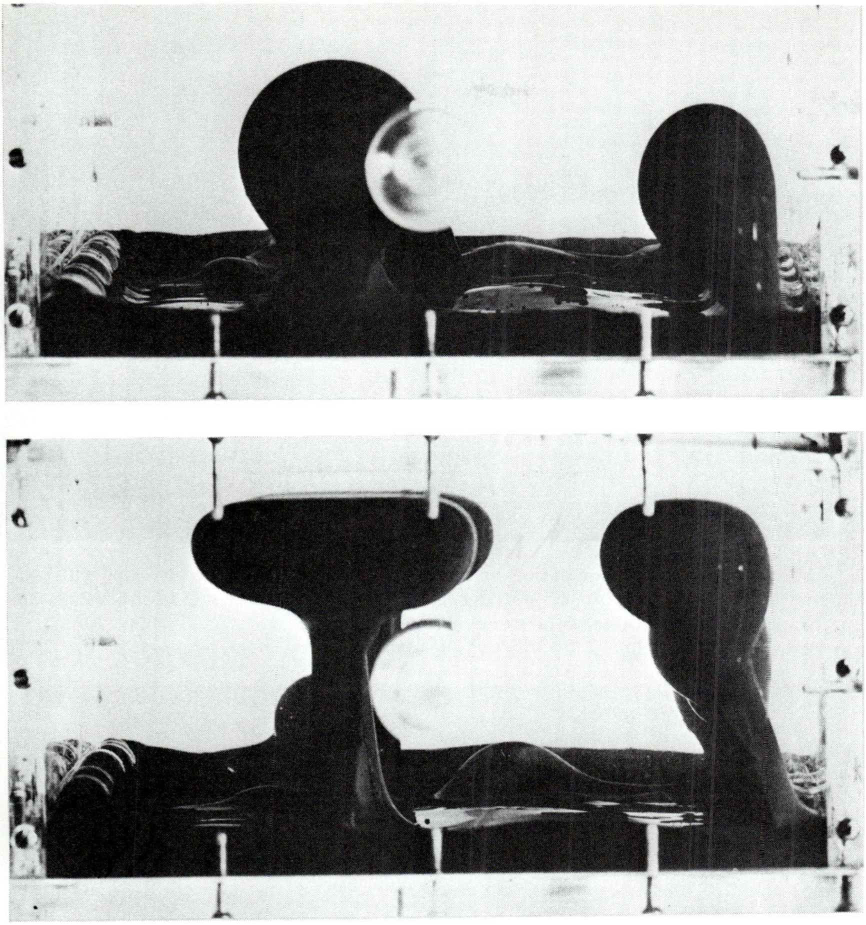

Fig. 7.4. Five stages of evolution of two rows of diapirs of viscous mineral oil rising through heavy syrup. Note that the diapir on the right tends to move away from the wall in attempt to decrease the shear-strain energy associated with the movement parallel to the wall.

initially generated anticlines and between subsequently developed culminations and domes on the ridges. Dome diameter, as well as the spacing between domes, appeared to vary linearly with source-layer thickness.

In some tests the source layer was made wedge-shaped in profile (by letting the box slope a little while the oil layer stabilized itself on the top before turning the box). The ridges and subsequent domes would then develop faster and rise more rapidly at the thick end of the layer, where the cross-section of the domes also would be larger (Fig. 7.5; see also discussion p. 107).

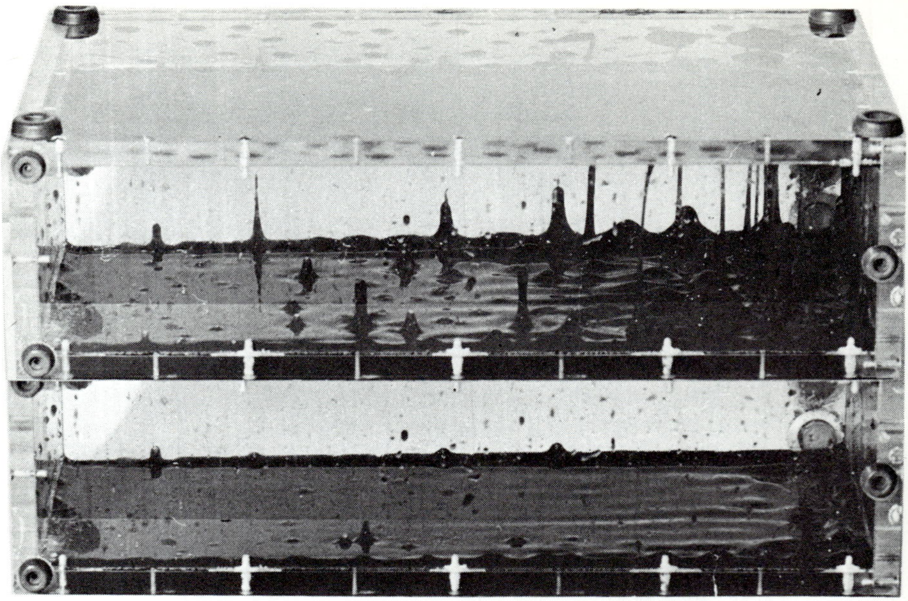

Fig. 7.5 Two stages of evolution of buoyant basal layer of oil in Plexiglass box filled with salt solution. The oil layer was originally wedge-shaped with the thicker side to the left in the figure. Bottom figure shows early stage of evolution, top figure later stage. Note continuously increased velocity of evolution towards the thick end of the oil sheet. See text.

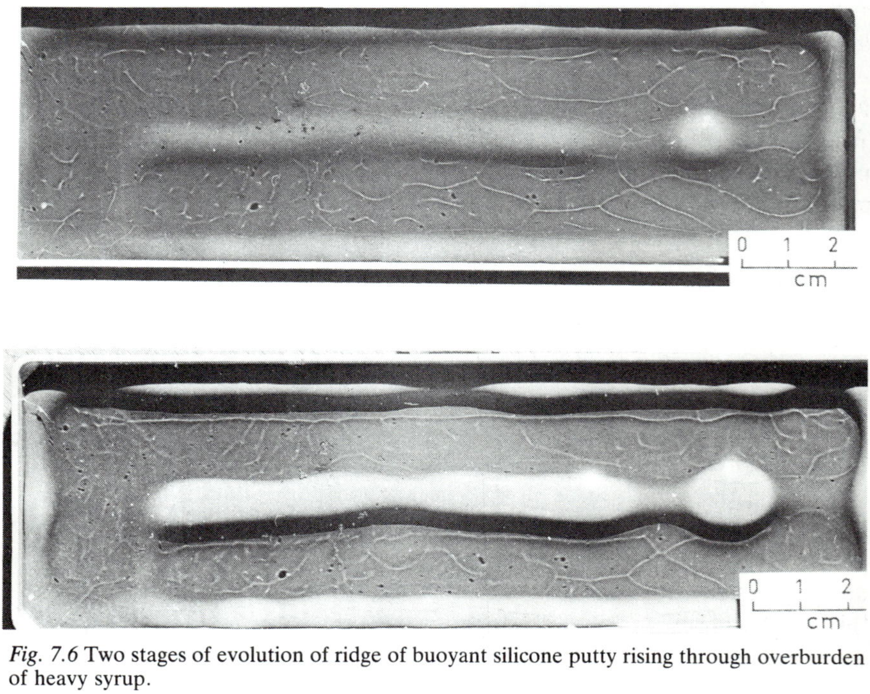

Fig. 7.6 Two stages of evolution of ridge of buoyant silicone putty rising through overburden of heavy syrup.

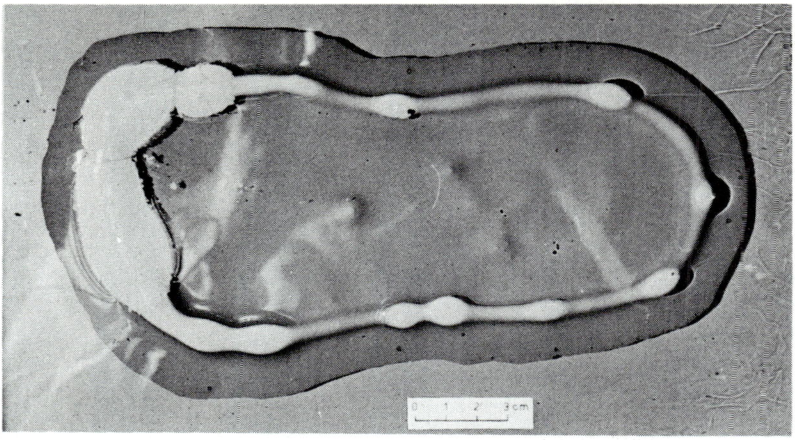

Fig. 7.7 Buoyant sheet of silicone putty below overburden of heavy syrup showing localization of anticlinal and domal rises controlled by shape of buoyant layer. See text.

Incidentally, rim synclines were characteristic features of domes formed in the tests above.

Recently, several students of salt structures have performed experimental tests (e.g. Heye, 1978). Diapir and mantle plume models have also been run experimentally by Whitehead and Luther (1975). Numerical experiments on salt diapirs have been performed by Berner *et al.* (1972) and Woidt (1978). Parmentier and Turcotte (1975) performed interesting numerical experiments on the flow pattern, as regards both heat and material in mantle plume models.

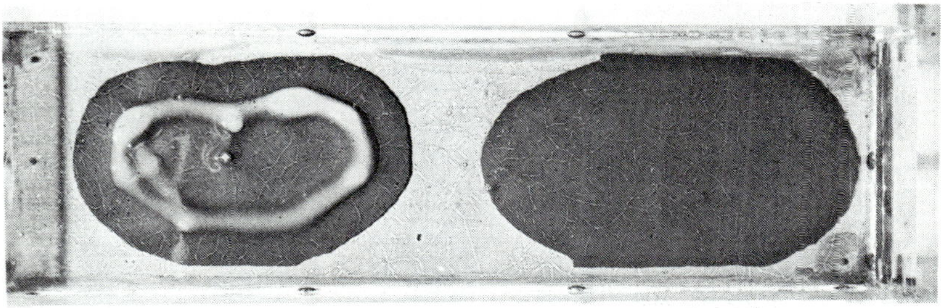

Fig. 7.8. Ridge parallel to the circumference of source layers of silicone putty rising through syrup overburden. Silicone-putty layer on the right more viscous (stiffer) than that on the left.

FLUID-DYNAMIC ANALYSIS

Introduction

In the dynamic analysis we shall treat a more general model than that exhibited by the above oil-syrup experiments in which the top boundary of the overburden is rigid and straight, as is also the bottom boundary of the buoyant stratum. In nature the overburden is limited by the surface of the Earth which is free to develop a wavy pattern reflecting the movement of the buoyant layer. The rocks subjacent to the buoyant layer—the substratum—may well be mobile enough to reflect the sinuous movement in the buoyant layer. Moreover, in nature the overburden often consists of several strata with different thicknesses, different densities and different rheological properties. Even the buoyant layer may be multiple. (These conditions are imitated in many of the centrifuged models to be described in later sections.) To make the analysis more applicable to both natural rock complexes and experimental models we shall accordingly consider multilayers in our fluid-dynamic treatment.

The objective of the analysis is primarily to obtain information on (1) the dominant wavelength (which controls the spacing of domes and anticlincal ridges), and (2) the velocity of amplitude growth—or decline—of sinusoidal deflections as a function of gemoetrical and rheological properties of the systems. The analysis that follows is mainly from Ramberg (1968b). See Lord Rayleigh (1900), Taylor (1950), Lewis (1950), Chandrasekhar (1955 and 1961), Hide (1955), Daneš (1964), Biot and Odé (1965), Selig (1965), Fletcher (1967), and Berner *et al.* (1972).

The theory is applicable equally well to the growth of deflections in systems with one or more buoyant layer as to the isostatic flattening of deflections in systems with normal density stratification. The theoretical treatment presented is rigorous under the following premises:

(1) The substances show Newtonian behaviour, i.e. their viscosity is independent of the magnitude of the deviatoric stresses applied, and their yield point is zero. This is probably not too unrealistic for crystalline rocks and mantle material under the condition of recrystallization flow and microcreep at deviatoric stresses less than those at which slip and fracture in mineral grains become significant. Observations from regional-metamorphic gneisses and schists show that recrystallization flow and microcreep account for much of the deformation structures in the Earth's crust. There are reasons to believe that such processes become increasingly important as the depth increases. Even if rocks in the process of slow creep at small

but long-lasting deviatoric stresses do not conform to the Newtonian model, one can at least qualitatively predict how the theory must be modified to fit the natural rocks. For example, if even under long-lasting action a minimum deviatoric stress is needed to make rocks flow, the perturbations on the interfaces must have a minimum amplitude and/or wavelength before growth, or flattening, begins.

(2) The inertial terms in the fluid-dynamic equations are neglected, implying that inertial forces are considered insignificant relative to the viscous force and the body force of gravity. That this is a realistic approximation for flowage of solid rocks even when their masses reach global dimensions can be shown quantitatively by comparing Chandrasekhar's theory which includes inertia with the theory in which inertia is neglected. This is discussed on pp. 77ff.

(3) The amplitude/wavelength ratio ($\bar{y}/\lambda$) of the sinusoidal waves at the boundaries between layers must be small. It is difficult to give an exact upper limit of the $\bar{y}/\lambda$ ratio permitted without rendering the theory inaccurate to the extent that it becomes useless. In this connection we note that the theory is applied to two major processes: (a) the development of diapirs and anticlinal rises in systems with inverted density stratification, and (b) the flattening of deflected boundaries in systems with normal density stratification. (Deflection means in this context deviation from the isopotential surfaces.) With respect to the doming process it is of interest to know both the geometrical distribution of domal bulges which rise from a given instable layer and the rate at which the domes ascend through the superincumbent strata. The distribution of the domes is determined at an early stage while the amplitude of the perturbations is still so small that the theory is applicable. This is actually verified by experimental checks. As the bulges rise, however, the theory becomes increasingly inaccurate for predicting the rate of upheaval.

(4) The theory applies to an orthogonal gravitational field.

In conclusion one can say that the theory is capable of predicting the spacing between domes and anticlinal rises, but that it is not valid for the velocity of rise when the amplitude exceeds say 5 to 10% of the wavelength of the perturbations. (A complete understanding of the entire inversion process of layers with inverted density stratification requires experimental scale-model work and numerical tests, e.g. by the finite element method, see Berner *et al.*, 1972; Woidt, 1978 and Hayashi and Kizaki, 1979.)

Isostatic adjustment means generally vertical movement of surface deflections whose amplitude is often small relative to the "wavelength" of

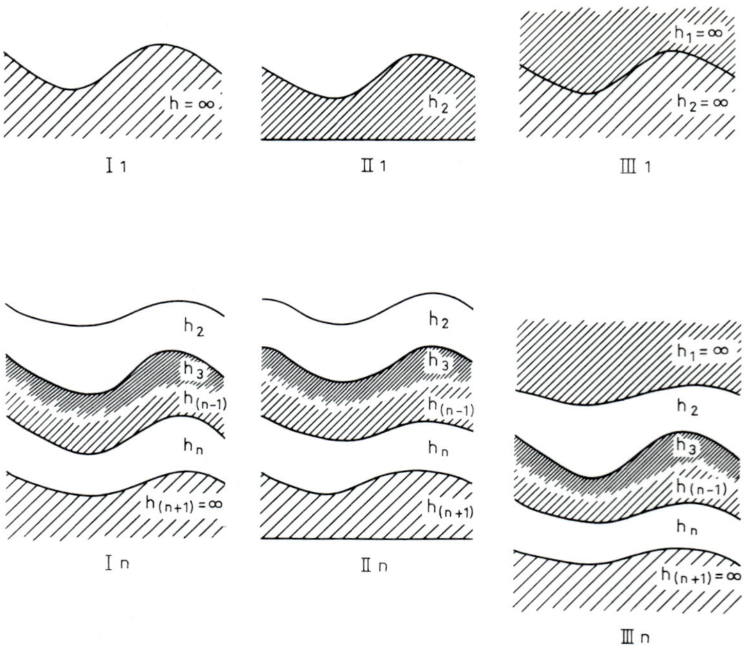

Fig. 7.9 Schematic profiles of models of types I to III; n is number of deformable boundaries and interfaces. The irregular space between layer 3 (h_3) and $n - 1$ (h_{n-1}) indicates presence of additional layers (from Ramberg, 1968b).

the affected region. For such processes the theory is valid with good approximation within the whole range of vertical movement, provided the deflections from the equipotential surfaces can be considered sinusoidal.

Figures 7.9 and 7.10 show schematic profiles across the fold axis of the models analysed, the simplest being an infinite half space with wavy surface and the more complex a multilayer with rigid top and base. Each layer is uniform with respect to density and effective viscosity, but in any one model the various layers may show different thicknesses, different densities and different viscosities. A dense layer below a less dense one means that the waviness of the interface between them is unstable in the sense that isostatic flattening tends to occur. If a layer is overlaid by a more dense one, perturbations on the unstable boundary tend to grow with time.

In the half-space model, I-1, and the single-layer model II-1, isostatic flattening is the only possible process, whereas both flattening and anticlinal or domal growth are possible in the other models depending upon whether the density stratification is normal or inverted.

The aim of the study is to determine how the rate of growth or flattening

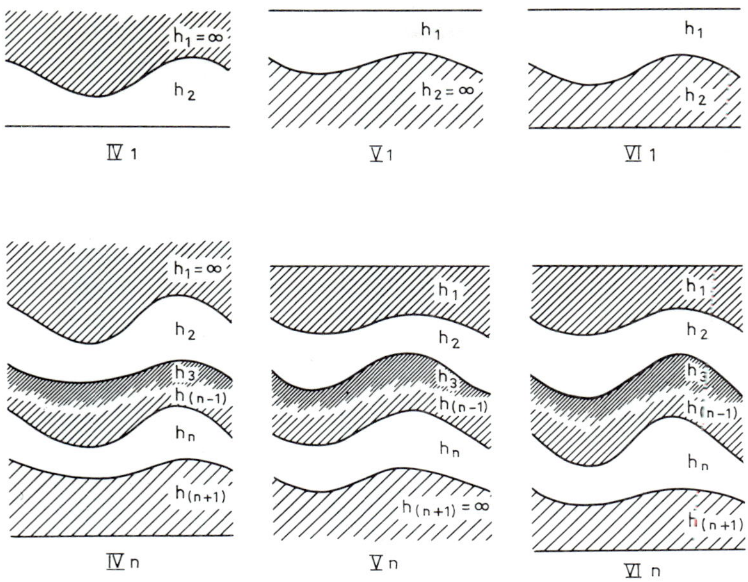

Fig. 7.10. Schematic profiles of models of types IV to VI. See legend to Fig. 7.9.

of the waves on the interfaces varies with the density and viscosity of the media, the thickness of the layers, and the wavelength of the perturbations for the various models. This study leads also to the determination of the dominant wavelength of buoyant systems.

For a solution of the problem we first determine the velocity distribution consistent with sinusoidal deflective movements of the interfaces or/and surfaces. The velocity pattern leads to a distribution of stresses according to standard fluid-dynamic relations. The conditions for dynamic equilibrium, viz. that the normal stress and the shear stress are continuous across the interfaces and vanish at the free surface, give solutions in terms of rate of change of amplitude as a function of viscosities, densities, number of layers in the complex, thickness of the layers, and wavelength of the sinusoidal perturbations.

For certain problems solutions in terms of rate of change of amplitude give sufficient information, for example when one wants to know the wavelength/thickness ratio of the dominant perturbation in a system with only two layers. The spacing between diapirs and between gravity-generated anticlines is chiefly controlled by the wavelength of the perturbation which grows with maximum speed. This also means maximum rate of dissipation of energy input.

For domal growth in structures with more than two layers and certain other geodynamic problems, the rate of change of amplitude is not sufficient information: it is often also necessary to know the magnitude of the amplitude after a certain lapse of time. This is particularly important for problems related to isostatic adjustment. The mathematical theory employed solves such integration problems, which for multilayers with several coupled movable interfaces involve the solution of simultaneous differential equations. The theory is limited to small amplitude/wavelength ratios so the integration must not be carried beyond an amplitude/wavelength ratio equal to about 0·1.

Velocity relations for a single layer

Consider a layer with thickness h whose upper surface coincides with the plane $y = 0$ in an orthogonal co-ordinate system, y being the vertical direction in the field of gravity, reckoned positive upward (Fig. 7.11). The

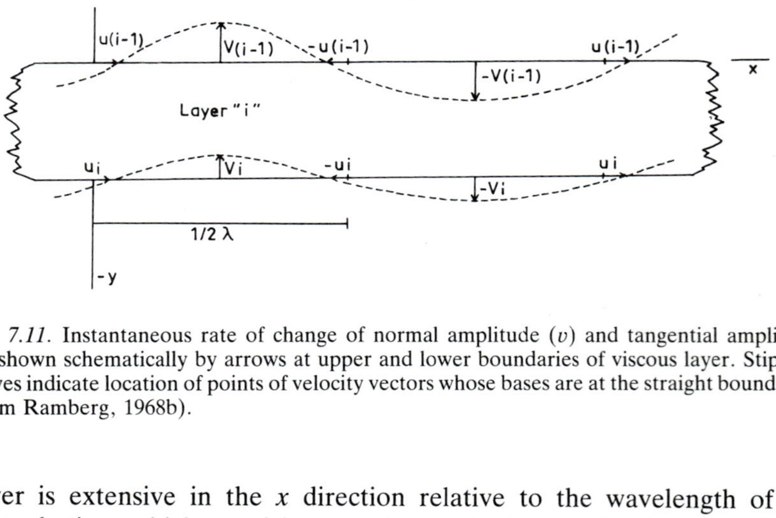

Fig. 7.11. Instantaneous rate of change of normal amplitude (v) and tangential amplitude (u) shown schematically by arrows at upper and lower boundaries of viscous layer. Stippled curves indicate location of points of velocity vectors whose bases are at the straight boundaries (from Ramberg, 1968b).

layer is extensive in the x direction relative to the wavelength of the perturbations which we wish to consider on its boundaries. The movement is two-dimensional in the x, y plane, and the initially very gentle perturbations are sinusoidal of type

$$y = \bar{y} \sin(\omega x), \tag{7.1}$$

as is the rate of change of amplitude of the deflections

$$v = \bar{v} \sin(\omega x), \tag{7.2}$$

where $v \equiv \partial y/\partial t$, $\bar{y}$ is amplitude, $\bar{v}$ rate of change of amplitude, $\omega \equiv 2\pi/\lambda$, and λ is wavelength.

It can be demonstrated by means of the stream function that the sine variation of the normal velocity, v, at a boundary leads to cosine variation of the tangential velocity, u, at the boundary and exponential variation of both velocity components with y within the layer. Also terms of type: $y \exp(\pm\omega y)$ occur. The general velocity equations are

$$u = [A \exp(\omega y) + B \exp(-\omega y) + Cy \exp(\omega y) + Dy \exp(-\omega y)] \cos(\omega x), \quad (7.3)$$

$$v = [A \exp(\omega y) - B \exp(-\omega y) + C(y - 1/\omega) \exp(\omega y) - D(y + 1/\omega) \exp(-\omega y)] \sin(\omega x). \quad (7.4)$$

A, B, C and D are coefficients later to be determined in terms of the amplitude velocities at the upper and lower boundaries of the layer.

For plane non-inertial strain of an incompressible Newtonian fluid the stream function, ψ, is a solution of the biharmonic equation

$$\frac{\partial^4\psi}{\partial x^4} + 2\frac{\partial^4\psi}{\partial x^2\partial y^2} + \frac{\partial^4\psi}{\partial y^4} = 0 \quad \text{in which} \quad \frac{\partial\psi}{\partial x} = v \quad \text{and} \quad \frac{\partial\psi}{\partial y} = -u.$$

For information on the stream function and its significance and properties in general, see e.g. Jaeger (1956) and Cole (1962). See also Chapter 9.

Determinations of the coefficients in the velocity equations

The most general boundary conditions for a layer considered in the present analyses of models (Figs 7.9, 7.10 and 7.11) are that sine waves occur at both the upper and the lower boundary, i.e. at $y = 0$ and $y = -h$. The conditions of two infinite half spaces joined along a flexible boundary, and of layers with one straight rigid boundary are special cases accounted for by modifications of the coefficients of the general layer.

At the upper boundary of a wavy layer let the rate of growth or flattening of the deflections be

$$v = \bar{v}_t \sin(\omega x), \quad (7.5)$$

and the horizontal velocity component

$$u = \bar{u}_t \cos(\omega x). \quad (7.6)$$

When $y = 0$ the general velocity equations (7.3) and (7.4) become identical to eqns (7.5) and (7.6) respectively, and we obtain

$$\bar{u}_{\mathrm{t}} = A + B \tag{7.7}$$

and

$$\bar{v}_{\mathrm{t}} = A - B - \frac{C}{\omega} - \frac{D}{\omega}. \tag{7.8}$$

Here subscript t indicates amplitude velocity at the top boundary of the layer.

At the lower boundary of the layer the velocity components are

$$v = \bar{v}_{\mathrm{b}} \sin(\omega x), \tag{7.9}$$

and

$$u = \bar{u}_{\mathrm{b}} \cos(\omega x), \tag{7.10}$$

the subscript b referring to amplitude velocities at the bottom boundary of the layer. Inserting $y = -h$ in the general velocity equations (7.3) and (7.4) makes them identical to eqns (7.9) and (7.10). Consequently

$$\bar{u}_{\mathrm{b}} = Aa + Ba^{-1} - Cha - Dha^{-1} \tag{7.11}$$

and

$$\bar{v}_{\mathrm{b}} = Aa - Ba^{-1} - C\left(h + \frac{1}{\omega}\right)a + D\left(h - \frac{1}{\omega}\right)a^{-1} \tag{7.12}$$

where $a \equiv \mathrm{e}^{-h\omega}$.

The amplitudes are considered small relative to h, the wavelength is the same on both boundaries, and the two trains of waves are "in phase", i.e. the inflection points of the two sets of waves occur initially at the same x-values.

Expressions for the coefficients A, B, etc., in terms of the boundary conditions $\bar{v}_{\mathrm{t}}$, $\bar{u}_{\mathrm{t}}$, $\bar{v}_{\mathrm{b}}$ and $\bar{u}_{\mathrm{b}}$ obtained by solving the simultaneous eqns (7.7), (7.8), (7.11) and (7.12) are found in Table 7.1. The amplitude velocities $\bar{v}_{\mathrm{t}}$, $\bar{u}_{\mathrm{t}}$, $\bar{v}_{\mathrm{b}}$ and $\bar{u}_{\mathrm{b}}$ have been replaced by $\bar{v}_{i(i-1)}$, $\bar{u}_{i(i-1)}$ $\bar{v}_{ii}$ and $\bar{u}_{ii}$, respectively.

The index i refers to the ith layer in a multilayer model, v_{ii} and u_{ii} being the normal and tangential velocity components at the lower boundary of layer i, and $v_{i(i-1)}$, $u_{i(i-1)}$ those at the upper boundary of the same layer. That is, the first index on the velocity signs refers to the layer i and the second index refers to the upper $(i - 1)$ or lower (i) boundary of layer i. The layers themselves in the models are indexed from 1 (or in some cases 2) to $n + 1$, counting from top to bottom. As we are only considering welded contacts and continuous velocities across contacts, it follows that

Table 7.1. Coefficients in equations (7.7, 7.8, 7.11 and 7.12)

$$A_i \equiv [-2\phi_i^2\bar{v}_{i(i-1)} + (e^{2\phi_i} - 1 - 2\phi_i - 2\phi_i^2)\bar{u}_{i(i-1)} + 2\phi_i \sinh\phi_i \bar{v}_{ii} + 2(\phi_i \cosh\phi_i - \sinh\phi_i)\bar{u}_{ii}]\, \nu_i^{-1}$$

$$B_i \equiv [2\phi_i^2\bar{v}_{i(i-1)} + (e^{-2\phi_i} - 1 + 2\phi_i - 2\phi_i^2)\,\bar{u}_{i(i-1)} - 2\phi_i \sinh\phi_i \bar{v}_{ii} + 2\,(\sinh\phi_i - \phi_i \cosh\varphi_i)\bar{u}_{ii}]\, \nu_i^{-1}$$

$$C_i \equiv [(\omega - 2\phi_i\omega - e^{2\phi_i}\,\omega)\,\bar{v}_{i(i-1)} + (e^{2\phi_i}\,\omega - 2\phi_i\omega - \omega)\,\bar{u}_{i(i-1)} + 2(\omega \sinh\phi_i + e^{\phi_i}\,\phi_i\,\omega)\,\bar{v}_{ii} - 2(\omega \sinh\phi_i - e^{\phi_i}\,\phi_i\omega)\,\bar{u}_{ii}]\, \nu_i^{-1}$$

$$D_i \equiv [(\omega + 2\phi_i\omega - e^{-2\phi_i}\omega)\,\bar{v}_{i(i-1)} + (\omega - e^{-2\phi_i}\,\omega - 2\phi_i\omega)\,\bar{u}_{i(i-1)} - 2(\omega \sinh\phi_i + e^{-\phi_i}\,\omega\phi_i)\,\bar{v}_{ii} - 2(\omega \sinh\phi - e^{-\phi_i}\,\phi_i\omega)\,\bar{u}_{ii}]\, \nu_i^{-1}$$

$$\nu \equiv 2\cosh 2\phi_i - 2 - 4\phi_i^2; \qquad \phi_i \equiv \omega h_i$$

$\bar{v}_t$, $\bar{u}_t$, $\bar{v}_b$ and $\bar{u}_b$ in eqns (7.7) to (7.12) have been replaced by $\bar{v}_{i(i-1)}$, $\bar{u}_{i(i-1)}$, $\bar{v}_{ii}$ and $\bar{u}_{ii}$ respectively.

the velocity components at the bottom of a layer bear the same second index as the velocity components at the top of the layer below.

Determination of the stresses at the boundaries of a deforming layer

The velocity distributions implied in the expressions for the coefficients *A*, *B*, etc., are related to the normal stress, σ_y, and the shear stress, τ_{xy}, acting at the boundaries. This is true since deviatoric compression or tension parallel to layering is not included in the models. Models exposed to deviatoric layer-parallel compression are treated in Chapter 8. The relationships follow from the general velocity equations (7.3 and 7.4) in combination with the following conventional fluid-dynamic formulas valid for two-dimensional non-inertial flow of incompressible Newtonian substances, viz.

$$\sigma_y = 2\mu\dot{\varepsilon}_y - p, \tag{7.13}$$

$$\tau_{xy} = \mu\dot{\gamma}_{xy}, \tag{7.14}$$

$$\dot{\varepsilon}_y \equiv \partial v/\partial y; \qquad \dot{\gamma}_{xy} = \partial u/\partial y + \partial v/\partial x, \tag{7.15}$$

$$\partial p/\partial x = \mu(\partial^2 u/\partial x^2 + \partial^2 u/\partial y^2), \tag{7.16}$$

$$\partial p/\partial y = \mu(\partial^2 v/\partial x^2 + \partial^2 v/\partial y^2) - \rho g, \tag{7.17}$$

where μ is viscosity coefficient, p pressure, ρ density and g acceleration due to gravity. This routine operation yields the sought relationships

$$\tau_{i(i-1)} = 2\mu_i\omega(A_i - B_i)\cos(\omega x) \tag{7.18}$$

and

$$\tau_{ii} = 2\mu_i\omega(A_i a_i - B_i a_i^{-1} - C_i h_i a_i + D_i h_i a_i^{-1})\cos(\omega x), \tag{7.19}$$

for the shear stresses at the upper and lower boundaries respectively, of layer i. The normal stresses become

$$\begin{aligned}\sigma_{i(i-1)} = {} & 2\mu_i(A_i\omega + B_i\omega - C_i + D_i)\sin(\omega x) \\ & + \rho_i\bar{y}_{i(i-1)}g\sin(\omega x) - p_{(i-1)},\end{aligned} \tag{7.20}$$

$$\begin{aligned}\sigma_{ii} = {} & 2\mu_i(A_i\omega a_i + B_i\omega a_i^{-1} - C_i(\omega h_i + 1)a_i \\ & - D_i(\omega h_i - 1)a_i^{-1})\sin(\omega x) + \rho_i\bar{y}_{ii}g\sin(\omega x) - p_i,\end{aligned} \tag{7.21}$$

respectively, for the upper and lower boundaries, where $\bar{y}_{i(i-1)}$ and $\bar{y}_{ii}$ are the amplitudes of the deflections at the upper and lower boundaries, $p_{i(i-1)}$ and p_i are pressures at the inflection points.

Introduction of the expressions for the coefficients A, B, etc.—as determined above—into eqns (7.18), (7.19), (7.20), and (7.21) gives directly the relationships between the stresses and the velocities at the upper and lower boundaries. For layer i the relationships are

$$\begin{aligned}\tau_{i(i-1)} = {} & 2\omega\mu_i(S_{i(i-1)}^{v(i-1)}\bar{v}_{i(i-1)} + S_{i(i-1)}^{u(i-1)}\bar{u}_{i(i-1)} \\ & + S_{i(i-1)}^{v_i}\bar{v}_{ii} + S_{i(i-1)}^{u_i}\bar{u}_{ii})\cos\omega x,\end{aligned} \tag{7.22}$$

$$\begin{aligned}\tau_{ii} = {} & 2\omega\mu_i(S_{ii}^{v(i-1)}\bar{v}_{i(i-1)} + S_{ii}^{u(i-1)}\bar{u}_{i(i-1)} \\ & + S_{ii}^{v_i}\bar{v}_{ii} + S_{ii}^{u_i}\bar{u}_{ii})\cos\omega x,\end{aligned} \tag{7.23}$$

$$\begin{aligned}\sigma_{i(i-1)} = {} & 2\mu_i h_i^{-1}(P_{i(i-1)}^{v(i-1)}\bar{v}_{i(i-1)} + P_{i(i-1)}^{u(i-1)}\bar{u}_{i(i-1)} \\ & + P_{i(i-1)}^{v_i}\bar{v}_{ii} + P_{i(i-1)}^{u_i}\bar{u}_{ii})\sin\omega x \\ & + \rho_i\bar{y}_{i(i-1)}g\sin\omega x - p_{(i-1)},\end{aligned} \tag{7.24}$$

$$\begin{aligned}\sigma_{ii} = {} & 2\mu_i h_i^{-1}(P_{ii}^{v(i-1)}\bar{v}_{i(i-1)} + P_{ii}^{u(i-1)}\bar{u}_{i(i-1)} \\ & + P_{ii}^{v_i}\bar{v}_{ii} + P_{ii}^{u_i}\bar{u}_{ii})\sin\omega x + \rho_i\bar{y}_{ii}g\sin\omega x - p_i,\end{aligned} \tag{7.25}$$

The indexing of the coefficients P and S follows the same rules as applied on the velocities. The first suffix refers to the layer to which the coefficient belongs and the second suffix associates the coefficient with the upper surface $(i - 1)$ or with the lower surface (i), respectively, of the layer. The affixes $v_{(i-1)}$, u_i, etc. indicate the velocity component to which the coefficients relate. So, for example, $S_{i(i-1)}^{v(i-1)}$ is a measure of how much the stress $\tau_{i(i-1)}$ is influenced by the velocity component $v_{i(i-1)}$. If $S_{i(i-1)}^{v(i-1)}$ is relatively large, a correspondingly large percentage of $\tau_{i(i-1)}$ is due to the velocity

Table 7.2 Coefficients in equations (7.22) to (7.25)

For finite thickness of general layer i:

$$S^{v(i-1)}_{i(i-1)} = S^{vi}_{ii} = \frac{-2\phi_i^2}{\cosh 2\phi_i - 1 - 2\phi_i^2}$$

$$S^{u(i-1)}_{i(i-1)} = -S^{ui}_{ii} = \frac{\sinh 2\phi_i - 2\phi_i}{\cosh 2\phi_i - 1 - 2\phi_i^2}$$

$$S^{vi}_{i(i-1)} = S^{v(i-1)}_{ii} = \frac{2\phi_i \sinh \phi_i}{\cosh 2\phi_i - 1 - 2\phi_i^2}$$

$$S^{ui}_{i(i-1)} = -S^{ui(i-1)}_{ii} = \frac{2(\phi_i \cosh \phi_i - \sinh \phi_i)}{\cosh 2\phi_i - 1 - 2\phi_i^2}$$

$$P^{v(i-1)}_{i(i-1)} = -P^{vi}_{ii} = \frac{\phi_i(\sinh 2\phi_i + 2\phi_i)}{\cosh 2\phi_i - 1 - 2\phi_i^2}$$

$$P^{u(i-1)}_{i(i-1)} = P^{ui}_{ii} = \frac{-2\phi_i^3}{\cosh 2\phi_i - 1 - 2\phi_i^2}$$

$$P^{vi}_{i(i-1)} = -P^{v(i-1)}_{ii} = \frac{-2\phi_i(\phi_i \cosh \phi_i + \sinh \phi_i)}{\cosh 2\phi_i - 1 - 2\phi_i^2}$$

$$P^{ui}_{i(i-1)} = P^{u(i-1)}_{ii} = \frac{-2\phi_i^2 \sinh \phi_i}{\cosh 2\phi_i - 1 - 2\phi_i^2}$$

For infinite thickness of overburden, layer 1, and substratum, layer $(n+1)$:

$S^{u1}_{11} = -1$, $S^{v1}_{11} = 0$, $P^{v1}_{11} = -\omega h_1$, $P^{u1}_{11} = 0$, $S^{un}_{(n+1)n} = 1$,

$S^{un}_{(n+1)n} = 0$, $P^{vn}_{(n+1)n} = \omega h_{(n+1)}$, $P^{un}_{(n+1)n} = 0$.

All other P- and S-coefficients are irrelevant since the velocities at infinite distance above boundary 1 and below boundary n vanish ($\bar{v}_o = \bar{u}_o = \bar{v}_{(n+1)} = \bar{u}_{(n+1)} = 0$). Note that the infinite thicknesses h_1 and $h_{(n+1)}$ in the P-coefficients vanish in the stress equations (7.27) and (7.29) because the coefficients are multiplied by $2\mu_1 h_1^{-1}$ or $2\mu_{(n+1)} h_{(n+1)}^{-1}$, cf. eqns (7.24) and (7.25).

$v_{i(i-1)}$; if the same coefficient is small, $v_{i(i-1)}$ contributes little to the stress, and so on.

As shown in Table 7.2, the coefficients S and P contain only the ratio between the layer thickness and the wavelength of perturbations (implied in the symbol ϕ). It follows that the shear stress at the two boundaries of a layer i with given thickness and consisting of a substance with given viscosity, is completely determined by the four amplitude velocity components $\bar{v}_{ii}$, $\bar{v}_{i(i-1)}$, $\bar{u}_{ii}$, $\bar{u}_{i(i-1)}$, by the wavelength of the perturbations and by the position vector x. The normal stress is controlled by the same factors in addition to the density and the magnitude of the amplitudes $\bar{y}_{i(i-1)}$ and $\bar{y}_{ii}$. The effect of the amplitude on the normal stress is due to the body force of gravity, which acts on the perturbations at the boundary between substances with different densities. The normal stress also depends on the hydrostatic pressure at the inflection points, but this pressure does not contribute to the deviatoric stress responsible for the evolution of the waves.

The top and bottom layers, layer 1 and $(n + 1)$ respectively, are different from the rest in the sense that layers 1 and $(n + 1)$ are either infinite half spaces or delimited by a rigid top or a rigid bottom respectively. In either case v_{10}, u_{10}, $v_{(n+1)(n+1)}$ and $u_{(n+1)(n+1)}$ vanish, for which reason the corresponding coefficients $S_{11}^{v_0}$, $S_{11}^{u_0}$, etc, are not included in the stress expressions. Moreover, for an overburden and a substratum that are infinitely thick (i.e. infinite half-spaces) the coefficients $S_{11}^{v_1}$, $S_{11}^{u_1}$, $S_{(n+1)n}^{v_n}$, $S_{(n+1)n}^{u_n}$ and the corresponding coefficients $P_{11}^{v_1}$, $P_{11}^{u_1}$ etc. change because ϕ_1 and $\phi_{(n+1)}$ are infinite for infinite half-spaces. Accordingly, the stresses for infinite top and bottom layers are:

$$\tau_{11}^{\infty} = -2\mu_1\omega\bar{u}_{11}\cos(\omega x) \tag{7.26}$$

$$\sigma_{11}^{\infty} = -2\mu_1\omega\bar{v}_{11}\sin(\omega x) + \rho_1\bar{y}_{11}g\sin(\omega x) - p_1 \tag{7.27}$$

$$\tau_{(n+1)n}^{\infty} = 2\mu_{(n+1)}\omega u_{(n+1)n}\cos(\omega x) \tag{7.28}$$

$$\sigma_{(n+1)n}^{\infty} = 2\mu_{(n+1)}\omega v_{(n+1)n}\sin(\omega x) + \rho_{(n+1)}\bar{y}_{(n+1)n}g\sin(\omega x) - p_n \tag{7.29}$$

For layer 1 delimited by a rigid top the stresses at its lower boundary are

$$\tau'_{11} = 2\mu_1\omega(S_{11}^{v_1}\bar{v}_{11} + S_{11}^{u_1}\bar{u}_{11})\cos(\omega x) \tag{7.30}$$

$$\sigma'_{11} = 2\mu_1 h_1^{-1}(P_{11}^{v_1}\bar{v}_{11} + P_{11}^{u_1}\bar{u}_{11})\sin(\omega x) + \rho_1\bar{y}_{11}g\sin(\omega x) - p_1 \tag{7.31}$$

The stresses at the upper boundary of layer $(n + 1)$ with rigid base, are

$$\tau'_{(n+1)n} = 2\mu_{(n+1)}\omega(S_{(n+1)n}^{v_1}\bar{v}_{(n+1)n} + S_{(n+1)n}^{u_1}\bar{u}_{(n+1)n})\cos(\omega x), \tag{7.32}$$

and

$$\begin{aligned}\sigma'_{(n+1)n} = {} & 2\mu_{(n+1)}h_{(n+1)}^{-1}(P_{(n+1)n}^{v_n}\bar{v}_{(n+1)n} + P_{(n+1)n}^{u_n}\bar{u}_{(n+1)n})\sin(\omega x) \\ & + \rho_{(n+1)}\bar{y}_{(n+1)n}g\sin(\omega x) - p_n.\end{aligned} \tag{7.33}$$

The symbol ∞ attached to the stresses indicates infinite thickness of the top and bottom layers, and a prime (′) refers to limited thickness and a straight rigid top or base of the layers. For further discussion it is convenient to distinguish between these two possible geometries of the two terminal layers in the models.

It is significant to note that the only variable in S and P is the ratio between the thickness of a layer and the wavelength of perturbations. Now, for any given layered system of the kind under consideration the thicknesses of the layers do not change appreciably with time as long as the amplitude is small relative to h. In such multilayers the only variable which can affect the coefficients A, B, C and D and thus also influence the coefficients S and P, is the wavelength of possible perturbations. Hence the rate of change of amplitude $\bar{v}_{ij}$ and $\bar{u}_{ij}$ of the perturbations in a given system will in general change with the wavelength, as will the boundary stresses, τ_{ij} and σ_{ij} ($j = i$ or $j = (i - 1)$).

Velocity–stress relations in composite models

As the relationships between the stresses and velocities at the boundaries of an individual layer are now determined we can proceed to study how a multilayer behaves under the condition of non-inertial dynamic equilibrium. Dynamic equilibrium requires that the normal stress and the shear stress are continuous at the boundaries. For multilayers with n flexible interfaces and $(n + 1)$ layers the following set of equations accordingly occurs:

$$\left.\begin{array}{ll}
\tau_{11} = \tau_{21} & \text{first interface} \\
\sigma_{11} = \sigma_{21} & \\
\tau_{22} = \tau_{32} & \text{second interface} \\
\sigma_{22} = \sigma_{32} & \\
\vdots & \\
\tau_{nn} = \tau_{(n+1)n} & n\text{th interface} \\
\sigma_{nn} = \sigma_{(n+1)n} &
\end{array}\right\} \qquad (7.34)$$

We consider generally welded contacts between the layers for which reason the velocities are continuous across the contacts. Consequently, the following equations are also taken into account in the models:

$$\left.\begin{array}{l}
\bar{v}_{11} = \bar{v}_{21} = \bar{v}_1 \\
\bar{u}_{11} = \bar{u}_{21} = \bar{u}_1 \\
\bar{v}_{22} = \bar{v}_{32} = \bar{v}_2 \\
\bar{u}_{22} = \bar{u}_{32} = \bar{u}_2 \\
\vdots \\
\bar{v}_{nn} = \bar{v}_{(n+1)n} = \bar{v}_n \\
\bar{u}_{nn} = \bar{u}_{(n+1)n} = \bar{u}_n
\end{array}\right\} \qquad (7.35)$$

Likewise the amplitudes of either side of a contact must be equal in magnitude, thus:

$$\left.\begin{array}{l} \bar{y}_{11} = \bar{y}_{21} = \bar{y}_1 \\ \bar{y}_{22} = \bar{y}_{32} = \bar{y}_2 \\ \vdots \\ \bar{y}_{nn} = \bar{y}_{(n+1)n} = \bar{y}_n \end{array}\right\} \tag{7.36}$$

The first suffix refers to the layer, the second to the interface.

Based on the expressions for τ and σ (eqns 7.22 to 7.25, 7.26 to 7.29 or 7.30 to 7.33) the above conditions (7.34 to 7.36) lead to a set of equations with $\bar{v}_i$ and $\bar{u}_i$ as unknown written in matrix form as follows:

$$\begin{array}{c|cccccccccc} & & 1 & & 2 & & 3 & & 4 & \\ \hline 1' & c_{11} & d_{12} & e_{13} & f_{14} & \cdot & \cdot & \cdot & \cdot & \cdot \\ & i_{21} & j_{22} & k_{23} & l_{24} & \cdot & \cdot & \cdot & \cdot & \cdot \\ \hline 2' & a_{31} & b_{32} & c_{33} & d_{34} & e_{35} & f_{36} & \cdot & \cdot & \cdot \\ & g_{41} & h_{42} & i_{43} & j_{44} & k_{45} & l_{46} & \cdot & \cdot & \cdot \\ \hline 3' & \cdot & \cdot & a_{53} & b_{54} & c_{55} & d_{56} & e_{57} & f_{58} & \cdot \\ & \cdot & \cdot & g_{63} & h_{64} & i_{65} & j_{66} & k_{67} & l_{68} & \cdot \\ \hline 4' & \cdot & \cdot & \cdot & \cdot & a_{75} & b_{76} & c_{77} & d_{78} & e_{7} \\ & \cdot & \cdot & \cdot & \cdot & g_{85} & h_{86} & i_{87} & j_{88} & k_{8} \\ \hline 5' & \cdot & \cdot & \cdot & \cdot & \cdot & \cdot & a_{97} & b_{98} & c_{9} \\ & \cdot & \cdot & \cdot & \cdot & \cdot & \cdot & g_{10,7} & h_{10,8} & i_{1} \\ \hline 6' & \cdot & \cdot & \cdot & \cdot & \cdot & \cdot & \cdot & \cdot & a_{1} \\ & \cdot & \cdot & \cdot & \cdot & \cdot & \cdot & \cdot & \cdot & g_{1} \end{array}$$

Here

$$q_i \equiv \frac{\rho_i - \rho_{(i+1)}}{2\mu_{(i+1)}} h_{(i+1)} g, \tag{7.38}$$

(but $q_n' \equiv \dfrac{\rho_n - \rho_{(n+1)}}{2\mu_{(n+1)}} h_n g$ if the basal layer $(n+1)$ is an infinite half-space) and $\bar{y}_i$ is amplitude of the interfacial waves. Full stops denote zero elements.

As it stands eqn (7.37) is valid for systems with 6 flexible boundaries. Systems with 1, 2, 3, etc., flexible boundaries are treated by employing the upper left submatrices limited by the dashed lines 1–1′, 2–2′, 3–3′, etc., whereas extension of the matrix elements along the diagonal zone makes the equation applicable to systems with any finite number of layers and flexible interfaces.

The elements in the matrix of eqn (7.37) as defined in Table 7.3 follow from the expression for S and P in Table 7.2. The reason for the somewhat unusual choice of element notation, i.e. that different letters are employed as symbols in addition to both row and column suffixes, is that elements symbolized by the same letter are of identical form with respect to the three parameters $\mu_{i:(i+1)}$, ϕ_i and $\phi_{(i+1)}$. In short, elements with the same letter are identical functions of $\mu_{i;(i+1)}$, ϕ_i and $\phi_{(i+1)}$ (except for layers 1 and $(n + 1)$). Thus the letter symbol shows the form of the expression for the element while the two suffixes give the position of the element in the matrix as well as the indices to be attached to μ and ϕ. The indices on the

$$
\begin{bmatrix}
5 & & 6 \\
 & \cdot & \cdot \\
 & \cdot & \cdot \\
 & \cdot & \cdot \\
 & \cdot & \cdot \\
 & \cdot & \cdot \\
 & \cdot & \cdot \\
\cdots_{7,10} & \cdot & \cdot \\
\cdots_{8,10} & \cdot & \cdot \\
\cdots_{9,10} & e_{9,11} & f_{9,12} \\
\cdots_{10,10} & k_{10,11} & l_{10,12} \\
\cdots_{11,10} & c_{11,11} & d_{11,12} \\
\cdots_{12,10} & i_{12,11} & j_{12,12}
\end{bmatrix}
\begin{bmatrix}
\bar{v}_1 \\ \bar{u}_1 \\ \bar{v}_2 \\ \bar{u}_2 \\ \bar{v}_3 \\ \bar{u}_3 \\ \bar{v}_4 \\ \bar{u}_4 \\ \bar{v}_5 \\ \bar{u}_5 \\ \bar{v}_6 \\ \bar{u}_6
\end{bmatrix}
=
\begin{bmatrix}
0 \\ q_1\bar{y}_1 \\ 0 \\ q_2\bar{y}_2 \\ 0 \\ q_3\bar{y}_3 \\ 0 \\ q_4\bar{y}_4 \\ 0 \\ q_5\bar{y}_5 \\ 0 \\ q_6\bar{y}_6
\end{bmatrix}
\qquad (7.37)
$$

latter two parameters, which associate them with a certain layer, stand in a simple relation to the row suffix of the matrix elements because there are two equations for each flexible interface. The layer-identification suffix, i, is related to the even-row suffix, l, and to the odd-row suffix, k, thus: $i = \frac{1}{2}(k + 1)$, $i = \frac{1}{2}l$ (see Table 7.3).

One premise of the theory is that the wavelength is the same for all interfaces and that any change of wavelength which we undertake in the numerical analysis (in order to study how the rate of change of amplitude varies with ϕ) must be made simultaneously at all flexible contacts throughout the system. From the condition of equal wavelength at all interfaces and the definition $\phi_i \equiv 2\pi h_i/\lambda$, follows that the ϕ-values for the various interfaces are proportional to the thicknesses. For any one given system the proportions between the ϕ-values for the different layers is therefore fixed, and we need only consider the variation of an arbitrarily selected

Table 7.3 Matrix elements in the equation systems (7.37) and (8.51)

The row suffix k is always odd, l always even. Suffix i refers to layer i in model. Layer suffix i and element suffixes k and l are related thus: $i = \frac{1}{2}(k+1)$, $i = \frac{1}{2}l$. For definition of P and S, see Table 7.2. Element expressions valid for systems with top and bottom layers with finite thickness and rigid surface and base respectively:[a]

$$
\begin{aligned}
a_{k(k-2)} &= -\mu_{i:(i+1)}S_{ii}^{v(i-1)} & g_{l(l-3)} &= -\mu_{i:(i+1)}h_{(i+1):i}P_{ii}^{v(i-1)} \\
b_{k(k-1)} &= -\mu_{i:(i+1)}S_{ii}^{u(i-1)} & h_{l(l-2)} &= -\mu_{i:(i+1)}h_{(i+1):i}P_{ii}^{u(i-1)} \\
c_{kk} &= S_{(i+1)i}^{vi} - \mu_{i:(i+1)}S_{ii}^{vi} & i_{l(l-1)} &= P_{(i+1)i}^{vi} - \mu_{i:(i+1)}h_{(i+1)i}P_{ii}^{vi} \\
d_{k(k+1)} &= S_{(i+1)i}^{ui} - \mu_{i:(i+1)}S_{ii}^{ui} & j_{ll} &= P_{(i+1)i}^{ui} - \mu_{i:(i+1)}h_{(i+1):i}P_{ii}^{ui} \\
e_{k(k+2)} &= S_{(i+1)i}^{v(i+1)} & k_{l(l+1)} &= P_{(i+1)i}^{v(i+1)} \\
f_{k(k+3)} &= S_{(i+1)i}^{u(i+1)} & l_{l(l+2)} &= P_{(i+1)i}^{u(i+1)}
\end{aligned}
$$

For infinite half space on top and/or bottom use following element expressions in the upper and lower two rows:[b]

$$
\begin{aligned}
c_{11}^{\infty} &= S_{21}^{v1} & i_{21}^{\infty} &= P_{21}^{\nu 1} + \mu_{1:2}\phi_2 \\
d_{12}^{\infty} &= S_{21}^{u1} + \mu_{1:2} & j_{22}^{\infty} &= P_{21}^{\mu 1}
\end{aligned}
$$

$$
\begin{aligned}
g_{2n(2n-3)}^{\infty} &= -\mu_{n:(n+1)}P_{nn}^{\nu(n-1)} \\
h_{2n(2n-2)}^{\infty} &= -\mu_{n:(n+1)}P_{nn}^{\mu_n-1)} \\
c_{(2n-1)(2n-1)}^{\infty} &= -S_{nn}^{\nu_n}\mu_{n:(n+1)} & i_{2n(2n-1)}^{\infty} &= -P_{nn}^{\nu_n}\mu_{n:(n+1)} + \phi_n \\
d_{(2n-1)2n}^{\infty} &= -S_{nn}^{\mu_n}\mu_{n:(n+1)} + 1 & j_{2n2n}^{\infty} &= -P_{nn}^{\mu_n}\mu_{n:(n+1)}
\end{aligned}
$$

$e_{13}, f_{14}, k_{23}, l_{24}, a_{(2n-1)(2n-3)}, b_{(2n-1)(2n-2)}$ and all other element expressions remain unchanged. For models terminated at the top by a free deformable surface put $\mu_{1:2} = 0$ in elements d_{12}^{∞} and i_{21}^{∞}; all other elements are unchanged.

---

[a] Since the wavelength must be the same for all boundaries in a model, the quantity $\phi_i = (2\pi h_i)/\lambda$ in all coefficients must be changed to a reference value $\phi_r \equiv \phi_i h_{r:i}$ where h_r is the thickness of the reference layer, usually layer Nr. 2.

[b] The element expressions in this table are not suited for systems consisting *only* of two infinite half-spaces joined along a deformable boundary, see text.

ϕ, termed ϕ_r; the change of the other ϕ-values follow automatically when the transformation $\phi_i = h_{i:r}\phi_r$ is made. In the computer program developed we have found it convenient to refer generally all ϕ-values to ϕ_2 (i.e. $\phi_r = \phi_2$) inasmuch as h_1 and thus ϕ_1—which would otherwise have been the natural reference quantity—are infinite in many of the models analysed.

(In a double-layer model with an infinite half-space as base and an overburden with finite thickness it is necessary to use ϕ_1 as reference because $\phi_2 = \infty$. Double-layer models consisting of two infinite half-spaces are treated separately, p. 77–79.)

In the element notation adopted the 22 different definitions expressed in Table 7.3 are sufficient to specify the coefficients in the equations for models with any number of layers.

If we wish to analyse numerically a model with a flexible free upper surface (types I and II, Fig. 7.9) it is possible to use the computer program based on the elements defined in Table 7.3 and let the first interface be the free surface. This is accomplished by putting $\mu_{1:2} = 0$ in the elements d_{12} and i_{21} at the same time as ρ_1 is put equal to zero in q_1. This means that the top layer has vanishing viscosity and density, in short, its dynamic effect is zero. Because the uppermost flexible boundary always bears the index unity (1), and furthermore a layer and its basal boundary bear identical index, it follows that layer number 1 does not exist in models with a free flexible surface and that the uppermost layer in such models bears the index 2.

The equation system can easily be extended to represent dynamic models with any number of layers. For each new layer one simply adds to the diagonal zone in the matrix a set of elements of the type

$$\begin{matrix} a & b & c & d & e & f \\ g & h & i & j & k & l, \end{matrix}$$

c and j lying on the principal diagonal.

Models with one flexible interface

We can select from the equation system models with any number of layers. We start the discussion of a double-layer model with one flexible interface. The significant problem to study is the rate of change of transversal amplitude, $\bar{v}_1$, as a function of viscosities, densities and wavelength. To find $\bar{v}_1$, (as well as $\bar{u}_1$) the upper 2×2 submatrix of eqn (7.37) is utilized, the result being

$$\frac{\bar{v}_1}{\bar{y}_1} \equiv \frac{\partial \bar{y}_1}{\bar{y}_1 \partial t} = \frac{-d_{12}}{c_{11} j_{22} - d_{12} i_{21}} q_1 \equiv k q_1 \tag{7.39}$$

where

$$q_1 \equiv \frac{(\rho_1 - \rho_2) h_2 g}{2\mu_2}$$

and the thickness, h_2, of the basal layer is finite. (If h_2 is infinite and h_1 finite, a modified expression for q_1 must be used, viz.: $q_1' \equiv (\rho_1 - \rho_2)h_1 g/2\mu_2$. In case both h_1 and h_2 are infinite, see pp. 77 ff.)

$\bar{v}_1/\bar{y}_1$ is the rate of change of relative amplitude, an important quantity in perturbation theory of dynamically instable systems. The quantity kq_1 may be called the growth factor because in systems with inverted density stratification the perturbations grow with time as $\exp(kq_1 t)$, see e.g. Chandrasekhar (1955). Since k is a constant independent of $\bar{y}$ and t for any given system with given wavelength—as long as the deflections are gentle—eqn (7.39) can readily be integrated to yield the amplitude at time t:

$$y_t = \bar{y}_0 \exp(kq_1 t). \tag{7.40}$$

The factor

$$k = \frac{-d_{12}}{c_{11}j_{22} - d_{12}i_{21}}$$

assumes four different general forms characterizing four different types of double-layer models. Type III-1 (Figs 7.9 and 7.10) consists of two infinite half-spaces; type IV-1 consists of a top half-space ($h_1 = \infty$) and a basal finite layer with rigid lower boundary; type V-1 has a finite top layer with rigid surface overlaying an infinite half-space, and VI-1 contains two finite layers with rigid surface and base. Models I-1 and II-1 contain only one layer with a flexible free surface in which case $\rho_1 = 0$ and $q_1 < 0$, meaning that surface perturbations tend always to flatten with time and never to grow. These models apply to certain kinds of isostatic adjustment.

Just as the elements c_{11}, d_{12}, i_{21} and j_{22} are of different forms depending upon whether the thicknesses of the two layers are finite or infinite the factors k^{III}, k^{IV}, k^{V} and k^{VI} are also of different forms. (The roman superiors associate the k-factors with their respective types of model as shown in Figs. 7.9 and 7.10.)

Let us focus the attention on a certain structure belonging to one of the general model types, say type VI-1. As $\mu_{1:2}$ is fixed by the viscosities of the two materials and $h_{1:2}$ by the thicknesses of the layers, the wavelength implied in ϕ_1 and ϕ_2 is the only variable in the factor k^{VI}.

The ratio $\phi_{1:2}$ is fixed by the thickness ratio. For this particular structure k^{VI} is accordingly a function of only one of the two parameters ϕ, say ϕ_2. Numerical tests have shown that $k^{\mathrm{VI}} = f^{\mathrm{VI}}(\phi_2)$ generally has a maximum at a certain ϕ_2-value. As q_1 is a constant for a given structure of given substances it is evident that the rate of change of relative amplitude has a maximum at the same ϕ_2-value as has the factor k^{VI}. For positive values of q_1, i.e. when $\rho_1 > \rho_2$, $\bar{v}_1/\bar{y}_1$ is positive and the perturbations whose wavelength gives maximum k^{VI}—and hence maximum rate of rise of relative amplitude—will sooner or later dominate the structure.

For all double-layer structures belonging to a certain model type, k as a function of ϕ_2 (or ϕ_1) is identical at identical values of the dimensionless parameters $\mu_{1:2}$ and $h_{1:2}$. Accordingly, instable structures of a given model type showing the same viscosity ratio and the same thickness ratio, develop identical dominant wavelength/thickness ratio quite irrespective of the absolute magnitude of the viscosities, the thicknesses or the densities.

If among these structures q_1 differs (because $\rho_1 - \rho_2$, h and/or μ differ) the absolute magnitude of $\bar{v}_1/\bar{y}_1$ also differs at any given ϕ, but the maximum rate of growth of relative amplitude coincides with identical ϕ-value for all these structures.

The factor k in eqn (7.39) gives the rate of change of relative amplitude measured in the unit q_1 which has the dimension of time^{-1} and is characteristic for the particular system. Therefore k may be appropriately called the specific growth factor in analogy to the growth factor kq_1 which measures $\bar{v}_1/\bar{y}_1$ in absolute units (sec^{-1}, hour^{-1} or year^{-1}, etc.). It is worth noting, moreover, that the ratio between the final amplitude and the initial

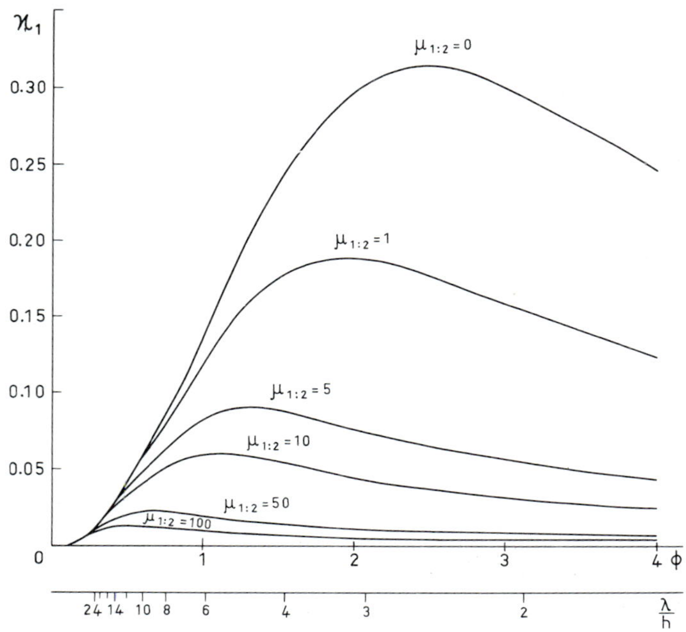

Fig. 7.12. Growth factor $k = v/yq$ plotted versus ϕ_2 and λ/h_2 for model type IV-1 at selected viscosity ratios, $\mu_{1:2}$. Maxima show ϕ_2- and λ/h_2-values of dominant perturbation (from Ramberg, 1968b).

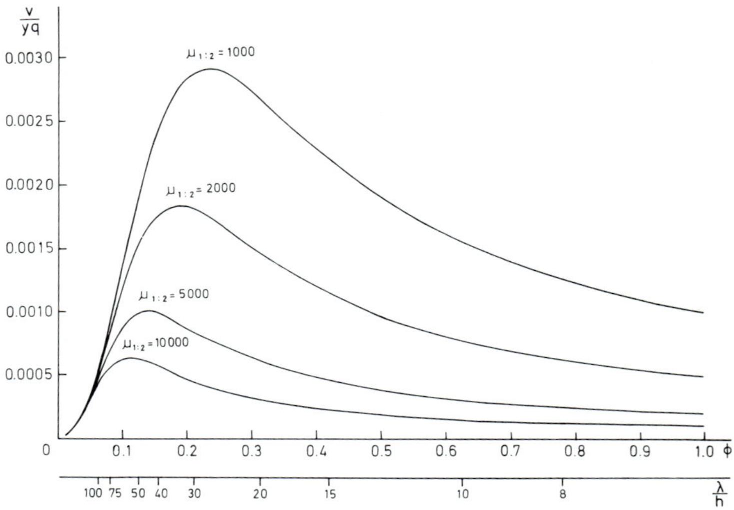

Fig. 7.13. Same as Fig. 7.12 but valid for different viscosity ratios.

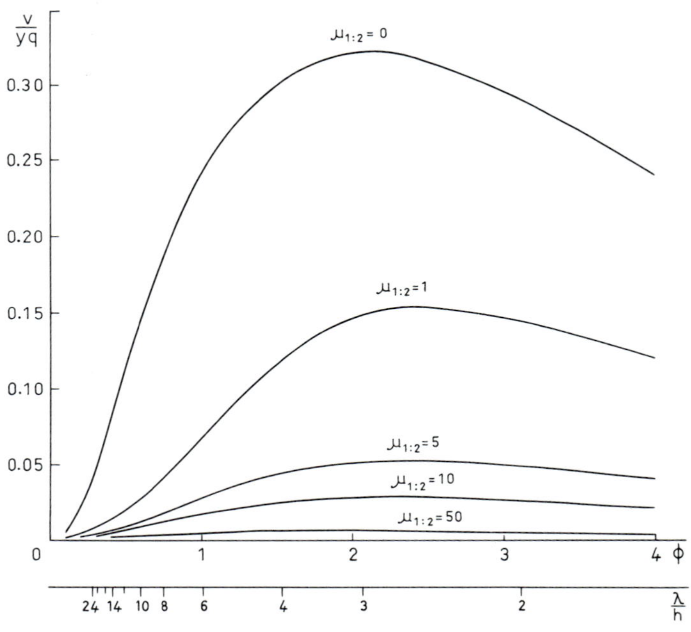

Fig. 7.14. The function $k \equiv v/yq = f(\lambda/h_2)$ for model type VI 1 for selected viscosity ratios, $\mu_{1:2}$.

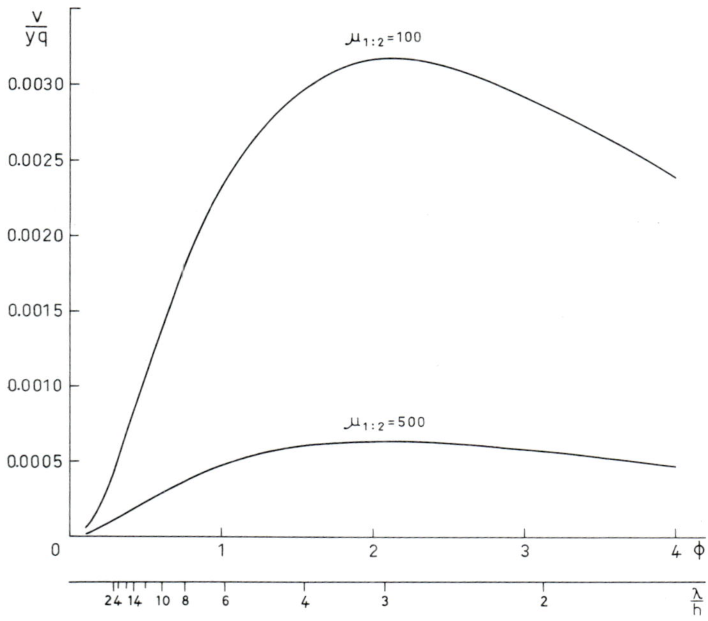

Fig. 7.15. Same as Fig. 7.14 but referring to different viscosity ratios.

amplitude, $\bar{y}_1/\bar{y}_0$, is the same for structures with the same specific growth factor provided that integration is performed over the same time interval as measured in the model unit of time, q_1^{-1}.

Plots of $k \equiv (\bar{v}_1/\bar{y}_1)q_1^{-1}$ versus ϕ for model types IV-1 and VI-1 at various values of $\mu_{1:2}$ are presented in Figs 7.12, 7.13, 7.14, and 7.15. See also Table 7.4, p. 90.

It is interesting to note the difference between the behaviour of models of type IV-1 and models of type VI-1. Models of type IV show a systematic increase of dominant wavelength/thickness ratio with increasing viscosity of the infinite overburden relative to that of the buoyant layer. This is a consequence of the greater mobility—relatively speaking—of the buoyant layer when the μ_o/μ_b ratio increases. By virtue of its increased relative mobility the buoyant layer is able to flow laterally over larger distances thereby augmenting the distance between rising anticlines or diapirs.

No such change of dominant wavelength with viscosity contrast is found in the models with top and bottom limited by straight rigid boundaries, provided the overburden and the buoyant layer are equally thick. Under this condition there is symmetry in the flow pattern in the sense that the downward flow of overburden material between the domes is identical to the upward flow of source material into the domes at least as long as the amplitude is small relative to the wavelength. Because of this symmetry in VI-1 models there should be no tendency for two models to develop different dominant wavelength in the special cases when the μ_o/μ_b ratio in the one is the same as the μ_b/μ_o ratio in the other. It is perhaps less obvious why the dominant wavelength should vary so little with changing viscosity contrast as the numerical calculation actually demonstrates, see Figs 7.14 and 7.15.

Model III-1 is of no direct geological significance because no global layers are infinitely thick. Indirectly, however, a model of two infinite half-spaces joined along a wavy interface is geodynamically interesting because it can be used to demonstrate quantitatively that inertia is insignificant in the evolution of domes and diapirs of crystalline rocks. Inertia can be shown to be of limited consequence even for global-scale structures, such as, for example, mantle diapirism, provided that magmas or other earth materials with low viscosity do not participate in the process.

The expression for rate of relative amplitude growth for model III-1 cannot be determined by the equation system (7.37) because both h_1 and h_2, and consequently ϕ_1, ϕ_2, q_1 and q_2, are infinite. Developed directly from eqns (7.26–29) based on the conditions

$$\tau_{11} = \tau_{21}, \qquad v_{11} = v_{21} = v$$

and

$$\sigma_{11} = \sigma_{21}, \qquad u_{11} = u_{21} = u$$

$$y_{11} = y_{21} = y$$

the expression for rate of relative amplitude growth becomes

$$\frac{\bar{v}}{\bar{y}} = \frac{(\rho_1 - \rho_2)g\lambda}{4\pi(\mu_1 + \mu_2)} \tag{7.41}$$

which shows that the amplitude velocity has no finite maximum as a function of the wavelength. Instead, the velocity of relative amplitude change increases linearly with the wavelength. This becomes physically unrealistic at large wavelengths, because the moving mass in the system increases with increasing wavelength and becomes infinite when the wavelength becomes infinite. In fact, the moving mass increases with the square of the wavelength because it can be shown that the perturbed motion penetrates on either side of the deflecting boundary to a depth which is proportional to the wavelength. So, when the wavelength of the instable deflections between the two half-spaces becomes sufficiently large the mass in the inertial term $\frac{1}{2}mv^2$ becomes dynamically significant and inertia prevents the amplitude for long-wavelength deflection to grow as fast as indicated by eqn (7.41).

The crucial questions are (1) At which wavelength does inertia become significant? and (2) Does inertia influence geologic structures? To answer these questions we compare eqn (7.41) valid for viscous systems in which inertia is disregarded, with a corresponding equation developed by Chandrasekhar (1955) valid for viscous systems with inertia included.

If the two velocity equations coincide over some regions of the wavelength it shows that the inertial force is of no consequence for systems with wavelength within those regions. On the other hand, inertia is significant within regions of the wavelength where the two equations do not coincide.

Chandrasekhar expresses the growth of amplitude for the infinite half-space system with inertia and viscosity included in an equation of form

$$\bar{y} = \bar{y}_0 \exp(nt) \tag{7.42}$$

in which n is a function of density, viscosity, acceleration due to gravity and wavelength. According to eqn (7.42) n in Chandrasekhar's notation is the same as $\partial\bar{y}/\bar{y}\partial t = \bar{v}/\bar{y}$, and may therefore be compared with the term $(\rho_1 - \rho_2)g\lambda/4\pi(\mu_1 + \mu_2)$ of eqn (7.41) in the non-inertial case.

Chandrasekhar has plotted $\bar{v}/\bar{y}$ versus $2\pi/\lambda$ as given by his expression for n, see wholedrawn curves in Fig. 7.16. The model does give a maximum on the amplitude velocity curve, and hence a dominant wavelength. This is a consequence of including inertia in the model.

In order to compare Chandrasekhar's $\bar{v}/\bar{y}$ versus $2\pi/\lambda$ curves with corresponding curves given by the function $(\rho_1 - \rho_2)g\lambda/4\pi(\mu_1 + \mu_2)$, we note that Chandrasekhar's treatment is limited to the special condition $\mu_1/\rho_1 = \mu_2/\rho_2 = \nu$, and that each graph refers to constant value of the ratio $(\rho_1 - \rho_2)/(\rho_1 + \rho_2)$. When the condition $\mu_1/\rho_1 = \mu_2/\rho_2 = \nu$ is applied to the expression $(\rho_1 - \rho_2)g\lambda/4\pi(\mu_1 + \mu_2)$ we obtain

$$\frac{\bar{v}}{\bar{y}} = \frac{(\rho_1 - \rho_2)g\lambda}{4(\rho_1 + \rho_2)\pi\nu}. \tag{7.43}$$

Further rearrangement is necessary because in Chandrasekhar's graphs $\bar{v}/\bar{y}$ is plotted in the unit $(g^2/\nu)^{1/3}$ and $2\pi/\lambda$ in the unit $(g/\nu^2)^{1/3}$. Taking these units into consideration, a rearrangement of eqn (7.41) yields

$$\frac{\bar{v}}{\bar{y}} = \frac{1}{2}\frac{\rho_1 - \rho_2}{\rho_1 + \rho_2}(g^2/\nu)^{1/3}\left(\frac{g}{\nu^2}\right)^{1/3}\frac{\lambda}{2\pi}. \tag{7.44}$$

Expressed in the units of Chandrasekhar, eqn (7.43) for the non-inertial case consequently becomes

$$\frac{\bar{v}}{\bar{y}} = \frac{1}{2}\frac{\rho_1 - \rho_2}{\rho_1 + \rho_2}\left(\frac{2\pi}{\lambda}\right)^{-1} \tag{7.45}$$

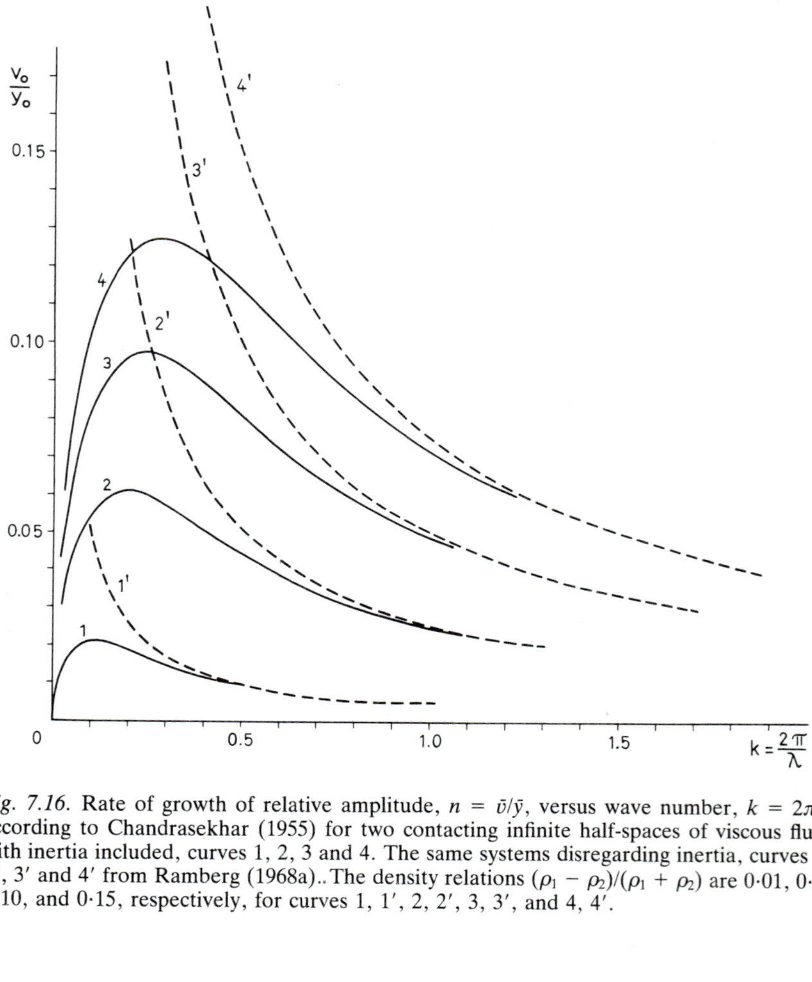

Fig. 7.16. Rate of growth of relative amplitude, $n = \bar{v}/\bar{y}$, versus wave number, $k = 2\pi/\lambda$, according to Chandrasekhar (1955) for two contacting infinite half-spaces of viscous fluids with inertia included, curves 1, 2, 3 and 4. The same systems disregarding inertia, curves 1′, 2′, 3′ and 4′ from Ramberg (1968a)..The density relations $(\rho_1 - \rho_2)/(\rho_1 + \rho_2)$ are 0·01, 0·05, 0·10, and 0·15, respectively, for curves 1, 1′, 2, 2′, 3, 3′, and 4, 4′.

which yields the $\bar{v}/\bar{y}$ versus $2\pi/\lambda$ plots as shown in Fig. 7.16 for the $(\rho_1 - \rho_2)/(\rho_1 + \rho_2)$ ratios 0·01, 0·05, 1·10, and 0·15, respectively.

The curves for the inertial-viscous model and the non-inertial-viscous model coincide for wavelengths less than certain limits which depend upon the ratio $(\rho_1 - \rho_2)/(\rho_1 + \rho_2)$.

To test whether inertia can be safely disregarded in our analysis of layered geodynamic systems the following reasonable parameters are selected for rocks: $\rho_1 = 3{\cdot}15$ g cm^{-3}, $\rho_2 = 2{\cdot}85$ g cm^{-3} and kinematic viscosity, $\nu = 10^{20}$ cgs units. For the selected density relations inertia is insignificant when $\lambda \leqslant 2\pi(\nu^2/g)^{1/3}$ as seen in Fig. 7.16, curves 2 and 2′.

With $\nu = 10^{20}$ cgs units and $g = 981$ cm sec^{-2} we find that inertia can safely be disregarded as a wavelength-controlling factor even for systems with wavelength as large as $1{\cdot}4\times10^{13}$ cm—far larger than any system on Earth! However, if the viscosity is less than assumed above, then the wavelength which delimits systems with and without significant inertial force will be shorter.

Rock salt is one of the most ductile and softest rocks which accordingly is most favourable for inertial forces. Its effective kinematic viscosity under favourable conditions may perhaps go down to 10^{16} cgs units (poise).

The density of salt is about 2·2 g cm^{-3}. An overburden with density 2·7 g cm^{-3} gives a value of $(\rho_1 - \rho_2)/(\rho_1 + \rho_2)$ equal to 0·098. From the curves 3 and 3′ in Fig. 7.16 we can see that the inertia-insignificant conditions occur for $2\pi/\lambda$-values greater than about 1·2 $(g/\nu^2)^{1/3}$ units. Hence

$$\lambda < \frac{2\pi}{1{\cdot}2}\left(\frac{10^{32}}{981}\right)^{1/3} \approx 2{\cdot}43\times10^{10}\ \text{cm}. \tag{7.46}$$

Even when using an exceptionally low viscosity for rock salt, inertia does not become significant as a wavelength-controlling agent unless the geometric dimensions go beyond the limit of the Earth.

This does not exclude inertial control of the late stages of mantle diapirism because as the amplitude/wavelength ratio increases the velocities become much faster than at the initial stage when the wavelength is determined.

The chief conclusion to be drawn from the comparison between the inertial and non-inertial systems is that the dominant wavelength of unstable layered global systems is not controlled by inertial forces, provided that the layers consist of crystalline rocks.

Multilayers with several flexible interfaces

To analyse a multilayer system with, say, two interfaces the upper 4×4 submatrix of eqn (7.37) is utilized. For a system with 3 interfaces the 6×6 submatrix limited by the dashed line 3—3′ serves, and so on. By thus employing larger and larger portions of the matrix of eqn (7.37) multilayers with increasing number of flexible interfaces and layers can be studied. Though the set of equations as stated represents a model with maximum 7 layers and 6 interfaces the set can easily be extended to cover any number of layers, the practical limitation being the capacity of the computer employed. The extension of the equation set requires only that elements of the kind already defined in Table 7.3 are added along the diagonal zone.

For our purposes it is practical to express the solution of eqn (7.37) with respect to the amplitude velocities as follows:

$$\begin{bmatrix} \bar{v}_1 \\ \bar{v}_2 \\ \vdots \\ \bar{v}_n \end{bmatrix} = \begin{bmatrix} k_{11} & k_{12}q_{2:1} \ldots k_{1n}q_{n:1} \\ k_{21} & k_{22}q_{2:1} \ldots k_{2n}q_{n:1} \\ \vdots & \vdots \qquad \vdots \\ k_{n1} & k_{n2}q_{2:1} \ldots k_{nn}q_{n:1} \end{bmatrix} \begin{bmatrix} q_1\bar{y}_1 \\ q_1\bar{y}_2 \\ \vdots \\ q_1\bar{y}_n \end{bmatrix} \tag{7.47}$$

or

$$\bar{v}_j = (k_{ji}q_{i:1})q_1\bar{y}_i$$

for the transverse amplitudes, and:

$$\begin{bmatrix} \bar{u}_1 \\ \bar{u}_2 \\ \vdots \\ \bar{u}_n \end{bmatrix} = \begin{bmatrix} k^u_{11} & k^u_{12}q_{2:1} \ldots k^u_{1n}q_{n:1} \\ k^u_{21} & k^u_{22}q_{2:1} \ldots k^u_{2n}q_{n:1} \\ \vdots & \vdots \qquad \vdots \\ k^u_{n1} & k^u_{n2}q_{2:1} \ldots k^u_{nn}q_{n:1} \end{bmatrix} \begin{bmatrix} q_1\bar{y}_1 \\ q_1\bar{y}_2 \\ \vdots \\ q_1\bar{y}_n \end{bmatrix} \tag{7.48}$$

$$u_j = (k^u_{ji}q_{i:1})q_1\bar{y}_i$$

for the tangential amplitude.

Presented in the above form the solution, with respect to $\bar{v}$ and $\bar{u}$, emphasizes the dynamic coupling among the various layers.

The reason for introducing q_1 in the equation system is to make all elements in the matrix dimensionless, thereby facilitating comparison among models with different absolute magnitude of density, viscosity or geometric dimensions, but with identical ratios between their various corresponding geometric and material properties.

The larger the magnitude of an element $k_{ji}q_{i:1}$ the stronger the dynamic coupling between interfaces j and i. A large numerical value of, say, k_{13}, for example, means that the amplitude velocity at interface number 1 is strongly influenced by deflections which exist at interface number 3; a small numerical value of say $k_{42}q_{2:1}$ means that the amplitude velocity on interface 4 is only weakly affected by deflections existing on interface 2, etc., etc.

The reason for using j as row index and i as column index rather than vice versa, the more natural order, is the wish to maintain index i to identify layer i. According to the summation convention the notation $k_{ji}q_{i:1}$ means the sum $k_{j1} + k_{j2}q_{2:1} + \ldots + k_{jn}q_{n:1}$. As q_i already is defined thus:

$$\frac{(\rho_i - \rho_{(i+1)})gh_{(i+1)}}{2\mu_{(i+1)}} \tag{7.49}$$

we see that index i is maintained as a reference to layer i. A notation of form $k_{ij}q_{i:1}$, on the other hand, would mean the sum $k_{1j} + k_{2j}q_{2:1} + \ldots + k_{nj}q_{n:1}$, which certainly is not what is intended because this form contains the wrong coefficients k. The index order in the form $k_{ij}q_{j:1}$ gives the correct summation, viz. $k_{i1} + k_{i2}q_{2:1} + \ldots + k_{in}q_{n:1}$ but only if the index referring to the layer is changed to j, which is exactly what we do not wish.

Application of matrix algebra shows that the coefficients k_{ji} of the above equations are related to the matrix of eqn (7.37) as follows:

$$\begin{aligned} k_{11} &= |A_{21}| : |A|; \quad k_{12} = |A_{41}| : |A|; \quad k_{13} = |A_{61}| : |A| \ldots k_{1n} = |A_{(2n)1}| : |A| \\ k_{21} &= |A_{23}| : |A|; \quad k_{22} = |A_{43}| : |A|; \quad k_{23} = |A_{63}| : |A| \ldots k_{2n} = |A_{(2n)3}| : |A| \end{aligned} \tag{7.50}$$

etc.

$$k_{n1} = |A_{2(2n-1)}| : |A|; \quad k_{n2} = |A_{4(2n-1)}| : |A|; \quad k_{n3} = |A_{6(2n-1)}| : |A| \ldots k_{nn} = |A_{2n(2n-1)}| : |A|.$$

Here $|A_{21}|$, etc., are the co-factors to the elements with the indices 2,1, etc., in the matrix of eqn (7.37), and $|A|$ is the determinant of the same matrix. In condensed form the coefficients which control the velocity $\bar{v}_j$ are written

$$k_{ji} \equiv |A_{(2i)(2j-1)}| : |A|. \tag{7.51}$$

The indices j and i take the values between 1 and n, n being the number of flexible boundaries in the multilayer model.

Similarily, for the coefficients in the equation for the tangential amplitude velocities we have:

$$\begin{aligned} k_{11}^u &= |A_{22}| : |A|; \quad k_{12}^u = |A_{42}| : |A|; \quad k_{13}^u = |A_{62}| : |A| \ldots k_{1n} = |A_{(2n)2}| : |A| \\ k_{21}^u &= |A_{24}| : |A|; \quad k_{22}^u = |A_{44}| : |A|; \quad k_{23}^u = |A_{64}| : |A| \ldots l_{2n} = |A_{(2n)4}| : |A| \end{aligned} \tag{7.52}$$

etc.

$$k_{n1}^u = |A_{2(2n)}| : |A|; \quad k_{n2}^u = |A_{4(2n)}| : |A|; \quad k_{n3}^u = |A_{6(2n)}| : |A| \ldots k_{(2n)(2n)} = |A_{(2n)(2n)}| : |A|.$$

A more condensely expressed form for the coefficients which control the velocity u_j, is

$$k_{ji}^u \equiv |A_{(2i)(2j)}| : |A|. \tag{7.53}$$

Let us for a moment discuss a system with two flexible interfaces. This is a relatively simple system, yet it demonstrates the principal difference between models with but one interface and models with several. Solutions with respect to the amplitude velocity $\bar{v}_1$ and $\bar{v}_2$ based upon the upper 4×4 submatrix of eqn (7.37) yields:

$$\begin{bmatrix} \bar{v}_1 \\ \bar{v}_2 \end{bmatrix} = \begin{bmatrix} k_{11}^{m2} & k_{12}^{m2} q_{2:1} \\ k_{21}^{m2} & k_{22}^{m2} q_{2:1} \end{bmatrix} \begin{bmatrix} q_1 \bar{y}_1 \\ q_1 \bar{y}_2 \end{bmatrix} \tag{7.54}$$

in which we have assumed that both q_1 and q_2 are finite, i.e. the basal layer

3 is not an infinite half-space. The superior $m2$ indicates that the coefficients refer to one of the double interface models I-2, II-2, . . . , or VI-2, Figs 7.9 and 7.10. The nature of the coefficients k_{ji}^{m2} depends upon the boundary conditions at the top and bottom of the models. If the bottom layer is infinitely thick q_2 becomes infinite and some modification of the equation system is necessary, see pp. 87–89.

For comparison with single-interface models, eqn (7.39), we form the rate of relative amplitude growth of one of the amplitudes, say, $\bar{y}_1$:

$$\frac{\bar{v}_1}{\bar{y}_1} = (k_{11}^{m2} + k_{12}^{m2} q_{2:1} \bar{y}_{2:1}) q_1 = K_1^{m2} q_1. \tag{7.55}$$

The expression

$$k_{11}^{m2} + k_{12}^{m2} q_{2:1} \bar{y}_{2:1} \equiv K_1^{m2} \tag{7.56}$$

plays the same role in double-interface models as does the specific growth factor k in single-interface models (eqn 7.39). It is evident from the nature of the elements of the matrix to eqns (7.54 and 7.37) that K_1^{m2} contains the same kinds of dimensionless parameters as embodied in k for the simpler models, viz. $\mu_{1:2}$, $\mu_{2:3}$, $h_{1:2}$ and $h_{2:3}$ as well as the wavelength/thickness ratio symbolized by the parameters ϕ_1 and ϕ_2. Numerical calculations demonstrate that the factors k_{11}^{m2} and k_{12}^{m2} as functions of ϕ generally have maxima, each factor reaching maximal value usually at different wavelengths in a given layered structure.

The ratio $q_{2:1} \equiv q_2/q_1$ is a new parameter not occurring in the specific growth factor for single interface models. The identity

$$q_{2:1} \equiv \frac{\rho_2 - \rho_3}{\rho_1 - \rho_2} h_{3:2} \mu_{2:3} \tag{7.57}$$

shows that the density-contrast ratio is the new parameter that enters the growth factor of the model with two flexible boundaries since the thickness ratio and the viscosity ratio already occurs in k_{11}^{m2} and k_{12}^{m2}.

Except for possible variation in the amplitude ratio, $y_{2:1}$, K_1^{m2}—as a function of ϕ_1 or ϕ_2—is identical for all structures that show the same viscosity ratios, the same density-contrast ratios and the same thickness ratios, provided that the structures belong to the same model class I-2, II-2 or III-2, etc. When these conditions for dynamic similarity are satisfied, any maximum on the function $K_1^{m2} = f(\phi)$ applies to all structures belonging to the same model class. The result is identical wavelength/thicknes ratio for the dominant perturbation quite independent of the absolute values of μ, h and ρ.

To compare the growth factors for multilayers with unlike numbers of flexible interfaces consider the following equations

$$\frac{\bar{v}_1}{\bar{y}_1} = k_1 q_1 \tag{7.58}$$

$$\frac{\bar{v}_1}{\bar{y}_1} = (k_{11}^{m2} + k_{12}^{m2} q_{2:1} \bar{y}_{2:1}) q_1 = K_1^{m2} q_1 \tag{7.59}$$

$$\frac{\bar{v}_2}{\bar{y}_2} = (k_{21}^{m2} \bar{y}_{1:2} + k_{22}^{m2} q_{2:1}) q = K_2^{m2} q_1 \tag{7.60}$$

$$\frac{\bar{v}_1}{\bar{y}_1} = (k_{11}^{m3} + k_{12}^{m3} q_{2:1} \bar{y}_{2:1} + k_{13}^{m3} q_{3:1} \bar{y}_{3:1}) q_1 = K_1^{m3} q_1 \tag{7.61}$$

$$\frac{\bar{v}_2}{\bar{y}_2} = (k_{21}^{m3} \bar{y}_{1:2} + k_{22}^{m3} q_{2:1} + k_{23}^{m3} q_{3:1} \bar{y}_{3:2}) q_1 = K_2^{m3} q_1 \tag{7.62}$$

$$\frac{\bar{v}_3}{\bar{y}_3} = (k_{31}^{m3} \bar{y}_{1:3} + k_{32}^{m3} q_{2:1} \bar{y}_{2:3} + k_{33}^{m3} q_{3:1}) q_1 = K_3^{m3} q_1. \tag{7.63}$$

These equations apply to models with one, two and three flexible interfaces respectively. By putting the superior m equal to I, II, etc., as the case may be, reference is made to the corresponding classes of model as visualized in Figs 7.9 and 7.10. Note the special definition of q_n if the basal layer $(n + 1)$ is an infinite half space, see eqn (7.71) and p. 89. In the above examples we have assumed that q_1 is finite. If it is not another reference factor, q_r, must be selected among the q-values which neither vanish (because $\rho_i - \rho_{(i+1)} = 0$) nor become infinite (because $h_{(i+1)} = \infty$).

The occurrence of the amplitude ratios as prominent parameters in the formulas for the growth factors for models with more than one mobile interface is significant because it means that the maximum of the growth factor, and hence maximum rate of amplitude growth, generally falls at different ϕ-value as the perturbations develop and the amplitude ratios presumably change.

Consider for example eqn (7.59). For a given structure k_{11}^{m2} and k_{12}^{m2} vary only with the wavelength contained in ϕ. The two coefficients have generally maxima at two different wavelengths as verified by many numerical examples studied by the writer. For a given system $q_{2:1}$ is constant, but the ratio $\bar{y}_{2:1}$ generally changes as the perturbations grow. Any maximum of the factor k_1^{m2} as a function of ϕ is due to the combined effect of k_{11}^{m2} and $(k_{12}^{m2} q_{2:1} \bar{y}_{2:1})$. The changing magnitude of the latter term as the amplitudes grow with different rate is apt to change the ϕ-value which maximizes K_1^{m2} and thus determines the wavelength. These conditions are well documented numerically in Ramberg (1968b).

Additional insight into the early phases of the evolution of buoyant multilayers is obtained from a study of the integrated versions of the amplitude-growth equations. Such a study shows how the amplitude ratio changes with time and how the wavelength of the dominant perturbation changes as the structures evolve.

A study of the integrated equations is not only important for the development of systems with inverted density stratification, but even more so for the flattening process of sinuous deviations from the equipotential surface in layered systems with normal density stratification. The analysis here is accordingly also of consequence for isostatic adjustment in a multilayered globe, even if the integration is valid only as long as the amplitude/wavelength ratio is small and the theory is limited to geometrically simple deflections.

In integrated form the velocity equations system eqn (7.47) is written:

$$\begin{bmatrix} \bar{y}_1 \\ \bar{y}_2 \\ \vdots \\ \bar{y}_n \end{bmatrix} = \begin{bmatrix} c_{11} & c_{12} \dots c_{1n} \\ c_{21} & c_{22} \dots c_{2n} \\ \vdots & \\ c_{n1} & c_{n2} \dots c_{nn} \end{bmatrix} \begin{bmatrix} \exp(\kappa_1 q_1 t) \\ \exp(\kappa_2 q_1 t) \\ \vdots \\ \exp(\kappa_n q_1 t) \end{bmatrix} \tag{7.64}$$

In this set of equations κ_i are the n eigenvalues of the n^2 matrix of eqn (7.47), t is time, and c_{ij} are coefficients determined partly by the eigenvectors that belong to the eigenvalues, partly by the initial amplitudes. To be specific, the coefficients c_{ij} are determined by the condition that the column of coefficients $c_{1j}, c_{2j}, c_{3j}, \dots, c_{(n-1)j}, c_{nj}$ are proportional to the n components of the eigenvector that belongs to the eigenvalue κ_j. There are n such eigenvectors associated with the eigenvalues $\kappa_1, \kappa_2, \kappa_3, \dots, \kappa_n$, respectively. This gives the relative values of the coefficients.

To obtain absolute values of the coefficients the following relationships are used:

$$\left.\begin{aligned} c_{11} + c_{12} + c_{13} \dots c_{1n} &= \bar{y}_{01} \\ c_{21} + c_{22} + c_{23} \dots c_{2n} &= \bar{y}_{02} \\ &\vdots \\ c_{n1} + c_{n2} + c_{n3} \dots c_{nn} &= \bar{y}_{0n} \end{aligned}\right\} \tag{7.65}$$

where $\bar{y}_{01}, \bar{y}_{02}, \dots, \bar{y}_{0n}$ are the amplitudes of the initial deflections at time zero. (For detailed explanation see books on differential equations, e.g. Ince (1956), Kreider *et al.* (1968) and others).

For a model with two deformable interfaces, for example, we have:

$$\begin{bmatrix} (k_{11} - \kappa_1) & k_{12} q_{2:1} \\ k_{21} & (k_{22} q_{2:1} - \kappa_1) \end{bmatrix} \begin{bmatrix} c_{11} \\ c_{21} \end{bmatrix} = 0$$

and

$$\begin{bmatrix} (k_{11} - \kappa_2) & k_{12} q_{2:1} \\ k_{21} & (k_{22} q_{2:1} - \kappa_2) \end{bmatrix} \begin{bmatrix} c_{12} \\ c_{22} \end{bmatrix} = 0. \tag{7.66}$$

These two sets of homogeneous equations give the ratios $c_{11}:c_{21}$ and $c_{12}:c_{22}$ which are the ratios between the two components of the eigenvectors that belong to κ_1 and κ_2.

Furthermore, at time zero the following condition exists:

$$c_{11} + c_{12} = \bar{y}_{01}, \tag{7.67}$$

and

$$c_{21} + c_{22} = \bar{y}_{02}$$

where $\bar{y}_{01}$ and $\bar{y}_{02}$ are the amplitudes of the initial deflections at the two interfaces.

Incidentally, it is not necessary that both interfaces are initially perturbed. That is, $\bar{y}_{01}$ and $\bar{y}_{02}$ need not both assume finite values. To initiate growth of the deflections it is sufficient that one of the initial amplitudes are finite; the other may vanish, implying that the corresponding interface coincides with the isopotential surface.

Among the eigenvalues one is more significant than the others. That is the one which together with q gives maximum value to the term $\exp(\kappa_1 q_1)$. This eigenvalue is given the index unity and referred to as the *principal eigenvalue*. For buoyant systems κ_1 assumes the sign which makes $\kappa_1 q_1$ positive. In case $q_1 < 0$—meaning that $\rho_1 < \rho_2$ and hence that the buoyancy of the system is not due to layer 2—numerical tests show that the principal eigenvalue is also negative. If $q_1 > 0$ the principal eigenvalue proves to be positive in buoyant systems.

As the product $\kappa_1 q_1$ occurs as exponents in the integrated equation the term $\exp(\kappa_1 q_1 t)$ becomes increasingly dominant as time passes, and the deflection whose wavelength corresponds to the numerical maximum on κ_1 grows faster than any other deflection. Sooner or later this deflection becomes the dominant wave of the system. The length of the wave which coincides with the maximum on the principal eigenvalue has accordingly been termed the *dominant wavelength* (see Biot and Odé, 1965; Ramberg, 1968b).

Under certain circumstances a buoyant system may be dominated by waves whose wavelength does not coincide with the maximum on κ_1. This is apt to happen when the amplitude is relatively large for initial perturbations whose wavelength is different from the one that coincides with the maximum on the principal eigenvalue. These buoyant perturbations will grow fast in the beginning and may well dominate the early evolution of a structure. As time passes, however, the wave which corresponds to κ_1 max. usually outgrows deflections with deviating wavelengths even if their initial amplitude were larger than that of the theoretical dominant wave. This can be demonstrated in numerical models, but too large amplitude of initial deflections with the "wrong" wavelength may well prevent the theoretical dominant wave developing altogether.

For these reasons the term *characteristic wavelength* is perhaps a more appropriate name for the wavelength which coincides with the maximum on the principal eigenvalue. In cases where misunderstanding is unlikely, dominant wavelength and characteristic wavelength are used interchangeably below.

In a series of dynamically similar systems which differ in terms of the thicknesses of corresponding layers, the characteristic wavelength is proportional to the thickness of any selected corresponding layer in the systems. For this reason we generally think of the characteristic wave in terms of the wavelength/thickness ratio rather than in terms of its absolute length.

In some layered structures we will find that none of the calculated eigenvalues yields a positive value of the exponents $\kappa_i q_1$. This means that none of the initial perturbations on the interfaces will grow. Instead they all tend to flatten, gradually establishing a layering which coincides with the equipotential surfaces in the gravitational field. This occurs when none of the layers is less dense than the superincumbent strata.

At the stage of development at which the characteristic wavelength has become the truly dominant one, the terms in the amplitude growth equation which contain the maximum eigenvalue are the only effective ones. Consequently, under this condition, the equation system becomes considerably simplified:

$$\left.\begin{aligned} \bar{y}_1 &= c_{11}\exp(\kappa_1 q_1 t) \\ \bar{y}_2 &= c_{21}\exp(\kappa_1 q_1 t) \\ &\vdots \\ \bar{y}_n &= c_{n1}\exp(\kappa_1 q_1 t) \end{aligned}\right\} \qquad (7.68)$$

Because $c_{11}/c_{21}/\ldots/c_{n1}$ are related as the components of the principal eigenvector, the above equations demonstrate the significant condition that when the characteristic wavelength dominates the structure the amplitudes at the different interfaces are related as the components of the principal eigenvector. During continued growth the amplitudes will remain in the same proportion to one another as long as the mathematical theory is applicable with sufficient approximation to real systems. Numerical tests indicate that this is true as long as none of the amplitudes exceeds 10% of the wavelength.

(Incidentally, it is interesting to note that the quantities K_1^{m2} and K_2^{m2} in eqns (7.59–7.60) and K_1^{m3}, K_2^{m3} and K_3^{m3} in eqns (7.61–7.63) are identical to any one of the eigenvalues, say κ_i, of the matrix of eqn (7.47) (as applied on double and triple interface systems respectively), provided that the ratios $\bar{y}_{2:1}$, $\bar{y}_{3:2}$, etc., equal the ratios between the components of the eigenvector belonging to the eigenvalue in question. This is in agreement with a general rule of matrix algebra, see Hammarling, 1970, p. 9.)

With the above restrictions on the amplitude ratios satisfied, therefore, $K^{m2} = K_1^{m2} = K_2^{m2}$ and $K^{m3} = K_1^{m3} = K_2^{m3} = K_3^{m3}$ are constants (for a given system at given ϕ-values) and eqns (7.59) to (7.63) are readily integrated to yield:

$$\left.\begin{aligned}\bar{y}_1 &= \bar{y}_{01}\exp(K^{m2}q_1t)\\ \bar{y}_2 &= \bar{y}_{02}\exp(K^{m2}q_1t)\end{aligned}\right\} \tag{7.69}$$

and

$$\left.\begin{aligned}\bar{y}_1 &= \bar{y}_{01}\exp(K^{m3}q_1t)\\ \bar{y}_2 &= \bar{y}_{02}\exp(K^{m3}q_1t)\\ \bar{y}_3 &= \bar{y}_{03}\exp(K^{m3}q_1t).\end{aligned}\right\} \tag{7.70}$$

When these sets of equations are compared with the set (7.68) it follows that K^{m2} and K^{m3} are identical to κ_1^{m2} and κ_1^{m3}.

In the continued discussion of the equations which express the amplitudes as functions of time, attention is drawn to the fact that the eigenvalues are completely determined by the dimensionless parameters $\mu_{i:(i+1)}$, $h_{i:(i+1)}$, $(\rho_i - \rho_{(i+1)})/(\rho_{(i+1)} - \rho_{(i+2)})$ and ϕ. This is a consequence of the forms of the elements in the matrix of eqn (7.37) from which κ is calculated.

The coefficients c_{ij} are determined by the same dimensionless parameters but also by the initial amplitudes as shown by eqn (7.65). Division of the coefficients by one of the initial amplitudes—say y_{0p}— results therefore in the dimensionless quantities $c_{ij}y_{0p}^{-1}$ in which the ratios of the initial amplitudes occur as factors. Equation (7.64) may accordingly be expressed in dimensionless form

$$\frac{\bar{y}_1}{\bar{y}_{0p}} = (c_{11}\bar{y}_{0p}^{-1})\exp(\kappa_1 q_1 t) + (c_{12}\bar{y}_{0p}^{-1})\exp(\kappa_2 q_1 t) \tag{7.71}$$

and

$$\frac{\bar{y}_2}{\bar{y}_{0p}} = (c_{21}\bar{y}_{0p}^{-1})\exp(\kappa_1 q_1 t) + (c_{22}\bar{y}_{0p}^{-1})\exp)\kappa_2 q_1 t) \tag{7.72}$$

if we for convenience consider structures with but two flexible boundaries. (In case q_1 vanishes—because $\rho_1 - \rho_2 = 0$—we may use q_2 as reference. If the latter becomes infinite—because $h_3 = \infty$— we may use a modified

$$q_2' = \frac{(\rho_2 - \rho_3)gh_2}{2\mu_3},$$

see below. If both qs are zero—i.e. $\rho_1 = \rho_2 = \rho_3$—the model is dynamically stable quite independent of the shape of the interface and therefore uninteresting from our point of view.) $\bar{y}_{0p}$ is either $\bar{y}_{01}$ or $\bar{y}_{02}$, referring to the non-vanishing initial amplitude if one interface happens to coincide with the equipotential surface at time zero.

From these comments it appears that structures belonging to the same

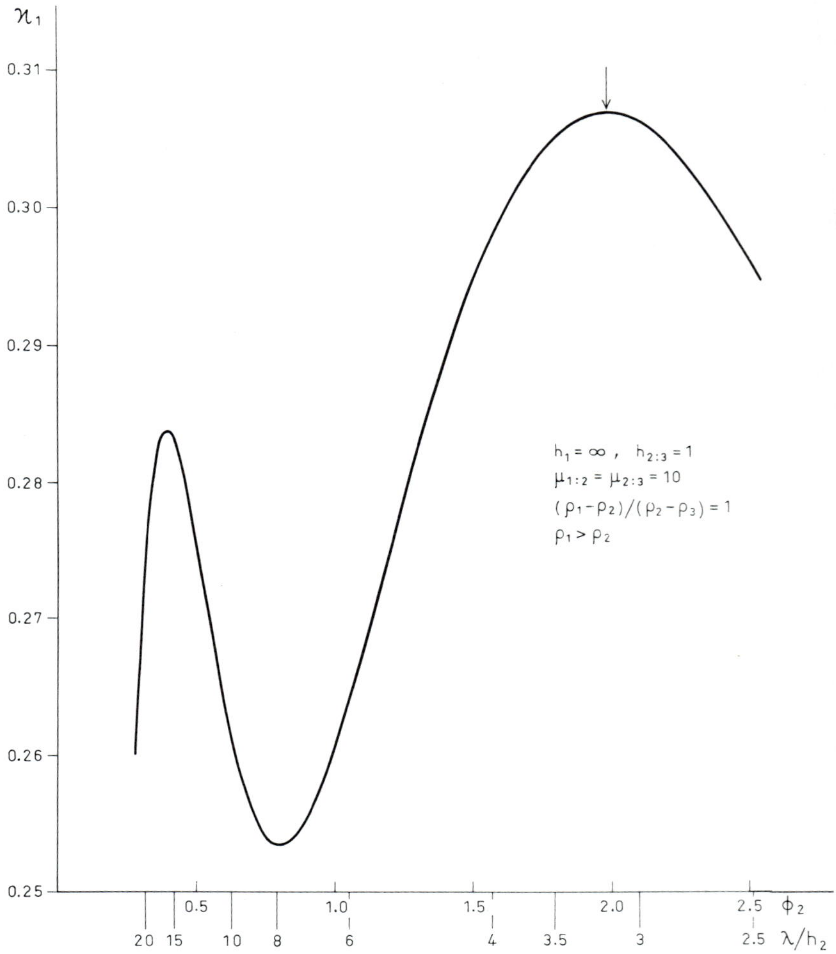

Fig. 7.17 Eigenvalue, κ_1, versus ϕ_2 and λ/h_2 exhibiting two maxima for a system consisting of two layers resting on a rigid basement and overlaid by viscous infinite half-space. Pertinent properties are recorded in the figure. See text.

general type of models and showing identical values of the corresponding parameters $\mu_{1:2}$, $\mu_{2:3}$, $h_{1:2}$, $h_{2:3}$, $(\rho_1 - \rho_2)/(\rho_2 - \rho_3)$ and $\bar{y}_{01:02}$, will develop identical ratio $\bar{y}_{1:2}$ after equal time as measured in the respective model units of time, which is q^{-1}. A further condition is that the comparison is made at equal ϕ-value. Corresponding conclusions can be made for systems with any number of deformable boundaries.

In some multilayer models with certain relations between viscosity contrasts, density contrasts and layer thicknesses the κ_1 versus ϕ function shows more than one maximum. In such models two or more dominant wavelengths develop simultaneously. An example is shown in Figs. 7.17 and 7.18. If the same conditions occur in three-dimensional models it would give rise to domes-in-domes structure, a situation which perhaps is applicable to multiple granite batholiths, see pp. 321 ff.

The even spacing predicted theoretically for the protuberances ascending from buoyant layers is seen in the distribution of salt diapirs and salt ridges (see Fig. 11.3), in the distribution of some granites (see Fig. 11.8) and even in the distribution of volcanic vents (see Figs 11.5–11.7).

For models whose basal layer is an infinite half-space the term q_n in the matrix of the amplitude-velocity equation system eqn (7.47) becomes infinite and a modification of some elements in the matrix is required. This can be accomplished by changing the definition of

$$q_n \text{ to } q_n' \equiv \frac{(\rho_n - \rho_{(n+1)})gh_n}{2\mu_{(n+1)}}$$

such that the last column in the amplitude velocity matrix changes to

$$(k_{1n}q_n' : q_1 \quad k_{2n}q_n' : q_1 \ldots k_{nn}q_n' : q_1).$$

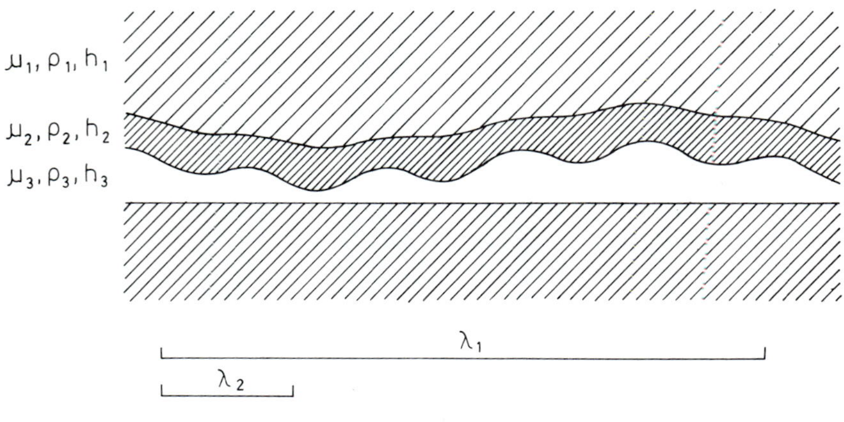

Fig. 7.18 Development of two sets of perturbations with different wavelengths giving rise to domes-in-dome structures. See Fig. 7.17 (from Ramberg, 1968b).

NUMERICAL MODELS

Models with one deformable boundary

Models consisting of a rigid basement, a viscous buoyant layer and a viscous infinite half-space as overburden belong to the model class IV-1 and are some of the least complicated buoyant structures from a fluid-dynamic point of view.

Figures 7.12 and 7.13 and Table 7.4 show some pertinent numerical results in terms of dominant wavelength and maximum values of κ_1. Note the continuous increase of the wavelength/thickness ratio with decreasing relative viscosity of the buoyant stratum relative to the overburden.

Table 7.4 Maximum principal eigenvalues and dominant wavelengths for buoyant layer with infinite half-space overburden and rigid base

$\mu_{1:2}$	ϕ_2	λ/h_2	κ_1
0	2·13	2·95	0·3214
1	1·7	3·70	0·1942
5	1·21	5·19	0·0900
10	0·99	6·35	0·0595
50	0·62	10·13	0·0212
100	0·50	12·56	0·0134
500	0·32	19·63	0·0046
1000	0·23	27·32	0·00291
2000	0·182	34·52	0·00183
5000	0·134	46·89	0·000995
10 000	0·107	58·70	0·000627

If the characteristic wavelength is less than the thickness of the overburden in natural occurrences, such that the deflections on the top of the buoyant layer do not affect the free surface, this class of model is applicable to some natural salt domes, granite domes, etc.

Models with two deformable boundaries

In numerous geologic structures, the perturbed strain at the boundary between the overburden and a buoyant layer penetrates up to the free surface which reflects the wavy pattern, but with reduced amplitude. This type of structures belongs to class II-2 in our system of classification (or

Table 7.5 Maximum principal eigenvalues, dominant wavelengths and amplitude ratios of models of type II-2

h_2/h_3	$(\rho_2-\rho_3)/\rho_2$	μ_2/μ_3	$-\kappa_1$	ϕ_2	λ/h_3	$\bar{y}_2/\bar{y}_1$
1	0·1	1	0·01753	2.05	3·06	30·93
1	0·1	10	0·0396	1·57	4·0	13·97
1	0·1	100	0·05074	1·06	5·92	10·94
1	0·1	1000	0·05437	0·75	8·37	10·4
0·01	0·1	1	0·03185	0·84	0·0748	12·98
0·01	0·1	100	0·05483	0·53	0·1185	10·07
0·01	0·1	1000	0·05545	0·33	0·1903	10·01
0·01	0·01	1	0·0029996	0·87	0·0722	126·9
0·01	0·01	100	0·004990	0·54	0·1163	100·43
0·01	0·01	1000	0·0050417	0·36	0·1744	100·05

to II-3 if the basement also is plastic, see below). Table 7.5 gives the value of κ_1 max, the ratio between the amplitude at the top of the buoyant layer and at the free surface and the dominant wavelength/thickness ratio for selected ratios between the viscosities, the thicknesses and the densities respectively, of overburden and buoyant layer.

The numerical examples demonstrate that—at otherwise identical conditions—the characteristic wavelength relative to the thickness of the buoyant layer decreases with decreasing thickness of the overburden. (Compare also the characteristic wavelength of models with infinite overburden, Table 7.4.) This is an interesting consequence of the change of the gravity potential of the system. When the thickness of the overburden is less than the dominant wavelength, then the deformation at the contact between overburden and buoyant layer penetrates all the way up to the free surface. As numerically demonstrated in Table 7.5 the thinner the overburden relative to λ—at otherwise identical conditions—the larger the amplitude of the deflections at the free surface relative to the deflections at the instable interface. This means increased gravity potential. If the wavelength now decreases at the buoyant layer–overburden boundary, the amplitude of the deflections on the free surfaces decreases also relative to those on the top of the buoyant layer, and the rise of the gravity potential becomes less. This reduction of the dominant wavelength with decreasing overburden thickness is exactly what happens in the numerical models.

In connection with the theoretical prediction presented here it is interesting to note the empirical relationships established between the spacing of volcanoes and crustal thickness in the East African Rift region.

The thinner the crust, the closer the volcanic vents, as demonstrated by Mohr and Wood (1976). (See Figs 11.6 and 11.7.)

The expected and well-documented condition that the amplitude ratio $\bar{y}_{2:1}$ decreases with diminishing overburden thickness (Table 7.5) is particularly striking when the overburden is relatively ductile, much less so when the overburden shows 100 and 1000 times higher viscosity than the buoyant layer.

It is interesting to note how the density ratio $(\rho_2 - \rho_3)/\rho_2$ affects the difference between the amplitude on the surface and on the buoyant layer–overburden interface. When all other variables are fixed a change in this density ratio from 0·1 to 0·01 implies a tenfold increase in the ratio $\bar{y}_{2:1}$ as recorded in the 6 lowest rows in Table 7.5. This trend is expected because the density contrast $(\rho_2 - \rho_3)$ is proportional to the driving force of the system while the resisting force increases with increasing density ρ_2 of the overburden owing to the deflections on the free surface.

The relationship between the characteristic wavelength and the thickness of the overburden is of interest for the theory of evolution of "greenstone belts" which appear to be a widespread type of tectonic structure in Precambrian shields, see e.g. Gorman *et al.* (1978) and Drury (1977). A dominant process responsible for the particular structure one encounters in greenstone-belt regions is the diapiric rise of a mobilized granitoid basement in and through a cover of metavulcanics and metasediments. Under such circumstances it is likely that the cover is thin relative to the granitoid basement in which case the wavelength would be smaller than one would expect on basis of the thickness of the granitoid basement. A buoyant layer that occurs deep in the Earth's mantle may develop waves whose length is not sufficient for the perturbed strain to penetrate up to the top of the mantle or down to the next discontinuous boundary below the layer. This corresponds to a buoyant layer sandwiched between two infinite half-spaces, that is model III-2 in our classification system. (See Figs 7.9 and 7.10.)

Numerical tests have shown that this model behaves quite unlike the models recorded above so some numerical results are presented.

Though geometrically and mathematically simple, the model nevertheless offers infinite possibilities of variations in terms of the density-contrast ratio, $(\rho_1 - \rho_2)/(\rho_2 - \rho_3)$ and the viscosity ratios $\mu_{1:2}$ and $\mu_{2:3}$. Because two of the three members of the model are infinitely thick, the thickness ratio does not enter the matrix of coupling coefficients. It is solely the ratios $\mu_{1:2}$, $\mu_{2:3}$ and $(\rho_1 - \rho_2)/(\rho_2 - \rho_3)$ that control the $\kappa = f(\phi_2)$ function and hence the characteristic wavelength.

To restrict the option of parameter variation in the numerical tests, the following conditions are unchanged from model to model: $\mu_{1:2} = \mu_{3:2}$ and

Table 7.6 Dynamic data of bouyant layer, 2, sandwiched between two infinite half-spaces, 1 and 3[a]

$\mu_{1:2} = \mu_{3:2}$ ($\mu_1 = 10^{22}$ poise)	ϕ_2	λ/h_2	κ_1	$\bar{y}_1/\bar{y}_2$	$q_1 = \frac{0{\cdot}3\times 981\times 10^6}{2\mu_2}$	$dy/dt = \kappa_1 q_1 \bar{y}_1$ at $y_1 = h_2$ (cm/year)	Time, for $\bar{y}_1/\bar{y}_{01}$ to reach 10^3 (years$\times 10^6$)
0·01	0·004	1570	43·3832	1/0·997	$1{\cdot}4715\times 10^{-16}$	0·198	34·91
0·02	0·002	3140	27·7504	1/0·998	$2{\cdot}943\times 10^{-16}$	0·253	27·28
0·1	0·002	3140	5·188	1/0·998	$1{\cdot}4715\times 10^{-15}$	0·237	29·18
0·2	0·002	3140	2·50119	1/0·998	$2{\cdot}943\times 10^{-15}$	0·228	30·27
1	0·005	1256	0·49933	1/0·994	$1{\cdot}4715\times 10^{-14}$	0·227	30·33
1·8	0·033	190·4	0·277875	1/0·961	$2{\cdot}6487\times 10^{-14}$	0·228	30·28
2	0·125	50·2	0·251227	1/0·886	$2{\cdot}943\times 10^{-14}$	0·229	30·14
2·5	0·3	20·9	0·205917	1/0·745	$3{\cdot}67875\times 10^{-14}$	0·235	29·42
3	0·4	15·7	0·177042	1/0·644	$4{\cdot}4145\times 10^{-14}$	0·242	28·51
4	0·5	12·5	0·141412	1/0·552	$5{\cdot}886\times 10^{-14}$	0·258	26·77
5	0·56	11·2	0·119670	1/0·490	$7{\cdot}3575\times 10^{-14}$	0·273	25·31
10	0·6	10·4	0·0727354	1/0·387	$1{\cdot}4715\times 10^{-13}$	0·332	20·82
20	0·55	11·4	0·0448675	1/0·336	$2{\cdot}943\times 10^{-13}$	0·409	16·88
30	0·5	12·5	0·0339433	1/0·324	$4{\cdot}4145\times 10^{-13}$	0·465	14·87
50	0·44	14·2	0·0239441	1/0·309	$7{\cdot}3575\times 10^{-13}$	0·546	12·65
100	0·36	17·4	0·0149618	1/0·299	$1{\cdot}4715\times 10^{-12}$	0·683	10·12
500	0·22	28·5	$5{\cdot}06883\times 10^{-3}$	1/0·284	$7{\cdot}3575\times 10^{-12}$	1·156	5·98
10^3	0·18	34·9	$3{\cdot}18765\times 10^{-3}$	1/0·1271	$1{\cdot}4715\times 10^{-11}$	1·454	4·751
10^4	0·08	78·5	$6{\cdot}84174\times 10^{-4}$	1/0·265	$1{\cdot}4715\times 10^{-10}$	3·121	2·21
10^5	0·04	157·1	$1{\cdot}47463\times 10^{-4}$	1/0·258	$1{\cdot}4715\times 10^{-9}$	6·727	1·03
10^6	0·02	314·2	$3{\cdot}15234\times 10^{-5}$	1/0·225	$1{\cdot}4715\times 10^{-8}$	14·425	0·48

[a] $(\rho_2 - \rho_3)/(\rho_1 - \rho_2) = -1, \rho_1 - \rho_2 = 0{\cdot}3,\ h_2 = 10$ km, $\mu_1 = \mu_3 = 10^{22}$ poise

$(\rho_1 - \rho_2)/(\rho_2 - \rho_3) = -1$. Within these constraints the ratio $\mu_{1:2} = \mu_{3:2}$ has been changed in the test, as have the absolute values of the viscosity, the density and the thickness of the buoyant stratum. The constraints mean that the two half-spaces are identical with respect to density and viscosity.

Table 7.6 gives pertinent data for a large number of viscosity ratios. In

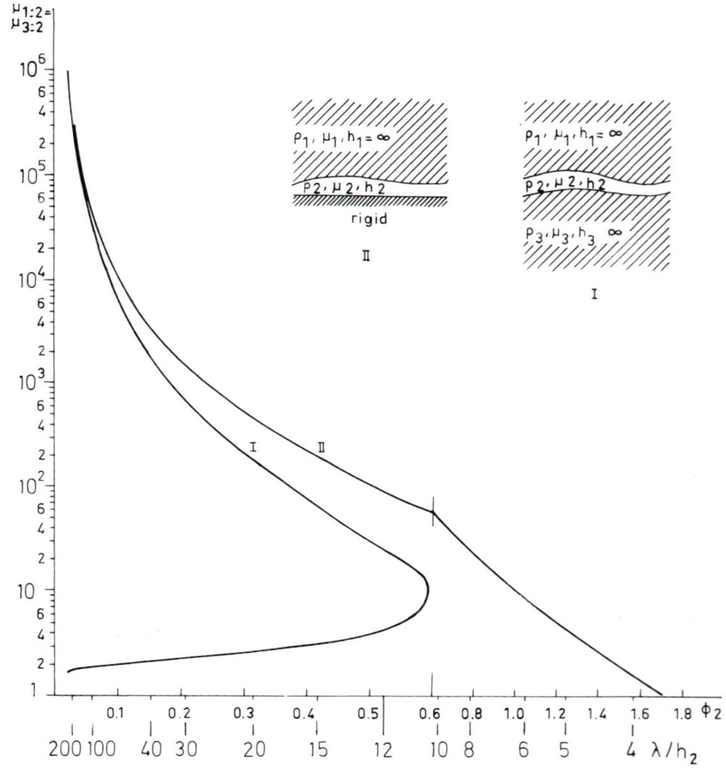

Fig. 7.19. Dominant wavelength as function of viscosity contrast for two single-layer models, one with infinite viscous overburden and rigid basement (II) and one with the layer sandwiched between two viscous infinite half-spaces with identical properties (Ramberg, 1972b). (Note break in scale at $\phi_2 \approx 0{\cdot}58$.)

two respects this type of model behaves unlike those presented above: (1) the characteristic wavelength reaches a minimum value at a certain $\mu_{1:2}$ ratio (around 10) and increases when the viscosity ratio changes in either sense from $\mu_{1:2} = 10$; and (2) the characteristic wavelength/thickness ratio reaches much larger values than in any of the other models.

Both these relationships show up in Fig. 7.19, where the characteristic wavelength is plotted as a function of the viscosity ratio for the model under discussion and a model with rigid base.

The condition that the characteristic wavelength/thickness ratio reaches values more than 1000 for the "layer between infinite half-spaces" model is significant for global-scale geodynamics. It means that the spacing between protuberances arising from even thin buoyant layers deep in the mantle may amount to several thousand kilometres. This is, of course, of consequence for the theories of mantle plumes, sea-floor spreading and continental drift (see Chapters 16 and 18).

Models with three deformable boundaries

During the evolution of dome and diapir structures in the Earth generally at least three layer contacts are deformed. In addition to the active boundary at the top of the buoyant stratum, not only the free surface may be deformed, as considered in the above examples, but also the contact between the basement and the buoyant layer. Structures of this kind belong to our class II-3 (see Fig. 7.9).

At the outset we repeat a minor formality which is adopted for practical reasons in the computer program but which might create some confusion when it comes to the indexing of the layers. We always define the uppermost *deformable* boundary as boundary number *one* in the models. Moreover, we also use the same index for a layer and for its *lowermost* boundary. Hence the layer immediately above the uppermost deformable boundary is defined as layer number one. Consequently, in a model whose uppermost boundary is free and deformable and therefore bears the index unity, layer number one (1) does not exist. (In the computer program the top layer vanishes simply by putting $\rho_1 = 0$ and $\mu_1 = 0$.) In models with a deformable free surface the uppermost layer therefore bears the index 2.

The crucial dimensionless parameters which characterize the dynamics of this type of structures are $h_{2:3}$, $h_{3:4}$, $\mu_{2:3}$, $\mu_{3:4}$, $(\rho_2 - \rho_3)/\rho_2$ and $(\rho_3 - \rho_4)/\rho_2$.

As the effect of varying density-contrast ratio has been demonstrated in the analysis of the double-boundary models above, the density-contrast ratios are kept fixed in the three-boundary models (with two exceptions) while the effect of varying thickness ratios and viscosity ratios are tested. The selected density-contrast ratios are $(\rho_2 - \rho_3)/\rho_2 = 0{\cdot}1$, and $(\rho_3 - \rho_4)/\rho_2 = -0{\cdot}1$. For two models the ratio $(\rho_3 - \rho_4)/\rho_2 = -1{\cdot}0$. These are reasonable figures for some tectonic models: they mean that both overburden and substratum are 10% more dense than the buoyant layer, a condition which would well prevail when a salt stratum is sandwiched

Table 7.7. Evolution of dominant wavelength in selected examples of model type II-3[a] (Fig. 7.9)

h_2/h_3	h_3/h_4	μ_2/μ_3	μ_3/μ_4	$-\kappa$	ϕ_2	λ/h_3	$\bar{y}_1/\bar{y}_2/\bar{y}_3$
1	1	100	100	0·052630	0·90	6·96	3·5/36·77/1·0
1	1	100	1	0·05170	1·0	6·20	2·10/23·16/1·0
1	1	100	0·01	0·050860	1·11	5·65	5·0/54·71/1·0
1	1	1	100	0·02685	1·35	4·65	0·148/2·64/1·0
1	1	1	1	0·022177	1·53	4·10	0·083/3/46/1·0
1	1	1	0·01	0·017585	2·05	3·06	4·10/126·7/1·0
1	0·1	100	100	0·05265	0·92	6·82	3·57/37·31/1·0
1	0·1	100	1	0·05198	0·94	6·68	3·14/33·53/1·0
1	0·1	100	0·01	0·051306	1·05	5·97	3·81/38·13/1·0
1	0·1	1	100	0·02685	1·35	4·65	0·148/2/63/1·0
1	0·1	1	1	0·02252	1·46	4·30	0·158/3·28/1·0
1	0·1	1	0·01	0·01759	2·04	3·07	3·37/104·0/1·0
0·5	1	100	100	0·05417	0·73	4·30	8·70/88·63/1·0
0·5	1	100	1	0·05380	0·77	4·08	7·25/74·40/1·0
0·5	1	100	0·01	0·05360	0·82	3·83	10·36/106·94/1·0
0·5	1	1	100	0·03056	1·03	3·05	0·28/4·05/1·0
0·5	1	1	1	0·02830	1·12	2·80	0·38/6·23/1·0
0·5	1	1	0·01	0·02598	1·35	2·33	10·56/193·2/1·0
0·1	1	100	100	0·05483	0·51	1·23	199·2/2000/1·0
0·1	1	100	1	0·05483	0·51	1·23	207·8/2090/1·0
0·1	1	100	0·01	0·055349	0·51	1·23	510/5000/1·0
0·1	1	1	100	0·03172	0·83	0·76	42·37/536·7/1·0
0·1	1	1	1	0·03172	0·82	0·77	51·6/653·6/1·0
0·1	1	1	0·01	0·03172	0·82	0·77	42·73/541/1·0
1	0·01	100	100	0·052650	9·92	6·82	3·56/37·04/1·0
1	0·01	100	1	0·051970	0·97	6·47	3·15/33·52/1·0
1	0·01	100	0·01	0·050530	1·05	5·98	3·50/38·02/1·0
1[b]	0·01	100	100	0·05260	0·92	6·82	34·4/360·6/1·0
1[b]	0·01	100	1	0·05193	0·98	6·41	30·5/325/1·0

[a] Density contrast ratio $(\rho_2-\rho_3)/\rho_2=0.1$; $(\rho_3-\rho_4)/\rho_2=-0.1$; $q_1=\dfrac{-\rho_2 g h_2}{2\mu_2}$
[b] $(\rho_2-\rho_3)/\rho_2=0.1$; $(\rho_3-\rho_4)/\rho_2=-1.0$

between shale, or when a granitic layer overlaid by metabasalt rests on amphibolite.

Five series of numerical examples of model type II-3 with different thickness ratios are recorded in Table 7.7. In series (1) all three layers are of equal thickness ($h_{2:3} = h_{3:4} = 1$). In series (2) each of the two uppermost layers is one tenth as thick as the basal layer ($h_{2:4} = h_{3:4} = 0{\cdot}1$). Series (3) contains a top layer half as thick as either of the two below ($h_{2:3} = 0{\cdot}5$,

$h_{3:4} = 1$). The top layer of series (4) is one tenth as thick as the two lower layers ($h_{2:3} = 0{\cdot}1$, $h_{3:4} = 1$), and finally series (5) has a basal layer not less than hundred times thicker than either of the two upper layers ($h_{2:3} = 1$, $h_{3:4} = 0{\cdot}01$). Layer 4 rests upon a flat rigid base. Maximum κ_1-values, dominant wavelength/thickness ratios and the steadied amplitude proportions are shown in the table.

The effect of the viscosity ratios upon the dominant wavelength at constant thickness ratios and constant density-contrast ratios is principally of the same nature as found for the simplier models studied above, viz. the larger the viscosity ratios $\mu_{2:3}$ and/or $\mu_{3:4}$ the greater the dominant wavelength/thickness ratio. This is shown in each of the five series in Table 7.7. (Note that the last two examples in series (5) have different density-contrast parameters.)

In series (1), (2) and (5) the ratio $h_{2:3}$ is the same but $h_{3:4}$ differs. A study of the table indicates little effect of varying substratum thickness (varying $h_{3:4}$) upon the characteristic wavelength. This is at least true when the thickness of the substratum equals or exceeds that of the buoyant layer.

In series (1), (3) and (4), on the other hand, the buoyant layer and substratum are equaly thick, whereas $h_{2:3}$ varies. The same trend appears as displayed in the two-layer model, regarding the dominant wavelength and the ratio between the thicknesses of overburden and buoyant layer. The less the thickness of the overburden is relative to the layer below, the smaller the dominant wavelength relative to the thickness of the buoyant layer at given viscosity ratios.

In the last two examples of series (5) the substratum is more dense than in the other examples in the table. The effect on the dominant wavelength is small but amplitude y_3 is quite drastically reduced relative to y_1 and y_2. This is expected on account of the increased density of the substratum.

As natural examples on models with three deformable boundaries we shall select the following simple geological structures: (1) A rock-salt stratum overlain by, and resting on, somewhat more dense sediments; (2) A buoyant low-velocity layer in the upper mantle; (3) A granitoid gneissic basement overlaid by somewhat more dense metasediments, basic lavas, etc; and (4) A low-density layer buried deep in the mantle.

Unstable rock-salt stratum

Let a salt stratum h_3 cm thick with effective Newtonian viscosity μ_3 and density ρ_3 be buried below an overburden with the following pertinent parameters: h_2, μ_2 and ρ_2. The salt rests on a rock whose parameters are h_4, μ_4 and ρ_4. The complex is supported by a rigid substratum as indicated in Fig. 7.20.

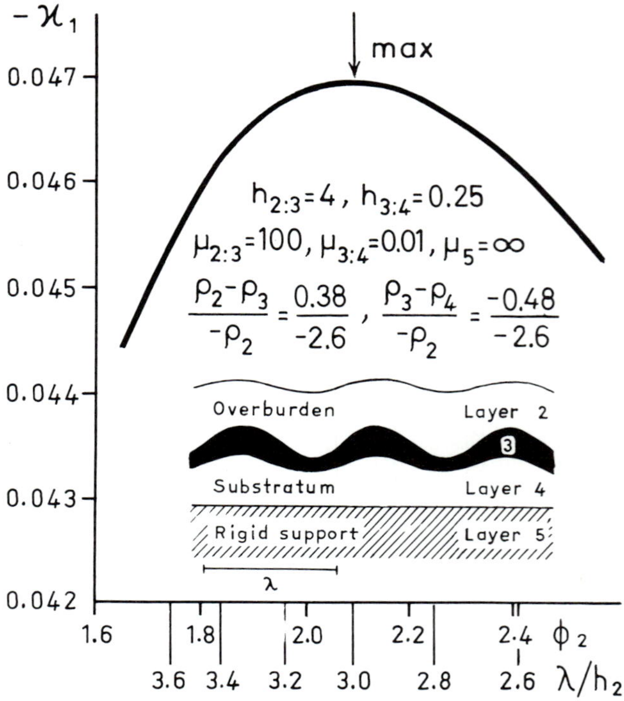

Fig. 7.20. Eigenvalue, κ_1, as function of ϕ_2 and λ/h_2 for buoyant salt stratum overlaid by overburden of limited thickness and resting upon viscous substratum whose lower boundary is rigid. Geometry of structure and pertinent properties shown in figure (from Ramberg, 1972a).

The dimensionless parameters needed for the calculation are selected thus: $h_{2:3} = 4$; $h_{3:4} = 0{\cdot}25$; $\mu_{2:3} = 100$; $\mu_{3:4} = 0{\cdot}01$; $(\rho_2 - \rho_3)/\rho_2 = 0{\cdot}146154$; $(\rho_3 - \rho_4)/\rho_2 = -0{\cdot}184615$. (The last two irregular values arise from the densities adopted, viz. $\rho_2 = 2{\cdot}6$ g cm^{-3}, $\rho_3 = 2{\cdot}22$ g cm^{-3} and $\rho_4 = 2{\cdot}7$ g cm^{-3}.)

Figure 7.20 shows the computed function $\kappa_1 = f(\phi_2)$ within a region around its numerical maximum. A dominant wavelength about 3 times the thickness of the overburden and 12 times the thickness of the buoyant layer is indicated. The computed eigenvector components are related as 0·0762:1:0·13025. This relation is also the proportions between the amplitude $\bar{y}_1$ to $\bar{y}_2$ to $\bar{y}_3$ after their steadied proportions have been reached. Thus the amplitude of the deflections on the surface is only 7·6% of the amplitude on the interface between the buoyant layer and the overburden. The heavy substratum is sucked up, as it were, to an amplitude corresponding to 13% of the amplitude on the top of the buoyant layer.

To determine $q_1 = -\rho_2 g h_2/2\mu_2$ it is necessary to specify values for h_2, μ_2, and g. For example, when $h_2 = 2\times10^5$ cm, $\mu_2 = 10^{20}$ poise, $g = 981$ cm sec^{-2} and $\rho_2 = 2{\cdot}6$ g cm^{-3} q becomes $-2{\cdot}55\times10^{-12}$ sec^{-1}. Insertion of this value of q_1, and $\kappa_1 = -4{\cdot}684\times10^{-2}$, in the equation

$$v_2 = \kappa_1 q_1 \bar{y}_2 \tag{7.73}$$

gives 0·221 cm year^{-1} at the moment $\bar{y}_2$ is 0·1 λ = 600 m. (From Fig. 7.20 we see that the dominant wavelength is 3 $h_2 = 6\times10^5$ cm = 6000 m.) As long as $\bar{y}_2 \leqslant 600$ m the velocity estimate is reliable. From the equation $\bar{y} = \bar{y}_0 \exp(\kappa_1 q_1 t)$ we may estimate the amplitude of the initial perturbations, whose amplitude has grown to 600 m after a selected lapse of time.

$$\bar{y}_{02} = \bar{y}_2 \exp(-\kappa_1 q_1 t) = 600 \exp(-11{\cdot}94\times10^{-14} t). \tag{7.74}$$

That is, a deflection whose amplitude now is 600 m was 14·79 m one million years ago, 8·99 mm three million years ago, and not more than 0·22 mm four million years ago.

Of course, initial perturbations only a few mm high do not create enough stress to initiate movement unless the materials have no strength. Though field observations of regional-metamorphic rocks indicate very low yield strength it is not likely that the rocks are entirely lacking in strength even considering the long time span of many tectonic events. Hence there must be a lower limit to the geometric dimensions of deflections which are able to initiate growth.

The growth of deflections may also be expressed as the change of amplification with time of the initial amplitude, i.e. $\bar{y}_i/\bar{y}_{0i}$ as a function of time. This is shown in Table 7.8 for the model under discussion. The growth is quite negligible during the first ten thousand years. Even after 100 000 years the amplitude is not more than 45% larger than initially. After a million years, however, the amplitude growth is explosive.

Table 7.8. Amplification of rock-salt stratum

Years	10	10^2	10^3	10^4	10^5
$\bar{y}_2/\bar{y}_{02}$	1·000037	1·00037	1·0037	1·038	1·4482
Years	10^6	2×10^6	3×10^6	4×10^6	5×10^6
$\bar{y}_2/\bar{y}_{02}$	40·57	1646	64769	$2{\cdot}71\times10^6$	$1{\cdot}1\times10^8$

Rise of an unstable low-velocity layer

Assume that the low-velocity layer in the upper mantle (Gutenberg, 1954, 1959; Anderson, 1965; Anderson *et al.*, 1970; Press, 1970; Lehman, 1967; Tarakanov and Leviy, 1968) is 10^7 cm thick (h_4), has a density $\rho_4 = 3{\cdot}35$ g cm^{-3}, and an effective viscosity $\mu_4 = 10^{20}$ poise. Assume, moreover, that a 10^7-cm thick (h_3) stratum with $\rho_3 = 3{\cdot}50$ g cm^{-2} and $\mu_3 = 10^{23}$ poise rests directly on the low-velocity layer, and that the Earth's crust is 2×10^6 cm thick (h_2), shows an average density $\rho_2 = 2{\cdot}80$ g cm^{-3} and possesses a viscosity $\mu_2 = 10^{23}$ poise. The low-velocity layer rests on a thick mantle with $\rho_5 = 3{\cdot}60$ g cm^{-3} and $\mu_5 = 10^{22}$ poise. The latter is an order-of-magnitude viscosity value commonly used for the mantle derived from Haskell's (1935) classic calculations of isostatic adjustment in Scandinavia.

The consequent dimensionless parameters are $\mu_{2:3} = 1$, $\mu_{3:4} = 1000$, $\mu_{4:5} = 10^{-2}$, $h_{2:3} = 0{\cdot}2$, $h_{3:4} = 1$, $h_5 = \infty$, $(\rho_2 - \rho_3)/(-\rho_2) = 0.25$, $(\rho_3 - \rho_4)/(-\rho_2) = -0{\cdot}05357$, and $(\rho_4 - \rho_5)/(-\rho_2) = 0{\cdot}08929$.

All these values are uncertain, particularly the viscosity values; nevertheless, it would be interesting to see how an upper mantle–crust structure would behave dynamically if it had the properties chosen.

From Fig. 7.21 we see that the dominant wavelength is 50 times the thickness of the crust and 10 times the thickness of the low-velocity layer. The computed eigenvector that belongs to κ_1 gives the proportion between the steadied amplitudes $\bar{y}_1:\bar{y}_2:\bar{y}_3:\bar{y}_4 = 0{\cdot}0078:0{\cdot}174:1:0{\cdot}007$. This relationship means that the amplitude of deflections created on the Earth's surface is only $0{\cdot}781\%$ of the amplitude of the deflection on the top of the active low velocity layer.

According to the parameters adopted q_1 assumes the value $-2{\cdot}7468 \times 10^{-14}$ sec^{-1}. Together with the calculated value of $\kappa_1 = -0{\cdot}1352606$, this value of q_1 results in the relations between time and amplification of initial perturbations as given in Table 7.9.

As the dominant wavelength is $10\ h_3 = 1000$ km the theory is applied until the amplitude $\bar{y}_3$ is 100 km. Although the amplitude velocity obtained at this large amplitude is questionable, it is interesting to see the result:

$$\begin{aligned}\bar{v}_3 &= \kappa_1 q_1 10^7 \text{ cm sec}^{-1}\\ &= 0{\cdot}13526\times2{\cdot}7468\times10^{-14}\times10^7 \text{ cm sec}^{-1} = 1{\cdot}152 \text{ cm year}^{-1}.\end{aligned}$$

The vertical movement at the Earth's surface is much less, viz.

$$\bar{v}_1 = 0{\cdot}13526\times2{\cdot}7468\times10^{-14}\times10^7\times0{\cdot}00781 = 0{\cdot}008995 \text{ cm year}^{-1}.$$

Here $7{\cdot}81\times10^4$ cm is the amplitude at the surface when $\bar{y}_3 = 10^7$ cm.

Even if the application of the theory is questionable at amplitudes as

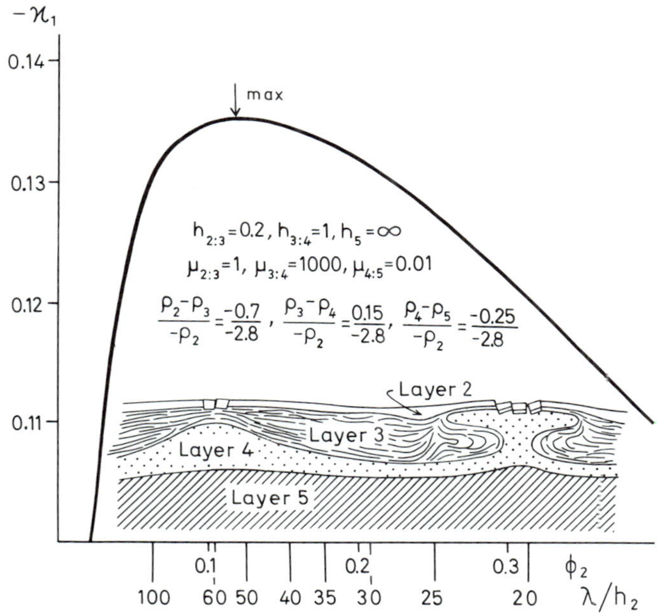

Fig. 7.21. Eigenvalue, κ, as function of ϕ_2 and λ/h_2 for low-density layer in upper mantle. Geometry of structure and pertinent properties shown in figure (from Ramberg, 1972a).

large as 100 km, the ratios between finite and initial amplitude given in Table 7.9 are good approximations when $\bar{y}_3 < 10$ km, which is only 1% of the dominant wavelength. The values presented in the table can therefore with confidence be interpreted as follows: a dominant deflection at the top of the low-velocity layer whose amplitude today is 10 km was 8·9 km 10^6 years ago, 3·16 km 10 million years ago, and not more than 9·9 cm 100

Table 7.9. Growth of amplitude of deflections on upper boundary of low-velocity layer

Years	10^6	10^7	2×10^7	3×10^7	4×10^7	5×10^7
$\bar{y}_3/\bar{y}_{03}$	1·122	3·164	10·009	31·658	100·183	316·714
Years	6×10^7	7×10^7	8×10^7	9×10^7	10^8	
$\bar{y}_3/\bar{y}_{03}$	1003·256	3171·627	$1{\cdot}004\times10^4$	$3{\cdot}176\times10^4$	$1{\cdot}01\times10^5$	

million years ago. This result is of course no better and no worse than the input parameters employed.

Upheaval of gneiss-granitic basement in orogens

Domal and anticlinal gneiss-granitic protrusions from the basement are well known in the Alps, the Scandinavian Caledonides, the Appalachians, and other orogenic belts (Wegmann, 1930, 1935; van Bemmelen, 1960; Ramberg, 1966, 1980; Zen *et al.*, 1968; Eskola, 1948; Haller, 1955, 1971; Brun, 1980). Gneiss-granitic domes and diapirs of various shapes are also dominant structures in Precambrian greenstone belts, see e.g. Gorman *et al.* (1978) and Fyson and Frith (1979).

Let us see whether a simple buoyancy model is at all quantitatively reasonable in terms of rate of tectonic processes, density contrasts of rocks, geometrical dimensions of structures, and effective viscosity.

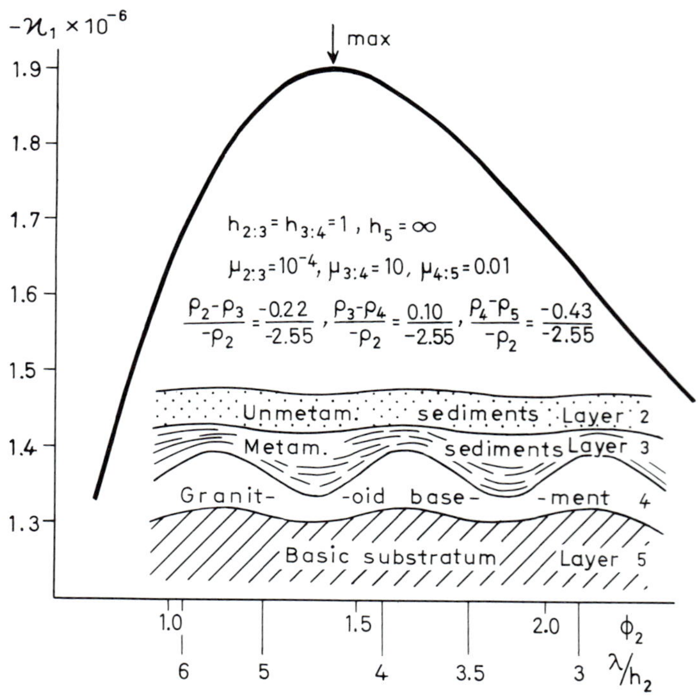

Fig. 7.22. Eigenvalue, κ_1, as function of ϕ_2 and λ/h_2 for unstable granitoid substratum buried below metasediments. Sketch of structure and pertinent properties shown in figure (from Ramberg, 1972a).

Resting on a simatic substratum with density $\rho_5 = 3{\cdot}1$ g cm^{-3} and viscosity $\mu_5 = 10^{22}$ poise are three main layers; on the bottom a granitoid layer with $\rho_4 = 2{\cdot}67$ g cm^{-3}, $\mu_4 = 10^{20}$ poise, and thickness $h_4 = 5\times10^5$ cm; next a 5×10^5-cm-thick (h_3) layer consisting of metamorphosed sediments and lavas with averge density $\rho_3 = 2{\cdot}77$ g cm^{-3} (a value corresponding to 10% amphibolite or greenstone and 90% schists, with densities as determined of alpine rocks, Wenk and Wenk, 1969) and an average viscosity $\mu_3 = 10^{21}$ poise; and on the top a layer of unmetamorphosed sediments with $h_2 = 5\times10^5$ cm, $\rho_2 = 2{\cdot}55$ g cm^{-3} and $\mu_2 = 10^{17}$ poise. The consequent dimensionless parameters are $\mu_{2:3} = 10^{-4}$, $\mu_{3:4} = 10$, $\mu_{4:5} = 10^{-2}$, $(\rho_2 - \rho_3)/-\rho_2 = 0{\cdot}0862745$, $(\rho_3 - \rho_4)/-\rho_2 = -0{\cdot}03921569$, $(\rho_4 - \rho_5)/-\rho_2 = 0{\cdot}16862745$, $h_{2:3} = 1$, and $h_{3:4} = 1$. Here h_5 is considerably larger than the dominant wavelength, which is practically the same as putting h_5 infinitely large.

In Fig. 7.22 the computed κ_1 values are plotted versus ϕ_2. Maximum κ_1 defines a dominant wavelength given by $\phi_2 = 1{\cdot}4$, i.e. $\lambda = 4{\cdot}485h_2 = 22{\cdot}5$ km. From the parameters defined above, $q_1 = -6{\cdot}2539 \times 10^{-9}$ sec^{-1}. Insertion of q_1 and $\kappa_1 = -1{\cdot}8996 \times 10^{-6}$ in exp $(\kappa_1 q_1 t)$ gives the ratio between finite and initial amplitudes at various times (Table 7.10). At the moment when $\bar{y}_3 = 2{\cdot}25\times10^5$ cm—which is 10% of λ—the velocity of amplitude growth is $\bar{v}_3 = \kappa_1 q \bar{y}_3 = 0{\cdot}08286$ cm year^{-1}. From these values we conclude that the buoyant rise of domes and anticlines from a gneiss granitic basement is a physically realistic process in orogenic belts whose lifetime usually is of the order 10^8 years.

Table 7.10. Growth of amplitude of deflections on upper boundary of granitic basement

Years	10^6	10^7	2×10^7	3×10^7	10^8
$\bar{y}_3/\bar{y}_{03}$	1·45	39·75	1580·0	62 830·0	$9{\cdot}86\times10^{15}$

Low-density layer located deep in the Earth's mantle

Theory shows that a buoyant stratum sandwiched between two rheologically identical half-spaces will develop much larger wavelengths than the same stratum with the same overburden, but resting on a rigid support (see p. 94).

In the search for a motor for continental drift and sea-floor spreading, it is important to study dynamically active systems which have characteristic wavelengths comparable to the width of oceans and continents. The "layer between infinite half-spaces" model is particularly promising in this respect,

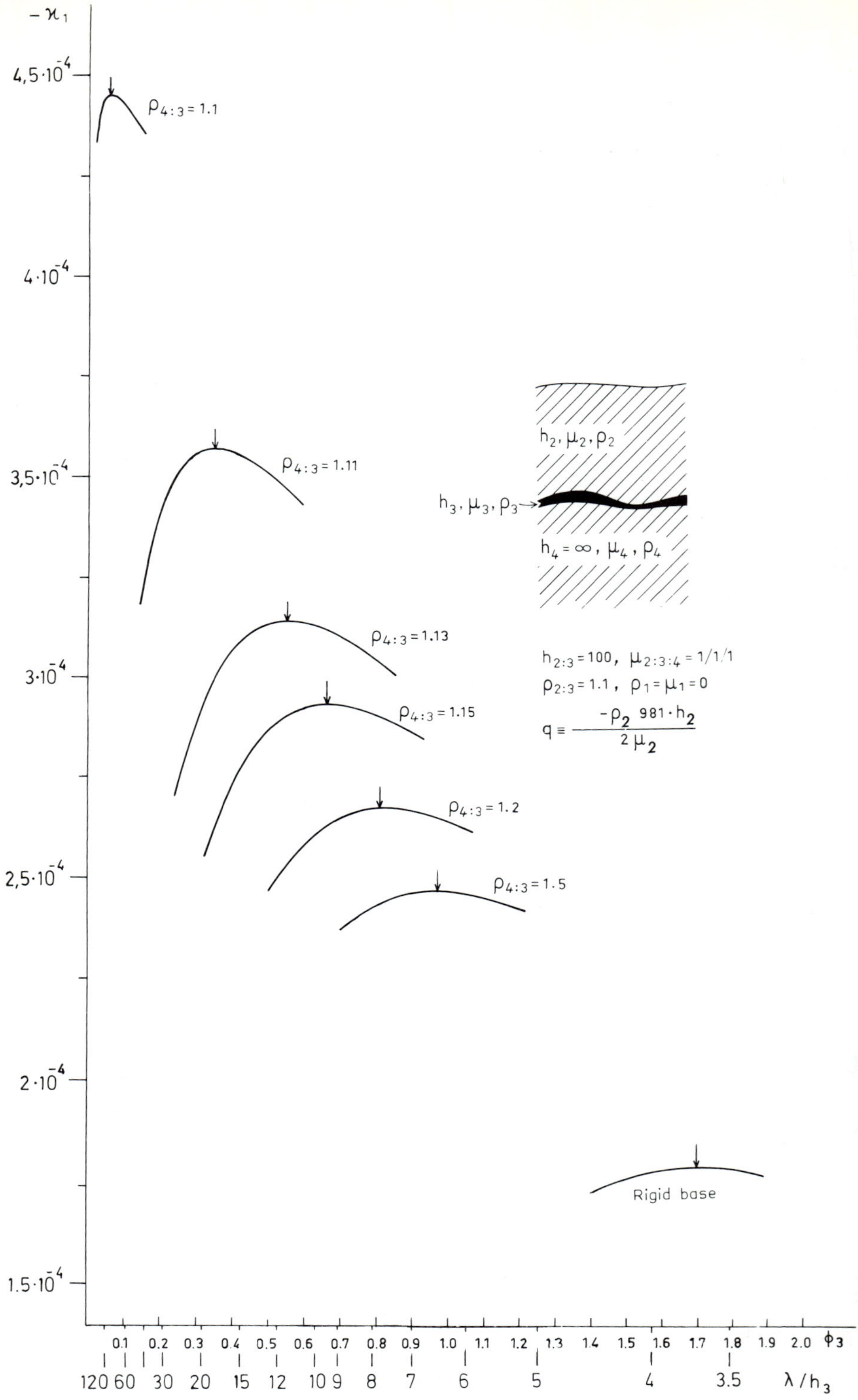

$-\varkappa_1$
4,5·10⁻⁴
4·10⁻⁴
3,5·10⁻⁴
3·10⁻⁴
2,5·10⁻⁴
2·10⁻⁴
1.5·10⁻⁴
$\rho_{4:3} = 1.1$
$\rho_{4:3} = 1.11$
$\rho_{4:3} = 1.13$
$\rho_{4:3} = 1.15$
$\rho_{4:3} = 1.2$
$\rho_{4:3} = 1.5$
Rigid base
h_2, μ_2, ρ_2
h_3, μ_3, ρ_3
$h_4 = \infty, \mu_4, \rho_4$
$h_{2:3} = 100, \ \mu_{2:3:4} = 1/1/1$
$\rho_{2:3} = 1.1, \ \rho_1 = \mu_1 = 0$
$q \equiv \frac{-\rho_2 \ 981 \cdot h_2}{2\mu_2}$
0.1 0.2 0.3 0.4 0.5 0.6 0.7 0.8 0.9 1.0 1.1 1.2 1.3 1.4 1.5 1.6 1.7 1.8 1.9 2.0
ϕ_3
120 60 30 20 15 12 10 9 8 7 6 5 4 3.5
λ / h_3

except for the unrealistic condition that the mantle is treated as infinitely thick. This causes only negligible errors if the characteristic wavelength of the buoyant layer does not exceed the thickness of the overburden.

As the wavelength for a layer a few kilometres thick may reach several thousand kilometres (Table 7.6), the error could be large if the infinite half-space model is used instead of models with limited overburden. When a model with a limited-thickness overburden is treated, account is taken of the potential energy involved in the sinuous deflections generated at the free surface.

Since the theory in this book is based on an orthogonal gravitational field, we unfortunately are not able to account for the effect caused by its actual spherical symmetry. The unknown effect may be large for systems spanning across the whole mantle.

The model now to be treated numerically is as follows (Fig. 7.23). A homogeneous upper layer 2 with free surface, thickness $h_2 = 1000$ km, density $\rho_2 = 3{\cdot}3$ g cm^{-3} at the surface, and effective viscosity $\mu_2 = 10^{22}$ poise. Below is the buoyant stratum with thickness h_3, density ρ_3 and effective viscosity μ_3. In the numerical tests a series of thicknesses of layer 3 and two viscosities are tested, viz. $\mu_3 = 10^{22}$ poise and $\mu_3 = 10^{16}$ poise. Layer 3 rests on the substratum which is an infinite half-space with viscosity $\mu_4 = \mu_2 = 10^{22}$ poise and density ρ_4 which varies from test to test.

The density differences at the boundaries between layers 2 and 3, and between layers 3 and 4 respectively, at the pressure at these levels are essential because they imply the driving power. In all tests we assume

$$\rho_{2:3} = 1{\cdot}1 \text{ and } (\rho_2 - \rho_3)/\rho_2 = 0{\cdot}09091.$$

Several values of ρ_4 are tested, determined by the conditions

$$\rho_{4:3} = 1{\cdot}1,\ 1{\cdot}11,\ 1{\cdot}13,\ 1{\cdot}15,\ 1{\cdot}20 \text{ and } 1.50 \text{ respectively.}$$

The first value specifies that the basal layer and the overburden are equally dense; the rest of the sequence of values gives a basal layer which is increasingly more dense than the overburden. The effect of varying basement density on the dominant wavelength is demonstrated numerically in Fig. 7.23 and in Table 7.11.

Fig. 7.23. Eigenvalue, κ_1, as function of ϕ_3 and λ/h_3 for buoyant stratum buried below a thick overburden (e.g. a layer deep in the earth's mantle) and resting on a half-space. Note how the dominant wavelength decreases with increasing density of the substratum, ρ_4. Note also that the dominant wavelength is much longer than for the same layer and the same overburden resting on a rigid base (from Ramberg, 1972b).

Table 7.11. Dynamic data for a buoyant stratum with finite overburden and infinite substratum. $\rho_{2:3} = 1{\cdot}10$; $\mu_{2:3} = \mu_{3:4} = 1$; $h_{2:3} = 100$, $h_4 = \infty$

$\rho_{4:3}$	ϕ_3	λ/h_3	$-\kappa_1$ ($\times 10^4$)	$\bar{y}_1/\bar{y}_2/\bar{y}_3$
1·10	0·0458	137	4·47	$4{\cdot}1\times10^{-4}/1/0{\cdot}9557$
1·11	0·34	18·5	3·5981	$2{\cdot}4\times10^{-15}/1/0{\cdot}69662$
1·13	0·54	11·63	3·1556	$1{\cdot}0\times10^{-23}/1/0{\cdot}53582$
1·15	0·66	9·52	2·9407	$8{\cdot}4\times10^{-29}/1/0{\cdot}44524$
1·20	0·82	7·66	2·6735	$1{\cdot}29\times10^{-35}/1/0{\cdot}32292$
1·50	0·97	6·47	2·4621	$4{\cdot}9\times10^{-42}/1/0{\cdot}21183$

In Table 7.12 various computer results are recorded. Here one finds the dominant wavelength/thickness ratio, the maximum principal eigenvalue, the proportion between the steadied amplitudes, the velocity of amplitude growth when the amplitude equals the thickness of the buoyant layer, and the time needed for 100-fold, 1000-fold and 10 000-fold amplification of the deflections. The overburden is the same in all tests, viz. $h_2 = 1000$ km, $\mu_2 = 10^{22}$ poise and ρ_2 at the surface $= 3{\cdot}3$ g cm^{-3}. In all examples in Table 7.12 basement and overburden are equally dense, viz. $\rho_{4:3} = \rho_{2:3} = 1{\cdot}1$.

When considering the velocity of amplitude growth and the time needed for n-fold amplification we have to watch lest the amplitudes should become so large that the theory ceases to be applicable. For example, in order that the 10 000-fold amplification shall not mean a final amplitude in excess of the arbitrary limit 10% of λ, the initital amplitude must not exceed one hundred thousandth of λ. Consider a wavelength equal to 1000 km (most wavelengths in Table 7.12 are larger than 1000 km). The maximum amplitude permitted without rendering the theory useless is then some 100 km. With 10 000-fold amplification this limits the intial amplitude to maximum 10^{-2} km $= 10$ m.

Another quantity to watch is the velocity $d\bar{y}_2/dt$ in the table. This is the velocity obtained when amplitude y_3 equals the thickness of layer 3; but since the characteristic wavelength is much larger than ten times the thickness of layer 3 the calculated velocity is well within the validity of the theory.

The time needed for the 10^n-fold amplifications of the amplitudes is based on the expression

$$10^n = y/y_0 = \exp(\kappa_1 q_1 t) \tag{7.75}$$

or

$$t = n\frac{\ln 10}{\kappa_1 q_1} \tag{7.76}$$

where $n = 2, 3$ and 4 for 100-fold, 1000-fold and 10 000-fold amplitication. $q_1 = -1{\cdot}6186\times10^{-11}$ sec^{-1} as obtained when $\rho_2 = 3{\cdot}3$ g cm^{-3}, $g = 981$ cm sec^{-2}, $h_2 = 10^8$ cm and $\mu_2 = 10^{22}$ poise are inserted in the formula $(-\rho_2 g h_2)/(2\mu_2)$.

The amplitude velocity at the moment $\bar{y}_2 = h_3$ is given by the expression $\bar{v}_2 = \kappa_1 q_1 h_3$ valid when the amplitude proportions have become steady.

It is interesting to see how the velocity increases with increasing thickness and diminishing viscosity of the buoyant layer. Note particularly that the amplitude velocity increases much faster than linearly with the thickness. In fact the velocity is closely proportional to the square of the thickness if systems with the same viscosity are compared. The reason for the $d\bar{y}/dt \approx h^2$ relationship is as follows: if the amplitude is kept constant while unlike thicknesses are tested, then the velocity of amplitude growth is directly proportional to the thickness of the layer. Similarly, if the layer thickness is kept fixed and the amplitude allowed to vary, the velocity is directly proportional to the amplitude as long as the deflections are gentle. When both thickness and amplitude are changed, so as to remain in constant proportion to one another, the velocity will be proportional to the square of the thickness or of the amplitude as the case may be. Moreover, since the characteristic wavelength is proportional to the thickness, the velocity of amplitude growth is also proportional to the square of the wavelength (for fixed viscosities and densities).

When the amplitudes have reached the state of steady proportions, the velocity of amplitude rise at the free surface, $\bar{v}_1$, and at the bottom of the buoyant stratum, $\bar{v}_3$, are related to $\bar{y}_1$ and $\bar{y}_3$, respectively, as $\bar{v}_2$ is related to $\bar{y}_2$.

One notes accordingly that the vertical movement on the Earth's surface is very slow as compared with the velocity at the depth of the active buoyant layer. In other words, considerable dynamic activity may occur at depth without noticeable surface disturbances.

The steadied amplitude proportion shown in Table 7.12, column 10, contains interesting information. Qualitatively, one should expect the contrast between $\bar{y}_1$ and $\bar{y}_2$ to decrease with increasing stiffness of the overburden relative to the buoyant stratum. The flowage within a stiff overburden—its gravitational spreading— is slow relative to the rate at which the mobile low-density stratum rises and lifts the overburden. It is also to be expected that the contrast between $\bar{y}_1$ and $\bar{y}_2$ should decrease with increasing wavelength relative to the thickness of the overburden, when compared at equal corresponding viscosities. This is exactly what the numerical model shows. Furthermore, the mathematical model brings quantitative support to the intuitive expectation that the difference between the amplitudes on the upper and lower boundaries of the buoyant layer

Table 7.12. Dynamic data of a buoyant layer buried deep in the mantle[a]

$h_{2:3}$	h_3 (km)	$\mu_{2:3} = \mu_{4:3}$	ϕ_2	λ (km)	$-\kappa_{1\,\text{max.}}$ ($\times 10^3$)	Time to reach $\bar{y}_t/\bar{y}_0 = 10^2$ (10^6 year)	Time to reach $\bar{y}_t/\bar{y}_0 = 10^3$ (10^6 year)	Time to reach $\bar{y}_t/\bar{y}_0 = 10^4$ (10^6 year)	$\bar{y}_1/\bar{y}_2/\bar{y}_3$	$d\bar{y}_2/dt$ when $\bar{y}_2 = h_3$ (cm year^{-1})
10	100	1	3·09	2032	3·98	2·31	3·46	4·62	0·0067/1/0·752	19·96
		10^6	0·63	9968	49·3	0·186	0·279	0·372	0·0995/1/0·0059	247·3
20	50	1	3·58	1754	2·113	4·34	6·51	8·68	0·0032/1/0·847	5·301
		10^6	0·92	6826	47·45	0·1934	0·290	0·387	0·0857/1/0·021	119·1
30	33·3	1	3·85	1631	1·44	6·37	9·56	12·8	0·00194/1/0·885	2·406
		10^6	1·13	5558	44·88	0·2049	0·307	0·409	0·0801/1/0·0375	74·98
40	25	1	4·01	1566	1·093	8·40	12·6	16·8	0·00134/1/0·9076	1·371
		10^6	1·32	4758	41·93	0·219	0·342	0·438	0·07413/1/0·0619	52·59
50	20	1	4·13	1521	0·8802	10·4	15·6	20·9	0·00104/1/0·9232	0·883
		10^6	1·50	4187	38·9	0·236	0·348	0·475	0·0673/1/0·0803	39·03
60	16·66	1	4·26	1474	0·737	12·5	18·7	24·9	0·00078/1/0·9323	0·617
		10^6	1·68	3738	35·93	0·256	0·383	0·511	0·06152/1/0·10954	30·04
70	14·3	1	4·35	1444	0·634	14·5	21·7	29·0	0·00068/1/0·9415	0·455
		10^6	1·80	3489	33·2	0·276	0·415	0·553	0·0549/1/0·1231	23·83
80	12·5	1	4·45	1411	0·556	16·5	24·8	33·0	0·00057/1/0·9474	0·3487

80		10^6	1·97	3188	30·7	0·299	0·449	0·598	0·0486/1/0·1342	19·34
90	11·11	1	4·52	1389	0·406	18·5	27·8	37·0	0·00048/1/0·9520	0·2765
		10^6	2·12	2962	28·4	0·323	0·485	0·647	0·04401/1/0·1578	15·93
100	10	1	4·58	1371	0·447	20·6	30·8	41·1	0·00041/1/0·9557	0·2243
		10^6	2·26	2779	26·33	0·349	0·523	0·697	0·0386/1/0·1648	13·21
110	9·09	1	4·63	1356	0·407	22·6	33·9	45·1	0·00037/1/0·9596	0·1856
		10^6	2·42	2595	24·5	0·375	0·562	0·7495	0·03487/1/0·18503	11·7
120	8·33	1	4·67	1345	0·373	24·6	36·9	49·2	0·00032/1/0·9621	0·1559
		10^6	2·54	2472	22·84	0·402	0·603	0·804	0·0315/1/0·2039	9·544
130	7·69	1	4·72	1331	0·345	26·6	39·9	53·2	0·00030/1/0·9649	0·1331
		10^6	2·67	2352	21·36	0·430	0·639	0·860	0·0274/1/0·2064	8.24
140	7·142	1	4·76	1319	0·3205	29·6	43·0	59·3	0·000262/1/0·9667	0·1118
		10^6	2·77	2267	20·03	0·4583	0·6874	0·9165	0·0248/1/0·2225	7·177
150	6·67	1	4·80	1310	0·2994	32·7	45·98	65·3	0·000244/1/0·969	0·0937
		10^6	3·0	2093	18·83	0·4874	0·7311	0·9748	0·0215/1/0·2233	6·30
500	2	1	5·4	1163	0·0905	101·7	152	203·3	0·000052/1/0·9893	0·0091
		10^6	9·1	690	5·777	1·59	2·383	3·18	0·00012/1/0·2724	0·5798
1000	1	1	5·7	1102	0·0454	202·2	303·2	404·3	0·0000228/1/0·9944	0·00228
		10^6	18·2	345	2·89	1·03	1·547	2·06	0·0000002/1/0·2733	0·145

[a] $(\rho_2-\rho_3)/\rho_2 = (\rho_4-\rho_3)/\rho_2 = 0{\cdot}3/3{\cdot}3$, $h_2 = 1000$ km, $\mu_2 = 10^{22}$ poise

(y_2 and y_3) should be large for a stratum which is soft relative to the overburden and substratum, and small for one that is more stiff.

Table 7.11 records the effect which varying density of layer 4 has on the characteristic wavelength and on the amplitude proportion. Both effects show up distinctly. When the density of the substratum is equal to the density of the overburden, then amplitude $\bar{y}_3$ is not much different from $\bar{y}_2$, $\bar{y}_3$, being only about 5% less than $\bar{y}_2$. An increment of the substratum density corresponding to a ρ_4/ρ_3 ratio equal to 1·5, compared to 1·1 for the ρ_2/ρ_3 ratio, produces a large difference between the two amplitudes, amplitude $\bar{y}_3$ is now only about 20% of $\bar{y}_2$. The same density change of the substratum also reduces the wavelength drastically, the characteristic wavelength/thickness ratio decreasing from 137 to 6·47.

These trends are intuitively expected, but only the mathematical treatment gives numbers to the qualitatively predicted variations.

Isothermal multilayer models simulating thermal convection in the Earth's mantle

The analytical theory of unstable layered systems described above is valid for layers that are uniform in terms of density and viscosity. In nature, however, both density and viscosity often change continuously with depth even if the constituent rock or magma is homogeneous in its mineralogical and chemical composition. The high pressure and temperature at deep levels in the Earth cause density and effective viscosity to change. Such continuous variation of density and viscosity may be approximated in theoretical models by selecting a large number of relatively thin layers whose density and/or viscosity change in small steps from layer to layer. By making the steps sufficiently small and the number of layers correspondingly large the results from the calculation will approach the condition of a model with continuous variation of density and/or viscosity in the vertical direction.

A few multilayer systems of this type have been calculated. As an example we shall discuss a multilayer model with downward decreasing density which to a certain extent may be regarded as an imitation of the overturn process in the Earth's mantle during thermal convection. The difference in density between the rising and sinking columns in thermal convection cells in the mantle is simulated by the variation of density from layer to layer in the isothermal multilayer model. The velocity of flow in the unstable multilayer model at certain defined values of the amplitudes will then compare well with the velocity and pattern of flow in a thermally convecting mantle.

For true thermal convection in the mantle see e.g. Runcorn (1965), Tozer (1967), Turcotte and Oxburgh (1972), McKenzie (1977), McKenzie *et al.* (1974), and a recent comprehensive study of mantle convection edited by Cassen (1979). See also the recent book by McElhinny (1979).

The first step towards constructing a multilayer suited to simulate the mechanics of a thermal convection cell is to determine how the driving forces in the convecting system can be replaced by corresponding forces in the unstable isothermal multilayer.

In thermal convection the driving force is the buoyancy due to density contrast between the rising branch and the descending branch of the convecting cell. The rising branch is hotter—and therefore less dense—than the descending branch when compared at the same horizontal level. This density contrast occurs when the vertical thermal gradient exceeds the adiabatic gradient. An obvious conclusion is that the thermal expansion results in a downward decreasing density, and that this is the reason for the unstability of the convecting cell. Inverted density stratification is, however, not needed: it is sufficient that the temperature gradient is greater than the adiabatic gradient of the material because when this condition is satisfied the rising branch will always be hotter than the descending branch when compared at equal level in the gravitational field. It does not matter that the high pressure at depth compresses the material more than it is expanded by the high temperature.

Knowing the temperature difference between the rising and descending currents, the thermal expansion coefficient of the medium, and its compressibility, one can calculate the density difference between the two oppositely moving currents. This can then be compared with the density difference between the rising and the descending currents in an isothermal multilayer with a stepwise decreasing density downward. The density difference in the isothermal multilayer can be adjusted by selecting appropriate amplitudes of the deflections. In this manner the driving force in the isothermal system can be matched quite accurate with the force which drives the thermal system.

Consider the multilayer with inverted density stratification in Fig. 7.24. We note that the amplitude is largest in the middle part ($\bar{y}_7$) and decreases both upward and downward in the model. This means that each of the layers from number 7 downward assumes maximum thickness in the anticlines—that is, in the region of rising flow—whereas each layer above number 7 assumes maximum thickness in the synclinal region where the flow is descending. A consequence of this is that the average density is greater in the descending branch of the flow than in the ascending branch. The *average* density may be estimated as the mass of a column through the entire multilayer pile, divided by the volume of the column. For columns

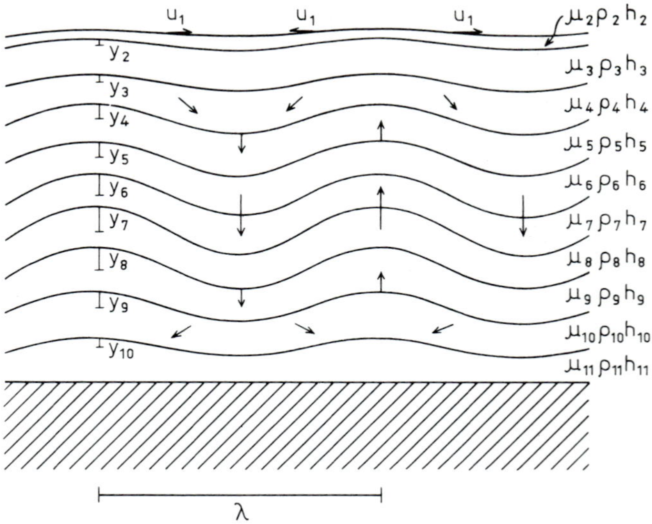

Fig. 7.24. A ten-layer buoyant model simulating a half-cycle of thermal convection in the upper mantle. Thin upper layer represents the earth's crust. Arrows indicate movement pattern.

with 1 cm^2 horizontal cross-section, the average density is the mass divided by the height.

For simplicity let all layers be of equal thickness, h, prior to the commencement of flow and let each layer be Δ g cm^{-3} less dense than the layer directly above. ρ_0 is the density of the top layer.

The mass of a column, 1 cm^2 in cross-section, through the centre of the anticline is accordingly:

$$\left.\begin{aligned}
&[h + (1 - \bar{y}_{2:1})\bar{y}_1]\rho_0 \\
&+ [h + (\bar{y}_{2:1} - \bar{y}_{3:1})\bar{y}_1]\,(\rho_0 - \Delta) \\
&+ [h + (\bar{y}_{3:1} - \bar{y}_{4:1})\bar{y}_1]\,(\rho_0 - 2\Delta) \\
&+ [h + (\bar{y}_{4:1} - \bar{y}_{5:1})\bar{y}_1]\,(\rho_0 - 3\Delta) \\
&\quad\vdots \\
&+ [h + (\bar{y}_{(n-1):1} - \bar{y}_{n:1})\bar{y}_1]\,[\rho_0 - (n-2)\Delta] \\
&+ [h + (\bar{y}_{(n:1)} - 0)\bar{y}_1]\,[\rho_0 - (n-1)\Delta] \\
&= nh\rho_0 - (\Delta + 2\Delta + 3\Delta \\
&\quad + \ldots + (n-1)\Delta)h + \rho_0\bar{y}_1 \\
&\quad - (\bar{y}_{2:1} + \bar{y}_{3:1} + \ldots + \bar{y}_{(n-1):1} + \bar{y}_{n:1})\bar{y}_1\Delta \\
&= M_{\text{anticl.}}
\end{aligned}\right\} \qquad (7.77)$$

The mass of a column with unit cross-section through the centre of the syncline is correspondingly:

$$\left.\begin{aligned}
&[h + (\bar{y}_{2:1} - 1)\bar{y}_1]\rho_0\\
&+ [h + (\bar{y}_{3:1} - \bar{y}_{2:1})\bar{y}_1]\,[\rho_0 - \Delta]\\
&+ [h + (\bar{y}_{4:1} - \bar{y}_{3:1})\bar{y}_1]\,[\rho_0 - 2\Delta]\\
&\quad\vdots\\
&+ [h + (\bar{y}_{n:1} - \bar{y}_{(n-1):1})\bar{y}_1][\rho_0 - (n-2)\Delta]\\
&+ [h + (0 - \bar{y}_{n:1})\bar{y}_1][\rho_0 - (n-1)\Delta]\\
&= nh\rho_0 - (\Delta + 2\Delta + 3\Delta + \ldots + (n-1)\Delta)h - \rho_0\bar{y}_1\\
&+ (\bar{y}_{2:1} + \bar{y}_{3:1} + \ldots + \bar{y}_{n:1})\bar{y}_1\Delta\\
&= M_{\text{syncl.}}
\end{aligned}\right\} \qquad (7.78)$$

The difference in mass of the columns is

$$M_{\text{syncl.}} - M_{\text{anticl.}} = 2[-\rho_0 + (\bar{y}_{2:1} + \ldots + \bar{y}_{n:1})\Delta]\bar{y}_1, \qquad (7.79)$$

and the average density difference is

$$\Delta\rho = 2\,\frac{[-\rho_0 + (\bar{y}_{2:1} + \bar{y}_{3:1} + \ldots + \bar{y}_{n:1})\Delta]\bar{y}_1}{nh}. \qquad (7.80)$$

Here we have assumed that $\bar{y}_1$ is small relative to nh.

This density contrast between the ascending and the descending columns in the unstable isothermal model can be compared with the difference in density of the hot and cool columns in a thermal convection cell.

A model of this type may be applied to the uppermost 900 km thick shell of the Earth's mantle. Let this part of the mantle consist of 9 layers with thickness 100 km and viscosity 10^{22} poise. To the top layer of the mantle we assign a density $\rho = 3{\cdot}4$ g cm^{-3}. The density of each layer below is 0·018 g cm^{-3} less than its directly overlying stratum. Floating on the top of the mantle occurs a 33-km-thick Earth's crust with density 2·8 g cm^{-3} and effective viscosity 10^{23} poise.

These figures yield the dimensionless parameters needed to calculate the characteristic wavelength of the system and the steadied proportions of the amplitudes. We have

$$\mu_{2:3} = 10;\ \mu_{3:4} = \mu_{4:5} = \ldots = \mu_{10:11} = 1;\ h_{2:3} = 0{\cdot}33,$$
$$h_{3:4} = h_{4:5} = \ldots = h_{10:11} = 1;$$

$$q_{2:1} = \frac{-0{\cdot}6\times100\times10^5\times10^{23}}{-2{\cdot}8\times44\times10^5\times10^{22}} = 6{\cdot}495$$

Table 7.13. Steadied amplitude proportions and growth factors for ten-layer convection imitation model with $\kappa_{max} = 0{\cdot}511$, $\phi_2 = 0{\cdot}164$, $\lambda/h_2 = 38{\cdot}3$

Subscript i	1	2	3	4	5
$\bar{y}_i$	1	12·22	76·11	164·76	240·05
$-k^u_{1i}$	3·1863	0·18946	0·32416	0·33181	0·28426
Subscript i	**6**	**7**	**8**	**9**	**10**
$\bar{y}_i$	277·80	267·78	212·40	127·47	41·88
$-k^u_{1i}$	0·22905	0·15589	0·0975	0·0491	0·0143

and

$$q_{3:1} = q_{4:1} = \ldots = q_{10:11} = \frac{0{\cdot}018\times100\times10^5\times10^{23}}{-2{\cdot}8\times33\times10^5\times10^{22}} = -0{\cdot}1929;$$

$$q_1 = \frac{2{\cdot}8\times33\times10^5\times981}{2\times10^{23}} = -4{\cdot}532\times10^{-14}\,\text{sec}^{-1}.$$

The calculated values of κ_{max}, λ/h_2 and the steadied amplitude proportions are recorded in Table 7.13 and partly in Fig. 7.25.

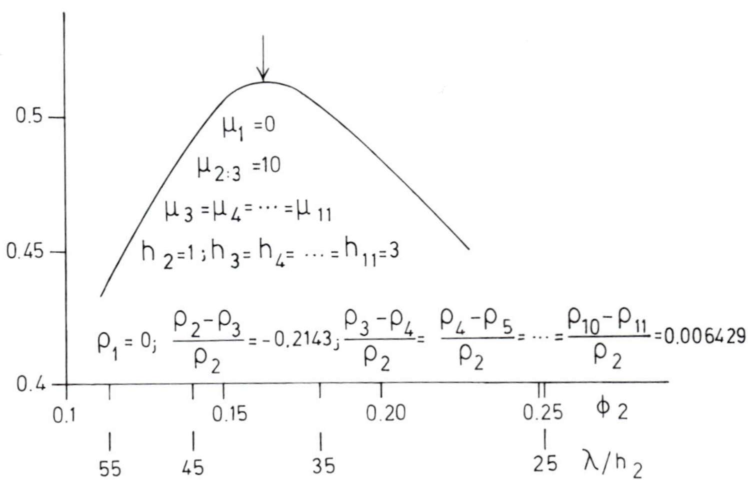

Fig. 7.25. Eigenvalue, κ_1, (vertical scale) versus ϕ_2 and λ/h_2 for ten-layer model simulating thermal convection in the earth's mantle. See text and Table 7.13.

Because of the presence of a crust which is less dense than the underlying mantle this Earth model does not conform exactly to the conditions specified for the theoretical model. It can be shown that the expression for the average density contrast between the descending and the ascending branches of the overturn flow now takes the form

$$\Delta\rho = 2\frac{(1-\bar{y}_{1:2})\rho_2 - \rho_3 + (\bar{y}_{3:2} + \bar{y}_{4:2} + \ldots + \bar{y}_{n:2})\Delta}{nh_3 + h_2}\bar{y}_2. \qquad (7.81)$$

Here ρ_2 and ρ_3 are the densities of the crust and the uppermost layer of the mantle, h_2 is the thickness of the crust and h_3 that of each of the n mantle layers, and Δ is the density difference between each layer in the mantle and the directly underlying one.

According to these data the average density contrast becomes

$$\begin{aligned}\Delta\rho &= 2[(1-0{\cdot}082)2{\cdot}8 - 3{\cdot}4 + (6{\cdot}23 + 13{\cdot}48 + 19{\cdot}64 + 22{\cdot}73 \\ &\quad + 21{\cdot}91 + 17{\cdot}38 + 10{\cdot}43 + 3{\cdot}43)0{\cdot}018]\bar{y}_2/(900+33) \\ &= 0{\cdot}00267\,\bar{y}_2\ \mathrm{g\ cm^{-3}}\end{aligned} \qquad (7.82)$$

when $\bar{y}_2$ is given in km.

To obtain a definite value for the density difference between the ascending and descending branches of a mantle whose uppermost 900 km undergoes thermal convection we assume a temperature difference of 100°C between the upper part of the two branches of flow. We take the density at the top of the mantle to be $3{\cdot}4\ \mathrm{g\ cm^{-3}}$ and the thermal expansion coefficient, α, to be $3\times10^{-5}\ ^\circ\mathrm{C}^{-1}$. At the top of the mantle the density difference between the descending and the ascending columns of flow is accordingly

$$\Delta\rho_\mathrm{T} = 3{\cdot}4\ \mathrm{g\ cm^{-3}} \times 3\times10^{-5} \times 100 = 0{\cdot}0102\ \mathrm{g\ cm^{-3}}. \qquad (7.83)$$

To make the calculation simple we assume that (1) the temperature difference between the two oppositely moving columns of flow, (2) the thermal expansion as a function of pressure and (3) the compressibility as a function of temperature, are so related that the density difference, $\Delta\rho_\mathrm{T} = 0{\cdot}0102\ \mathrm{g\ cm^{-3}}$, is also the average density difference between the two columns of flow at any level down in the mantle.

In order for the driving force in the isothermal model to equal the driving force in the thermal model $\Delta\rho$ must equal $\Delta\rho_\mathrm{T}$. Equating the two density differences (eqns 7.82 and 7.83) defines the $\bar{y}_2$ value which equalizes the buoyancy force in the two models, viz.

$$\bar{y}_2 = \frac{0{\cdot}0102}{0{\cdot}00267} = 3{\cdot}827 \text{ km} \tag{7.84}$$

This value yields for the largest amplitude, $\bar{y}_6$, the value $3{\cdot}824 \times 22{\cdot}73 = 86{\cdot}9$ km. The number 22·73 is the ratio $\bar{y}_6/\bar{y}_2$, see Table 7.13. Accordingly, the largest amplitude in the system is less than 10% of the dominant wavelength—which is 1264 km—and we feel that the model gives a reasonably correct velocity field for a mantle whose ascending limbs are 100°C hotter than the descending limbs.

The vertical component of the velocity of flow at the moment $\bar{y}_2 = 3{\cdot}824$ km follows from equation $\bar{v}_i = \kappa_1 q_1 \bar{y}_i$ which gives the vertical flow component at any level i. The maximum velocity is at interface 6 which is obtained when values for κ_1, q_1 and $\bar{y}_6$ are inserted in the equation above, thus

$$\begin{aligned}\bar{v}_6 &= 0{\cdot}511 \times 4{\cdot}532 \times 10^{-14} \times 8{\cdot}69 \times 10^6 \\ &= 2{\cdot}01 \times 10^{-7} \text{ cm sec}^{-1} \text{ or } 6{\cdot}32 \text{ cm year}^{-1}.\end{aligned} \tag{7.85}$$

The vertical amplitude velocity at the Earth's surface will be much less, viz.

$$\begin{aligned}\bar{v}_1 &= 0{\cdot}511 \times 4{\cdot}532 \times 10^{-14} \times 3{\cdot}13 \times 10^4 = 7{\cdot}25 \times 10^{-10} \text{ cm sec}^{-1} \\ &= 2{\cdot}277 \times 10^{-2} \text{ cm year}^{-1}.\end{aligned} \tag{7.86}$$

(Here $3{\cdot}13 \times 10^4$ cm is the amplitude at the surface.)

It is interesting for the question of sea-floor spreading and continental drift to calculate the velocity of horizontal flow at the top of our theoretical model. At the inflection points, i.e. in the middle between the crest of the anticline and the keel of the adjacent syncline, the horizontal velocity component $\bar{u}_1$ is given by the equation

$$\bar{u}_1 = (k^u_{11} + k^u_{12} q_{2:1} \bar{y}_{2:1} + k^u_{13} q_{3:1} \bar{y}_{3:1} + \ldots + k^u_{1.10} q_{10:1} \bar{y}_{10:1}) q_1 \bar{y}_1. \tag{7.87}$$

For the system under study the k^u_{ji}-values have been computed and presented in Table 7.13. Inserted in eqn (7.87), together with the values of $q_{i:1}$, $\bar{y}_{i:1}$ and q_1, we obtain

$$\bar{u}_1 = \pm 5{\cdot}03 \times 10^{-8} \text{ cm sec}^{-1} = \pm 1{\cdot}58 \text{ cm year}^{-1}. \tag{7.88}$$

The $\pm$-sign indicates that the horizontal velocity component changes direction as one passes the crest of the anticline: on the left side of the anticline the velocity vector points toward the left, on the right side of the anticline the velocity is directed toward the right. In other words, the flow is directed away from the anticlines.

In the chapter on experimental models the case of a six-layer system believed to simulate convection is tested experimentally, see p. 415.

Ten-layer systems with stepwise changing viscosity from top to bottom

The viscosity of the Earth's mantle is hardly constant along the vertical direction. Below the base of the low-velocity layer the viscosity probably increases with depth, but within and above the low-velocity layer an opposite variation of viscosity with depth possibly occurs. Depending upon the local pressure and temperture relationships, regions may also exist deep in the mantle where viscosity stratification is inverted.

A vertical variation of the mantle's viscosity must influence the flow pattern in a convecting mantle and in and around mantle diapirs. For example, is it possible that only the low-viscosity part of the mantle will convect even if the thermal gradient is superadiabatic throughout the entire mantle? This is a problem which relates to the horizontal dimension of convective cells and therefore is of interest in plate tectonics and continental drift.

To gain some insight into this question we tested numerically two models, both consisting of 10 equal-thickness viscous layers, each layer being less dense than the immediate overlying one, as defined by the relationships

$$\frac{\rho_2 - \rho_3}{\rho_2} = \frac{\rho_3 - \rho_4}{\rho_2} = \ldots = \frac{\rho_9 - \rho_{10}}{\rho_2} = \frac{\rho_{10} - \rho_{11}}{\rho_2} = 0{\cdot}1.$$

The thickness ratios are $h_{2:3} = h_{3:4} = h_{4:5} = \ldots = h_{9:10} = h_{10:11} = 1$. Resting on a rigid support and exhibiting a free, deformable surface, the complex belongs to model type II-10 (Figs 7.9 and 7.10). In accord with the notation used in this book the top layer bears the index 2 (Fig. 7.26).

In a model with the boundary conditions described above let the viscosity ratios be $\mu_{2:3} = \mu_{3:4} = \ldots = \mu_{9:10} = \mu_{10:11} = 0{\cdot}5$. It follows from this that the viscosity ratio $\mu_{2:11}$ is 0·001953125, or that the basal layer is 512 times more viscous than the top layer (Fig. 7.26).

The calculated principal eigenvalue is negative and has a maximum at $\phi_2 = 0{\cdot}78$ corresponding to $\lambda/h_2 = 8{\cdot}05$ (Fig. 7.27). Since h_2 is 10% of the thickness of the complete multilayer it means that the wavelength/composite thickness ratio is only 0·805, indicating that the width of the overturning cells is about 20% less than their height.

Incidentally, the negative sign of the eigenvalue must be seen in context

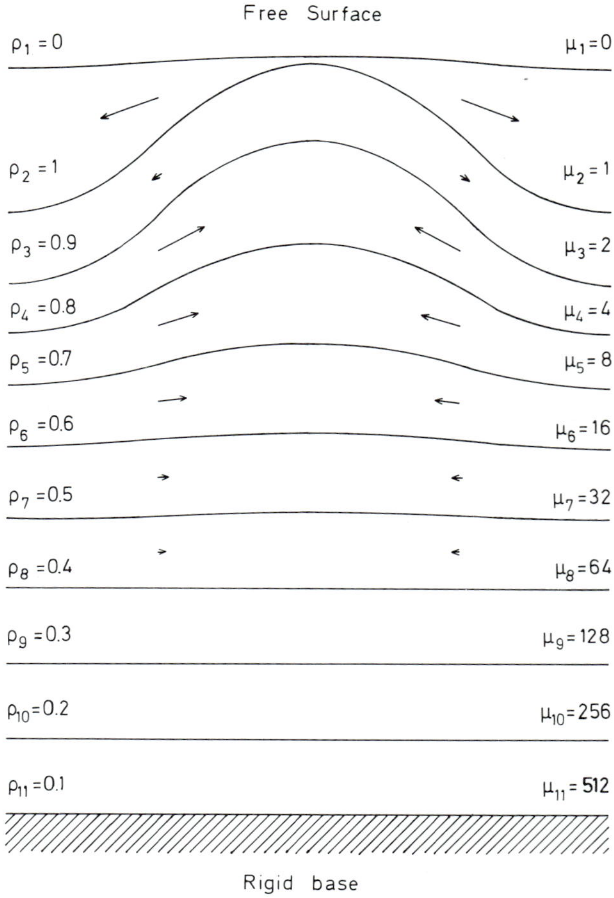

Fig. 7.26. Relative amplitudes indicated in ten-layer buoyant model with viscosity decreasing stepwise upward. Only one wavelength shown.

with the sign of q_1 which also is negative $\left(q_1 = \frac{-\rho_2 g h_2}{2\mu_2}\right)$, signifying that the product $\kappa_1 q_1$ is positive and hence that the amplitudes grow with time.

It is interesting to note the relationship between the components of the eigenvectors, and hence the relationship between the steadied amplitudes, indicated in Fig. 7.26. It appears that the overturn flow from the unstable to the stable density stratification is much more pronounced in the upper half of the complex than in its lower half. This theoretical result makes good sense in view of the difference in viscosity between the upper and lower layers. A comparison with the following model is informative.

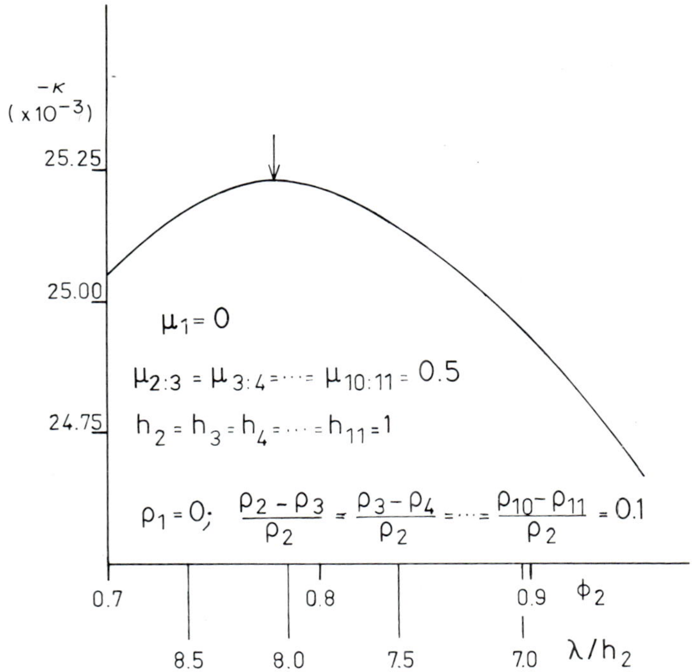

Fig. 7.27. Eigenvalue, κ_1, versus ϕ_2 and λ/h_2 for the model shown in Fig. 7.26.

In a model whose viscosity decreases from top to bottom all conditions are identical to those specified above except the reversal of the viscosity ratios, thus $\mu_{2:3} = \mu_{3:4} = \ldots = \mu_{9:10} = \mu_{10:11} = 2$. Consequently, the top layer is now $2^9 = 512$ times more viscous than the bottom layer (Fig. 7.28).

The negative principal eigenvalue has a numerical maximum at $\phi_2 = 0{\cdot}935$ or at $\lambda/h_2 = 6{\cdot}71$ (Fig. 7.29). Considering the composite thickness it means that the width of the overturning cell is only 67% of the height.

It is significant that the characteristic wavelength in this system is somewhat less than in the corresponding model with viscosity decreasing upward. In the present model the flow takes essentially place within the lowermost half of the model as seen from the calculated amplitude proportions exhibited in Fig. 7.28. The upper part of the system behaves a little like a rigid plate which causes much shear-strain in the horizontally flowing medium below. This effect is apt to decrease the horizontal distances between the anticlinal rises relative to the model with the most mobile layers in the upper part where there is no shear strain at the free surface.

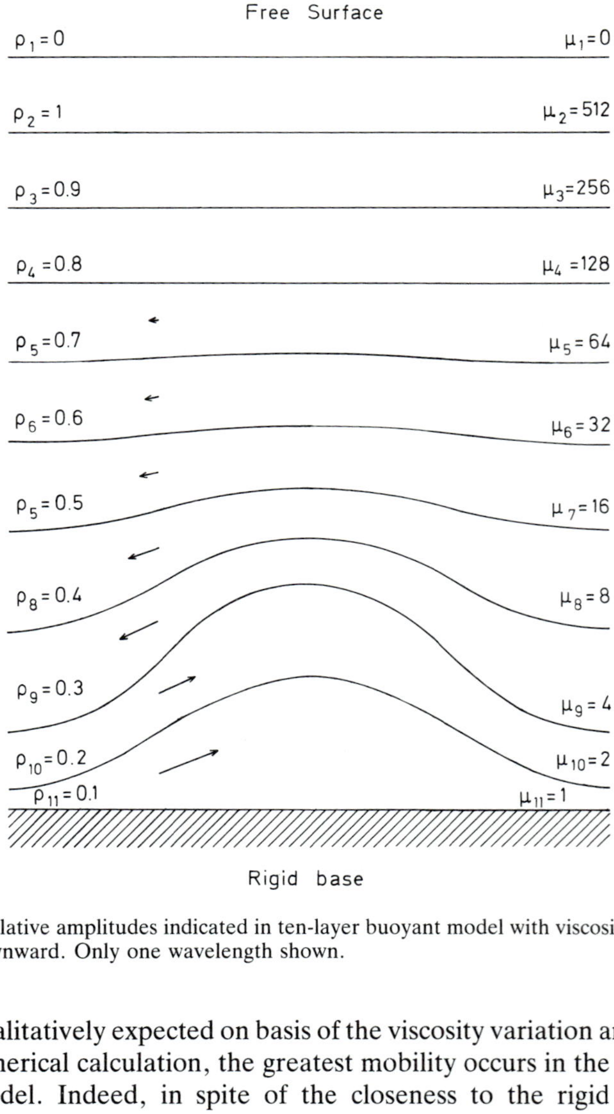

Fig. 7.28. Relative amplitudes indicated in ten-layer buoyant model with viscosity decreasing stepwise downward. Only one wavelength shown.

As is qualitatively expected on basis of the viscosity variation and verified by the numerical calculation, the greatest mobility occurs in the lower half of the model. Indeed, in spite of the closeness to the rigid base, the amplitude at the second interface from the bottom, $\bar{y}_9$, is rising with maximum speed whereas at the surface waves hardly develop at all.

Several models of the kind described in this section have been run in the centrifuge and will be described in later chapters.

The great difference in principal eigenvalues of the two systems may at first glance be taken to indicate greatly different velocity of evolution. However, this is not so if we compare the viscosities of the two systems.

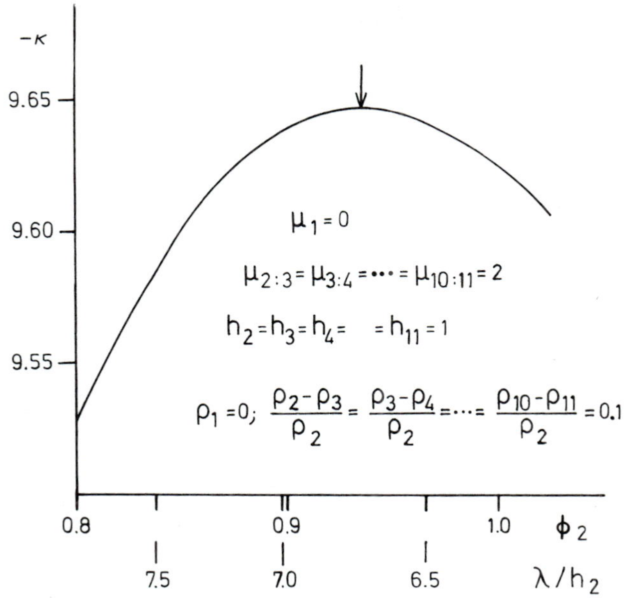

Fig. 7.29. Eigenvalue, κ_1, versus ϕ_2 and λ/h_2 for the model shown in Fig. 7.28.

In the system with the viscosity ratio $\mu'_{2:11} = 0{\cdot}001953125 = 1/512$ the growth factor $\kappa'_1 q'_1$ is

$$-2{\cdot}5212 \times 10^{-2} \times \frac{-\rho_2 h_2 g}{2\mu'_2}.$$

In the system with $\mu''_{2:11} = 512$, the growth factor $\kappa''_1 q''_1$ is

$$-9{\cdot}64 \times \frac{-\rho_2 h_2 g}{2\mu''_2}.$$

Here the prime and double prime refer to the two systems in which viscosities differ but corresponding densities and thicknesses are the same.

In order that the two systems should be comparable with respect to the absolute values of their viscosities, the viscosity of the bottom layer in the prime system must equal the viscosity of the top layer in the double-prime system. When this condition is satisfied the relations $\mu'_{2:3} = \mu'_{3:4} = \ldots = \mu'_{10:11} = 0{\cdot}5$ and $\mu''_{2:3} = \mu''_{3:4} = \ldots = \mu''_{10:11} = 2$, shows that for each layer in the one system there will be one, and only one, layer in the other system with the same viscosity. The average viscosity is also the same in both systems.

Let $\mu_2' = 1$, $\mu_{11}' = 512$, $\mu_2'' = 512$ and $\mu_{11}'' = 1$. Under this condition the two growth factors become

$$\left.\begin{aligned} \kappa_1' q_1' &= -2{\cdot}5212\times10^{-2}\times\frac{-\rho_2 h_2 g}{2\times1} = 1{\cdot}2606\times10^{-2}\times\rho_2 h_2 g, \\ \text{and} \qquad & \\ \kappa''_1 q''_1 &= 09{\cdot}64\times\frac{-\rho_2 h_2 g}{2\times512} = 0{\cdot}9414\times10^{-2}\times\rho_2 h_2 g. \end{aligned}\right\} \tag{7.89}$$

In spite of the great difference in principal eigenvalues it appears accordingly that the growth factors—which control the velocity of amplitude growth in the expression

$$\bar{v}_i = \kappa_1 q_1 \bar{y}_i \tag{7.90}$$

—are not much different in the two systems when based on comparable viscosities.

It makes good sense in view of mechanics that the model with the most mobile layers in the top half has a somewhat larger growth factor than the model whose lower half contains the most mobile layers. This can be explained by the same argument used when the difference in characteristic wavelength of the two systems was discussed. The flow in the model with the high viscosity in the upper half can to some extent be compared with the flow between two rigid plates, the rigid base and the rather stiff overburden. This is apt to lead to a slower rate of development than in the model, whose least viscous layers are in the upper half where there is no shear strain along the surface.

FINITE-ELEMENT METHOD APPLIED ON DIAPIRS

The analytical theory discussed in the above sections determines the characteristic wavelength of layered systems and the velocity at which the amplitude of deflections changes as long as the deflections are gentle. The maximum amplitude/wavelength ratio for which the analytical theory is valid with acceptable approximation to reality, is somewhat arbitrarily put equal to 0·1. For systems with deflections of this magnitude the velocity of amplitude change is quite accurately given by the theory in the forms of eqns (7.47) and (7.48).

If one wishes to study the evolution of domes or diapirs beyond the stage mentioned, numerical and experimental methods are available. The latter will be treated extensively in Chapter 11 as one of the main objectives of the book. A treatment of numerical methods applicable to advanced stages of diapir evolution is outside the scope of this book, but the usefulness of numerical methods in connection with the increasing availability of large computers demands that some references to relevant papers be given.

Fig. 7.30. Six stages on the evolution of an oil diapir rising through syrup. Experimental model run to test numerical model and analytical theory (from Berner *et al.*, 1972).

Berner *et al.* (1972) used the finite element method to study the evolution of two diapirs whose initial stages were treated by the analytical method. The finite element results were also compared with the experimental evolution of oil diapirs rising into an over-burden of glucose (Figs 7.30–7.32).

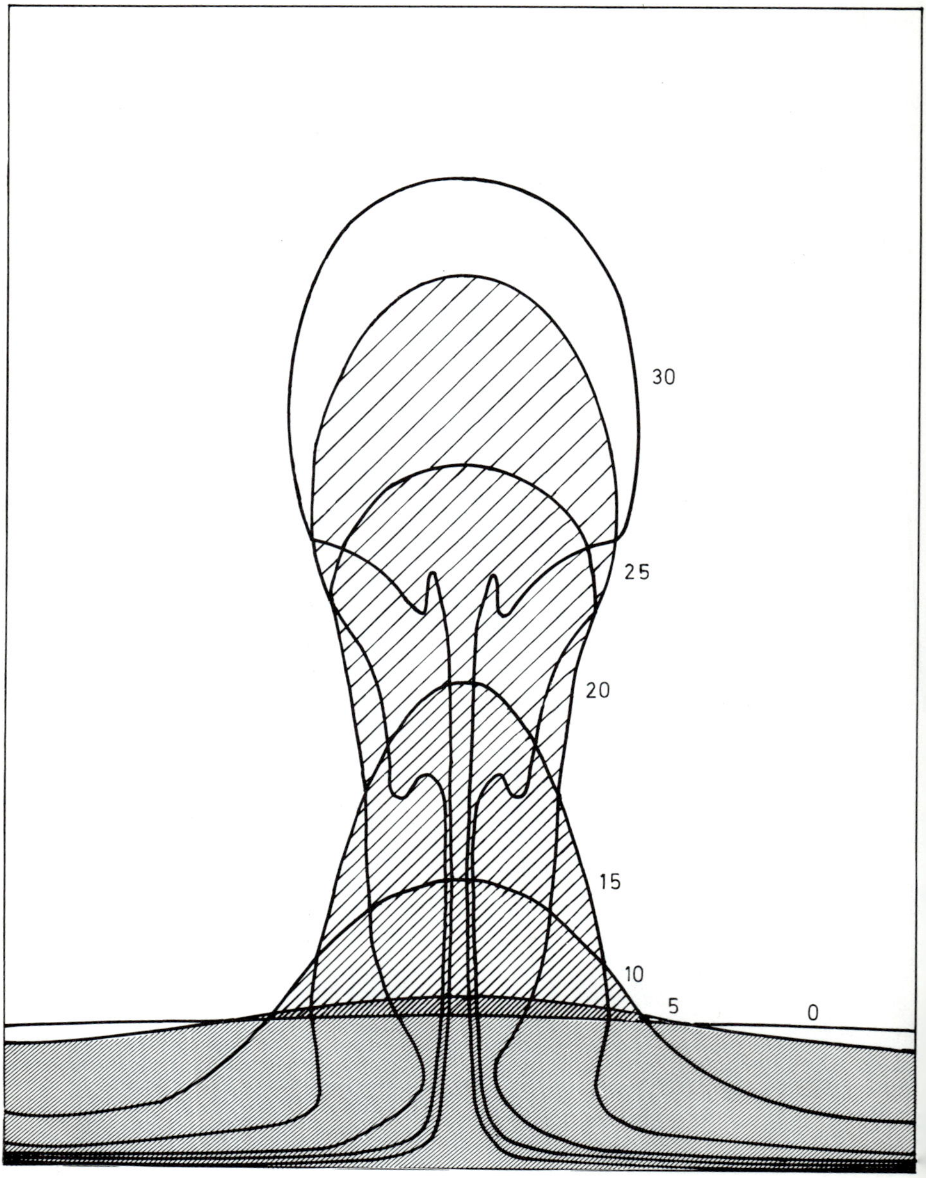

Fig. 7.31. Finite-element model of diapiric growth. Numbers indicate relative time of evolution (from Berner *et al.*, 1972).

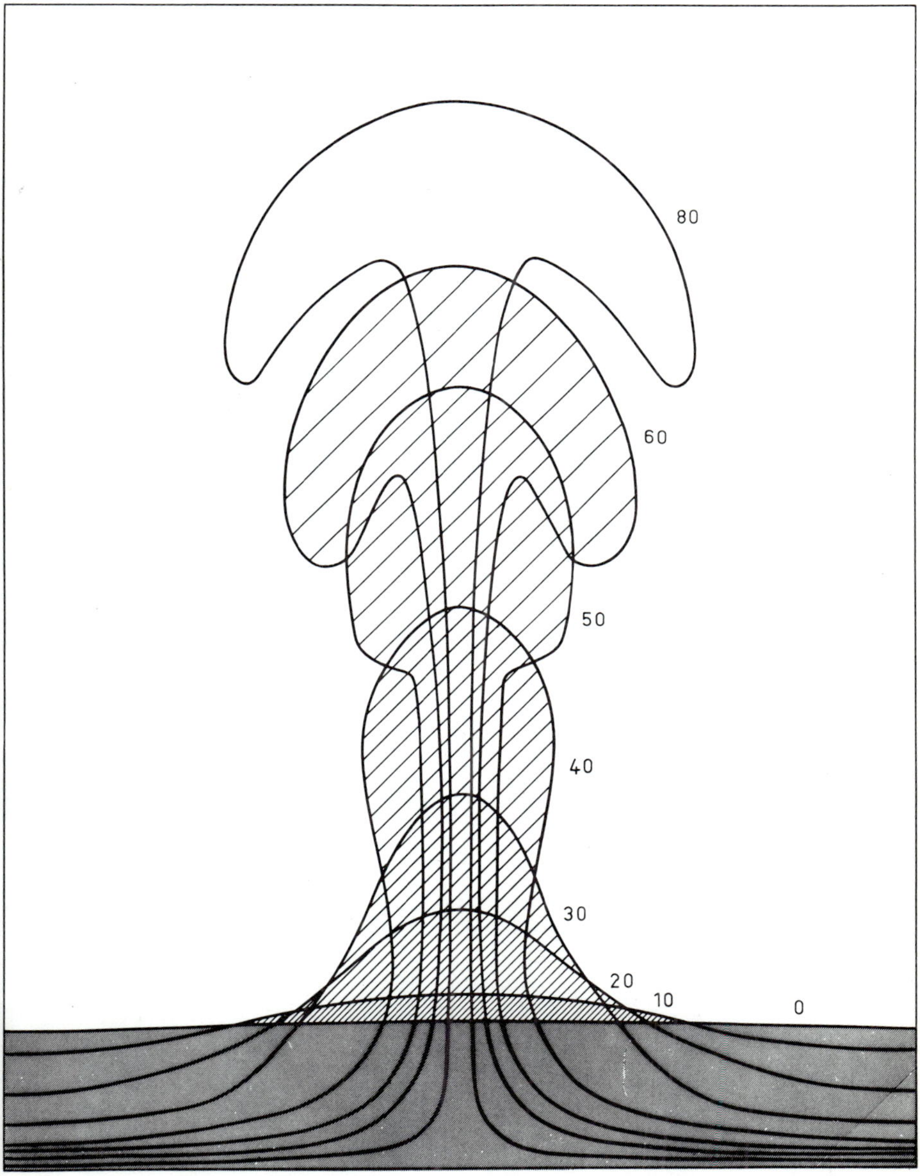

Fig. 7.32. Finite-element model of diapiric growth. Numbers indicate relative time of evolution (from Berner *et al.*, 1972).

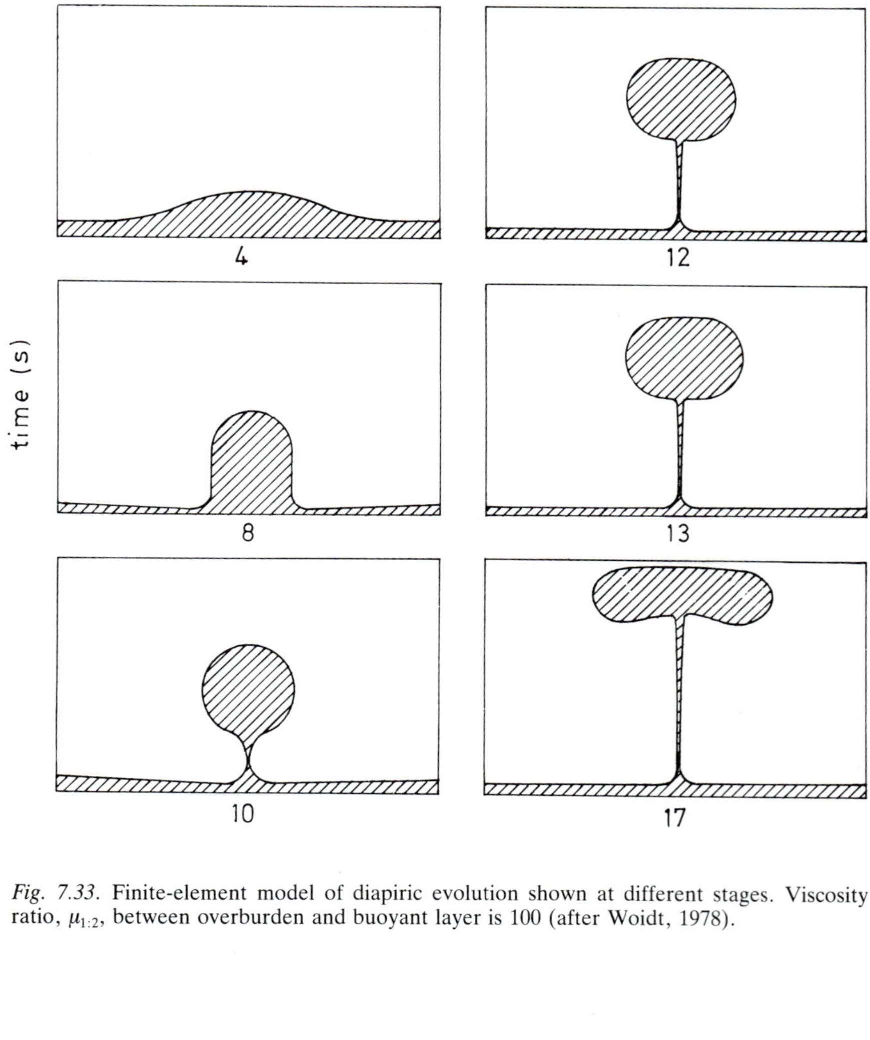

Fig. 7.33. Finite-element model of diapiric evolution shown at different stages. Viscosity ratio, $\mu_{1:2}$, between overburden and buoyant layer is 100 (after Woidt, 1978).

Recently, Woidt (1978) in an important contribution, simulated numerically a number of salt diapirs by the finite element method (Figs 7.33–7.35). His results are not essentially different from those of Berner *et al.* They compare also quite well with the experimental results. The papers by Berner *et al.* and Woidt demonstrate that much valuable information

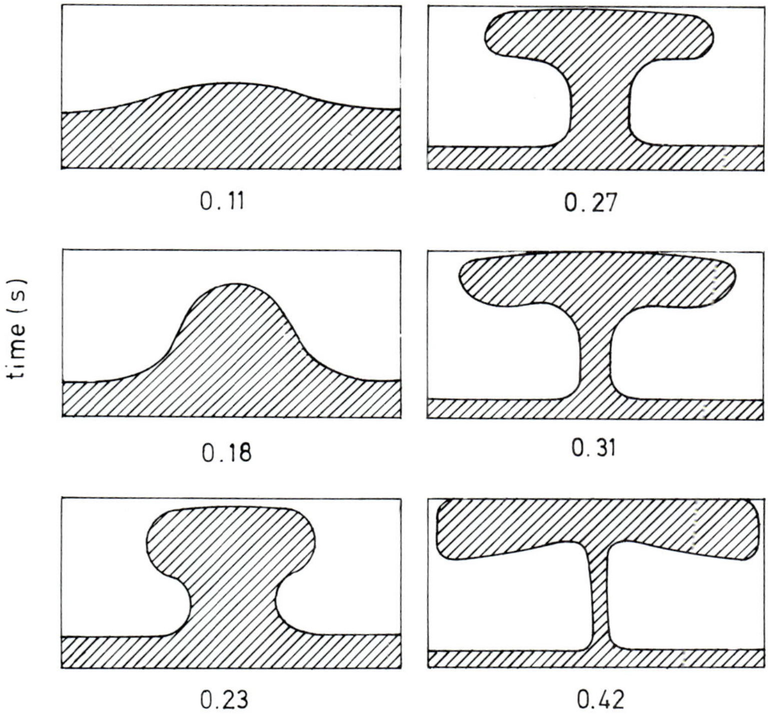

Fig. 7.34. Same as shown in Fig. 7.23, but viscosity ratio $\mu_{1:2}$ is here 1 (after Woidt, 1978).

on the evolution of geometrically complicated structures can be obtained by the finite element method. That the method is useful for a large variety of tectonic structures is documented by the many contributions in the special issue edited by Toksöz: Numerical Modelling in Geodynamics, *Tectonophysics* **50**, 1978. See also the paper by Hayashi and Kizaki (1979).

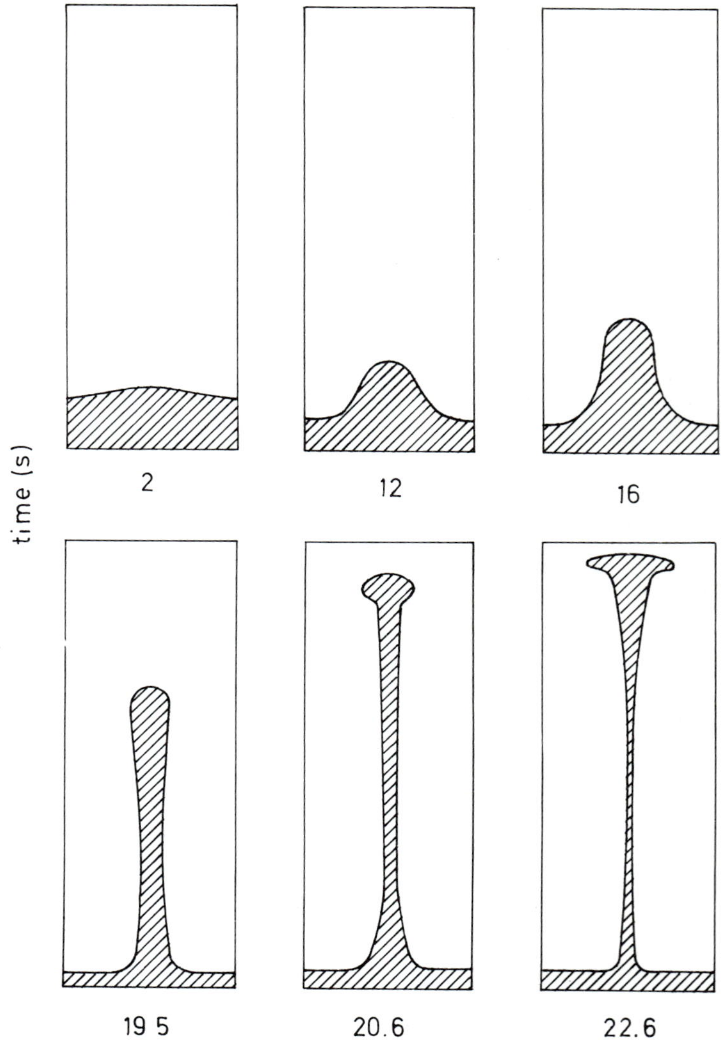

Fig. 7.35. Same as shown in Figs 7.33 and 7.34, but here the viscosity ratio, $\mu_{1:2} = 0{\cdot}01$ (after Woidt, 1978).

APPLICATION OF THE MULTILAYER THEORY ON ISOSTATIC ADJUSTMENT IN A LAYERED GLOBE

The theory of dynamics of layered systems in the field of gravity presented above is valid independent of whether the density stratification is inverted or normal. If there is no mobile layer that is less dense than any of the strata in the overburden, there is no buoyant force in the system, and initial deflections from corresponding equipotential surfaces of the interfaces or the free surface tend to flatten until the layering becomes parallel to the equipotential surfaces in the gravitational field. Were it not for the fact that at least the upper part of the crust possesses a certain yield strength, this tendency of smearing out all perturbations would ultimately lead to complete isostatic equilibrium.

In the mathematical theory the lack of buoyant layers in a system shows up in the signs of the eigenvalues of the matrix of the amplitude-growth equation (7.47). In non-buoyant systems all eigenvalues assume the signs which make the products $\kappa_i q_1$ negative. Accordingly, all exponential terms in the integrated amplitude equation (7.64) go to zero as time increases.

Numerical tests show that in detail the flattening of the deflections—the isostatic adjustment— may be quite complicated in multilayers with two or more deformable boundaries even if the deflections are simple sine functions. So, for example, the diminishing of the amplitude of deflections on one interface—say interface i— may even cause deflections to grow on neighbouring boundaries above or below. In the course of time the intitial deflections on boundary i vanish, as does their effect upon the neighbouring interfaces, and the evolution of the deflections created there changes from growth to flattening.

Even if thus in the early stage of isostatic adjustment dynamic coupling between adjacent layers may create deflections on originally unperturbed interfaces, in the end all deviations from the equipotential surfaces will level out in layered system consisting of Newtonian substances.

To show how the dynamic coupling between layers functions, consider the velocity equations valid for a non-buoyant system with two flexible boundaries, for example a free surface and an interface.

$$\bar{v}_1 = k_{11}q_1\bar{y}_1 + k_{12}q_2\bar{y}_2 \tag{7.91}$$

$$\bar{v}_2 = k_{21}q_1\bar{y}_1 + k_{22}q_2\bar{y}_2. \tag{7.92}$$

That the system is non-buoyant implies negative sign of q_1 and q_2 (because $\rho_1 = 0$ and $\rho_2 < \rho_3$), while k_{11}, k_{12}, k_{21} and k_{22} all are positive.

Let amplitude $\bar{y}_1$ initially assume a finite positive value (i.e. positive at $x = \frac{1}{4}\lambda$ and $(n+\frac{1}{4})\lambda$, but of course negative when $x = (n-\frac{1}{4})\lambda)$, while boundary number 2 coincides with the equipotential surface, i.e. $\bar{y}_2 = 0$ at all x. Applied in eqns (7.91) and (7.92), these initial values of $\bar{y}_1$ and $\bar{y}_2$ make both amplitude velocities $\bar{v}_1$ and $\bar{v}_2$ negative. While amplitude $\bar{y}_1$ diminishes, amplitude $\bar{y}_2$ grows from zero to greater and greater negative values (at $x = \frac{1}{4}\lambda$ and $(n+\frac{1}{4})\lambda)$. The deflections forming on interface 2, as caused by the spontaneous flattening at the surface, are accordingly in antiphase with the surface deflections. But the negative value which amplitude $\bar{y}_2$ gradually assumes means that the newly formed deflections on interface 2 act back, as it were, on the surface deflections. This feedback from interface 2 to boundary 1 is expressed in the term $k_{12}q_2\bar{y}_2$ in eqn (7.91). $\bar{y}_2$ has become negative, consequently the whole term $k_{12}q_2\bar{y}_2$ is positive and opposes the negative term $k_{11}q_1\bar{y}_1$, though generally not sufficiently to prevent the diminution of amplitude $\bar{y}_1$. The newly formed deflections on interface 2 merely reduce the rate at which the deflections on the surface flatten.

This simple analysis indicates that, in a layered Earth, isostatic adjustment in terms of surface movement does not follow a simple exponential expression.

The dynamic coupling between the deflections on several boundaries is demonstrated in a numerical model believed to simulate an idealized model of the Earth (Ramberg, 1968b). In this model the crust consists of a 40 km thick layer with density 2·7 g cm^{-3} and effective viscosity 10^{21} poise. The next layer below is 200 km thick and shows a density = 3·3 g cm^{-3} and an effective viscosity = 10^{20} poise. This layer rests on a basement which is regarded rigid.

The data selected yield the following dimensionless quantities to feed into the computer:

$$h_{2:3} = 0{\cdot}5; \quad h_{3:4} = 0{\cdot}4; \quad \mu_{2:3} = 10; \quad \mu_{3:4} = 10;$$

$$q_{2:1} = \frac{1{\cdot}1777\times10^{-12}}{q_1}; \; q_{3:1} = \frac{-2{\cdot}943\times10^{-11}}{q_1};$$

$$q_1 = \frac{-2{\cdot}7 \times 4\times10^{6} \times 981}{2\times10^{22}} = -5{\cdot}2974\times10^{-13}\,\text{sec}^{-1}.$$

Because the velocity of isostatic adjustment depends upon the wavelength of the deflections, two different wavelengths are tested. In one case the selected ϕ_2-value is 0·32, meaning λ = 785·2 km; in the other the ϕ_2-value is taken as 1·80, which corresponds to λ = 139·6 km (Figs 7.36 and 7.37).

The amplitude of the initial deflections on the surface was 500 m, i.e. $\bar{y}_1 = 500$ m. Interfaces number 2 and 3, at depths 40 and 120 km respectively, were initially coincident with the equipotential surfaces, i.e. $\bar{y}_2 = \bar{y}_3 = 0$, and the basal boundary at 320 km depth is rigid.

Fig. 7.36. Change of amplitudes with time of sinusoidal deflections at the free surface (y_1) and at two interfaces (y_2 and y_3) of a two-layer model crust resting on an infinite half-space. See explanation in text.

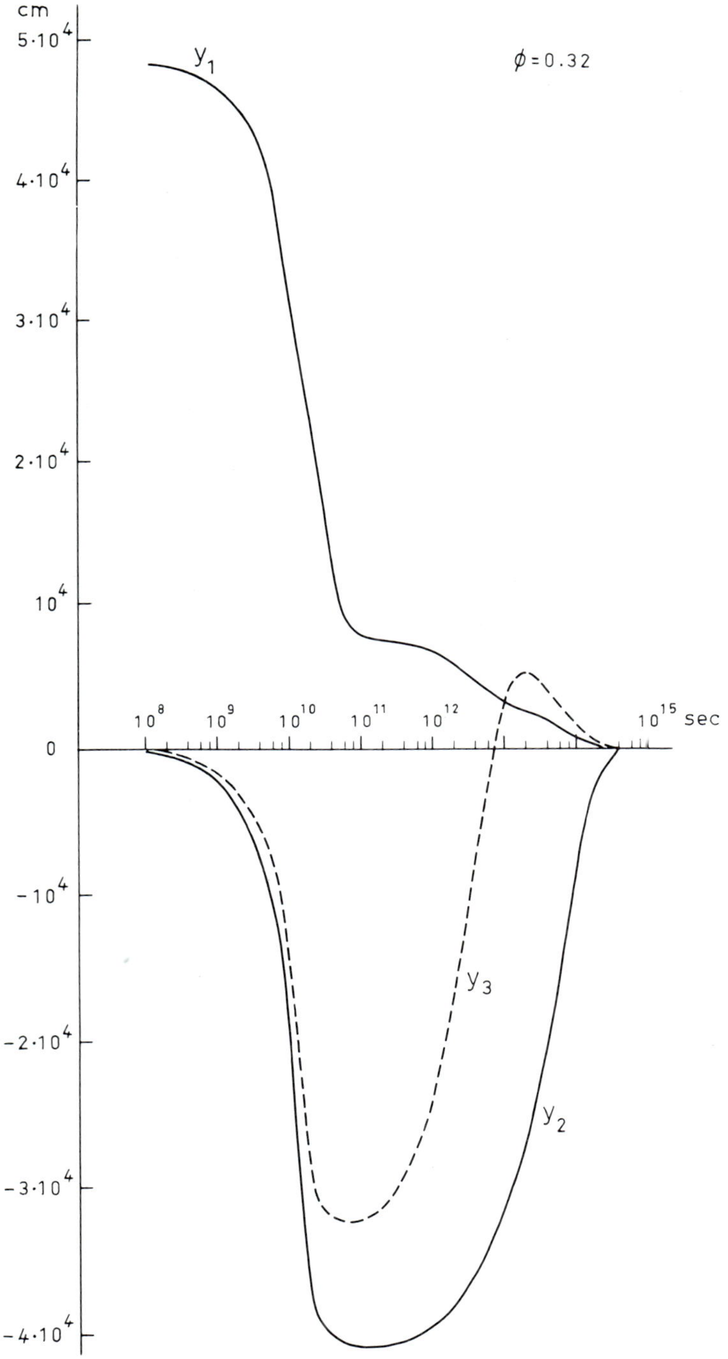

cm
5·10^4
4·10^4
3·10^4
2·10^4
10^4
0
-10^4
-2·10^4
-3·10^4
-4·10^4
y1
y2
y3
ϕ=0.32
10^8
10^9
10^10
10^11
10^12
10^15 sec

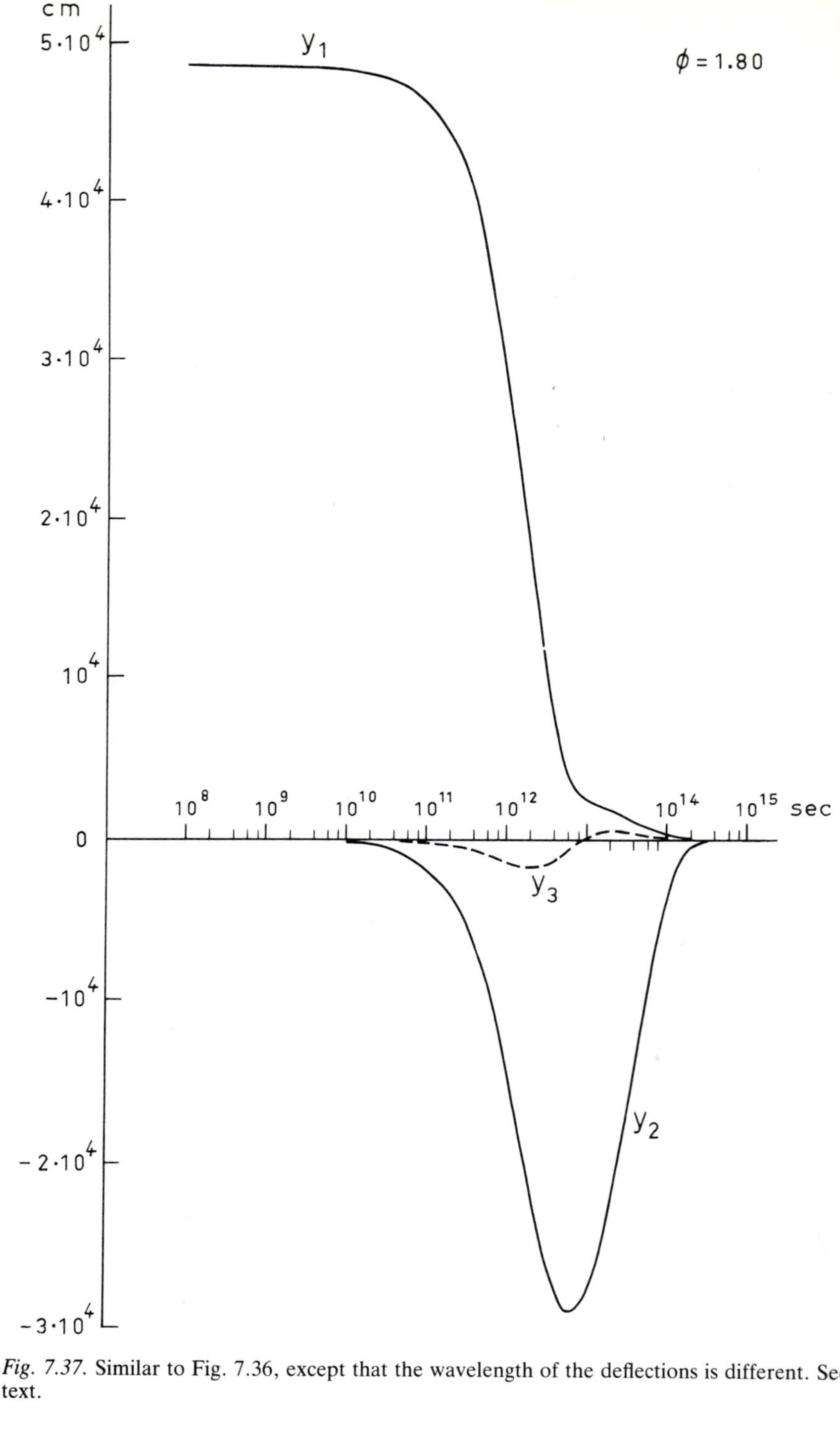

Fig. 7.37. Similar to Fig. 7.36, except that the wavelength of the deflections is different. See text.

Figures 7.36 and 7.37 show how the amplitudes on all three deformable boundaries change with time in the two examples. It is particularly interesting to note how deflections grow on the initial equipotential interfaces 2 and 3. The downward movement of the bulge on the free surface is, so to speak, transmitted to the two deep interfaces which therefore gradually develop downward-facing bulges or keels. As time passes, the amplitude of the surface bulge diminishes, thereby decreasing the tendency to deepen the keels on interfaces 2 and 3. At a certain moment the keels accordingly reach their maximum size, and from then on they begin to flatten. This reversed movement of the keels has a detectable retarding effect upon the rate at which the amplitude of the initial surface bulge decreases, as shown in the figures. It is particularly interesting to follow the deformation of the deepest interface. That it becomes depressed to a keel together with the second interface is expected, so is the much larger effect on interface 3 ($\bar{y}_3$) in the long wave case (Fig. 7.36), than in the case of the short waves (Fig. 7.37). Likewise one expects the keel subsequently to flatten. What is not immediately obvious, perhaps, is why interface 3 should cross the equipotential surface and develop an antikeel (i.e. positive amplitude $\bar{y}_3$) before it ultimately flattens to reach complete isostatic equilibrium.

This numerical model demonstrates that even on the simple premises of Newtonian viscosity, homogeneous layers, even thickness and sinusoidal deflections, isostatic adjustment in a layered globe is not a straightforward process of smoothing out unstable deflection on the Earth's surface or on discontinuous boundaries at depth. Instead, the flattening of initial deflections on one boundary is apt to create new deflections on other boundaries, and the deflections in a multilayered globe may move up and down in a complicated fashion before isostatic equilibrium is ultimately attained.

Other recent studies of isostatic adjustment and its relation to the properties of the Earth's mantle include Crittenden (1963), Cathles (1975), Peltier (1976), and Peltier and Andrews (1976). [See also Mörner (1980).]

EXPERIMENTAL TESTS OF ISOSTATIC ADJUSTMENT

The implication of the above theoretical reasoning for isostatic adjustment was visualized in two simple experimental tests described in Ramberg (1968b).

In a Plexiglass container a heavy syrup was supporting a less dense layer of silicone putty whose free surface had a wavy shape (Fig. 7.38). The boundary between the floating silicone putty and the substratum of syrup was initially straight and horizontal.

For a layered system of this nature the theory predicts how the initially

(*a*)

(*b*)

(*c*)

(*d*)

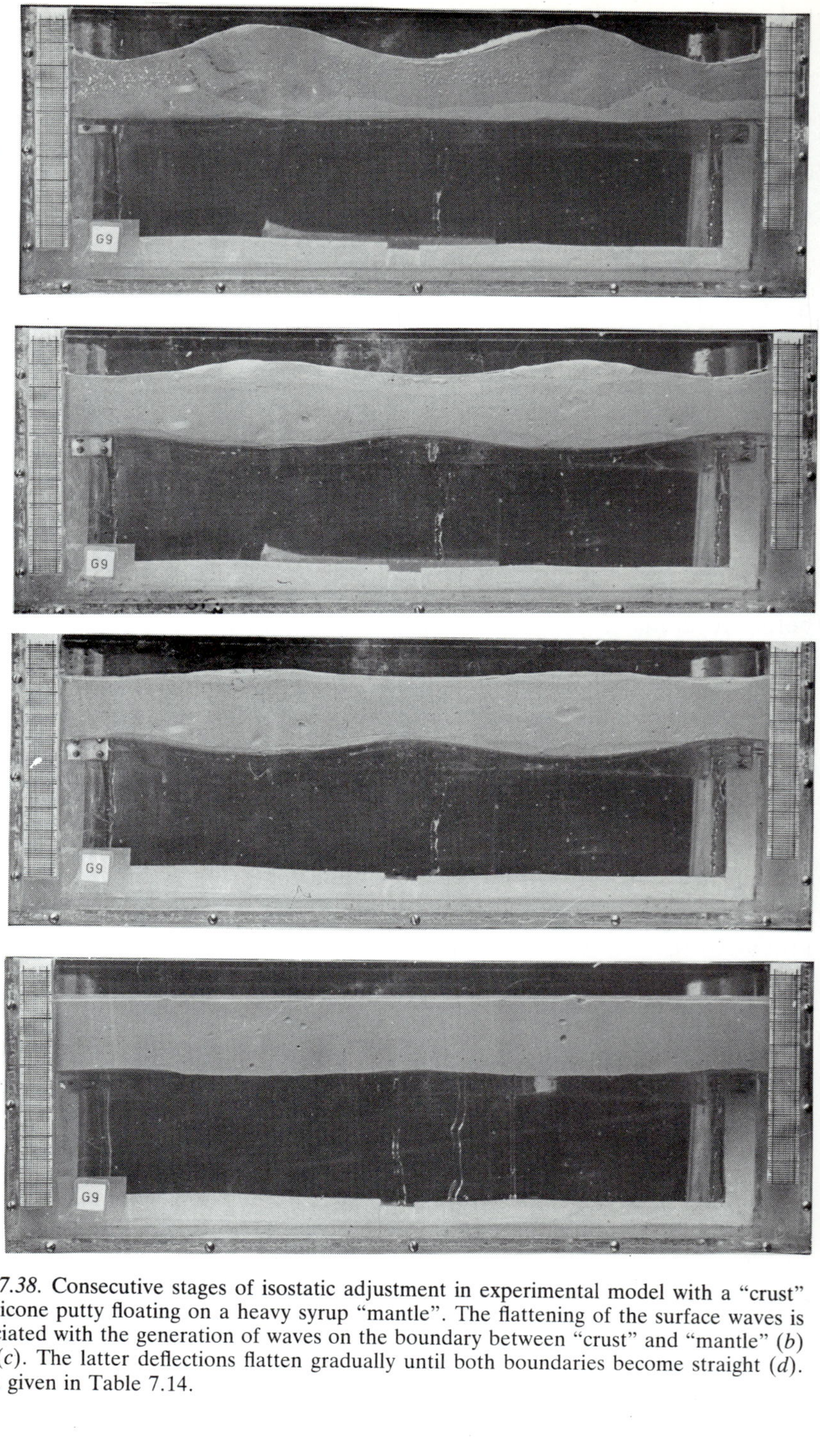

Fig. 7.38. Consecutive stages of isostatic adjustment in experimental model with a "crust" of silicone putty floating on a heavy syrup "mantle". The flattening of the surface waves is associated with the generation of waves on the boundary between "crust" and "mantle" (*b*) and (*c*). The latter deflections flatten gradually until both boundaries become straight (*d*). Data given in Table 7.14.

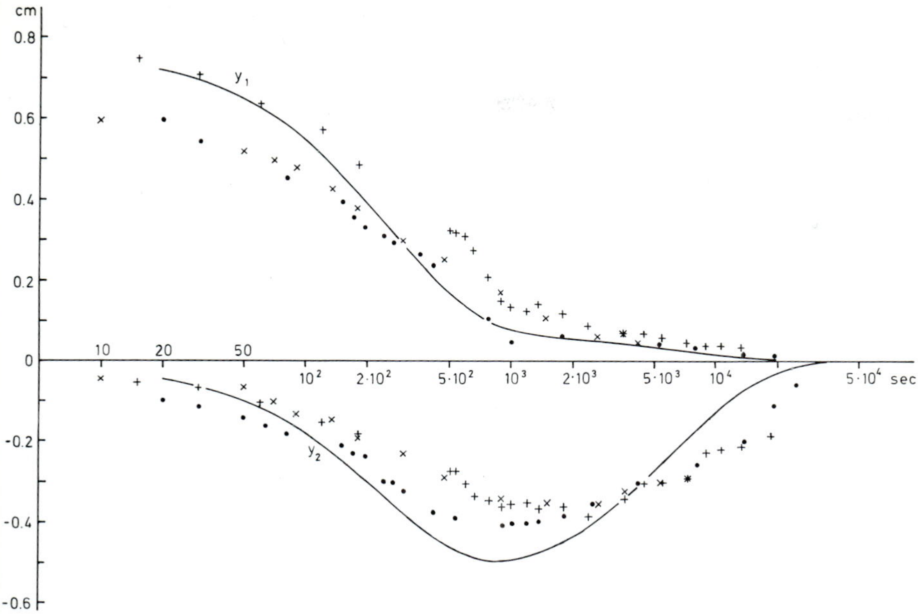

Fig. 7.39. Theoretical and experimental results from "isostatic" test no. 2, Table 7.14. Points, orthogonal- and diagonal crosses represent three different runs with same model; solid curve shows theoretical values based on data given in Table 7.14.

wavy surface tends to flatten and how in the process the boundary between the crust and the substratum is deflected into waves that are in antiphase with the surface waves. The amplitude of the secondary waves at the interface grows to a certain maximum value controlled by the viscosities, the thicknesses, the densities, the wavelength and the amplitude of the waves. After passing a maximum amplitude the secondary waves at the interface begin to flatten until ultimately stable equilibrium is attained at which both interface and surface have levelled out.

Table 7.14. Data for experimental isostatic tests

	Test 1	Test 2
λ	23·0 cm	13·6 cm
h_2	3·2 cm	3·5 cm
h_3	7·2 cm	7·0 cm
ρ_2	1·256 g cm^{-3}	1·25 g cm^{-3}
ρ_3	1·441 g cm^{-3}	1·44 g cm^{-3}
μ_2	$2\cdot8\times10^5$ poise	$8\cdot40\times10^5$ poise
μ_3	$2\cdot0\times10^3$ poise	$4\cdot20\times10^4$ poise
y_{01}	0·55 cm	0·77 cm
y_{02}	0·00 cm	0·00 cm

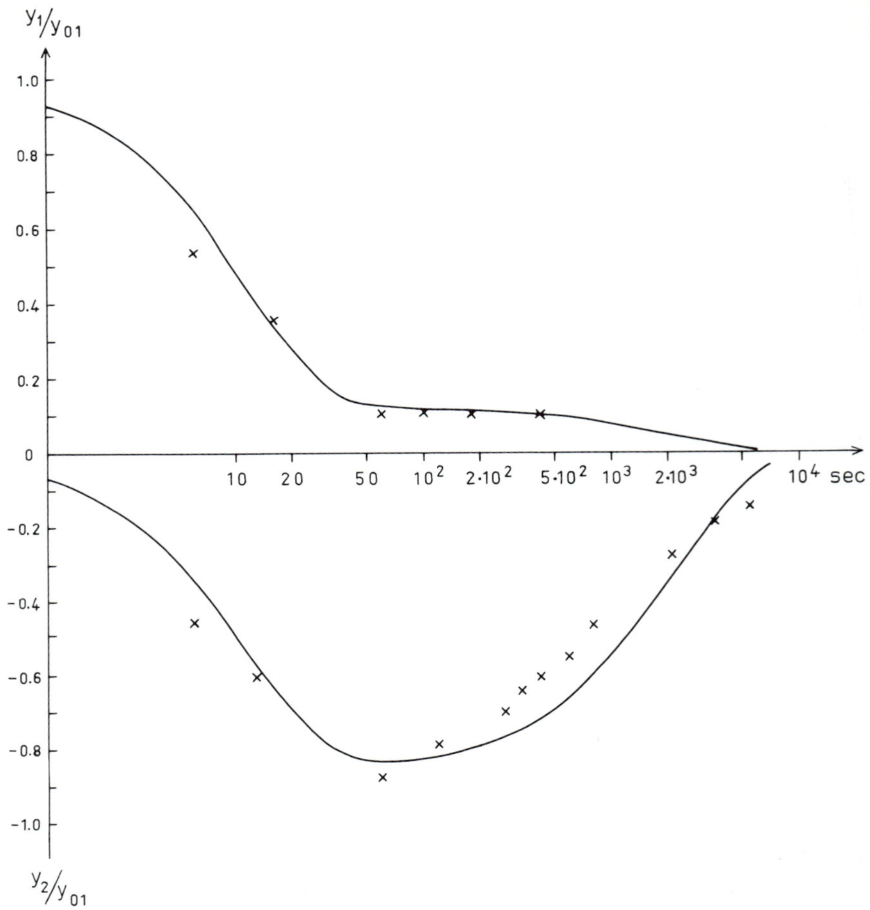

Fig. 7.40. Theoretical and experimental results from "isostatic" model test No. 1, Table 7.14. Crosses represent experimental observations, solid curve theoretical values based on data in Table 7.14.

The pertinent experimental parameters of two sets of tests with unlike initial wavelength are shown in Table 7.14. In Figs 7.39 and 7.40 the measured relationships between amplitudes and time in the two sets are compared with the theoretically calculated relationships. The correspondence between experiments and theory is considered quite satisfactory.

To determine the amplitudes as functions of time the equation below is used.

$$\begin{bmatrix} \bar{y}_1 \\ \bar{y}_2 \end{bmatrix} = \begin{bmatrix} c_{11}\, c_{12} \\ c_{21}\, c_{22} \end{bmatrix} \begin{bmatrix} \exp(\kappa_1 q_1 t) \\ \exp(\kappa_2 q_1 t) \end{bmatrix} \tag{7.93}$$

In this equation c_{ij} are coefficients determined by the initial amplitudes and other properties of the system as explained on p. 84.

8 Instability of Horizontally Compressed Layered Systems in the Field of Gravity

INTRODUCTION

During the growth of domes, buoyant anticlines and other rises, gravity is a force which produces folds on an originally horizontal interface because of unstable density stratification. On the other hand, in layered systems with a stable density stratification gravity tends to flatten any deflection from the equipotential surfaces which may exist, as discussed in the section on isostatic adjustment. Because of its tendency to smooth out geometric perturbation, gravity is a controlling factor in the evolution of some buckle folds even if these are produced by a horizontal compressive stress acting on layered systems with normal density stratification.

The compressive force tends to throw the layers into folds, a process which gravity acts against because the surface folds represent a deviation from the equipotential surfaces in the field of gravity and accordingly increased gravity potential. Generally, the increment of gravity potential, counted per unit length of the system, is greater the longer the wavelength. (For a given, small relative buckle-shortening of the layers the amplitude change is proportional to the wavelength, hence the mentioned increase of gravity potential with increasing length of the waves for a given relative shortening.) It follows that the general tendency of gravity is to make the characteristic wavelength shorter than it would have been if gravity did not act. (This is not necessarily true if the system is buoyant in the sense that it contains a layer which is less dense than overlying layers. An example is treated on p. 187 ff.).

Because of the rapidly diminishing efficiency of the force of gravity with diminishing geometric size of systems, gravity only affects wavelengths which are larger than a certain limit. This limit is not fixed but depends upon the rheological properties of the rocks involved, their density and the strength of the compressive horizontal stress.

Before commencing on the quantitative analysis it may be useful to point out one essential difference in the dynamics of deflections existing in systems with normal and systems with inverted density stratifications, respectively.

In the inverted system the pressure in the buoyant layer at a given level is high underneath the synclines and low underneath the anticlines. Flow in the buoyant layer is accordingly directed from underneath the synclines toward the anticlines, causing the anticlines to rise and the synclines to subside.

In systems with normal density stratification pressure at a given level is high underneath the anticlines and low underneath the synclines. Flow in a plastic basement therefore extends from the anticlines toward the synclines, while flow in the cover is directed in the opposite sense. The result is a flattening of the waviness.

The horizontal compressive force which tends to buckle layered complexes must not only reverse the direction of the flow just mentioned, and hence augment the gravity potential, but must also overcome the strength and the viscous resistance of the rocks.

BUCKLING OF ELASTIC PLATES FLOATING ON A FLUID

The theory of the effect of gravity on the buckling of elastic plates when floating on a fluid was developed in two important papers which unfortunately are too little known to geologists (Smoluchowski, 1909, 1910). We shall presently see that simple tests confirm Smoluchowski's theory surprisingly well.

Smoluchowski used plates of gelatine and gutta-percha floating on mercury, and tin-foil as well as gold foil floating on water. In response to a small lateral stress buckles would form whose wavelength was controlled according to the theoretical formula

$$\lambda = 2\pi \sqrt[4]{\left[\frac{Eh^3}{12(1-\nu^2)\rho g}\right]}. \tag{8.1}$$

E is Young's modulus, ν Poisson's ratio, ρ density of the fluid substratum (Hg or H_2O) and h thickness of the layer.

For scale-model consideration it is convenient to make the expression explicit in the dimensionless ratio λ/h, because structures with the same λ/h ratio and proportional amplitude are geometrically similar independent of their size. Thus transformed, the expression becomes

$$\frac{\lambda}{h} = 2\pi \sqrt[4]{\left[\frac{E}{12(1-\nu^2)\rho g h}\right]} \tag{8.2}$$

which shows that structures of the above type, in which the ratio $E/(1 - \nu^2)\rho gh$ is constant, are geometrically similar in the sense that the ratio λ/h is constant.

One notes that buckles are not generated unless the lateral stress reaches a certain limit which is a function of E and other parameters. As shown by Smoluchowski (1910) and Ramberg and Stephansson (1964) the deviatoric stress necessary to produce the stable elastic buckles is related to the elastic parameters E and ν, and to the thickness and density of the layer through the formula

$$\Delta\sigma = \sqrt{\frac{Eh\rho g}{3(1 - \nu^2)}} \tag{8.3}$$

A combination of the two last equations yields

$$\frac{\lambda}{h} = \pi\sqrt{\frac{2\Delta\sigma}{h\rho g}}. \tag{8.4}$$

In this equation $\Delta\sigma$ is not an independent variable but instead takes the value specified in eqn (8.3) if buckles are to form at all. Weaker stress does not produce buckles; stronger stress increases the amplitude and/or causes ruptures.

The effect of gravity is implied in the term $h\rho g$ in the above equations. Under otherwise similar conditions, increasing density of the supporting fluid decreases the wavelength/thickness ratio. We note that $\Delta\sigma/h\rho g$ is the force ratio between the pressure or stress force and the gravitational force, i.e. the inverse of the Smoluchowski number in Table 6.1. For dynamic similarity to prevail in systems the Smoluchowski number must be constant.

Equation (8.4) shows the noteworthy fact that, for a surface layer of given material, the compressive stress needed to produce a certain wavelength/thickness ratio increases with the thickness of the layer and hence also increases with the wavelength. In view of the finite mechanical strength of solids, therefore, an upper limit must exist for the length of waves which can be produced by elastic compression of a global crust such as the granitic or gneissic shell (Smoluchowski, 1910; Ramberg, 1965a, 1970). This fact will be considered in more detail in connection with plastic buckling, p. 144 ff.

Ramberg and Stephansson (1964) performed several experiments with buckling of elastic rubber sheets floating on mercury ($\rho = 13{\cdot}5$ g cm^{-3}) and on a saturated solution of KI ($\rho = 1{\cdot}68$ g cm^{-3}). The arrangement of these experiments is shown in Fig. 8.1.

The buckling formula for the elastic case (elastic layer on a fluid substratum with negligible viscosity) was developed by Ramberg and Stephansson (1964) along a line of argument different from that employed

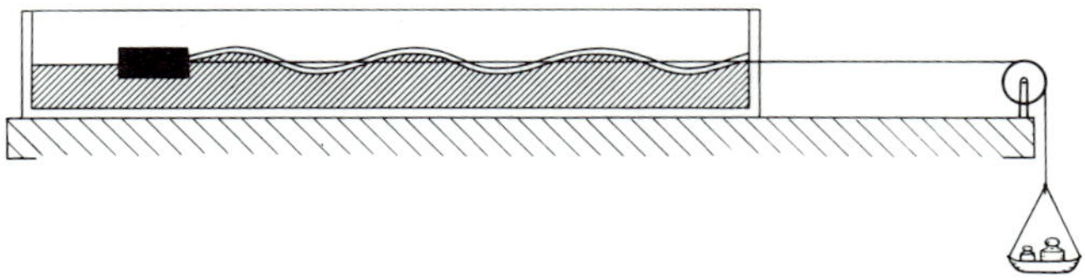

Fig. 8.1 Arrangement of buckling experiments.

by Smoluchowski, for which reason the identity of the equations obtained was quite a satisfactory result.

Figures 8.2 and 8.3 give the experimental results, including two from Smoluchowski (1910), all showing remarkable agreement with theory.

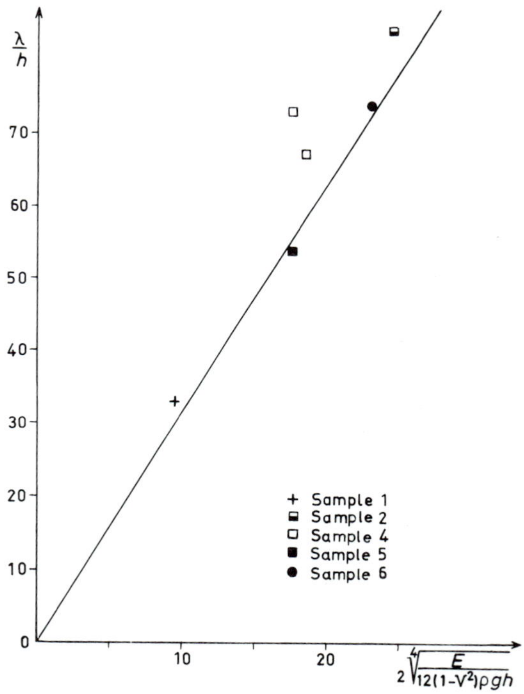

Fig. 8.2. Results from experimental buckling of elastic rubber plates floating on a solution of *KI* (density 1.68 g cm^{-3}). Straight line represents the relation

$$\frac{\lambda}{h} = 2\pi \sqrt[4]{\left(\frac{E}{12(1-v^2)\rho g h}\right)}$$

(Ramberg and Stephansson, 1964).

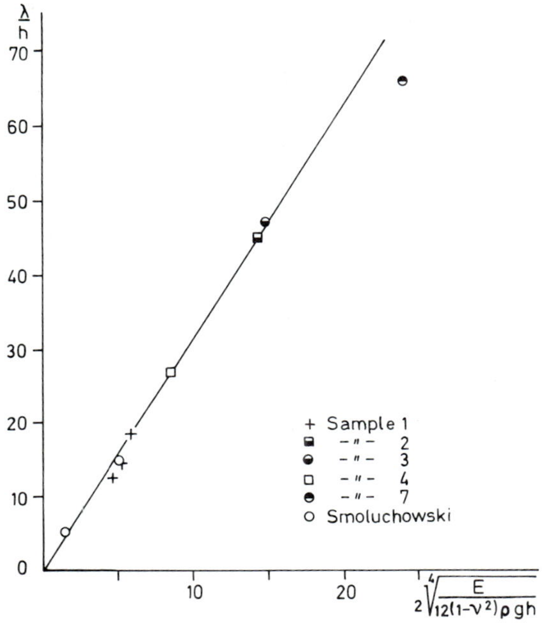

Fig. 8.3. Results from experimental buckling of elastic plates of rubber and gelatine floating on mercury. Straight line has same significance as in Fig. 8.2 (Smoluchowski, 1910. Ramberg and Stephansson, 1964).

According to a simple model the Earth's crust above the Moho may be treated as an elastic plate floating on a viscous substratum. Although the model is not very realistic (see p. 175), it is interesting to apply Smoluchowski's elastic theory on this model.

The largest deviatoric compressive stress we may apply on crustal rocks is of course given by their strength. A realistic value of the strength of crystalline rocks in deviatoric compression is say 5×10^9 dyne cm^{-2}. A reasonable value for Young's modulus, E, is 5×10^{11} dyne cm^{-2} and Poisson's ratio is taken as 0·2. The density of the asthenosphere just below the crust is, say, 3·3 g cm^{-3}.

When applied in eqn (8.3) these quantities put a limit on the thickness which an elastic crust is permitted to have if it is to be buckled by a horizontal compressive stress not exceeding the strength 5×10^9 dyne cm^{-2}.

$$h \leqslant \sigma^2 \frac{3(1-0{\cdot}04)}{5\times10^{11}\times3{\cdot}3\times981} = 4{\cdot}45\times10^4 \text{ cm} = 445 \text{ m}.$$

According to this formula an elastic crust thicker than 445 m cannot be buckled by compressive forces unless the strength is greater than 5×10^9 dyne cm^{-2} and/or the Young modulus is less than 5×10^{11} dyne cm^{-2}. (Incidentally, even if the strength of the crust is greater than 5×10^9 dyne cm^{-2}, a crust thicker than the above value will not buckle if the deviatoric horizontal compressive stress in the crust does not exceed 5×10^9 dyne cm^{-2}.)

Evidently the thickness of the part of the Earth's crust which may be considered elastic is much greater than 445 m. It therefore seems that elastic buckling of the Earth's crust is not possible.

Aspects of the viscous buckling of the crust of the globe will be discussed below.

THEORY OF VISCOUS BUCKLING IN THE GRAVITATIONAL FIELD

Introduction

Biot (1959, 1961) and Ramberg (1963b, 1964b, 1970) have developed a theory for viscous buckling of Newtonian materials under the simultaneous

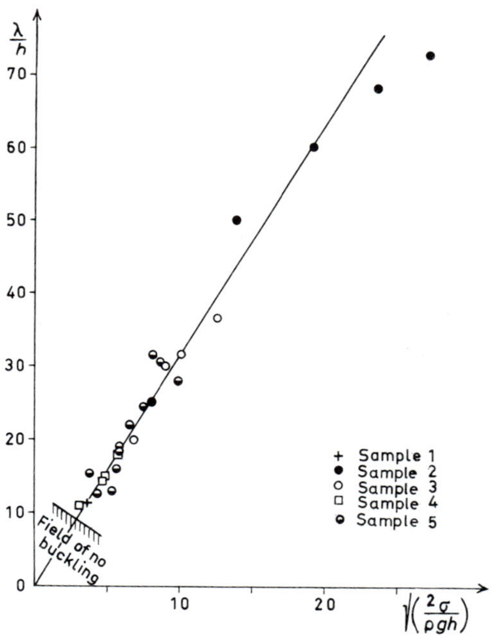

Fig. 8.4. Relationship between wavelength/thickness ratio and the term $\sqrt{(2\sigma/\rho gh)}$ in experiments on buckling of viscous resin plates floating on mercury. Straight line represents the theoretical relation $\lambda/h = \pi\sqrt{(2\sigma/\rho gh)}$ (Ramberg and Stephansson, 1964).

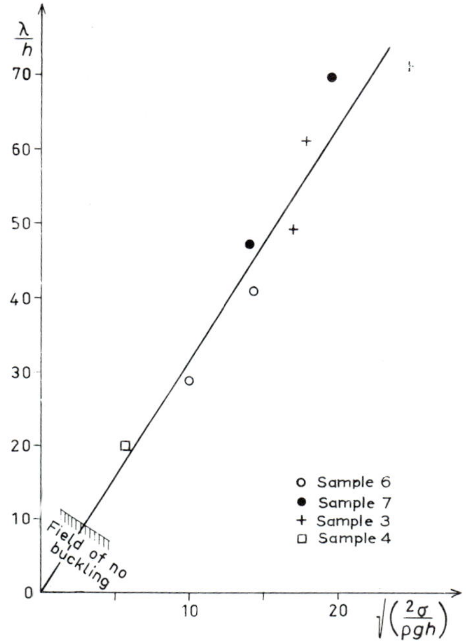

Fig. 8.5 Relationship between wavelength/thickness ratio and the term $\sqrt{(2\sigma/\rho gh)}$ in experiments on buckling of viscous resin plates floating on a solution of KI (ρ = 1·68 g cm^{-3}). Theoretical relation $\lambda/h = \pi\sqrt{(2\sigma/\rho gh)}$ shown by straight line (Ramberg and Stephansson, 1964).

action of gravity and horizontal compression. A part of the theory of the viscous case has been quite satisfactorily checked by experiments with plates of stitching wax floating on mercury and on a solution of KI (Ramberg and Stephansson, 1964). Some results are recorded in Figs 8.4 and 8.5.

These experimental tests relate only to the special model consisting of a single viscous plate floating on a heavy fluid with negligible viscosity. The treatment of a more general model consisting of several layers with unlike densities, unlike effective viscosities and unlike thicknesses being simultaneously exposed to layer-parallel compressive forces and to the pull of gravity, will be considered below.

The behaviour of such systems exposed solely to the force of gravity has been treated in Chapter 7. We shall follow a similar procedure when compressive forces are added to the system. The aim of the study is chiefly to develop equations that relate the characteristic wavelength of multilayered systems to their rheological–geometrical properties, to the compressive forces and to the body force of gravity. The study also furnishes information on the velocity of the folding process, especially the rate of amplitude growth or decline as long as the buckles are gentle.

Viscous buckling of a single free layer

Buckling of a layer of viscous, plastic or elastic materials is associated with both longitudinal stress and shear stress parallel to, and across, the sheet. For a viscous sheet these stresses are related to the velocity of folding (rate of growth of amplitude or rate of change of curvature of the buckles); for an elastic sheet the stresses are functions of the static magnitude of the curvature.

In the discussion of the viscous case it is advantageous first to treat separately the two above-mentioned kinds of stresses and their associated strains. Superposition of these strains and stresses gives the resultant velocity of buckling in terms of rate of growth of transversal and tangential amplitudes (see Fig. 8.6) as functions of the force moment represented by the product of the compressive force and the transversal deflection.

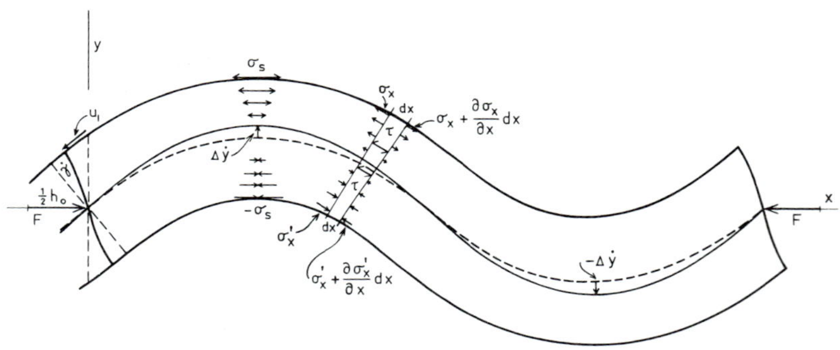

Fig. 8.6. Stresses and strains in a buckling layer. Amplitude exaggerated; description in text relates to more gentle deflections than shown in the figure (from Ramberg, 1970).

The relation between rate of change of radius of curvature and the longitudinal strain parallel to the layer is given by an equation analogous to one much used in the strength-of-material theory of buckling of plates (see Fig. 8.7):

$$\dot{\varepsilon}_{xs} = \tfrac{1}{2}h_0\dot{r}^{-1} \approx -\tfrac{1}{2}h_0\, \mathrm{d}(\mathrm{d}^2y/\mathrm{d}x^2)/\mathrm{d}t, \qquad (8.5)$$

an acceptable approximation for small curvature whose approximate expression is $\mathrm{d}^2y/\mathrm{d}x^2$. $\dot{\varepsilon}_{xs}$ is the rate of change of layer-parallel longitudinal strain in the surface fibres of the layer, h_0 its thickness, r the radius of curvature of the centroid, and t time.

In a sheet that is long in the direction of the fold axis relative to its thickness, the buckling and associated strains are two-dimensional. The strain rate is accordingly related to the longitudinal layer-parallel stress in the surface fibres by the general expression (see e.g. Jaeger, 1956) valid for pure shear

$$\sigma_{xs} = 4\mu\dot{\varepsilon}_{xs} \approx -2\mu h_0 \dot{y}''_{\varepsilon}. \tag{8.6}$$

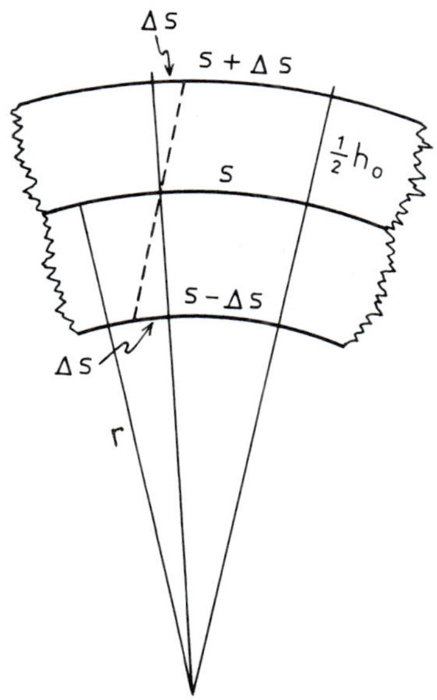

Fig. 8.7. Part of bent layer without layer-parallel shear. Marker lines which were normal to the layer prior to bending are parallel to radii after bending. Longitudinal strain in convex surface fibres is $\varepsilon_{xs} = \Delta s/s = h_0/2r\rho$ in concave surface fibres $\varepsilon'_{xs} = -\Delta s/s = h_0/2r$. Stippled line is parallel to radius shown at distance s from the stippled line (from Ramberg, 1970).

(If the stress σ_{ys} normal to the surface does not vanish, σ_{xs} must be exchanged by $\Delta\sigma_s = \sigma_{xs} - \sigma_{ys}$). The viscosity coefficient of the material is μ and

$$\dot{y}''_{\varepsilon} \equiv \mathrm{d}(\mathrm{d}^2y/\mathrm{d}x^2)/\mathrm{d}t,$$

the subscript ε emphasizing that this particular part of the rate of change of curvature is due solely to the longitudinal strain parallel to the layer.

Accepting a linear variation of σ_x with distance from the centroid of the layer and vanishing longitudinal stress in the y direction throughout, the stress σ_x yields a bending moment, M, when integrated at any cross-section of the layer, from the centroid to the two surfaces

$$M = 4\frac{\sigma_{xs}}{h_0}\int_0^{\frac{1}{2}h_0} h^2\,\mathrm{d}h = \tfrac{1}{6}\,\sigma_{xs}h_0^2. \tag{8.7}$$

(σ_{xs} is the stress at the surface fibres exclusively due to the bending of the layer. The total stress at a cross-section is $\sigma_x + \sigma_b$ where σ_b is the layer-parallel compressive stress (=buckling stress) to which the layer is exposed. As distinct from σ_x, σ_b is constant in magnitude and sign over any given cross-section. The moment formed by integrating $\sigma_b h\mathrm{d}h$ on both sides of the centroid therefore cancels.)

M is reckoned per unit length in the direction of the fold axis. At dynamic equilibrium this moment is balanced by the buckling moment Fy (see Fig. 8.6):

$$\tfrac{1}{6}\sigma_{xs}h_0^2 = Fy = -\sigma_b h_0 y \tag{8.8}$$

where $\sigma_b h_0$ is the compressive force (=buckling force) acting on the layer at the inflection points. Hence we obtain the differential equation

$$\dot{y}''_\varepsilon = -3\frac{Fy}{\mu h_0^3}, \tag{8.9}$$

the solution of which gives the rate of change of amplitude of gentle buckles as related exclusively to the longitudinal strain.

It is also necessary to consider the effect of layer-parallel shear upon the folding process. To this end we first calculate the rate of change of curvature of the folds as if they were solely due to layer-parallel shear strain. The radius of curvature would then be related to the variation of the shear angle along the sheet by the approximation

$$\frac{\mathrm{d}\dot{\gamma}}{\mathrm{d}x} = -\dot{r}^{-1} \approx \dot{y}''_\gamma, \tag{8.10}$$

provided the amplitude/wavelength ratio is small. This equation is realized by inspection of Fig. 8.8, where $\mathrm{d}\dot{a}$ is related to $\frac{1}{2}h_0$ as $\mathrm{d}s$ is related to $\dot{r}$. The rate of change of shear angle is $\dot{\gamma}$, and the subscript γ emphasizes that this component of the buckling process is due exclusively to shear strain parallel to, and across, the layer.

The next step is to find how the rate of change of shear, $\dot{\gamma}$, is related to the longitudinal stress σ_x in the layer, and thereby arrive at a relation between layer-parallel shear and the buckling force. To reach this goal we

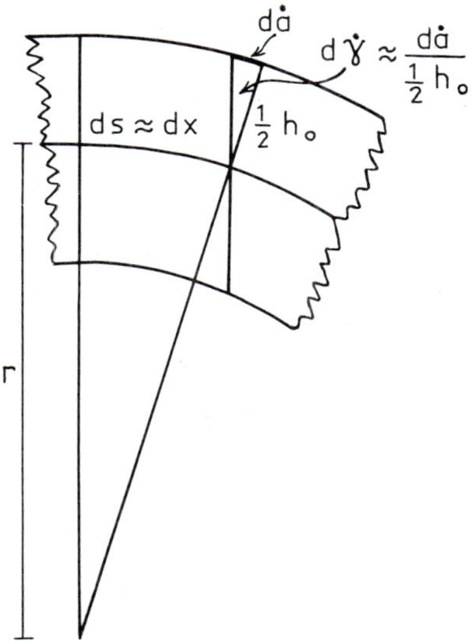

Fig. 8.8 Part of bent layer without layer parallel longitudinal strain. Marker line normal to layer prior to bending remains parallel to one another during gentle bending. Shear angle varies with distance along the gently bended layer thus $\partial\gamma/\partial x \approx r^{-1}$ since $\partial\gamma = \partial a/\frac{1}{2}h_0 = \partial s/r$ (from Ramberg, 1970).

first note from Fig. 8.6 that the shear stress τ_{xy} at distance h from the centroid is related to the gradient of the longitudinal stress in the x direction, thus

$$\tau_{xy} = \frac{2}{h_0}\frac{\mathrm{d}\sigma_{xs}}{\mathrm{d}x}\int_h^{\frac{1}{2}h_0} h\,\mathrm{d}h = \frac{\mathrm{d}\sigma_{xs}}{\mathrm{d}x}\left[\tfrac{1}{4}h_0 - \frac{h^2}{h_0}\right], \tag{8.11}$$

because the longitudinal stress as well as its gradient in the x direction are assumed to vary linearly with the distance from the centroid. The rate of change of shear strain at distance h is accordingly

$$\dot{\gamma}_h = \mu^{-1}\frac{\mathrm{d}\sigma_{xs}}{\mathrm{d}x}\left[\tfrac{1}{4}h_0 - \frac{h^2}{h_0}\right]. \tag{8.12}$$

This equation expresses quantitatively the expected condition that the shear assumes a maximum value at the centroid and vanishes at the free surfaces ($h = \pm\frac{1}{2}h_0$).

For the calculation of the contribution of layer-parallel shear to the rate of buckling we make use of the average shear rate across the considered cross-sections of the sheet. This average is

$$\dot{\gamma}_{av} = \frac{2}{h_0}\int_0^{\frac{1}{2}h_0} \dot{\gamma}_h \, dh = \frac{1}{6\mu}\frac{d\sigma_{xs}}{dx} h_0. \tag{8.13}$$

Differentiating eqn (8.13) with respect to x and utilizing eqn (8.10) yields

$$\frac{d\dot{\gamma}_{av}}{dx} = \frac{h_0}{6\mu}\frac{d^2\sigma_{xs}}{dx^2} \approx \dot{y}''_{\gamma}. \tag{8.14}$$

At dynamic equilibrium the moments M and Fy must be equal, cf. eqn (8.8). This leads to an expression for σ_{xs}:

$$\sigma_{xs} = 6h_0^{-2}Fy, \tag{8.15}$$

from which is obtained

$$\frac{d^2\sigma_{xs}}{dx^2} = 6h_0^{-2}F\frac{d^2y}{dx^2}$$

to insert in eqn (8.14). A differential equation relating $\dot{y}''_{\gamma}$ to the buckling force follows:

$$\dot{y}''_{\gamma} \approx \mu^{-1}h_0^{-1}Fy''. \tag{8.16}$$

The contributions to the buckling process both from the layer-parallel longitudinal strain and from the layer-parallel shear strain may now be combined:

$$\dot{y}''_{\varepsilon} + \dot{y}''_{\gamma} = \dot{y}'' = -F\mu^{-1}h_0^{-1}(3h_0^{-2}y - y''). \tag{8.17}$$

For the boundary conditions $y = 0$ at $x = 0$ and at $x = \frac{1}{2}\lambda$ for $t \geqslant 0$, and $y = \bar{y}$ at $x = \frac{1}{4}\lambda$ for $t = 0$ a solution of this equation is

$$y = \bar{y}_0 \exp[-\sigma_b \mu^{-1}(3\phi^{-2} + 1)t] \sin(\omega x) \tag{8.18}$$

in which the buckling force is replaced by the buckling stress according to $F = -\sigma_b h_0$; $\phi \equiv 2\pi h_0/\lambda$ and $\omega \equiv 2\pi/\lambda$ have been introduced. (Equation (8.18) is found by trying a solution of type $y = e^{(mx+nt)}$ for eqn (8.17), and then determining m and n from the relations $y'' = m^2y$, $\dot{y}'' = m^2ny$, and $\sqrt{[(3F\mu^{-1}h_0^{-3})/(n - F\mu^{-1}h_0^{-1})]} = 2\pi/\lambda$.).

The rate of growth of the sinusoidal deflections is consequently

$$\frac{\partial y}{\partial t} \equiv v = -(3\phi^{-2} + 1)\mu^{-1}\sigma_b\bar{y}_0 \exp[-\sigma_b\mu^{-1}(3\phi^{-2} + 1)t] \sin(\omega x), \tag{8.19}$$

the term $-(3\phi^{-2}+1)\mu^{-1}\sigma_b\bar{y}_0 \exp[-\sigma_b\mu^{-1}(3\phi^{-2}+1)t]$ being the rate of growth of amplitude.

As already pointed out the rate of growth of amplitude here calculated refers to the deflection of the plane of particles which prior to buckling was located in the middle of the layer as indicated in Fig. 8.6. This plane of particles deflects somewhat faster than the two surfaces because the material on the concave side of the centroid is compressed parallel to the layer and the material on the convex side is stretched. For materials with Newtonian viscosity under small strain $\dot{\varepsilon}_{xs}$ at the concave surface is approximately equal to $-\dot{\varepsilon}_{xs}$ at the opposite point on the convex surface. Therefore, were it not for the uniform shortening of the laterally compressed layer, it would retain its thickness during gentle buckling and the rate of increase of amplitude would be the same for the two surfaces.

As long as we are concerned with the problem of determining the characteristic wavelength produced during buckling, the homogeneous strain which is related to the stress σ_b is of less consequence. To find the rate of change of amplitude due to the homogeneous strain we reason as follows. The layer is compressed homogeneously at a strain rate $\dot{\varepsilon}_x = (\sigma_b - \sigma_y)/4\mu$ and thickened homogeneously at a strain rate $\dot{\varepsilon}_y = (\sigma_y - \sigma_b)/4\mu$.

In the model under study we have assumed that σ_y vanishes. Accordingly, the homogeneous strain contributes to the growth rate of the transversal amplitude by a velocity

$$\bar{v} = \dot{\varepsilon}_y \bar{y} = -\frac{\sigma_b}{4\mu}\bar{y} \tag{8.20}$$

assuming plane strain in the x, y plane.

The amplitude velocity expressed in eqn (8.20) is a mere consequence of the homogeneous strain. It will be referred to as the *homogeneous amplitude* velocity.

The amplitude velocities expressed in eqns (8.19) and (8.26) are results of the active buckling alone; these are termed the *perturbed amplitude* velocity.

In the analysis of the folding process we are chiefly concerned with the perturbed amplitude velocities. The true amplitude velocity is the sum of the homogeneous and the perturbed velocities.

Returning to the rate of perturbed amplitude growth let $\Delta\dot{\bar{y}}$ be the difference in amplitude velocity of the centroid and of either of the two surfaces. From the remarks above follows

$$\Delta\dot{\bar{y}} = 2\dot{\varepsilon}_{ys} h_0^{-1} \int_{-\frac{1}{2}h_0}^{0} h \, \mathrm{d}h = \tfrac{1}{4}\dot{\varepsilon}_{ys} h_0, \tag{8.21}$$

where $\dot{\varepsilon}_{ys}$ is the rate of change of the normal strain at the surface, and a linear variation of $\dot{\varepsilon}$ with distance from the centroid is assumed. By eqns (8.6), (8.8) and (8.7), combined with the condition $\dot{\varepsilon}_{ys} = -\dot{\varepsilon}_{xs}$, $\Delta\dot{y}$ becomes related to the buckling stress by the function

$$\Delta\dot{y} = -\tfrac{3}{8}\mu^{-1}\sigma_b\bar{y}_0. \tag{8.22}$$

The rate of growth of amplitude of the surfaces 1 and 2 follows from eqns (8.19) and (8.22):

$$\dot{\bar{y}} \equiv v_1 = v_2 = v - \Delta\dot{y} = -(3\phi^{-2} + \tfrac{5}{8})\mu^{-1}\sigma_b\bar{y}. \tag{8.23}$$

Here $\bar{y}$ is the amplitude of either of the two surfaces.

For the continued analysis a relation between the rate of change of the tangential amplitude (i.e. the perturbed velocity component in the x direction, u, at both surfaces at $x = 0$ and $x = n\lambda$, n being an integer) and the buckling force is also required. Aided by Fig. 8.6 in which we imagine that surface 1 is the upper boundary and 2 the lower, the following equations can be deduced:

$$u_1 = -u_2 \text{ (because of symmetry)},$$
$$u_1 = \int_{\frac{1}{4}\lambda}^{0} \dot{\varepsilon}_{xs}\mathrm{d}x \tag{8.24}$$

where $\dot{\varepsilon}_x$ is the perturbed strain rate along the upper surface within the region $0 \leqslant x \leqslant \frac{1}{4}\lambda$.

For plane strain we have

$$\dot{\varepsilon}_{xs} = \frac{\sigma_{xs}}{4\mu} = -\frac{3}{2}\frac{\sigma_b\bar{y}}{\mu h_0}\sin(\omega x) \tag{8.25}$$

from eqn (8.8). Accordingly, after inserting for $\dot{\varepsilon}_{xs}$ in eqn (8.24) and integration, we obtain:

$$\bar{u}_1 = -\bar{u}_2 = \frac{3}{2}\frac{\sigma_b}{\omega\mu h_0}\bar{y} = \frac{3}{2}\frac{\sigma_b\bar{y}}{\phi\mu}. \tag{8.26}$$

These are the perturbed tangential amplitude velocities.

With compressive stress and initial amplitude fixed for a given layer, the only variable which can influence the rate of buckling is the wavelength as implied in the symbol ϕ. It follows from eqns (8.19) and (8.26) that the rate of buckling in terms of the velocity of both the transversal and the tangential amplitudes increases with decreasing ϕ, that is, with increasing wavelength. v is proportional to the square of the wavelength and u increases linearly with the length of the waves.

An embedded layer, of course, behaves differently because of the resistance against the sidewise deflections offered by the surrounding material. Since this resistance generally increases as the waves become longer, there will be a wave with intermediate length which grows with optimal rate. This condition is of paramount geotectonic consequence and shall be treated in detail in the section on multilayers.

Buckling of the free surface of a compressed infinite half-space

The buckling eqns (8.19) and (8.26) become increasingly inaccurate the thicker the layer relative to the wavelength. It is not possible to give a definite limit for the thickness/wavelength ratio beyond which the equations are useless, but clearly they are not valid for sinusiodal deflections on the free surface of an infinite half-space which is compressed normal to the axial plane of the wavy surface. If we wish to treat the buckling of a layer that is sandwiched between infinite half-spaces, therefore, we could either adopt a model in which the half-spaces are not compressed—the compressive stress acts only on the layer—or we could attempt to develop relationships between the compressive stress acting on a half-space and the perturbed rate of growth of waves on its free surface. Although not exact, the relationships arrived at in the following between growth velocity of surface buckles and compressive stress acting on a half-space are useful approximations for our purpose.

For the special case of a single embedded layer not affected by gravity the question of mathematical rigour has been discussed by Fletcher (1977, 1979) and Ramberg (1979).

Valuable information on the dynamics of the surface waves on a compressed half-space is obtained if we compare the evolution of these waves with the evolution of a passive sinusiodal marker-surface within a body undergoing homogeneous strain with the principal compressive strain normal to the axial plane of the sinuous marker.

We wish in particular to find how the movement of the buckles on the free surface of the compressed half-space deviates from the movement of the passive marker whose sinuous shape initially was identical to the shape of the surface buckles (see Figs 8.9 and 8.10). The differences between corresponding velocities at the buckling surface and at the passive marker are the perturbed velocities in which we are interested. The passive marker by definition is deformed only by the homogeneous strain.

The essential dynamic differences between the two sets of waves is that the shear stress and the normal stress vanish at the buckles along the free surface whereas these stresses exist at the passive sinusoidal marker along which the stresses vary periodically. The stresses at the marker vary with the magnitude of the angle between the tangent to the marker curve and

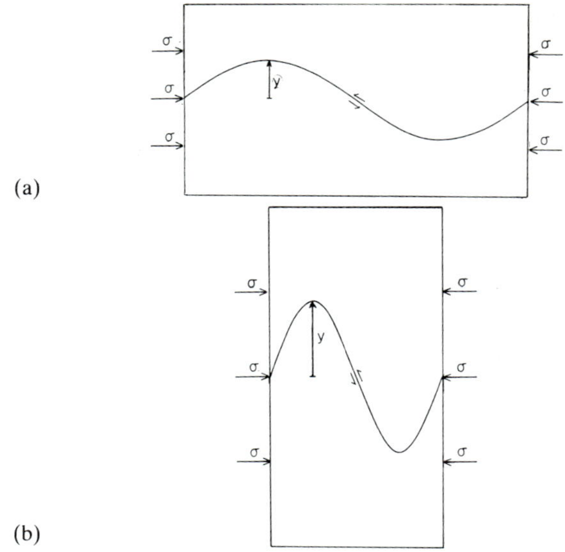

Fig. 8.9. Homogeneous compressive strain parallel to the trend of a gentle sinuous marker: (*a*) initial shape, (*b*) compressed shape. Both curves are sine functions.

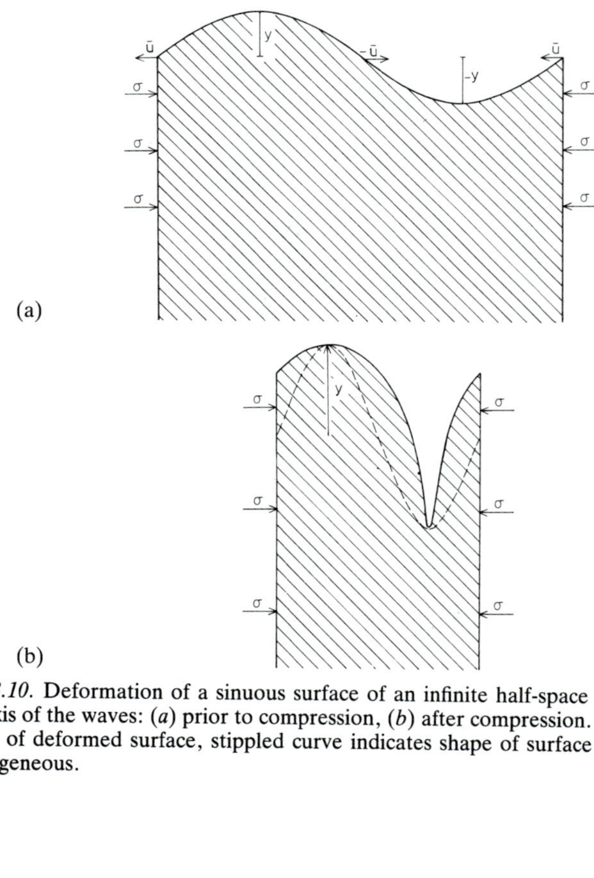

Fig. 8.10. Deformation of a sinuous surface of an infinite half-space compressed normal to the axis of the waves: (*a*) prior to compression, (*b*) after compression. Solid-line curve shows shape of deformed surface, stippled curve indicates shape of surface if the strain had been homogeneous.

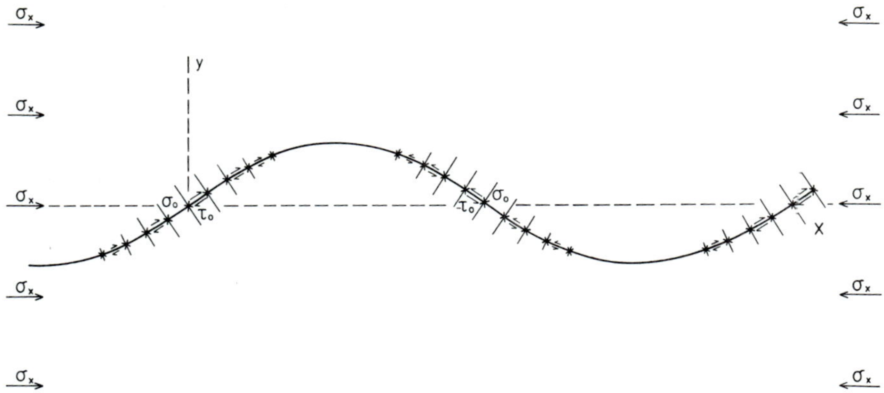

Fig. 8.11. Distribution of normal stress and shear stress at a sinuous passive marker in a homogeneous body compressed homogeneously parallel to a line containing the inflection points.

the direction of the principal compressive stress σ_x (Fig. 8.11).

We take σ_y to be zero and consider the homogeneous strain as irrotational and two-dimensional in the x, y-plane.

On these premises the component of shear stress parallel to the sinusoidal marker is

$$\tau_\alpha = -\sigma_x \sin\alpha \cos\alpha, \tag{8.27}$$

and the normal stress across the marker is

$$\sigma_\alpha = \sigma_x \sin^2 \alpha. \tag{8.28}$$

Here α is the angle between the tangent to the marker curve and the direction of the principal compressive stress σ_x. These are basic stress relationships, see e.g. Jaeger (1959).

The angle α follows from the equation for the sinusoidal marker

$$y = \bar{y} \sin(\omega x) \tag{8.29}$$

which yields

$$\operatorname{tg} \alpha = \mathrm{d}y/\mathrm{d}x = \omega\bar{y} \cos(\omega x). \tag{8.30}$$

For gentle sinusoidal deflections the angle α is small even at the inflection points. Hence we put $\cos\alpha \approx 1$ and $\sin\alpha \approx \operatorname{tg}\alpha$, and obtain from eqns (8.27), (8.28) and (8.30)

$$\tau_\alpha \approx -\sigma_x \omega\bar{y} \cos(\omega x), \tag{8.31}$$

and

$$\sigma_\alpha \approx \sigma_x \omega^2 \bar{y}^2 \cos^2(\omega x). \tag{8.32}$$

First consider the shear stress. According to eqn (8.31) the shear stress parallel to the marker assumes numerical maxima with alternating signs at the inflection points, and vanishes at the crests of the anticlines and at the keels of the synclines.

Since the marker moves passively with the homogeneously strained material we conclude that τ_α and σ_α as given above, are exactly the stresses along and across a sinusoidal marker which are required in order to keep the strain homogeneous. If the stresses at the sinusoidal marker were different from those given by eqns (8.27) and (8.28) the deflection of the marker would deviate from the movement pattern characteristic of homogeneous strain.

Now compare this with the buckles on the free surface of the compressed half-space. Here the shear stress vanishes at all x, so does the normal stress. Obviously, then, the movement of the buckled surface is not consistent with homogeneous strain and consequently different from the movement of the marker curve: the movement at the buckled surface is perturbed.

It is known that if the shear stress at the free surface of an infinite half-space varies proportional to $\cos(\omega x)$, then the velocity component u at the surface varies proportionally to the shear stress and hence also proportionally to $\cos(\omega x)$, according to the expression:

$$u = \tfrac{1}{2}\mu^{-1}\omega^{-1}\tau = \tfrac{1}{2}\mu^{-1}\omega^{-1}\bar{\tau}\cos(\omega x). \tag{8.33}$$

The shear stress at the inflection point at $x = 0$ is $\bar{\tau}$, and the stress at any x is τ. This equation is valid for half-spaces with their free surface facing upward, i.e. substratum half-spaces. For half-spaces whose free surface faces downward, i.e. overburden half-spaces, the equation is

$$u = -\tfrac{1}{2}\mu^{-1}\omega^{-1}\tau = -\tfrac{1}{2}\mu^{-1}\omega^{-1}\bar{\tau}\cos(\omega x). \tag{8.34}$$

In the analysis of buckling problems we are concerned with the perturbed velocities. These are defined as the difference between the actual velocities at the buckling boundary, and the velocities which would have occurred had the strain been homogeneous.

The perturbed velocity component u follows from the difference in shear stress at the buckling free surface and at the homogeneously strained sinusiodal marker. This shear-stress difference is

$$\tau = 0 - \tau_\alpha = \sigma_x \omega \bar{y} \cos(\omega x). \tag{8.35}$$

Insertion of this expression for τ in eqn (8.33) gives the perturbed velocity

u at the buckling surface of a laterally compressed substratum half-space

$$u_\sigma = \tfrac{1}{2}\sigma_x\mu^{-1}\bar{y}\cos(\omega x). \qquad (8.36)$$

Correspondingly, the perturbed surface velocity of an overburden half-space is

$$u_\sigma = -\tfrac{1}{2}\sigma_x\mu^{-1}\bar{y}\cos(\omega x). \qquad (8.37)$$

The tangential amplitude velocities are + and $-\tfrac{1}{2}\sigma\mu^{-1}\bar{y}$ in the two situations.

The velocities in eqns (8.36) and (8.37) have been marked with the subscript σ to emphasize that these are perturbed tangential velocities caused by the lateral compressive stress, σ_x, acting on the half-spaces.

The tangential velocities here described give rise to a widening of the bulges and a narrowing of the depression at the free surface of a compressed half-space. That is, the antiforms will widen and the synforms narrow on a substratum half-space, whereas the opposite is true for an overburden half-space. Thus the well known *cusp-and-arc* fold structure is formed (Ramberg, 1970b, p. 86; Smith, 1979, p. 285).

It is evident from the above equations that the tangential velocity at the buckling surface of an infinite half-space is independent of the wavelength. This is in noteworthy contrast to the conditions characterizing buckling layers, see eqn (8.26).

The normal stress across the sinusoidal marker must also be considered. By virtue of the quadratic cosine term in eqn (8.32) the period of the normal stress is only half the period of the shear stress. The normal stress does not change sign, and reaches its maximum value at $x = 0$, $x = \tfrac{1}{2}\lambda$, $x = \lambda$, etc., that is, at the inflection points at the flanks of the waves. The minimum value, which for the normal stress is zero, is reached at the crests of the anticlines and the keels of the synclines. Since σ_x is compressive in our model it follows from eqn (8.28) that the stress σ_α is also compressive. The value given by eqn (8.28) is exactly the magnitude of compressive stress across the sinusoidal marker which is required to keep the marker to deform passively in the homogeneous strain field.

At the buckling free surface of a half-space the normal stress vanishes at all points. It is evident, then, that the normal velocity must be different from the velocity normal to a sinusoidal passive marker. The normal-stress difference between the wavy free surface and the passive marker as expressed by eqn (8.32) indicates that the normal velocity at the free surface also varies with a period only half that of the surface buckles.

Evidently, the lack of constraint from a compressive stress normal to the surface at the flanks of the buckles allows the flanks to bulge out to make the anticlines wide and rounded and the synclines narrow and pointed for half-spaces whose surface face upward. For half-spaces with a downward-facing surface the anticlines become narrow and pointed and the synclines wide and rounded. In other words, the normal velocity has a similar effect on the shape of the surface buckles as has the tangential velocity (Fig. 8.10), accentuating the *cusp-and-arc* form of the folds.

The movement pattern here described means that the shape of initially gentle sinusoidal folds on the free surface of a laterally compressed half-space deviates from the sinusoidal shape increasingly as the compression of the half-space proceeds. Sinusoidal deflections change gradually to a *cusp-and-arc* pattern as the compression proceeds. This is in contrast to the passive marker which maintains the sinusoidal shape during the compression, notwithstanding the increase of the amplitude and the shortening of the wavelength. Since the normal stress across the passive marker and at the buckling free surface is the same at the crests and keels we assume that the transversal amplitude velocity also is the same for the marker and the free surface. Accordingly, we consider the perturbed transversal amplitude velocity of the surface buckles to be zero:

$$\bar{v}_\sigma = 0. \tag{8.38}$$

Buckling of multilayers

The velocity components v and u in the previous equations and in those that follow are the perturbed velocities that belong exclusively to the periodically varying strain which constitutes the buckling process. The velocities associated with the homogeneous strain—i.e. the uniform shortening and thickening of the compressed strata and half-spaces—are not included in the analysis. Shortening and thickening are not insignificant in fold structures—in fact, this deformation is of prime consequence in the evolution, for example, of different orders of folds in multilayers and of schistosity in such rock complexes (Ramberg, 1964b, 1970; Ramberg and Ghosh, 1968)—rather it need not be considered in an analysis especially aimed at determining the initial wavelength/thickness ratio in buckling systems. The effect of the homogeneous strain can be added after the dominant wavelength has been determined.

When treating buckling of enclosed layers, of composite stratified complexes and the effect of gravity on such structures, an efficient method of attack is to apply the condition of continuity of the normal stress and the shear stress at the contacts between the strata. The method requires equations that express the stresses at the boundaries of a sheet or a half-space as functions of three parameters: (1) the periodically varying velocity components at the boundaries, (2) the buckling stress, and (3) the force of gravity. Functions of this nature may be obtained by combining eqns

(7.22–7.33), valid for a folding layer or half-space not laterally compressed, with eqns (8.23), (8.26), (8.36), (8.37) and (8.38), referring to a laterally compressed buckling layer or half-space under vanishing stresses at the surfaces.

Referring first to a layer with finite thickness, let $\bar{v}'$, $\bar{u}'$, $\bar{v}'$ and $-\bar{u}'$ be the rate of growth of transversal and tangential amplitudes at the upper and lower surfaces respectively, as generated solely by the buckling force according to eqns (8.23) and (8.26). Let on the other hand $\bar{v}''_{(i-1)}$ and $\bar{u}''_{(i-1)}$ be the amplitude velocities on the upper surface as caused exclusively by periodically varying stresses at the upper and lower boundaries, and by the gravity force. $\bar{v}''_i$ and $\bar{u}''_i$, are the corresponding velocities at the lower surface caused solely by the last-mentioned stresses and forces in accordance with eqns (7.22–7.25).

The total perturbed velocities at the surfaces of a layer simultaneously acted upon by a lateral compressive stress, by sinusoidal varying stresses at the surfaces and by the force of gravity, are accordingly

$$\left.\begin{aligned} \bar{v}_{(i-1)} &= \bar{v}' + \bar{v}''_{(i-1)} & \bar{u}_{(i-1)} &= \bar{u}' + \bar{u}''_{(i-1)} \\ \bar{v}_i &= \bar{v}' + \bar{v}''_i & \bar{u}_i &= -\bar{u}' + \bar{u}''_i. \end{aligned}\right\} \tag{8.39}$$

When we replace $\bar{v}'$ and $\bar{u}'$ in the above equations by their forms as expressed in the buckling equations (8.23) and (8.26), and insert $\bar{v}''_{(i-1)} = \bar{v}_{(i-1)} - \bar{v}'$, $\bar{v}''_i = \bar{v}_i - \bar{v}'$, $u''_{(i-1)} = u_{(i-1)} - u'$ and $u''_i = u_i + u'$ in eqns (7.22–7.25) the sought functions follow, viz. expressions for each of the boundary stresses $\sigma_{i(i-1)}$, σ_{ii}, $\tau_{i(i-1)}$ and τ_{ii} as related to the resultant perturbed amplitude velocities, to the buckling stress and to the gravity force. The resulting functions are

$$\begin{aligned} \sigma_{i(i-1)} = {} & 2\mu_i h_i^{-1}(P_{i(i-1)}^{v(i-1)}\bar{v}_{i(i-1)} + P_{i(i-1)}^{u(i-1)}\bar{u}_{i(i-1)} \\ & + P_{i(i-)}^{v_i}\bar{v}_{ii} + P_{i(i-1)}^{\bar{u}_i}\bar{u}_{ii}) \sin \omega\bar{x} \\ & + N_{i(i-1)}h_i^{-1}\sigma_{bi}(\bar{y}_{i(i-1)} + \bar{y}_{ii}) \sin \omega x \\ & + \rho_i g \bar{y}_{i(i-1)} \sin \omega x + \sigma'_{(i-1)}, \end{aligned} \tag{8.40a}$$

$$\begin{aligned} \sigma_{ii} = {} & 2\mu_i h_i^{-1}(P_{ii}^{v(i-1)}\bar{v}_{i(i-1)} + P_{ii}^{u(i-1)}\bar{u}_{i(i-1)} \\ & + P_{ii}^{v_i}\bar{v}_{ii} + P_{ii}^{u_i}\bar{u}_{ii}) \sin \omega x \\ & + N_{ii}h_i^{-1}\sigma_{bi}(\bar{y}_{i(i-1)} + \bar{y}_{ii}) \sin \omega x \\ & + \rho_i g \bar{y}_{ii} \sin \omega x + \sigma'_i, \end{aligned} \tag{8.40b}$$

$$\begin{aligned} \tau_{i(i-1)} = {} & 2\mu_i\omega(S_{i(i-1)}^{v(i-1)}\bar{v}_{i(i-1)} + S_{i(i-1)}^{u(i-1)}\bar{u}_{i(i-1)} \\ & + S_{i(i-1)}^{v_i}\bar{v}_{ii} + S_{i(i-1)}^{u_i}\bar{u}_{ii}) \cos \omega x \\ & + M_{i(i-1)}\omega\sigma_{bi}(\bar{y}_{i(i-1)} + \bar{y}_{ii}) \cos \omega x, \end{aligned} \tag{8.40c}$$

$$\begin{aligned} \tau_{ii} = {} & 2\mu_i\omega(S_{ii}^{v(i-1)}\bar{v}_{i(i-1)} + S_{ii}^{u(i-1)}\bar{u}_{i(i-1)} \\ & + S_{ii}^{v_i}\bar{v}_{ii} + S_{ii}^{u_i}\bar{u}_{ii}) \cos \omega x \\ & + M_{ii}\omega\sigma_{bi}(\bar{y}_{i(i-1)} + \bar{y}_i) \cos \omega x. \end{aligned} \tag{8.40d}$$

The coefficients S and P are defined in Table 7.2. The buckling stress acting on layer i is σ_{bi}. M and N are terms which assume the forms given in Table 8.1. In eqns (8.40) layer-index i has been added to the velocities.

The four stress equations developed above, together with the corresponding equations (8.44), (8.45), (8.46) and (8.47) valid for half-spaces, are the basic ones in the present analysis of multilayers exposed simultaneously to lateral compression and to the force of gravity. The equations specify the effect upon the shear stress and the normal stress at the boundaries, as caused by the amplitude velocities, by the buckling stress and by the force of gravity.

The coefficients S and P show how the boundary stresses are controlled by the amplitude velocities, the coefficients N and M determine the extent to which the buckling stress controls the stresses at the boundaries, and finally the term $\rho g\bar{y}$ shows how gravity affects the stresses.

When the overburden or the substratum is thick relative to the wavelength of the buckles the sinusoidal contact strain does not penetrate through to the surface of the overburden or to the bottom of the substratum. In such cases the overburden and the substratum are dynamically equivalent to infinite half-spaces.

Accordingly, it is necessary to determine the stresses at the bottom face of an infinite half-space occurring as overburden, and at the top face of an infinite half-space occurring as substratum. These stresses are functions of (1) the sinusoidal velocities at the overburden base or at the substratum top, (2) the compressive stress acting on the overburden or on the substratum and (3) the body force of gravity.

At the basal face of an infinite overburden the perturbed tangential velocity as caused solely by a periodically varying shear stress of type $\tau = \bar{\tau}\cos(\omega x)$ is

$$u = \bar{u}\cos(\omega x) = -\tfrac{1}{2}\omega^{-1}\mu^{-1}\tau. \tag{8.41}$$

A compressive stress acting on the overburden normal to the axial plane of the buckles augments the perturbed tangential velocity by an amount equal to

$$u_\sigma = \bar{u}_\sigma\cos(\omega x) = -\tfrac{1}{2}\sigma\mu^{-1}\bar{y}\cos(\omega x). \tag{8.42}$$

The total perturbed tangential velocity at the basal face of an infinite overburden is accordingly

$$\begin{aligned} u_{11}^{\infty} &= (\bar{u} + \bar{u}_\sigma)\cos(\omega x) = \bar{u}_{11}^{\infty}\cos(\omega x) \\ &= -\tfrac{1}{2}\omega^{-1}\mu^{-1}\tau - \tfrac{1}{2}\mu^{-1}\sigma\bar{y}_{11}\cos(\omega x), \end{aligned} \tag{8.43}$$

and the shear stress at the base becomes

$$\tau_{11}^{\infty} = -2\omega\mu_1\bar{u}_{11}^{\infty}\cos(\omega x) - \omega\sigma\bar{y}_{11}\cos(\omega x). \tag{8.44}$$

Here we have attached superscript ∞ to indicate association with an infinite half-space rather than a layer with finite thickness. The subscript 11 indicates that the stress and the velocity refer to the basal boundary of the uppermost member—i.e. member 1 and boundary 1—of the system.

Equation (8.44) relates the shear stress at the buckling base of an infinite half-space to the tangential amplitude velocity $\bar{u}_{11}^{\infty}$ and to the compressive stress σ to which the half-space has been exposed.

To determine the normal stress at the buckling base of the compressed half-space we recall that the normal stress vanishes at the crests and keels of the buckles, and that the compression of the half-space presumably does not give rise to a significant perturbed transversal amplitude velocity. The normal stress at the base is therefore not affected by the horizontal compression. Let it therefore suffice to consider the relationship between the normal stress, the periodically varying velocity at the wavy base and the force of gravity. The expression is the same as the one valid for an infinite overburden not exposed to compression, viz.:

$$\sigma_{11}^{\infty} = -2\omega\mu_1\bar{v}_{11}^{\infty}\sin(\omega x) + \rho_1\bar{y}_{11}g\sin(\omega x) - p_1 \tag{8.45}$$

(see also p. 66).

An analogous argument applied to an infinite half-space occurring as substratum results in the following expressions for the stresses at the buckling surface:

$$\tau_{(n+1)n}^{\infty} = 2\omega\mu_{(n+1)}\bar{u}_{(n+1)n}^{\infty}\cos(\omega x) - \omega\sigma\bar{y}_{(n+1)n}\cos(\omega x), \tag{8.46}$$

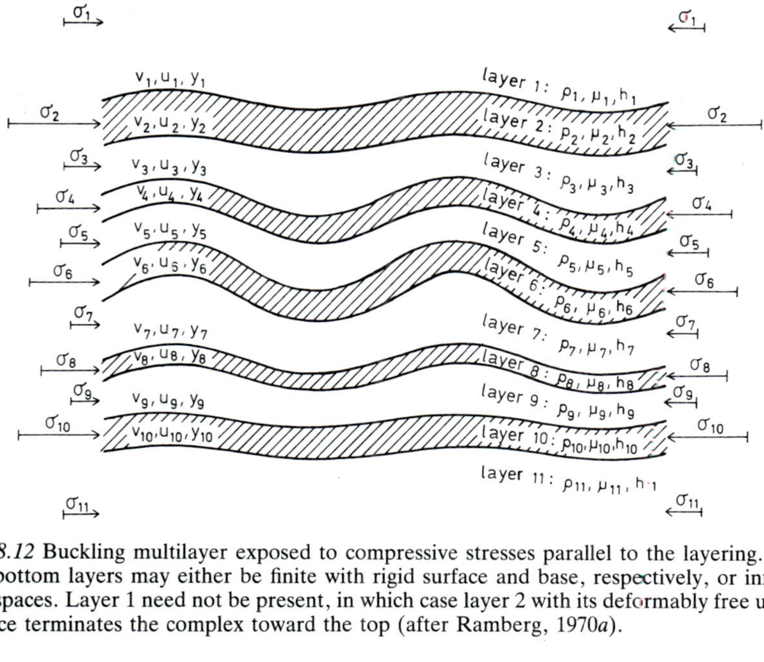

Fig. 8.12 Buckling multilayer exposed to compressive stresses parallel to the layering. Top and bottom layers may either be finite with rigid surface and base, respectively, or infinite half-spaces. Layer 1 need not be present, in which case layer 2 with its deformably free upper surface terminates the complex toward the top (after Ramberg, 1970*a*).

and for the normal stress across the buckling face:

$$\sigma^{\infty}_{(n+1)n} = 2\omega\mu_{(n+1)}\bar{v}^{\infty}_{(n+1)n}\sin(\omega x) + \rho_{(n+1)}\bar{y}_{(n+1)n}g\sin(\omega x) - p_n. \quad (8.47)$$

Subscript $(n + 1)$ refers to the substratum and subscript $(n + 1)n$ to the upper surface—surface n— of the substratum.

For a multilayer with welded contacts and $(n + 1)$ members the condition for dynamic equilibrium in terms of continuity of other stresses is (see Fig. 8.12)

$$\begin{aligned} &\left.\begin{aligned}\tau_{11} &= \tau_{21}\\ \sigma_{11} &= \sigma_{21}\end{aligned}\right\} \text{ for the first interface,}\\ &\left.\begin{aligned}\tau_{22} &= \tau_{32}\\ \sigma_{22} &= \sigma_{32}\end{aligned}\right\} \text{ for the second interface,}\\ &\qquad\vdots\\ &\left.\begin{aligned}\tau_{nn} &= \tau_{(n+1)n}\\ \sigma_{nn} &= \sigma_{(n+1)n}\end{aligned}\right\} \text{ for the } n\text{th interface.}\end{aligned} \quad (8.48)$$

The first subscript refers to the layer, the second to the interface. Following the notation defined on p. 62 the lower boundary of a layer bears the same index as the layer, the upper boundary is indexed by one unit less.

$$\left[\begin{array}{cccccccccc}
c_{11} & d_{12} & e_{13} & f_{14} & & & & & & \\
i_{21} & j_{22} & k_{23} & l_{24} & & & & & & \\
a_{31} & b_{32} & c_{33} & d_{34} & e_{35} & f_{36} & & & & \\
g_{41} & h_{42} & i_{43} & j_{44} & k_{45} & l_{46} & & & & \\
 & & a_{53} & b_{54} & c_{55} & d_{56} & e_{57} & f_{58} & & \\
 & & g_{63} & h_{64} & i_{65} & j_{66} & k_{67} & l_{68} & & \\
 & & & & \cdot\ \cdot\ \cdot & & & & & \\
 & & & & & \cdot\ \cdot\ \cdot & & & & \\
 & & & & & & \cdot\ \cdot\ \cdot & & & \\
 & & & & & & & a_{(2n-3)(2n-5)} & b_{(2n-3)(2n-4)} & c_{(2n-3)(2n-\ldots} \\
 & & & & & & & g_{(2n-2)(2n-5)} & h_{(2n-2)(2n-4)} & i_{(2n-2)(2n-\ldots} \\
 & & & & & & & & & a_{(2n-1)(2n-\ldots} \\
 & & & & & & & & & g_{2n(2n-3)}
\end{array}\right.$$

The necessary condition of continuity of the velocities at the interfaces is satisfied by introducing the velocities $\bar{v}_k$ and $\bar{u}_k\,(k = 1, 2 \ldots n)$ by means of the equations

$$\begin{aligned}
\bar{v}_{11} &= \bar{v}_{21} = \bar{v}_1 \\
\bar{u}_{11} &= \bar{u}_{21} = \bar{u}_1 \\
\bar{v}_{22} &= \bar{v}_{32} = \bar{v}_2 \\
\bar{u}_{22} &= \bar{u}_{32} = \bar{u}_2 \\
&\vdots \\
\bar{v}_{nn} &= \bar{v}_{(n+1)n} = \bar{v}_n .
\end{aligned} \tag{8.49}$$

Because of material coherence across the interfaces we can also introduce the transversal amplitude, $\bar{y}_k$, one for each interface according to eqns (8.50).

$$\begin{aligned}
\bar{y}_{11} &= \bar{y}_{21} = \bar{y}_1 \\
\bar{y}_{22} &= \bar{y}_{32} = \bar{y}_2 \\
&\vdots \\
\bar{y}_{nn} &= \bar{y}_{(n+1)n} = \bar{y}_n
\end{aligned} \tag{8.50}$$

$$\begin{bmatrix}
 & & \\
 & & \\
 & & \\
 & & \\
 & & \\
 & & \\
 & & \\
 & & \\
 & & \\
\ldots n-3)(2n-2) & e_{(2n-3)(2n-1)} & f_{(2n-3)2n} \\
\ldots n-2)(2n-2) & k_{(2n-2)(2n-1)} & l_{(2n-2)2n} \\
\ldots n-1)(2n-2) & c_{(2n-1)(2n-1)} & d_{(2n-1)2n} \\
\ldots (2n-2) & i_{2n(2n-1)} & j_{2n2n}
\end{bmatrix}
\begin{bmatrix}
\bar{v}_1 \\ \bar{u}_1 \\ \bar{v}_2 \\ \bar{u}_2 \\ \bar{v}_3 \\ \bar{u}_3 \\ \cdot \\ \cdot \\ \cdot \\ \bar{v}_{(n-1)} \\ \bar{u}_{(n-1)} \\ \bar{v}_n \\ \bar{u}_n
\end{bmatrix}
=
\begin{bmatrix}
Q_1 \\ Q_2 \\ Q_3 \\ Q_4 \\ Q_5 \\ Q_6 \\ \cdot \\ \cdot \\ \cdot \\ Q_{(2n-3)} \\ Q_{(2n-2)} \\ Q_{(2n-1)} \\ Q_{2n}
\end{bmatrix} \tag{8.51}$$

Table 8.1. Coefficient Q on right-hand side of equation system (8.51)[a]

Valid when $3 \leqslant k \leqslant (2n-3)$ and $4 \leqslant l \leqslant (2n-2)$:

$$Q_{\mathrm{k}} \equiv M_{ii}\frac{\sigma_i}{2\mu_{(i+1)}}(y_{(i-1)}+y_i) - M_{(i+1)i}\frac{\sigma_{(i+1)}}{2\mu_{(i+1)}}(y_i+y_{(i+1)})$$

$$Q_l \equiv h_{(i+1):i}N_{ii}\frac{\sigma_i}{2\mu_{(i+1)}}(y_{(i-1)}+y_i) + \frac{(\rho_i-\rho_{(i+1)})gh_{(i+1)}y_i}{2\mu_{(i+1)}} - N_{(i+1)i}\frac{\sigma_{(i+1)}}{2\mu_{(i+1)}}(y_i+y_{(i+1)})$$

Valid when $k = 1$ and/or $(2n-1)$, $l = 2$ and/or $2n$[b]:

$$Q_1^{\infty} \equiv -\frac{\sigma_1 y_1}{2\mu_2} - M_{21}\frac{\sigma_2(y_1+y_2)}{2\mu_2}$$

$$Q_2^{\infty} \equiv \frac{(\rho_1-\rho_2)gh_2y_1}{2\mu_2} - N_{21}\frac{\sigma_2(y_1+y_2)}{2\mu_2}$$

$$Q_{(2n-1)}^{\infty} \equiv M_{nn}\frac{\sigma_n(y_{(n-1)}+y_n)}{2\mu_{(n+1)}} + \frac{\sigma_{(n+1)}y_n}{2\mu_{(n+1)}}$$

$$Q_{2n}^{\infty} \equiv N_{nn}\frac{\sigma_n(y_{(n-1)}+y_n)}{2\mu_{(n+1)}} + \frac{(\rho_n-\rho_{(n+1)})gh_ny_n}{2\mu_{(n+1)}}$$

$$Q_1^{\sim} \equiv -M_{21}\frac{\sigma_2(y_1+y_2)}{2\mu_2}$$

$$Q_2^{\sim} \equiv -\frac{\rho_2gh_2y_1}{2\mu_2} - N_{21}\frac{\sigma_2(y_1+y_2)}{2\mu_2}$$

$$M_{i(i-1)} = M_{ii} \equiv (3\phi_i^{-2} + \tfrac{5}{8})\,(S_{i(i-1)}^{\mathrm{v}(i-1)} + S_{i(i-1)}^{\mathrm{v}i}) - \tfrac{3}{2}\phi_1^{-1}\,(S_{i(i-1)}^{u(i-1)} - S_{i(i-1)}^{ui})^{c}$$

$$N_{i(i-1)} = -N_{ii} \equiv (3\phi_i^{-2} + \tfrac{5}{8})\,(P_{i(i-1)}^{\mathrm{v}(i-1)} + P_{i(i-1)}^{\mathrm{v}i}) - \tfrac{3}{2}\phi_1^{-1}\,(P_{i(i-1)}^{u(i-1)} - P_{i(i-1)}^{ui})^{c}$$

---

[a] Row index k is always odd, l always even. Index i refers to layer i. Row indices k and l and layer index i are related thus: $i = \frac{1}{2}(k+1)$; $i = \frac{1}{2}l$. Definitions of P and S are given in Table 7.2.

[b] Superscript ∞ signifies infinite thickness of overburden (layer 1) and/or substratum (layer $n+1$); superscript $\sim$ signifies free, deformable surface of top layer which then carries index 2.

[c] These formulas for M and N not valid for infinite overburden or inifinite base.

Application of the stress eqns (8.40) for the individual strata in a multilayer and eqns (8.44) to (8.47) for infinite half-spaces as overburden and substratum, lead to the equation system (8.51), when the stresses are equated at the interfaces according to eqn (8.48). The elements of the matrix of the equation system are defined in Tables 7.3 and 8.1.

The expressions for the matrix elements a to l in eqn (8.51) are the same as for the elements in eqn (7.37), p. 68. The column of constants, Q_1 to Q_n is, however, different.

The maximum number of strata—including possible half-spaces as overburden and/or basement—that can be handled depends on the capacity of the computer available. The extension is accomplished by filling in elements along the diagonal zone in sets of two rows of type

$$\begin{matrix} a & b & c & d & e & f \\ g & h & i & j & k & l, \end{matrix}$$

c and j lying on the principal diagonal. Two rows are needed to account for each additional layer, because each new layer brings two new unknown velocities into the system.

Solves for $\bar{v}_1$, $\bar{v}_2$, $\bar{v}_3$, etc., the equation system (8.51) shows how the rate of change of transversal amplitude of the folds of each non-rigid interface varies with certain characteristic dimensionless parameters of the composite structure, such as thickness ratios, viscosity ratios, density-contrast ratios, and with the wavelength and the buckling stress exerted on the various sheets.

With the analysis to follow in mind it is especially informative to express the solutions in a form that shows the rate of change of transversal amplitude at any one of the boundaries as a function of the buckling stress and amplitude of each of the constituent strata. In other words we wish to express the solution of eqns (8.51) in the form

$$\begin{bmatrix} \bar{v}_1 \\ \bar{v}_2 \\ \vdots \\ \bar{v}_n \end{bmatrix} = \begin{bmatrix} k_{11} k_{12} \ldots k_{1n} \\ k_{21} k_{22} \ldots k_{2n} \\ \vdots \quad \vdots \quad\quad \vdots \\ k_{n1} k_{n2} \ldots k_{nn} \end{bmatrix} \begin{bmatrix} (\sigma_1/\mu_1)\bar{y}_1 \\ (\sigma_2/\mu_2)\bar{y}_2 \\ \vdots \\ (\sigma_n/\mu_n)\bar{y}_n \end{bmatrix}, \tag{8.52}$$

or more compactly:

$$\bar{v}_j = \sum_{i=1}^{n} k_{ji}\bar{y}_i \frac{\sigma_i}{\mu_i}. \tag{8.53}$$

(Note that σ_1/μ_1 must be replaced by σ_2/μ_2 for models terminated by a deformable free surface because layer 1 then vanishes in our formal math-

Table 8.2. Elements in the matrix of equation system (8.52)[a]

When $i = 1$ and $n > 2$:

$$k_{j1}^{\infty} \equiv \tfrac{1}{2}|A|^{-1}\Big[(-\mu_{1:2}-M_{21}\sigma_{2:1}\mu_{1:2})|A_{1(2j-1)}| - (N_{21}\sigma_{2:1}\mu_{1:2}-\frac{\rho_1-\rho_2}{\sigma_1}gh_2\mu_{1:2})|A_{2(2j-1)}|$$
$$+M_{22}\sigma_{2:1}\mu_{1:3}|A_{3(2j-1)}| + N_{22}h_{3:2}\mu_{1:3}\sigma_{2:1}|A_{4(2j-1)}|\Big]$$

$$k_{j1}^{\sim} \equiv \tfrac{1}{2}|A|^{-1}\Big[-M_{21}|A_{1(2j-1)}| - (N_{21}+\frac{\rho_2}{\sigma_2}gh_2)|A_{2(2j-1)}| + M_{22}\mu_{2:3}|A_{3(2j-1)}|$$
$$+N_{22}h_{3:2}\mu_{2:3}|A_{4(2j-1)}|\Big]$$

When $2 \leqslant i \leqslant (n-1)$ and $n > 2$:

$$k_{ji} \equiv \tfrac{1}{2}|A|^{-1}\Big[-M_{i(i-1)}|A_{(2i-3)(2j-1)}| - N_{i(i-1)}|A_{(2i-2)(2j-1)}|$$
$$+(M_{ii}\mu_{i:(i+1)} - M_{(i+1)i}\sigma_{(i+1):i}\mu_{i:(i+1)})|A_{(2i-1)(2j-1)}|$$
$$+\Big(N_{ii}h_{(i+1):i}\mu_{i:(i+1)} - N_{(i+1)i}\sigma_{(i+1):i}\mu_{i:(i+1)} +$$
$$+\frac{\rho_i-\rho_{(i+1)}}{\sigma_i}gh_{(i+1)}\mu_{i:(i+1)}\Big)|A_{2i(2j-1)}| + M_{(i+1)(i+1)}\sigma_{(i+1):i}\mu_{i:(i+2)}$$
$$|A_{(2i+1)(2j-1)}| + N_{(i+1)(i+1)}h_{(i+2):(i+1)}\sigma_{(i+1):i}\mu_{i:(i+2)}|A_{(2i+2)(2j-1)}|\Big]$$

When $i = n$ and $n > 2$ (n being the number of flexible boundaries in the system):

$$k_{jn} \equiv \tfrac{1}{2}|A|^{-1}\Big[-M_{n(n-1)}|A_{(2n-3)(2j-1)}| - N_{n(n-1)}|A_{(2n-2)(2j-1)}|$$
$$+(M_{nn}\mu_{n:(n+1)} + \sigma_{(n+1):n}\mu_{n:(n+1)})|A_{(2n-1)(2j-1)}| + (N_{nn}\mu_{n:(n+1)}$$
$$+\frac{\rho_n-\rho_{(n+1)}}{\sigma_n}\mu_{n:(n+1)}gh_n)|A_{2n(2j-1)}|\Big]$$

When $n = 2$:

$$k_{11}^{\infty n=2} \equiv \tfrac{1}{2}|A|^{-1}\Big[(-\mu_{1:2}-M_{21}\sigma_{2:1}\mu_{1:2})|A_{11}| - (N_{21}\sigma_{2:1}\mu_{1:2} - \frac{\rho_1-\rho_2}{\sigma_1}gh_2\mu_{1:2})|A_{21}|$$
$$+M_{22}\sigma_{2:1}\mu_{1:3}|A_{31}| + N_{22}\sigma_{2:1}\mu_{1:3}|A_{41}|\Big]$$

$$k_{12}^{\infty n=2} \equiv \tfrac{1}{2}|A|^{-1}\Big[-M_{21}|A_{11}| - N_{21}|A_{21}| + (M_{22}\mu_{2:3}+\sigma_{3:2}\mu_{2:3})|A_{31}|$$
$$+ (N_{22}\mu_{2:3}+\frac{\rho_2-\rho_3}{\sigma_2}\mu_{2:3}h_2g)|A_{41}|\Big]$$

$$k_{11}^{\sim n=2} \equiv \tfrac{1}{2}|A|^{-1}\left[-M_{21}|A_{11}|-(N_{21}+\frac{\rho_2 g h_2}{\sigma_2})|A_{21}|+M_{22}\mu_{2:3}|A_{31}|+N_{22}\mu_{2:3}|A_{41}|\right]$$

$$k_{12}^{\sim n=2} \equiv \tfrac{1}{2}|A|^{-1}\Big[-M_{21}|A_{11}|-N_{21}|A_{21}|+(M_{22}\mu_{2:3}+\sigma_{3:2}\mu_{2:3})|A_{31}| + (N_{22}\mu_{2:3}+\frac{\rho_2-\rho_3}{\sigma_2}\mu_{2:3}h_2 g)|A_{41}|\Big]$$

$k_{21}^{\infty n=2}$, $k_{22}^{\infty n=2}$, $k_{21}^{\sim n=2}$, $k_{22}^{\sim n=2}$ are formed from the above coefficients by changing the second subscript in the cofactors from 1 to 3, thus $|A_{13}|$, $|A_{23}|$ etc. $|A_{pq}|$ are the cofactors to the corresponding elements of the matrix of eqn (8.51).

---

[a] ~ indicates model terminated upward by free, deformable boundary; ∞ indicates model with infinite half-space as overburden.

ematical system. The top layer with the free surface is then indexed by the digit 2. See also the treatment of buoyant systems, p. 95.)

We shall presently see that k_{ji} contains the ratios between the buckling stresses and the density contrast between layers in addition to the same dimensionless parameters that occur in the elements of the matrix of eqn (8.51).

The use of j as row and i as column index rather than *vice versa* in the array of k_{ji} may seem awkward. However, this arrangement allows us to continue to use i as the index for the general layer in the formulas to follow. See Table 8.2 and explanation p. 80.

Presenting the solution of $\bar{v}_j$ in the above form emphasizes the dynamic coupling between the various strata. The larger the numerical value of a certain coefficient k_{ji} the tighter the coupling between interface j and layer i. A large value of k_{13}, for example, means that the stress on layer 3, the viscosity of layer 3 and the amplitude on interface 3 have a significant influence on the rate of deflection of interface 1. On the other hand, if a certain k-value is small, say, k_{2n}, it means that the rate of deflection on the second interface is not much affected by what goes on at the nth layer and interface.

To determine the coefficients k_{ji} in terms of the elements in the matrix of eqn (8.51), that is, in terms of the characteristic dimensionless parameters of the system, we combine Cramer's rule (e.g. Hadley, 1965, p. 166) for solving simultaneous equations with the method of expanding a determinant by its cofactors. This operation provides the following expression for the velocity of the transversal amplitude

$$\bar{v}_j = |A|^{-1} \sum_{p=1}^{2n} Q_p |A_{pq}|. \tag{8.54}$$

Here $|A|$ is the determinant of the matrix to eqn (8.51), $|A_{pq}|$ the cofactors to the element a_{pq} in the same matrix and Q_p the column of coefficients on the right-hand side of eqn (8.51) as defined in Table 8.1. In the above expression the index q is related to the index j thus: $q = (2j - 1)$.

When the expressions for Q from Table 8.1 are inserted in the set of equations (8.51) and the terms $(\sigma_i/\mu_i)\bar{y}_i$ are factored out, formulas for the coupling coefficients, k_{ji}, are obtained based on the identity of eqns (8.52) and (8.51). See Table 8.2.

In this section of the book we consider only layered systems whose basal member is an infinite half-space—or at least a layer very thick relative to the wavelength—while the top member may either be an infinite half-space or a layer with a free flexible surface.

The general formula for the first column of coupling coefficients, i.e. the coefficients k_{j1}, depends on whether the model is terminated upward by an infinite half-space or by a free surface. This distinction is indicated by ∞ and $\sim$ respectively, in the list of coupling coefficients (Table 8.2). The coupling coefficients belonging to the internal columns in the matrix of coefficients, i.e. coefficient from k_{j2} to $k_{j(n-1)}$, are defined by a single general formula which is independent of the boundary conditions at the top and bottom of the model under study. The last column of coefficients, k_{jn}, are given by a general formula somewhat different from the others because of the infinite half-space nature of the basement.

Models consisting of only one buckling layer resting on an infinite half-space and either being overlaid by an infinite half-space or having a free upper surface are controlled by coupling coefficients whose formulas deviate somewhat from those of the multilayer models. The formulas for monolayer systems are indicated by the superscript $(n = 2)$ in agreement with the order of the matrix $[k_{ji}]$ which is 2 for a single-layer model (there are two flexible boundaries and consequently two transversal amplitude velocities).

The equation system (8.52) demonstrates that the dynamic evolution of a laterally compressed layered system in the field of gravity is controlled partly by the nature of the coupling coefficients, partly by the magnitude of the compressive stress acting on the layers in relation to their viscosity, and partly by the amplitude of the geometric perturbations of the interfaces and/or surfaces.

As we start to analyse the dynamics of the systems we first note that the coupling coefficients are dimensionless. For several similar systems with the same number of layers and the same boundary conditions at the top and bottom, the coupling coefficients are identical provided the following dimensionless ratios are the same:

(1) the ratios between corresponding layer thicknesses,

(2) the ratios between corresponding viscosities,

(3) the ratios between corresponding densities,
(4) the ratios between corresponding buckling stresses acting on the layers,
(5) the ratios between corresponding gravity-induced stress, $\rho g\bar{y}$, and corresponding buckling stress, σ_i, and
(6) the ratios between wavelength and corresponding layer thicknesses (note that the wavelength during its variation is kept the same for all deflected boundaries in a given model).

The latter ratio is expressed in the symbol $\phi_i = 2\pi(h_i/\lambda)$. These statements follow from the definitions expressed in Tables 7.2, 7.3, 8.1 and 8.2.

The above comments may also be expressed as follows: for systems with conditions 1, 2, 3, 4 and 5 satisfied the corresponding coupling coefficients are identical functions of the wavelength/thickness ratio as implied in the symbol ϕ, entirely independent of the absolute values of the parameters contained in the coupling coefficients.

The fact that the coupling coefficients contain only dimensionless quantities economizes the treatment of the systems, since systems with unlike geometric size, unlike viscosities, densities, etc., will behave dynamically similar if only the corresponding dimensionless ratios are the same. This is of paramount importance for the experimental model study of geodynamic systems as we shall see later.

In order to find the characteristic wavelength of a model, and also as a base for further discussion of its dynamic behaviour, it is informative to integrate the amplitude-velocity equation (8.52). For any given layered system the coupling coefficients are determined by the material properties of the constituent layers, their thicknesses, the stress acting on them and the wavelength. The thicknesses of the layers and their material properties are fixed for the system, but it is possible to vary the wavelength as well as the compressive stresses on the layers. Such variations will change the coupling coefficients which, however, are constants as long as the wavelength is arbitrarily fixed and the compressive stresses kept constant. It is practical, though not necessary to make the stresses proportional to the viscosities, such that the ratio σ_i/μ_i is the same for all layers.

Under the conditions of constant coupling coefficients the equation system (8.52) can be integrated. Before integrating this system of amplitude-velocity equations, a minor modification is practical by which a reference stress/viscosity ratio, σ_r/μ_r, is introduced, thus:

$$\begin{bmatrix} \bar{v}_1 \\ \bar{v}_2 \\ \vdots \\ \bar{v}_n \end{bmatrix} = \begin{bmatrix} k_{11}\sigma_{1:r}\mu_{r:1} & k_{12}\sigma_{2:r}\mu_{r:2} & \dots & k_{1n}\sigma_{n:r}\mu_{r:n} \\ k_{12}\sigma_{1:r}\mu_{r:1} & k_{22}\sigma_{2:r}\mu_{r:2} & \dots & k_{2n}\sigma_{n:r}\mu_{r:n} \\ \vdots & \vdots & & \vdots \\ k_{1n}\sigma_{1:r}\mu_{r:1} & k_{n2}\sigma_{2:r}\mu_{r:2} & \dots & k_{nn}\sigma_{n:r}\mu_{r:n} \end{bmatrix} \begin{bmatrix} (\sigma_r/\mu_r)\bar{y}_1 \\ (\sigma_r/\mu_r)\bar{y}_2 \\ \vdots \\ (\sigma_r/\mu_r)\bar{y}_n \end{bmatrix} \tag{8.55}$$

or

$$\bar{v}_j = \sum_{i=1}^{n} k_{ji}\sigma_{i:r}\mu_{r:i}\bar{y}_i \frac{\sigma_r}{\mu_r}. \tag{8.56}$$

Here k_{ji} are identical to the coefficients in eqn (8.52). In case the ratios σ_i/μ_i are kept equal in all layers, the matrix elements reduce to k_{ji} because $\sigma_{i:r}\mu_{r:i} = 1$. (Incidentally, for models terminated by a free deformable surface σ_1 and μ_1 must be replaced by σ_2 and μ_2.) Integration yields

$$\begin{bmatrix} \bar{y}_1 \\ \bar{y}_2 \\ \vdots \\ \bar{y}_n \end{bmatrix} = \begin{bmatrix} c_{11}c_{12} \ldots c_{1n} \\ c_{21}c_{22} \ldots c_{2n} \\ \vdots \quad \vdots \quad \vdots \\ c_{n1}c_{n2} \ldots c_{nn} \end{bmatrix} \begin{bmatrix} \exp[\kappa_1(\sigma_r/\mu_r)t] \\ \exp[\kappa_2(\sigma_r/\mu_r)t] \\ \vdots \\ \exp[\kappa_n(\sigma_r/\mu_r)t] \end{bmatrix}. \tag{8.57}$$

Here κ_i are the eigenvalues of the coefficient matrix of eqn (8.55); c_{ij} are coefficients which partly depend on the eigenvectors belonging to the set of eigenvalues, partly on the initial amplitudes; $\bar{y}_i$ are amplitudes at time t and σ_r/μ_r the ratio between the stress applied on the reference layer and its viscosity.

The analysis is the same as that applied to gravity-driven systems (p. 84). Were it not for the buckling stress, eqn (8.57) would in fact be identical to eqn (7.64) valid for layered systems affected only by the force of gravity.

It has been demonstrated in connection with the evolution of diapirs (p. 85) that the dynamics of layered systems is controlled by the eigenvalue which gives the maximum positive value of the exponents in the integrated equation. Since σ is compressive and hence negative for buckling systems, it is the numerically largest negative eigenvalue which controls the evolution. This is termed the *principal eigenvalue* and given the index unity, κ_1. Numerical examples show that generally the eigenvalues, or at least the principal eigenvalue, run through a numerical maximum when the wavelength varies. For some multilayer systems the principal eigenvalue goes through more than one maximum.

The maximum on the function $\kappa_1 = f(\phi_r)$ determines the characteristic wavelength of systems; ϕ_r implies the wavelength/thickness ratio of the reference layer, r, usually layer number two from the top. (It is impractical to use layer number one as a reference layer because it is either an infinite half-space, i.e. $h_1 = \infty$, or it does not exist at all, i.e. $h_1 = 0$.)

A practical point to observe when one seeks the maximum on the $\kappa_1 = f(\phi_r)$ function is that the wavelength must at all times be the same on the several interfaces. Moreover, the folds must be kept synchronous when the wavelength is changed in the process of seeking the maximum. This synchronization is readily obtained by introducing the same ϕ, say ϕ_r, in all coupling coefficients by means of the operation $\phi_i = \phi_r . h_i/h_r$. In most numerical models tested $r = 2$.

Similarly as described in the section on diapirs (p. 86) the amplitudes of the deflections on the several boundaries tend to develop in proportion to the components of the eigenvector which belongs to the principal eigenvalue at its maximum value.

Experience from buckling tests of model materials indicates that the characteristic wavelength is picked out, as it were, at an early stage of the buckling process, at a stage when the deflections are still very gentle. Therefore, at later stages in the evolution of folds the wavelength becomes shorter than the true characteristic wavelength for two reasons; (1) because of the buckling process *sensu stricto*— the so-called *buckle-shortening*, and (2) because of the homogeneous strain which we refer to as *homogeneous shortening*.

Pure buckle-shortening is characterized by unchanged thickness of the layer and unchanged length of arc of the folds during the compression of the wavelength. In contrast, homogeneous shortening is associated with a thickening of the layer as long as the inclination of the flanks of the folds is less than 45°. The true shortening is then the sum of the buckle-shortening and the homogeneous shortening. As long as the folds are gentle there is little difference between the homogeneous shortening here defined and the so-called layer-shortening, a term often used in structural geology.

For the correct interpretation of folds it is necessary to study the relationships between the buckle-shortening and the homogeneous shortening a little further. Even though the treatment unfortunately must be limited to folds whose amplitude/wavelength ratio is less than say 0·1, the analysis is believed to be informative.

Buckle-shortening

At small $\bar{y}/\lambda$ ratios, the arclength, S, of a sine wave is found by integrating

$$\mathrm{d}S \approx \left[1 + \tfrac{1}{2}\left(\frac{\mathrm{d}y}{\mathrm{d}x}\right)^2\right] \mathrm{d}x \tag{8.58}$$

from $x = 0$ to $x = \lambda$.

Here

$$\left(\frac{\mathrm{d}y}{\mathrm{d}x}\right)^2 = \bar{y}^2\omega^2 \cos^2(\omega x). \tag{8.59}$$

The integration yields

$$S \approx \lambda + \pi^2\frac{\bar{y}^2}{\lambda}, \tag{8.60}$$

by differentiation of which we obtain a relation between the rate of change of amplitude and the rate of change of wavelength (S is assumed constant):

$$\frac{\mathrm{d}\lambda}{\mathrm{d}t} \approx -2\pi^2 \frac{\bar{y}}{\lambda}\frac{\mathrm{d}\bar{y}}{\mathrm{d}t}. \tag{8.61}$$

According to the buckling theory the amplitude of the buckles at boundary i changes as a function of time as follows

$$\bar{y}_i = \bar{y}_{0i} \exp\left(\kappa_1 \frac{\sigma_r}{\mu_r} t\right) \tag{8.62}$$

where $\bar{y}_{0i}$ is the initial amplitude at boundary i, and κ_1 is the maximum numerical value of the principal eigenvalue of the matrix of coupling coefficients for the particular system; σ_r and μ_r are the buckling stress and viscosity, respectively, of reference layer r. For multilayers with more than one flexible boundary, this form of the equation holds if the proportions between the initial amplitudes $\bar{y}_{01}/\bar{y}_{02}/\ldots/\bar{y}_{0(n-1)}/\bar{y}_{0n}$ equal the proportions between the components of the eigenvector that belongs to the eigenvalue κ_1.

The rate of change of amplitude follows from eqn (8.62)

$$\frac{\mathrm{d}\bar{y}}{\mathrm{d}t} = \bar{y}_0\kappa_1 \frac{\sigma_r}{\mu_r} \exp\left(\kappa_1 \frac{\sigma_r}{\mu_r} t\right), \tag{8.63}$$

From eqns (8.61) and (8.63) one obtains

$$\frac{\mathrm{d}\lambda}{\mathrm{d}t} \approx -2\pi^2 \frac{\bar{y}_0^2}{\lambda} \kappa_1 \frac{\sigma_r}{\mu_r} \exp\left(2\kappa_1 \frac{\sigma_r}{\mu_r} t\right), \tag{8.64}$$

in which also $\bar{y} = \bar{y}_0 \exp(\kappa_1 \mu_r^{-1} \sigma_r t)$ has been incorporated, $\bar{y}$ being the amplitude at time t.

The rate of relative buckle-shortening may also be defined, thus:

$$\frac{1}{\lambda}\frac{\mathrm{d}\lambda}{\mathrm{d}t} \approx -2\pi^2 \left(\frac{\bar{y}_0}{\lambda}\right)^2 \kappa_1 \frac{\sigma_r}{\mu_r} \exp\left(2\kappa_1 \frac{\sigma_r}{\mu_r} t\right). \tag{8.65}$$

The most significant fact to note is that the velocity of relative buckle-shortening at a given deviatoric compressive stress is proportional to the eigenvalue of the system and proportional to the square of the amplitude/wavelength ratio as long as the deflections are gentle.

Homogeneous shortening (layer-shortening)

The mathematical definition of relative buckle-shortening may be compared with the expression for the rate of relative homogeneous shortening, that is the rate of uniform compressive strain normal to the axial plane of a gently folded layer, r. This strain rate is

$$\dot{\varepsilon} \equiv \frac{d\varepsilon}{dt} = \frac{1}{4}\frac{\sigma_r}{\mu_r}, \tag{8.66}$$

assuming plane strain in the plane normal to the fold axis. σ_r is compressive stress parallel to the general trend of the buckling layer. The stress normal to the layer is put equal to zero. The ratio between the two kinds of shortening is consequently

$$\begin{aligned}\frac{\text{rate of buckle-shortening}}{\text{rate of homogeneous shortening}} &= -8\pi^2\left(\frac{\bar{y}_0}{\lambda}\right)^2 \kappa_1 \exp\left(2\kappa_1 \frac{\sigma_r}{\mu_r} t\right) \\ &= -8\pi^2\left(\frac{\bar{y}}{\lambda}\right)^2 \kappa_1,\end{aligned} \tag{8.67}$$

remembering that κ_1 is a negative quantity. Expressed in words, this equation states that the numerically larger the quantity κ_1 (which increases numerically with increasing viscosity contrast between the layers and the surroundings) and the greater the ratio between the amplitude and the wavelength at any time, the more dominant is buckle-shortening relative to homogeneous shortening. For our purpose sufficiently accurate informations being obtained from the equations for $\bar{y}/\lambda$ ratios up to 0·1.

It is a condition of considerable consequence for structural geology that the buckle-shortening part of a deformation relative to the homogeneous-shortening part decreases in importance with decreasing numerical value of the principal eigenvalue of the system. The eigenvalue reflects both the viscosity contrast between layers and the ratio between the stress force and the gravity force. The less the viscosity contrast and—for non-buoyant systems—the less the compressive stress relative to the gravity-induced stress, $\sigma/\Delta\rho g h$, the smaller is the maximum on the principal eigenvalue in numerical terms, and consequently the less important is shortening due to buckling *sensu stricto* in comparison with shortening due to homogeneous strain. This effect shows up in small-scale structures as well as in structures of regional scale. In small- and medium-scale structures, layers which are not sufficiently competent relative to adjacent rocks will shorten and thicken rather than buckle in response to layer-parallel compressive stress. On regional and global scale we shall find that unless the lateral compressive force is large relative to the gravity force, crustal layers tend to shorten and thicken rather than buckle in response to compressive forces in the Earth's crust.

It is geodynamically significant that in some laterally compressed systems containing buoyant layer(s), increasing $\sigma/\Delta\rho g h$ ratio causes the principal eigenvalue to decrease and accordingly makes buckle-shortening less important. This will be demonstrated in a numerical test on p. 187 ff.

To study quantitatively the relationship between buckle-shortening and homogeneous shortening let us compare systems with unlike maximal principal eigenvalue on the following basis. In response to layer-parallel compression we let the amplitude in all cases increase by a certain defined amount, and then calculate how much shortening this defined amplitude increment requires in the various systems. To keep the mathematical approximation reasonably accurate the final amplitude must not be allowed to exceed 10% of the wavelength. On the other hand, it is not possible to start the buckling with ideally straight layers: it is necessary to commence the process with deflections whose amplitudes are finite but small. We select 0·01 for the initial amplitude/wavelength ratio and 0·1 for the final ratio in the buckling models to be compared.

The treatment below is not valid for systems containing buoyant strata, because in such systems waves develop without any kind of lateral shortening.

The rate of amplitude growth due to buckling *sensu stricto* is

$$\frac{\mathrm{d}\bar{y}_b}{\mathrm{d}t} = \bar{y}_0 \kappa \frac{\sigma}{\mu} \exp\left(\frac{\sigma}{\mu} t\right) = \kappa \frac{\sigma}{\mu} \bar{y} \tag{8.68}$$

where $\bar{y}$ is amplitude at time t and $\bar{y}_0$ the initial amplitude. Subscript b indicates that the equation refers solely to the buckling part of the deformation; σ is the compressive stress parallel to the layering. The equation is valid for single-layer systems or for multilayers provided that the amplitudes at the various boundaries are related as the components of the eigenvector that belongs to the principal eigenvalue.

As a consequence of the homogeneous strain alone, on the other hand, the rate of amplitude growth is

$$\frac{\mathrm{d}\bar{y}_H}{\mathrm{d}t} = \dot{\varepsilon}_y \bar{y} = -\dot{\varepsilon}_x \bar{y} = -\frac{\sigma}{4\mu} \bar{y}, \tag{8.69}$$

assuming non-rotational plane strain in the x, y plane and vanishing stress in the y-direction. Subscript H indicates that the expression refers solely to the homogeneous strain. Together the two different deformations add up to the resultant amplitude velocity

$$\frac{\mathrm{d}\bar{y}}{\mathrm{d}t} = \bar{y} \frac{\sigma}{\mu} (\kappa - \tfrac{1}{4}). \tag{8.70}$$

Integration yields

$$\ln \frac{\bar{y}}{\bar{y}_0} = (\kappa - \tfrac{1}{4}) \frac{\sigma}{\mu} t. \tag{8.71}$$

Since we wish to compare systems with final amplitute 10 times the initial amplitude (the latter being 1% of the wavelength), the following relationship is obtained

$$\frac{\sigma}{\mu}t = \frac{\ln 10}{\kappa - \frac{1}{4}} = \frac{2{\cdot}30}{\kappa - \frac{1}{4}}. \tag{8.72}$$

The systems are compared on basis of their principal eigenvalues. For systems with known κ-values the above equation specifies the magnitude of the term $(\sigma/\mu)t$ which is needed to produce a tenfold increase of the amplitude.

For any given system the viscosity is fixed, but the compressive stress as well as the time in the term $(\sigma/\mu)t$ can change without affecting the

Table 8.3. Wavelength shortening for systems with different κ-values associated with a tenfold increase of amplitude starting at a y_0/λ_0 ratio 0·01 (see text)

$-\kappa$	$-\frac{\sigma}{\mu}t$	$-\pi^2\left(\frac{y_0}{\lambda}\right)^2\left[\exp\left(2\kappa\frac{\sigma}{\mu}t\right)-1\right]$	$-\frac{1}{4}\frac{\sigma}{\mu}t$	$-\Sigma$ exp.[c]	λ_t/λ_0
0	9·20	0	2·30	2·30	0·100[a]
1	1·84	0·0381	0·4675	0·5056	0·603
2	1·022	0·0579	0·2555	0·3134	0·731
3	0·7076	0·0679	0·1769	0·2448	0·7828
4	0·541	0·0738	0·1352	0·20905	0·8113
5	0·438	0·0778	0·1095	0·1873	0·829
6	0·368	0·0807	0·0920	0·1727	0·8414
7	0·317	0·0825	0·07925	0·16175	0·8507
8	0·2788	0·0844	0·06970	0·15410	0·8570
9	0·2486	0·0856	0·06215	0·14775	0·8626
10	0·2244	0·0868	0·0561	0·1429	0·8668
15	0·15082	0·09007	0·03771	0·12777	0·8801
20	0·1136	0·091852	0·02840	0·120252	0·8867
30	0·07603	0·09353	0·019008	0·112537	0·89356
60	0·038174	0·095343	0·0095435	0·104886	0·90043
100	0·02294	0·096029	0·005735	0·101764	0·90324
200	0·01149	0·096800	0·0028725	0·099680	0·90512
500	0·004598	0·0970039	0·0011495	0·098153	0·9065
1000	0·0022994	0·0970873	0·0005745	0·097662	0·906955
10 000	0·00023	0·0971016	0·0000575	0·097159	0·90741
100 000	0·000023	0·0971016	0·000006	0·971016	0·90746
					0·90602[b]

[a] Wavelength shortening at pure homogeneous strain.
[b] Wavelength shortening at pure buckling according to equation (8.73) calculated for a change of amplitude from $\bar{y}_0 = 0{\cdot}01\lambda$ to $\bar{y}_t = 0{\cdot}1\lambda$.
[c] sum of columns 3 and 4.

eigenvalue. To produce the desired amplitude growth and thereby the required magnitude of the term $(\sigma/\mu)t$ we could either apply a high stress for a short time or apply a small stress for a long time because it is only the product σt which is fixed by the κ-value and the viscosity of each particular system. Table 8.3 lists corresponding values of κ and $(\sigma/\mu)t$ which, according to eqn (8.72), are required for a tenfold increase of the initial amplitude. There is a clear trend in the values: the smaller the numerical value of κ, the larger the numerical value of the term $(\sigma/\mu)t$. It will become clear from the treatment below that the reason for the need of higher stress—or longer time—to generate the tenfold amplitude rise in systems with numerical small κ-values is that in such systems greater shortening is required—due to the homogeneous-strain part—to produce the specified amplitude growth. (A point to note is that the eigenvalue is a function of the ratio $\sigma/\Delta\rho gh$ for gravity-controlled systems. To keep the eigenvalue unchanged while σ is altered, therefore, the gravity stress $\Delta\rho$gh must be changed proportional to the change in σ.)

For gentle buckles the rate of buckle-shortening in terms of relative shortening of the wavelength is

$$\frac{\mathrm{d}\lambda_b}{\lambda \mathrm{d}t} \approx -2\pi^2\left(\frac{\bar{y}_0}{\lambda_0}\right)^2 \kappa\frac{\sigma}{\mu}\exp\left(2\kappa\frac{\sigma}{\mu}t\right). \tag{8.73}$$

Here we have disregarded the small change of wavelength in the ratio between the initial amplitude and the instantaneous wavelength, i.e. $(\bar{y}_0/\lambda)^2$, see eqn (8.65). The latter has accordingly been replaced by the initial characteristic wavelength in the term $(\bar{y}_0/\lambda_0)^2$. This approximation is introduced to facilitate the later integration.

Due to homogeneous strain the rate of shortening of relative wavelength is

$$\frac{\mathrm{d}\lambda_H}{\lambda \mathrm{d}t} = \dot{\varepsilon} = \frac{\sigma}{4\mu}. \tag{8.74}$$

Adding the two velocities of relative wavelength-shortening and integrating between λ_0 at time zero and λ at time t, yield

$$\ln\frac{\lambda}{\lambda_0} \approx -\pi^2\left(\frac{\bar{y}_0}{\lambda_0}\right)^2\left[\exp\left(2\kappa\frac{\sigma}{\mu}t\right) - 1\right] + \frac{\sigma}{\mu}t. \tag{8.75}$$

To accomplish the required tenfold increment of the amplitude from an initial y_0/λ_0 ratio $= 0{\cdot}01$, there is a certain relation between the κ-value of the various systems and the term$(\sigma/\mu)t$ as given by eqn (8.72) and recorded in Table 8.3. By inserting corresponding κ and $(\sigma/\mu)t$ values in the above integrated wavelength equation (8.75) we find the change of wavelength associated with the tenfold increment of the amplitude. The relevant numbers are shown in Table 8.3. Here one also finds the wavelength-shortening in terms of the ratio λ/λ_0 for pure buckling and for pure homogeneous strain respectively, based on the tenfold amplitude growth, starting at an initial amplitude/wavelength ratio 0·01. For pure buckle-shortening the λ/λ_0 ratio is 0·91, and for pure homogeneous shortening the λ/λ_0 ratio if 0·10. It follows from the table that the folding is in the form of almost pure buckle-shortening when the κ-value numerically is say 60 and more, while the folding process is chiefly in the form of homogeneous shortening and thickening of layers when the κ-value approaches unity.

In a book devoted to the effect of gravity in tectonics we shall not discuss folding not controlled by gravity—important as this may be—but devote the remaining part of the chapter to a few numerical examples of gravity-controlled folding.

When studying numerical examples we realize that all theoretical buckling models are approximations to the real physical process. There exist several approximations which give slightly different results in terms of characteristic wavelength/thickness ratio (see discussion in Fletcher, 1977, 1979; Ramberg, 1979).

NUMERICAL MODELS OF GRAVITY-CONTROLLED VISCOUS BUCKLING*

Model 1: A single-layer Earth's crust

This model is chosen so as to imitate the crust of the Earth in its most simple motion. It consists of a viscous plate of thickness h_2, density ρ_2 and viscosity μ_2 overlying an infinitely thick substratum whose density is ρ_3 and viscosity μ_3. The top surface of the plate is free.

We have selected the following dimensionless ratios needed for the computation of the dominant wavelength:

$$\mu_{2:3} = 10^3 \qquad \rho_2 < \rho_3$$

$$\sigma_{2:3} = \mu_{2:3} \qquad \rho_2/(\rho_2 - \rho_3) = -9.$$

The latter value is a consequence of putting ρ_2 equal to 2·7 g cm^{-3} and $\rho_3 = 3$ g cm^{-3}. Other reasonable densities which give the same ratio include $\rho_2 = 2{\cdot}8$ g cm^{-3} and $\rho_3 = 3{\cdot}111$ g cm^{-3}.

* The numerical models in this section are based on velocity formulas for an infinite half-space slightly different from eqns (8.19) and (8.26).

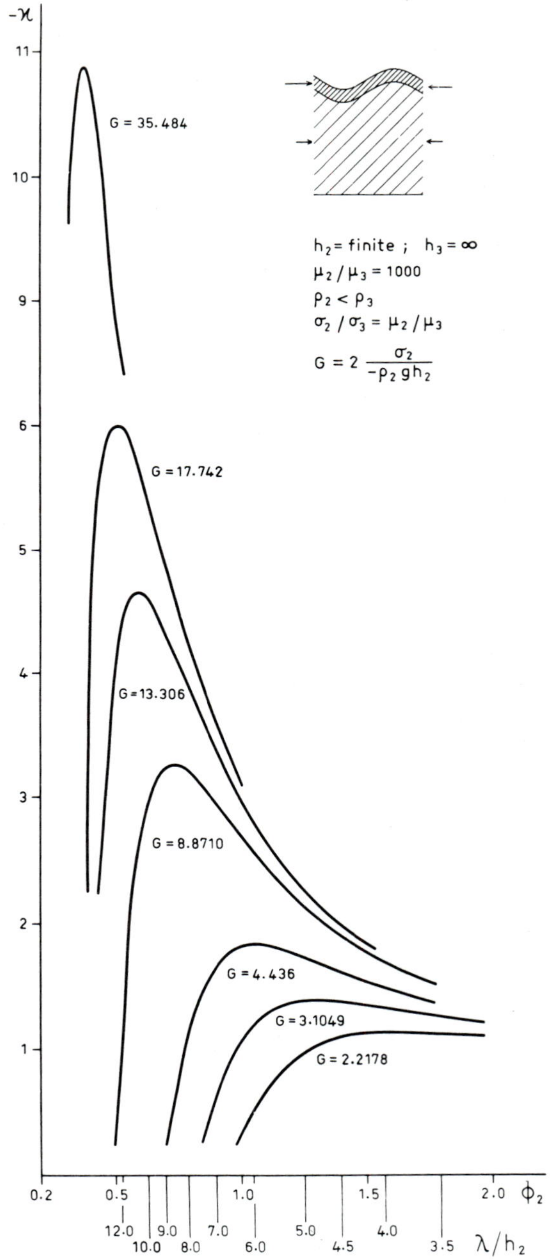

Fig. 8.13. Eigenvalue, κ_1, as a function of ϕ_2 and λ/h_2 at selected G-values for a simplified model of the Earth's crust. See text (after Ramberg, 1970b).

The ratio between the stress force and the gravity force in the model is defined thus

$$G = \frac{2\sigma_2}{-\rho_2 g h_2}, \tag{8.76}$$

since ρ_1 vanishes in this particular structure. See expressions for the coupling coefficients, Table 8.2.

Figure 8.13 contains the results of the κ v. ϕ calculations for a number of versions of the model in which all the dimensionless ratios are fixed at the values specified above except the force ratio G. As expected, the maximum on the function $\kappa = f(\phi_2)$ increases and becomes displaced toward larger wavelength/thickness ratios as the stress increases relative to the gravity force. In this system there is only one maximum at any given stress situation, signifying the occurrence of only one characteristic wavelength.

The results may be applied on a simple Earth's crust–mantle model consisting of a relatively competent crust say 16 km thick with density 2·7 g cm^{-3} and viscosity 10^{23} poise, resting on a simatic mantle whose upper part—which is affected by the sinusoidal strain—has a density 3·0 g cm^{-3} and viscosity 10^{20} poise, the latter selected to satisfy the ratio $\mu_{2:3}$ = 1000. The compressive stress—in excess of the lithostatic pressure—in the crust cannot exceed the ultimate strength if we wish the system to buckle rather than shatter and rupture. Thus, we have an upper limit on σ_2, say some 5×10^9 dyne cm^{-2}, being the deviatoric stress when the confining pressure is around 3000 bar. The G-value is determined accordingly

$$G = \frac{10^{10}}{2{\cdot}7\times10^3 \times 1{\cdot}6\times10^6} = 2{\cdot}315$$

(g is put equal to 1000 cm sec^{-2}).

The κ v. ϕ_2 curve to use in Fig. 8.13 is accordingly situated between the lowermost two curves, in fact very close to the lowest one. Hence the rather flat maximum falls on $\phi_2 \approx 1{\cdot}57$ and $\lambda/h_2 \approx 4{\cdot}0$. With 16 km as crustal thickness, the dominant wavelength becomes about 64 km. In accordance with the low numerical κ-value ($\approx -1{\cdot}0$) the compressing of the folds by homogeneous shortening rather than buckling is large, see below.

To obtain a larger characteristics wavelength, the G-value must be increased. An increment of G has also the effect of increasing the κ-value and accordingly making shortening by homogeneous strain less prominent. Increased G-value can theoretically be obtained by varying one or all of the quantities σ_2, ρ_2 and h_2, assuming that the variation in g is negligible.

It is hardly possible to do much with the density, because the values given cannot be significantly decreased if they are to represent the crust and the upper mantle. Anyway, if one wishes to change the density in order to obtain a larger value of G and consequently greater wavelength, one must remember that ρ_3 must be changed in proportion to ρ_2 if the information contained in Fig. 8.3 is still to be applicable to the model.

Similarly, a variation in σ_2 in the expression for G must be coupled to a proportional change of σ_3 because the ratio $\sigma_{2:3}$ must be constant in order to maintain dynamic similarity and to keep the information in Fig. 8.13 valid.

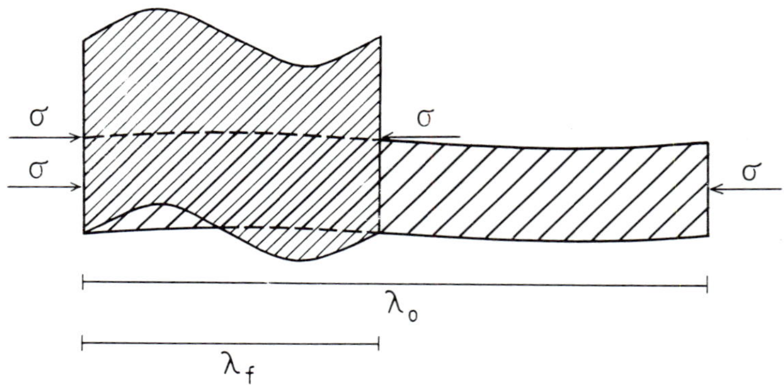

Fig. 8.14. Buckle shortening and layer shortening of an embedded layer when the viscosity of the layer is somewhat less than ten times that of the surroundings. Wide hatching: initial stage. Dense hatching: final state (after Ramberg, 1970b).

The situation is different with respect to h_2 in the expression for G. We can change h_2 as much as we like without the need of correspondingly changing other thicknesses in the model, because in the model considered there is only one finite thickness. In multilayers with several strata with finite thicknesses, any arbitrary change of h_2 in G must be associated with changes in all other finite thicknesses in order to maintain constant thickness ratios and unchanged dynamic similarity.

We have discussed earlier in this chapter how the shortening of buckling layers by homogeneous strain is increasingly important when the κ-value becomes smaller. At numerically small κ-values shortening and thickening mask much of the buckling process. When κ is unity, for example, the layer is compressed to only about 60% of its original length and correspondingly thickened, before the amplitude reaches 10% of the wavelength when initially being 1% of the wavelength of the initiating buckles (cf. Table 8.3 and Fig. 8.14).

From Table 8.3 we assume that κ should not be numerically less than 3 for well-developed buckles to form. From Fig. 8.13 we conclude then that G must exceed 8 in order that the crust should buckle rather than mainly thicken in response to a lateral compressive stress.

Since the lateral deviatoric compressive stress cannot be increased beyond the ultimate strength of the crust, and a lowering of crustal density below $2{\cdot}7$ g cm^{-3} is unrealistic, the only possible means of increasing G is to adopt a smaller crustal thickness. For example, putting $h_2 = 4$ km instead of the original 16 km, but maintaining ρ_2 and σ_2 as before, yields $G = 9{\cdot}26$. This value places maximum κ_1 at a ϕ_2-value somewhere between 0·6 and 0·7, corresponding to a characteristic λ/h_2 ratio between 9 and 11, i.e. more than twice the earlier wavelength/thickness ratio. Nevertheless, the dominant wavelength is considerably less than the previous one, viz. 40 km, because the crustal thickness is now only 4 km.

Incidentally, it is highly improbable that we can assume anything near 5×10^9 dyne cm^{-2} as the lateral compressive stress in excess of the lithostatic pressure in the Earth's crust. Weaker stresses give smaller G and shorter characteristic wavelengths and less buckling relative to uniform shortening and thickening.

The variation of maximum principal κ with G, as demonstrated by Fig. 8.13, has been treated by Biot (1959, 1961) and Ramberg (1964b, 1970) in other contexts. As these investigators have employed different theoretical methods it is particularly valuable to compare the results.

In Ramberg (1964b, 1970) the relationship between the ϕ-value for the characteristic wave and the stress-force/gravity-force ratio is given by

$$\tfrac{1}{2}G\rho_{2:3} = \frac{2\phi + 3\phi^{-2}\mu_{2:3}}{\phi - 3\mu_{3:2}} \tag{8.77}$$

in which we have introduced the symbols applied in this book and made some slight rearrangement of terms. In the cited work, the square root of $\frac{1}{2}G\rho_{2:3}$ was plotted versus the ϕ-value which coincided with maximum κ such as reproduced in Fig. 8.15. In this figure we also find the relationship between $\sqrt{(\frac{1}{2}G\rho_{2:3})}$, which is identical to $\sqrt{(\sigma_2/-\rho_3 g h_2)}$, and ϕ at maximum κ as taken from the graphs in Fig. 8.13.

Although the agreement between the results of the two quite different theoretical methods is excellent within the limited region of comparison, namely for $\mu_{2:3} = 1000$ and $\sqrt{(\frac{1}{2}G\rho_{2:3})}$ within the limits 1 and 4, there are reasons to expect greater deviation between the two methods at smaller $\mu_{2:3}$ ratios.

It is useful to consider the relationship between G and the ratio between the amplitudes, $\bar{y}_1$, at the free surface and $\bar{y}_2$ at the crust–mantle interface,

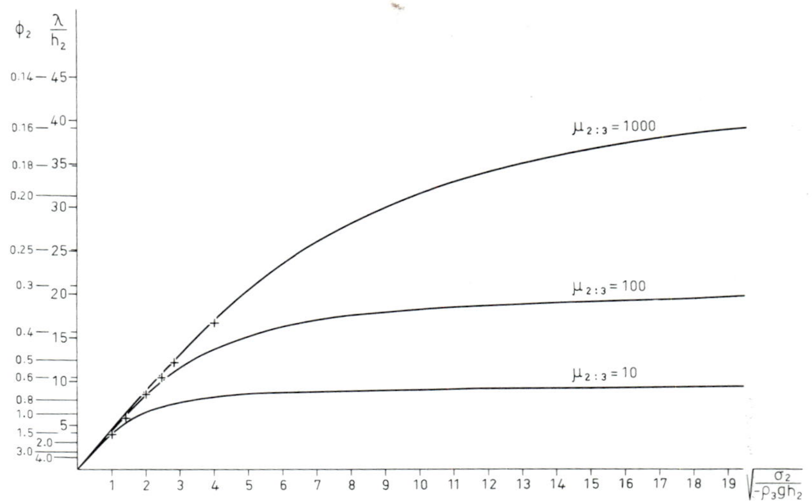

Fig. 8.15. Relationship between dominant wavelength and lateral-stress/gravity-stress ratio for buckling crustal layer shown at three selected $\mu_{2:3}$ ratios. Solid curves by eqn (8.77) crosses given by the maxima of the function $\kappa_1 = f(\phi_2)$ at various G-values as shown in Fig. 8.13. Note $\sqrt{\sigma_2/-\rho_3 g h_2} \equiv \sqrt{\frac{1}{2}G_{2:3}}$ (after Ramberg, 1970b).

as determined numerically and recorded in Table 8.4. Small G-values—which indicate low ratio between the horizontal stress force and the gravity force—denote low amplitude of the surface folds relative to that of the folds at the bottom of the crust. The amplitudes at the two boundaries become practically equal when $G \geqslant 8$. The amplitude ratio here considered

Table 8.4. Steadied amplitude ratio of buckled crustal layer

G	$\bar{y}_1/\bar{y}_2$	
	$\lambda/h_2 = 7{\cdot}85$	$3{\cdot}93$
0·2218	—	0·1785
0·4436	—	0·2863
1·3306	—	0·8204
1·7742	—	0·8894
2·2178	—	0·9210
3·1049	0·7132	0·9500
4·436	0·9603	0·9679
8·8710	0·9923	0·9854
13·306	0·9957	0·9906
17·742	0·9971	0·9931

is the steadied ratio which corresponds to the ratio between the components of the eigenvector that belongs to the principal eigenvalue.

This increase of $\bar{y}_1/\bar{y}_2$ from a small ratio to almost unity when G increases, occurs at any ϕ-value at which comparison is made. The relationship seems reasonable because of the large difference in density contrast at the two boundaries, $(\rho_1 - \rho_2) = -2{\cdot}7$ g cm^{-3} at the free surface and $(\rho_2 - \rho_3) = -0{\cdot}3$ g cm^{-3} at the crust–mantle interface. At small G-values, gravity is more important than lateral stress, and so the large negative density contrast at the surface prevents a rapid rise of the surface folds. At large G-values, the lateral stress-force overshadows the gravity-force, thus minimizing the consequence of any difference in density-contrast at the two boundaries and allowing the two amplitudes to rise with almost equal rate. The difference in amplitude ratio for the long waves and for the short waves is also informative. The increased importance of gravity with increasing wavelength shows up in the amplitude difference.

Model 2: A multilayer displaying two orders of folds

Model 2 exhibits a behaviour somewhat different from Model 1 inasmuch as it develops two orders of folds with different characteristic wavelengths. This is a variety of dynamic behaviour which is not rare in stratified columns of rocks.

The model consists of a 9-member multilayer with free surface, resting on an infinite half-space. The complex is exposed both to lateral compression and to the pull of gravity. The characteristic parameters are:

$$h_2 = h_4 = h_6 = h_8 = h_{10},$$

$$h_3 = h_5 = h_7 = h_9,$$

$$h_{11} = \infty,$$

$$h_{2:3} = \tfrac{1}{6},$$

$$\mu_2 = \mu_4 = \mu_6 = \mu_8 = \mu_{10},$$

$$\mu_3 = \mu_5 = \mu_7 = \mu_9,$$

$$\mu_{2:3} = 10, \; \mu_{2:11} = 100,$$

$$\rho_2 = \rho_3 = \ldots = \rho_{10},$$

$$(\rho_{10} - \rho_{11})/(-\rho_2) = 0{\cdot}111,$$

$$\sigma_2/\sigma_3/\ldots/\sigma_{10} = \mu_2/\mu_3/\ldots/\mu_{10}, \; \sigma_{11} = 0,$$

$$G \equiv \frac{2\sigma_2}{-\rho_2 g h_2}.$$

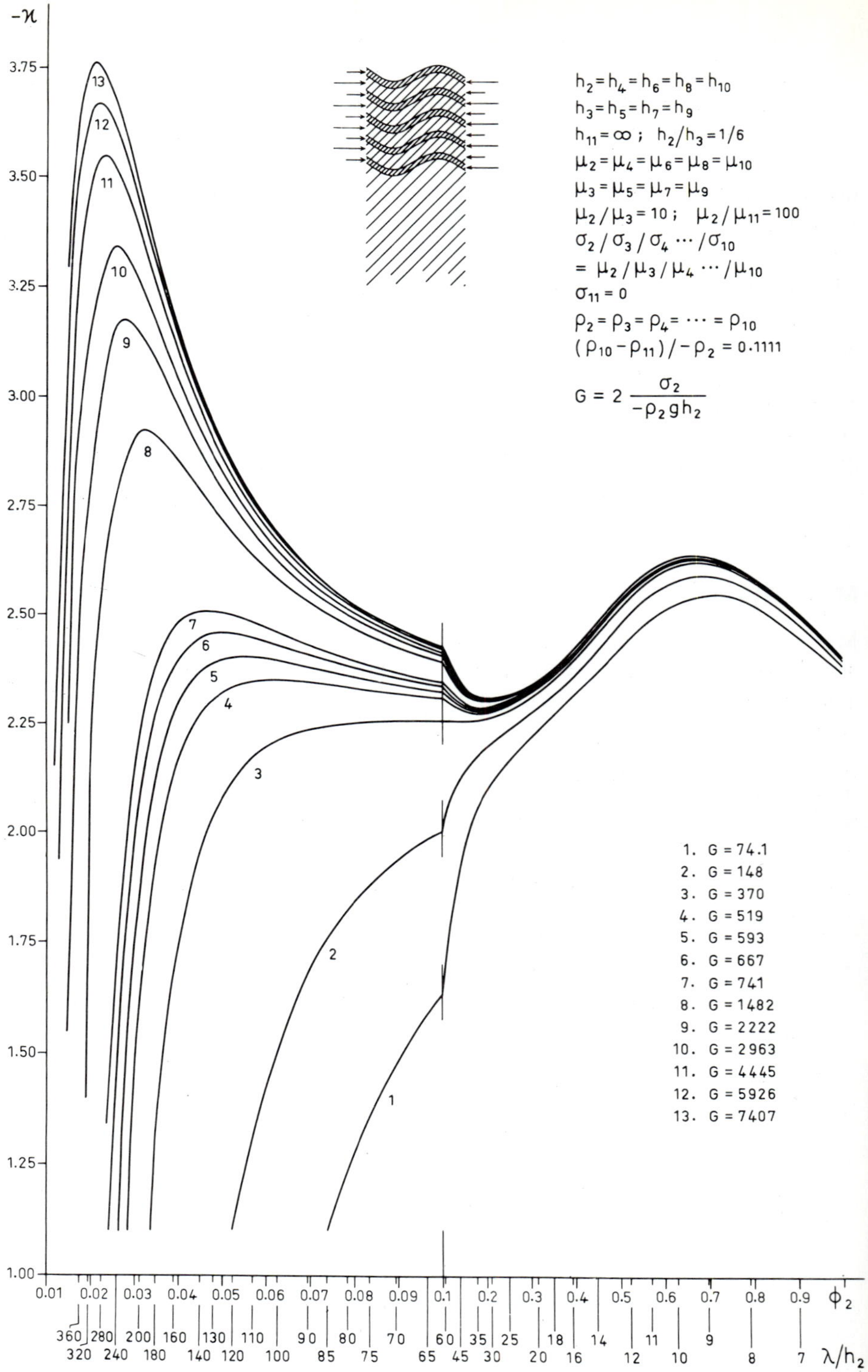

-κ
3.75
3.50
3.25
3.00
2.75
2.50
2.25
2.00
1.75
1.50
1.25
1.00
$h_2 = h_4 = h_6 = h_8 = h_{10}$
$h_3 = h_5 = h_7 = h_9$
$h_{11} = \infty$; $h_2/h_3 = 1/6$
$\mu_2 = \mu_4 = \mu_6 = \mu_8 = \mu_{10}$
$\mu_3 = \mu_5 = \mu_7 = \mu_9$
$\mu_2/\mu_3 = 10$; $\mu_2/\mu_{11} = 100$
$\sigma_2/\sigma_3/\sigma_4 \cdots /\sigma_{10}$
$= \mu_2/\mu_3/\mu_4 \cdots /\mu_{10}$
$\sigma_{11} = 0$
$\rho_2 = \rho_3 = \rho_4 = \cdots = \rho_{10}$
$(\rho_{10} - \rho_{11})/-\rho_2 = 0.1111$
$G = 2 \dfrac{\sigma_2}{-\rho_2 g h_2}$
1. G = 74.1
2. G = 148
3. G = 370
4. G = 519
5. G = 593
6. G = 667
7. G = 741
8. G = 1482
9. G = 2222
10. G = 2963
11. G = 4445
12. G = 5926
13. G = 7407
0.01 0.02 0.03 0.04 0.05 0.06 0.07 0.08 0.09 0.1 0.2 0.3 0.4 0.5 0.6 0.7 0.8 0.9 ϕ_2
360 280 200 160 130 110 90 80 70 60 35 25 18 14 11 9
320 240 180 140 120 100 85 75 65 45 30 20 16 12 10 8 7 λ/h_2

In this model member number 11—the basal infinite half-space—is not exposed to horizontal compression. However, numerical tests indicate that it ordinarily makes little difference in terms of dominant wavelength whether or not the basal half-space is exposed to a compressive stress.

Numerically determined κ v. ϕ_2 graphs are reproduced in Fig. 8.16. At low G-values only one maximum occurs, placing the characteristic λ/h_2-ratio at about 8·7 ($\phi_2 = 0{\cdot}72$). At higher ratios between the lateral stress force and the gravity force, i.e. at higher G-values, a second maximum occurs.

When $G = 370$, the second maximum is very flat and cannot be detected in the figure, but the digital result shows that it conincides with $\phi_2 = 0{\cdot}095$ and $\lambda/h_2 = 66{\cdot}1$. With increasing G the second maximum becomes more pronounced simultaneously as it moves toward larger wavelengths. The highest G-value studied places the maximum at about $\phi_2 = 0{\cdot}021$ which corresponds to $\lambda/h_2 \approx 300$. Contrary to this variation of the maximum at low values of ϕ_2, the maximum at $\phi_2 \approx 0{\cdot}7$ does not change markedly with changing G, neither in height nor in position on the ϕ_2-axis.

From the output data we conclude that at values of G below 370 only one order of buckles tends to form: a set with a wavelength/thickness ratio equal to about 8·7. At intermediate values of G, two orders of folds develop simultaneously: one set with $\lambda/h_2 \approx 8{\cdot}7$ and one with λ/h_2 larger than 66. When G is very large, say greater than 3000, the maximum at the long waves is so pronounced relative to that at the short waves that it is doubtful whether the latter actually would develop in experiments or in nature.

The variation of the amplitudes across the multilayer is particularly informative, cf. Table 8.5. Directing the attention first to the short buckles, we note that those on the top layer have by far the larger amplitude, the lowest layers being hardly folded at all (Fig. 8.17).

This distribution of amplitudes is physically plausible: since the wavelength is short, the flattening effect of gravity is weak, for which reason the density differences between layers or between air and the top layer, is of little consequence. The relative rate of short-wavelength buckling of the various layers is therefore chiefly controlled by the viscous resistance met when the more competent layers (i.e. numbers 2, 4, 6, 8, 10) deflect into the intercalated incompetent strata (i.e. 3, 5, 7, 9). This resistance is least for the top layer simply because its upper boundary is a free surface. An explanation of this nature goes well with the numerical results which show that the difference in amplitude between the top layer and those

---

Fig. 8.16. κ_1 as a function of ϕ_2 and λ/h_2 of compressed multilayer resting on an infinite half-space in the gravitational field. Note break in scale at $\phi_2 \approx 0.1$ (after Ramberg, 1970b).

Table 8.5. Steadied amplitude proportions of buckled multilayer

	$G = 74{\cdot}1$		$G = 741$		$G = 2963$		$G = 7407$	
	$\lambda/h_2 = 317$	9·67	$\lambda/h_2 = 317$	9·67	$\lambda/h_2 = 317$	9·67	$\lambda/h_2 = 317$	9·67
$\bar{y}_1$	0·032627	0·995128	0·892927	0·996945	0·969016	0·997082	0·969997	0·997109
$\bar{y}_2$	0·128562	1·000000	0·901246	1·000000	0·973262	1·000000	0·974174	1·000000
$\bar{y}_3$	0·655697	0·398540	0·944002	0·346477	0·991937	0·343123	0·992426	0·342463
$\bar{y}_4$	0·727062	0·390276	0·950880	0·337679	0·993935	0·334304	0·994355	0·333641
$\bar{y}_5$	0·988128	0·186379	0·980425	0·122552	0·999976	0·119610	1·000000	0·119041
$\bar{y}_6$	1·000000	0·184240	0·984110	0·119900	1·000000	0·116959	0·999963	0·116391
$\bar{y}_7$	0·877926	0·173529	0·998827	0·065398	0·994289	0·061907	0·993910	0·061249
$\bar{y}_8$	0·828232	0·175159	1·000000	0·065158	0·992359	0·061653	0·991929	0·060987
$\bar{y}_9$	0·422805	0·341874	0·999266	0·095902	0·974278	0·088781	0·973561	0·087451
$\bar{y}_{10}$	0·345998	0·340117	0·997838	0·095273	0·970167	0·088188	0·969407	0·086863

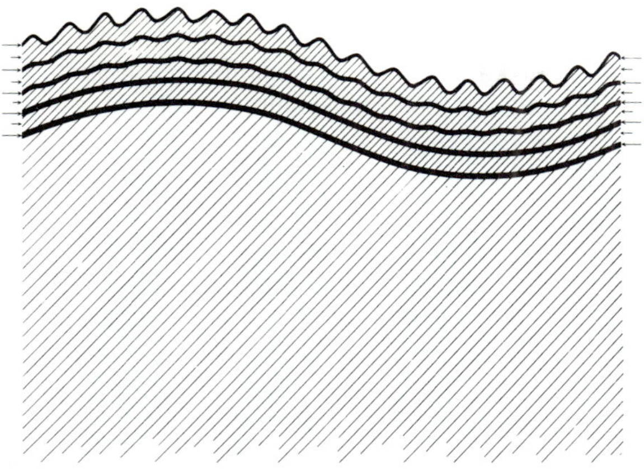

Fig. 8.17 Details of the multilayer shown in Fig. 8.16 and described in text. Note the change in amplitude of the second-order folds from top to bottom, see Table 8.5 (after Ramberg, 1970b).

below increases as the compressive stress is intensified as recorded by larger G-values, cf. Tables 8.5, columns $\lambda/h_2 = 9{\cdot}67$.

The long-wavelength buckles exhibit the same tendency as encountered in many other multilayers tested numerically, namely maximal amplitude at the central stata and a stepwise decrease toward the outermost layers. A certain detail is particularly revealing in this multilayer: the fact that the amplitude of the long waves at the top layer is less than the amplitude of the bottom layer (except at very large values of G), and, moreover, that this difference in amplitudes at the top and bottom diminishes and even becomes reversed when the lateral stress force increases relative to the gravity force, cf. Table 8.5. These relationships stand in marked contrast to the situation developed for the short-wavelength buckles. An explanation suggests itself: as a flattening force gravity is more effective for the long waves, hence the difference in density contrast, at the free surface and at internal boundaries, becomes a controlling factor. In spite of the relatively low viscous resistance offered against buckling of the top layer, the tendency of gravitational flattening is large enough to make the top layer buckle less rapidly than the deeper strata. Only if the lateral stress force becomes very large relative to the gravity force does the top layer buckle slightly faster than the layer at the bottom, Table 8.5, column $G = 7407$, $\lambda/h_2 = 317$.

Informative as the present model is with respect to dynamic behaviour it is too symmetric to simulate realistic rock complexes or global shells. Yet, it is interesting to attempt an application to a schematic and simplified pile of rocks, say a column of five competent sandstone strata interlayered with four less competent micaceous arkosic strata resting on a thick substratum of, say, shale or marl. Let the sandstone assume the properties h_2, μ_2 and ρ_2, the arkose the properties h_3, μ_3 and ρ_3, and finally the substratum the properties $h_{11} = \infty$, μ_{11} and ρ_{11}. The dimensionless ratios between these properties are as defined for the model, see above. Reasonable densities are $\rho_2 = \rho_3 = \ldots = \rho_{10} = 2{\cdot}52$ g cm^{-3} and $\rho_{11} = 2{\cdot}8$ g cm^{-3}, which agree with the condition $(\rho_{10} - \rho_{11})/(-\rho_2) = 0{\cdot}111$. The absolute values of the thicknesses are decided below.

Let us test how this structure behaves when exposed to a deviatoric lateral stress close to the ultimate strength of the rocks. The sandstone strata are exposed to the highest stress since they are the most competent members of the complex and the condition $\sigma_i/\sigma_{(i+1)} = \mu_i/\mu_{(i+1)}$ prevails. The ultimate strength of the sandstone is therefore decisive in determining the highest deviatoric stress which may be put on the system. An ultimate strength of the order 10^9 dyne cm^{-2} is reasonable for a sandstone at moderate confining pressure.

To avoid that buckling being masked by the homogeneous strain in response to a lateral compressive stress the eigenvalue must be numerically larger than about 3. According to the data presented in Fig. 8.16, the corresponding value for G is 1482. This puts an upper limit on the thicknesses which the multilayer can assume without loosing the tendency to buckle, thus:

$$1482 \leqslant G = \frac{2\sigma_2}{-\rho_2 gh} = \frac{2\times10^9}{-2{\cdot}52\times10^3 . h_2}, \tag{8.78}$$

which gives

$$h_2 \leqslant 5{\cdot}36\times10^2 \text{ cm} = 5{\cdot}36 \text{ m}.$$

Here g is put equal to 1000 cm sec^{-2}. The maximum thickness of the composite multilayer above the substratum is then

$$\Sigma h = 5 \times 5{\cdot}36 + 4 \times 6 \times 5{\cdot}36 = 155{\cdot}44 \text{ m}.$$

For $G = 1482$ the two maxima on the κ v. ϕ_2 function fall on $\phi_2 = 0{\cdot}032$ and $\phi_2 = 0{\cdot}65$, or $\lambda/h_2 = 196$ and $\lambda/h_2 = 9{\cdot}68$, respectively. The two characteristic wavelengths of the sandstone–arkose complex accordingly are 1051 m and 52 m when exposed to a lateral stress close to the ultimate strength of the rocks. A considerable homogeneous shortening parallel to the layering will also occur in agreement with the relatively small numerical value of κ. Thus, the final wavelengths will be well below the values arrived at above.

If we wish to reduce the shortening due to homogeneous strain, higher numerical values of κ must be sought.

However, this also means higher values of G and correspondingly smaller thicknesses, because the ultimate strength does not permit greater deviatoric compressive stress if cataclastic collapse is to be prevented. Nor is it realistic to apply different densities.

In real tectonic situations it is unlikely that the deviatoric horizontal compressive stress is as high as 10^9 dyne cm^{-2}, hence the G value will be less than 1482 and the wavelengths less than calculated above while homogeneous shortening and thickening of the complex becomes more dominating. If the horizontal deviatoric compressive stress is only around 3×10^8 dyne cm^{-2}, for example, and the thickness h_2 still 5·36 m, G assumes the value 444·2. Figure 8.16 shows that at this low G-value only the short-wavelength folds develop.

Below the depth of contact-strain penetration from the surface sheet of buckling sandstone the strain is essentially in the form of homogeneous shortening and thickening.

Model 3: A triple-layer system containing a buoyant sheet

As the last numerical buckling model, one with rather complicated dynamic behaviour has been selected. In this model the buckling process due to horizontal compression is combined with the buoyant rise of a low-density layer.

Three layers with thicknesses h_2, h_3 and h_4, effective viscosities μ_2, μ_3 and μ_4 and densities ρ_2, ρ_3 and ρ_4 overlie a thick buoyant layer whose relevant properties are h_5, μ_5 and ρ_5. The complex rests on a rigid base (Fig. 8.18).

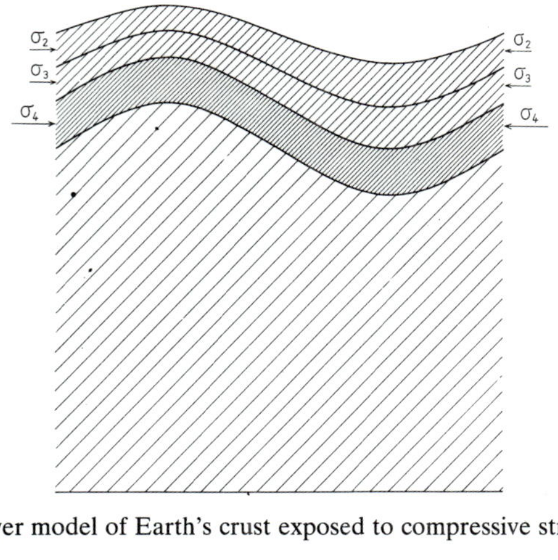

Fig. 8.18. Triple-layer model of Earth's crust exposed to compressive stress (after Ramberg, 1970b).

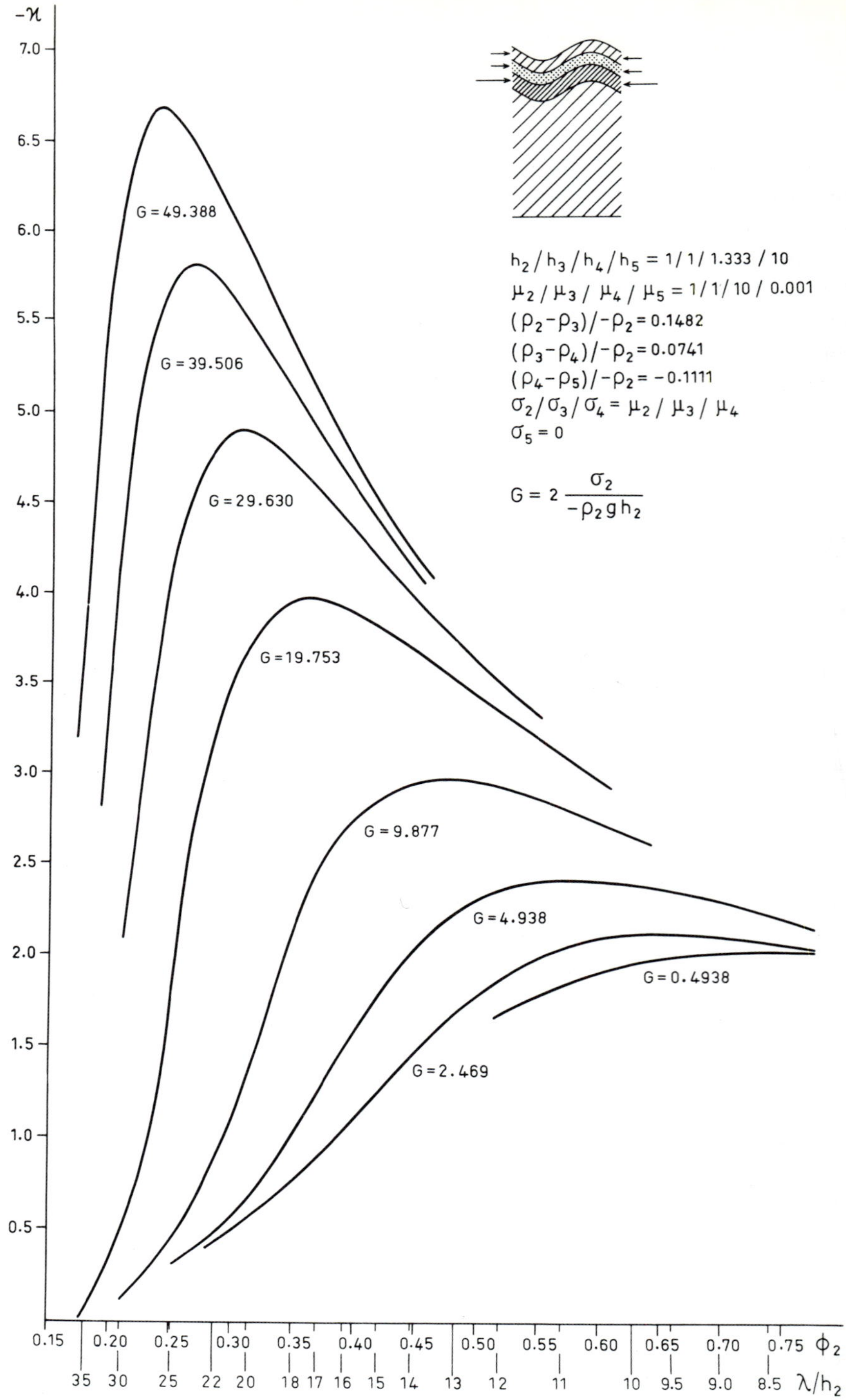

$-\varkappa$
7.0
6.5
6.0
5.5
5.0
4.5
4.0
3.5
3.0
2.5
2.0
1.5
1.0
0.5
G = 49.388
G = 39.506
G = 29.630
G = 19.753
G = 9.877
G = 4.938
G = 0.4938
G = 2.469
$h_2 / h_3 / h_4 / h_5 = 1/1/1.333/10$
$\mu_2 / \mu_3 / \mu_4 / \mu_5 = 1/1/10/0.001$
$(\rho_2 - \rho_3)/-\rho_2 = 0.1482$
$(\rho_3 - \rho_4)/-\rho_2 = 0.0741$
$(\rho_4 - \rho_5)/-\rho_2 = -0.1111$
$\sigma_2/\sigma_3/\sigma_4 = \mu_2/\mu_3/\mu_4$
$\sigma_5 = 0$
$G = 2 \frac{\sigma_2}{-\rho_2 g h_2}$
0.15 0.20 0.25 0.30 0.35 0.40 0.45 0.50 0.55 0.60 0.65 0.70 0.75 ϕ_2
35 30 25 22 20 18 17 16 15 14 13 12 11 10 9.5 9.0 8.5 λ/h_2

To select appropriate values for h, μ and ρ the model is regarded a simple imitation of the upper part of the Earth: a top granitic shell with free surface (layer 2), a basaltic shell (layer 3), the mantle above the low-velocity layer (layer 4) and the low-velocity layer itself (layer 5) resting on a mantle which is treated as rigid in the model.

The dimensionless ratios given below are based on equal thickness of the two crustal strata which also exhibit equal effective viscosity. The layer between the crust and the low-velocity layer is 30% thicker and shows a tenfold higher viscosity than the crustal layers. The low-velocity layer 5 is 10 times thicker than the granitic layer and exhibits a much smaller viscosity. The effect of the compressive stress on the low-velocity layer has been disregarded because of its low viscosity.

$$h_2/h_3/h_4/h_5 = 1/1/1{\cdot}33/10$$

$$\mu_2/\mu_3/\mu_4/\mu_5 = 1/1/10/0{\cdot}001$$

$$\sigma_2/\sigma_3/\sigma_4 = \mu_2/\mu_3/\mu_4; \qquad \sigma_5 = 0$$

$$-\rho_2/(\rho_2 - \rho_3)/(\rho_3 - \rho_4)/(\rho_4 - \rho_5) = -1/-0{\cdot}148/-0{\cdot}074/0{\cdot}11.$$

The irregular density-contrast proportions stem from the densities selected in agreement with reasonable global values, viz.

$$\rho_2 = 2{\cdot}7;\ \rho_3 = 3{\cdot}1;\ \rho_4 = 3{\cdot}3;\ \rho_5 = 3{\cdot}0,$$

all in g cm^{-3}. Of course, the value 3 g cm^{-3} for the low-velocity layer is questionable, but in lack of an empiric value it was arbitrarily selected in order to make the low-velocity layer buoyant. In all likelihood, however, the value is too low, giving too strong buoyant force in the model if it were to imitate a realistic Earth structure.

Figures 8.19 and 8.20 show the calculated principal eigenvalue as a function of ϕ_2 for a number of selected G-values. As in the previous models in this chapter the function $\kappa = f(\phi_2)$ is based on dimensionless parameters and is accordingly valid for any absolute magnitude of h, μ ρ and σ as long as their proportions are as specified.

If we wish to apply the model to the Earth appropriate absolute values must be choosen. With the densities already selected, reasonable values for the thicknesses, the viscosities and the horizontal compressive stresses must be choosen; 15 km each for the thicknesses of the granitic and the basaltic shells seems reasonable. Layer 4 then becomes 20 km thick, and

---

Fig. 8.19. Eigenvalue, κ_1, as a function of ϕ_2 and λ/h_2 of a compressed triple-layer Earth's crust. Shown for the region of G within which the dominant wavelength increases with increasing G-value (after Ramberg, 1970b). See text.

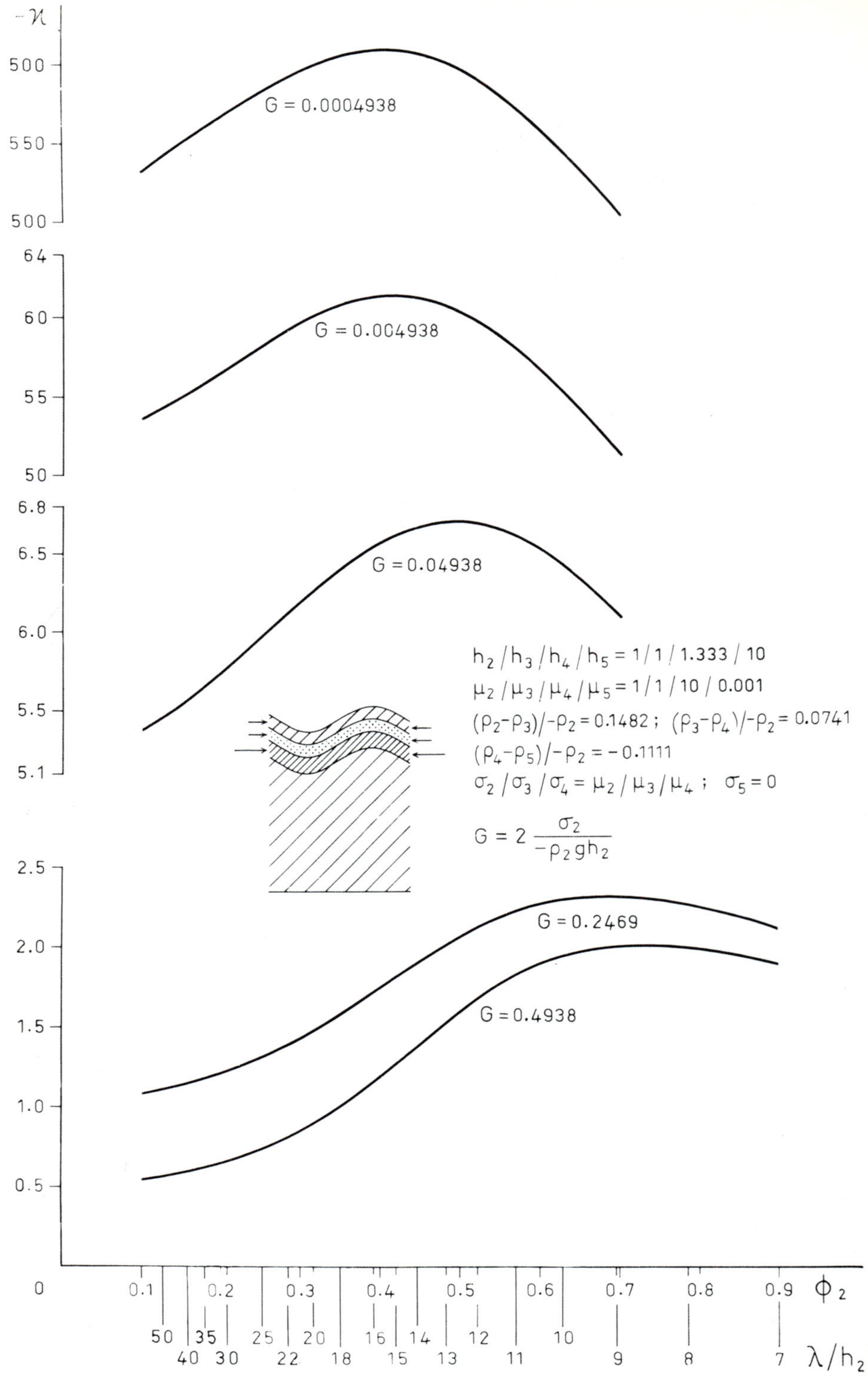

-κ
500
550
500
G = 0.0004938
64
60
55
50
G = 0.004938
6.8
6.5
6.0
5.5
5.1
G = 0.04938
h2 / h3 / h4 / h5 = 1 / 1 / 1.333 / 10
μ2 / μ3 / μ4 / μ5 = 1 / 1 / 10 / 0.001
(ρ2-ρ3)/-ρ2 = 0.1482 ; (ρ3-ρ4)/-ρ2 = 0.0741
(ρ4-ρ5)/-ρ2 = -0.1111
σ2 / σ3 / σ4 = μ2 / μ3 / μ4 ; σ5 = 0
G = 2 σ2 / -ρ2gh2
2.5
2.0
1.5
1.0
0.5
0
G = 0.2469
G = 0.4938
0.1 0.2 0.3 0.4 0.5 0.6 0.7 0.8 0.9
φ2
50 35 25 20 16 14 12 10
40 30 22 18 15 13 11 9 8 7
λ/h2

the low-velocity layer 150 km thick according to the thickness ratios specified.

Any selection of viscosity values will be very uncertain, but the following are as good as any: $\mu_2 = 10^{23}$, $\mu_3 = 10^{23}$, $\mu_4 = 10^{24}$ and $\mu_5 = 10^{20}$, all in poise. The horizontal deviatoric stress is equally uncertain; perhaps the following values are acceptable: $\sigma_2 = \sigma_3 = 5\times10^7$ dyne cm^{-2} (=50 bar). $\sigma_4 = 5\times10^8$ dyne cm^{-2} (=500 bar) and $\sigma_5 = 0$ as required by the dimensionless ratios.

Inserted in the formula for G these values give

$$G = 2\frac{\sigma_2}{-\rho_2 g h_2} = 2\frac{-5\times10^7}{-2{\cdot}7\times10^3\times15\times10^5} = 0{\cdot}0247. \tag{8.79}$$

This value of G coincides with a characteristic λ/h_2 ratio $\approx$ 14, or a characteristic wavelength equal to 210 km, according to the data in Fig. 8.21.

In the preceding buckling models the characteristic wavelength always increased when the G-value was augmented either by increasing the compressive stress, or by decreasing the density, respectively the layer thickness. The present model behaves less regularly. As long as the G-value is less than some 0·01, its effect on the wavelength is negligible, the length of the characteristic wave deviating very little from $14h_2$ according to Fig. 8.21. If the G-value is increased from 0·01 to about 0·5 the characteristic wavelength decreases from $14h_2$ to about $8{\cdot}5h_2$, while further increment of G makes λ rise to about $60h_2$.

At first glance the irregular behaviour seems puzzling. When considering the geometrical and mechanico-rheological properties of the model, however, its behaviour makes more sense.

First one notes that as the compressive stress vanishes the rise of layer 5 through the composite overburden is solely due to buoyancy. When $G = 0$, numerical calculation shows that the maximum on the $\kappa = \mathrm{f}(\phi_2)$ function coincides with λ/h_2 about 15. This, then, corresponds to the spacing between spontaneously rising diapirs or ridges in case no horizontal deviatoric stress acts on any of the layers.

Considered in relation to the thickness of the buoyant layer, the characteristic wavelength is only 1·50 times the thickness. This is noteworthy when compared with a somewhat similar model treated on p. 100. The latter exhibits a characteristic ratio between wavelength and buoyant-layer thickness not less than 10. This considerable difference in characteristic

---

Fig. 8.20. Eigenvalue, κ_1, as a function of ϕ_2 and λ/h_2 at different values of G of a compressed triple-layer Earth's crust. Shown for the region of G within which the dominant wavelength decreases with increasing G-value (after Ramberg, 1970b). See text.

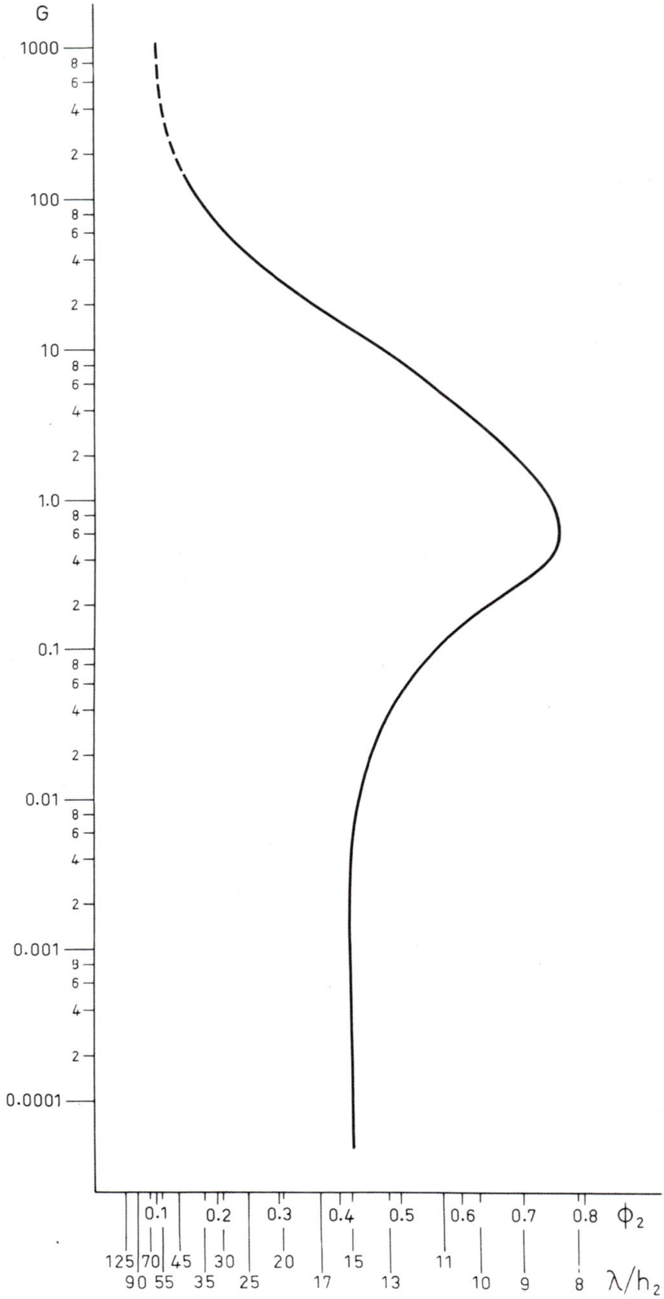

Fig. 8.21. Relationship between *G* and dominant wavelength of the model shown in Figs 8.19 and 8.20 (after Ramberg, 1970b).

wavelength is probably related to the unlike thickness of the overburden of the two models. In the present model the composite overburden is only one third as thick as the buoyant layer (viz. 3·333/10), while in the previous model the corresponding ratio is 1·2/1.

Suppose that the wavelength was the same in the two models; then the perturbed strain originating at the base of the overburden would have penetrated more strongly to the surface—produced larger amplitude at the surface—in the thin-overburden model than in the one with thick overburden. The gravity potential associated with the surface waves tends always to shorten the wavelength, but much more so for the thin-overburden model than for the one whose overburden is thick.

Although the condition here emphasized is important, there are other dissimilarities between the two models which contribute to the difference in characteristic wavelength, such as the viscosity contrast and the density contrast. A further circumstance of consequence is the relatively high mobility of the mantle below the low-velocity layer in the model treated on p. 100, in contrast to the present model whose mantle is rigid.

Returning to the discussion of the relationship between G and the characteristic wavelength for the present model it is seen in Figs 8.19 and 8.21 that at very large G-value the characteristic wavelength tends asymptotically to a certain λ/h_2-ratio which digital data show to be about 60. Large G-values mean that the effect of the compressive stress overshadows the gravity effect irrespective of whether the latter is of a buoyant or flattening nature. It is only to be expected, then, that the characteristic wavelength increases from $\lambda/h_2 \approx 15$, when the compressive stress is negligible, to $\lambda/h_2 \approx 60$, when the gravity force is negligible.

It is less clear why increasing G-value in the region $G = 0{\cdot}01$ to $0{\cdot}4$ is associated with a shortening of the characteristic wavelength. We find it reasonable that this is an effect of layer 4 which is the most competent member of the structure. The curves in Figs 8.19–8.21 can be interpreted to indicate that the laterally compressed model, if not under the influence of gravity, tends to develop two sets of buckles, one with a λ/h_2 ratio around 8 and one with a λ/h_2 ratio around 60. The short characteristic wavelength is essentially due to the high-viscosity layer number 4, while the long characteristic wavelength is due to the concerted tendency of layers 2, 3 and 4 to buckle as a unit because all three layers are more competent than the subjacent thick low-velocity layer.

What happens when gravity and a lateral compressive stress act together on the system is as follows. At moderate value of the stress, giving G-values in the neighbourhood of 0·1, an increase of the lateral stress tends to produce the short-wavelength folds characteristic for the buckling of the competent layer 4. Being short, these folds generate only gentle deflection on the surface and thus do not increase significantly the gravity potential. For this reason the folds develop even at moderate compressive stress as expressed in the form of moderate G-values.

Because the wavelength characteristic for these buckles is less than the wavelength for the buoyant domes or ridges, an augmented G-value is here locally associated with a shortening of the waves, from $\lambda/h_2 \sim 15$ to $\lambda/h_2 \sim 8$.

As the stress increases, the long waves, which are characteristic for the concerted buckling of all three uppermost layers as a unit, start to form but not until the stress has reached considerable intensity because of the large increment of the gravity potential associated with the surface deflection caused by the long waves. Numerical calculation demonstrates that as G changes from about 0·4 to about 400 the long folds gradually change their wavelength to the value characteristic for the system if it was not affected by gravity.

9 Spreading and Recumbent Folding under the Force of Gravity

INTRODUCTION

Bucher (1956) reported a series of interesting experiments simulating the role of gravity in certain phases of orogenic evolution, particularly the evolution of alpine-type recumbent anticlines such as the Pennine nappes.

He utilized the ability of stitching wax to flow and spread under its own weight even when occurring in pieces smaller than a few cubic millimetres. In one set of experiments horizontal layers of stitching wax alternating with layers of soft petrolatum would buckle in front of an expanding lobe of a spreading block of wax (Figs 9.1 and 9.2). This buckling of surface layers

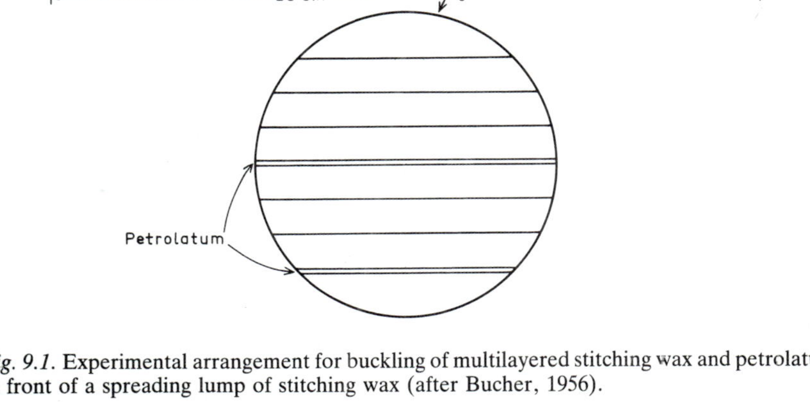

Fig. 9.1. Experimental arrangement for buckling of multilayered stitching wax and petrolatum in front of a spreading lump of stitching wax (after Bucher, 1956).

in front of spreading bodies is similar to the results of some centrifuged model experiments (Chapter 15). Whereas Bucher started with an elevated body of wax to give a mechanical "head" to drive the flow and folding, the elevated part is itself spontaneously produced in the centrifuged models; in these experiments domes and diapirs rise spontaneously, thus giving the "head" needed to generate the folding. That is, the centrifuged models have the mechanical instability—the mechanical "head"—originally located within the interior of the models in the form of unstable density distribution.

In a second series of tests Bucher had multilayered bars of stitching wax of different colours compressed slowly. (For example, $0{\cdot}6\,\mathrm{cm\,hr^{-1}}$ = rate of displacement of one end of a 50-cm-long bar.)

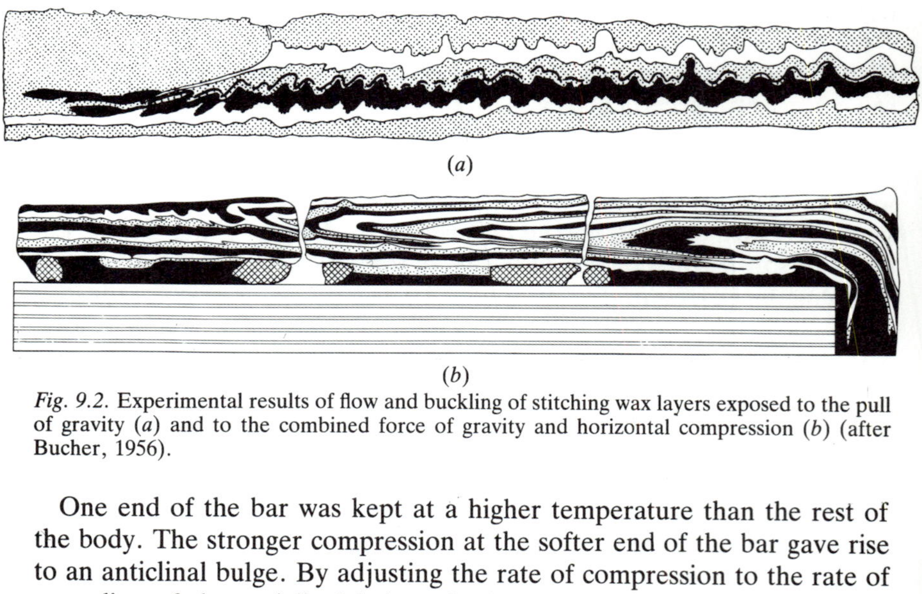

Fig. 9.2. Experimental results of flow and buckling of stitching wax layers exposed to the pull of gravity (*a*) and to the combined force of gravity and horizontal compression (*b*) (after Bucher, 1956).

One end of the bar was kept at a higher temperature than the rest of the body. The stronger compression at the softer end of the bar gave rise to an anticlinal bulge. By adjusting the rate of compression to the rate of spreading of the anticlinal bulge, the latter never rose much but rather spread horizontally in pace with the rising movement. A recumbent fold was thus produced by simple lateral compression (Fig. 9.2). Evolutions of this kind seem likely to occur on a large scale in tectonics. Bucher considered that the huge nappes so typical for many orogenic belts were produced by a mechanism akin to the above.

Again, the similarity to some of the centrifuged models is quite remarkable (see, e.g., Figs 15.13 and 15.14). In the centrifuged models the motivating power of the evolution is, however, in a sense more fundamental since no extra lateral stress is added.

The very rise of the anticline—not only its lateral spreading—is a spontaneous process in the centrifuged models, the propelling force being related to the density contrast between the dissimilar model materials. In Bucher's experiments the production of the anticline in the first place had to be accomplished by a lateral compressive stress whose origin remains to be explained. In the author's models, to be described below, there was no net lateral shortening.

In the course of the progress of the Swedish contribution to the international geodynamics project, The Caledonian Research Project, I became engaged in the question of the mechanisms of large-scale nappe-and thrust-sheet motion and the power behind these truly hard-to-understand phenomena.

The mechanical dilemma has long been recognized (cf. the discussion by Reade, 1908 and 1909; Fisher, 1909; Smoluchovski, 1909; Hubbert and Rubey, 1959; Oftedahl, 1943; Ramberg, 1977b; Chapman, 1979; and the compilation by Voight, 1976). The dilemma is that a pushing force in the back of a thrust sheet has to be so strong that it crushes the hind part of the sheet rather than causing it to move *en bloc* across its base. See, however, Wiltschko (1979) for a divergent opinion.

Under certain conditions the friction at the base may perhaps be small enough to allow *en bloc* movement without demanding a pushing stress in excess of the crushing strength. This may happen when the sole is lubricated by a layer of rock salt, a graphite-rich schist or some other exceptionally soft rocks. The Hubbert–Rubey model of high pore pressure, which gives the thrust sheet a sort of a "Hovercraft" lift, may occasionally also be applicable, but these will be exceptional situations unlikely to exist wherever thrust sheets are moving. It seems that the field experience from a number of orogenic belts demands a theory that is independent of these special circumstances.

Observations through many field seasons in the Scandinavian Caledonides by numerous workers, amongst whom we mention Törnebohm (1888, 1896), Oftedahl (1943, 1948, 1956), Kautsky (1953), Kvale (1945). Gee (1975 a and b, 1977), Zwart (1974), Zachrisson (1964), Williams and Zwart (1977), Ramberg (1977a, 1980)—have revealed structures which in my view must be taken as evidence that plastic flowage has played a major role during the movement of the nappes, not only in the mylonite zones of strain concentration, but, what is more significant, also within the bulk of some of the nappes, especially the hugh Seve nappe.

Within large regions of the Seve nappe the structure does not indicate the kind of strain to be expected if the nappe has been forced forward by a push from behind. To the contrary, the structure shows that there has been a shortening in the vertical direction and a volume-preserving lengthening in the horizontal plane in a sense which parallels the direction of movement of the nappe. This information from field observations, coupled with the results of numerous centrifuged models of spreading lobes of diapirs (e.g. Ramberg and Sjöström, 1973) and Bucher's (1956) inspiring experiments on collapsing anticlines, suggests that gravity collapse and complementary spreading must constitute an essential part of the dynamics of thrust sheets and recumbent-fold nappes. In an important paper Price (in De Jong and Scholten, 1973) treats large-scale gravity flow in the Canadian Rockies. The recent comparison between ice structures and structures in rocks by Huddleston (1977), is also very suggestive. The model is well known among glaciologists, who have accepted it as a natural explanation of the movement of ice sheets since the early 1950s (Nye, 1952).

Elliott (1976) applied the ice-sheet model to orogenic thrust sheets. He was mainly concerned with the shear stress that would theoretically develop at the base of thrust sheets depending upon the thickness of the sheet, its density and the slope angle of its surface. This formula is the same as the ice-sheet formula for the basal shear stress:

$$\tau = \rho g h_0 tg\alpha. \tag{9.1}$$

Here τ is the stress at the contact between the ice and the bedrock, ρ is the density of the ice, h_0 its thickness, α the slope angle at the surface, and g the acceleration due to gravity. In the above formula the base of the ice is considered horizontal, but by a small modification the formula also becomes applicable to an inclined base, emphazising the noteworthy condition that the ice may also move uphill. It is the slope on the free surface which determines the direction of motion, not so much the slope of the base.

DEVELOPMENT OF THE BASAL SHEAR-STRESS EQUATION

The basal-stress equation is unexpectedly simple in its form, and it is equally straightforward to develop, following the argument of the glaciologists, e.g. Nye (1952). We shall develop the formula in order to gain some understanding of the dynamics of the model.

Let a thin vertical column, Δx units thick, be marked out in the ice sheet (or nappe) whose surface is inclined, the body being very extensive in the direction normal to the x,y plane in relation to the dimensions l and h_0 in the x,y plane, Fig. (9.3). We consider plane strain in x,y.

Assuming lithostatic conditions, i.e. $\sigma_x = \sigma_y = \sigma_z$, and incompressibility, the horizontal normal stresses vary as the vertical normal stress down

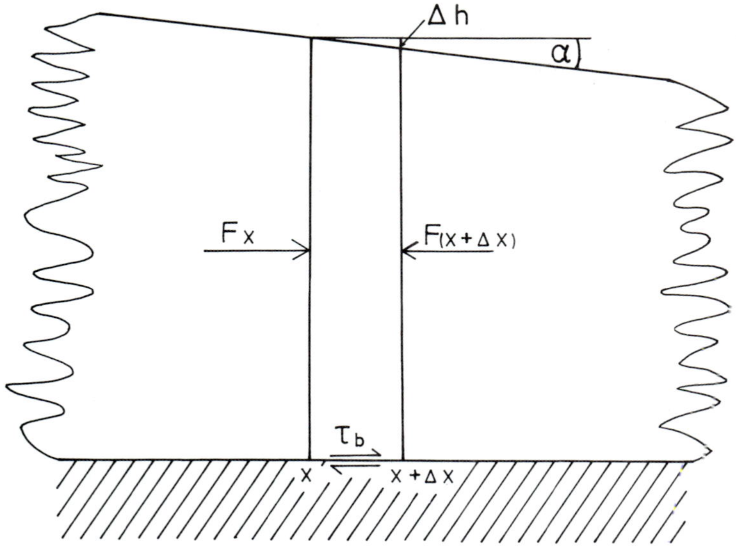

Fig. 9.3. Forces acting on an imaginary column through part of nappe (or ice sheet) with horizontal base and inclined surface.

through the bulks of the ice, hence:

$$\sigma_x = \sigma_z = \sigma_y = \rho g h \tag{9.2}$$

where h is the depth from the surface to the level in question.

The total horizontal force acting on the left-hand side of the column is then

$$F_x = \rho g \int_0^{h_0} h \mathrm{d}h = \tfrac{1}{2}\rho g h_0^2 \tag{9.3}$$

Here h_0 is the thickness of the ice sheet (or orogenic nappe) at x. On the right-hand side of the column the horizontal force is

$$F_{(x+\Delta x)} = \rho g \int_0^{(h_0-\Delta h)} h/\mathrm{d}h = \tfrac{1}{2}\rho g\,[h_0^2 - 2(\Delta h)\,h_0 + (\Delta h)^2]. \tag{9.4}$$

The resultant force which tends to push the column toward the right in the direction of the dip of the surface is then

$$\Delta F = F_x - F_{(x+\Delta x)} = \tfrac{1}{2}\rho g\,[2(\Delta h_0)h_0 - (\Delta h_0)^2]. \tag{9.5}$$

This force is balanced by the shear stress acting at the base along the segment Δx; consequently

$$\tau \cdot \Delta x = \tfrac{1}{2}\rho g[2(\Delta h)h_0 - (\Delta h)^2]. \tag{9.6}$$

When Δh_0 and Δx approach zero the above equation reduces to:

$$\tau = \rho g h_0 \frac{\partial h_0}{\partial x} = \rho g h_0 tq\alpha \qquad (9.7)$$

which is the celebrated ice-sheet formula.

Attractive though the equation is in its simplicity, it unfortunately cannot be entirely correct as is readily detected by considering the distribution of normal stresses down in the plastically collapsing ice sheet or nappe. The reason for the forward motion of the body along a horizontal base or uphill against a sloping base is the plastic shortening in the vertical direction and the complementary horizontal lengthening. In other words, the forward motion stems from the gravity collapse and the gravity spreading. But this means that the compressive stress in the vertical direction must be greater than the compressive horizontal stress in the direction of movement. The latter may even be tensile, as it in fact often is near the surface of the ice sheet (see discussion below). Since now σ_x, when compressive, is numerically less than σ_y, the condition of hydrostatic stress down in the ice is not satisfied, instead the following inequality prevails

$$|\sigma_x| < |\sigma_y| = \rho g h. \qquad (9.8)$$

Here the vertical lines signify numerical values.

This leads to the conclusion that the basal shear stress must be numerically less than given by formula (9.7) which is based on hydrostatic conditions. It is not possible to give an exact expression for the deviatoric stress down in the ice sheet, but it seems to be related to the length/thickness ratio (the slenderness ratio) of the ice sheet in such a fashion that the longer the sheet in relation to its thickness the closer to hydrostatic conditions are the stresses in the central part of the sheet and the more exact is formula (9.7) as an expression for the average shear stress at the base.

SEMIQUANTITATIVE THEORY OF VISCOUS COLLAPSE

Let us consider qualitatively what happens in a viscous body which spreads under the force of gravity while it rests on a rigid substratum, e.g. a pile of nappes in an orogenic belt. Experimental models which support the analysis are described on pp. 219–226.

Because the vertical compressive stress increases with depth, the velocity of horizontal lengthening also tends to increase with depth. It follows that shear is created parallel to the horizontal plane in the pile of nappes, because of drag between the actively spreading deep layers and the more passive layers above.

In the semiquantitative analysis of the processes here outlined, we shall assume that the rocks behave as Newtonian substances during the orogenic evolution, or at least that their yield strength is very small in response to stresses which act over millions of years. For stresses of magnitude between the small yield stress and the ultimate strength of the rocks the rate of change of strain is considered nearly proportional to the deviatoric stress. In the theoretical analysis we moreover assume that the deviatoric stresses in the nappes are limited to the interval between the small yield strength and the ultimate strength of the rocks. In natural thrust sheets, of course, the stress not infrequently reaches the ultimate strength of the rocks in question. The latter is defined as the maximum stress which a rock can withstand without discontinuous disruption.

In connection with the analysis that follows the papers by Weertman (1957, 1962 and 1979) should be consulted.

Model with free slip along base

First the effect of pure shear alone will be discussed. This condition is obtained if the model rests on an ideally frictionless sole. The situation is unrealistic as a tectonic model, but the analysis is important for the understanding of the more realistic and more complex situation encountered when the nappes are welded to the basement along the sole and consequently simple shear becomes an important part of the strain.

(For an experimental study of the free-slip model we have found it practical to let a plastic or viscous mass (e.g. silicone putty) rest on a layer of mercury. The great contrast in density causes the silicone putty to float almost on the top of the mercury, and the low viscosity of the mercury means that the friction at the base is negligible relative to the viscous internal friction in the spreading mass.)

Consider a rectangular block which is long in the z direction in relation to the width in the direction x; y is the vertical co-ordinate (Fig. 9.4). Because of the elongate shape of the parallelepiped the strain will be two-dimensional in the plane x,y. The contact between the body and its base is frictionless, and there is no shear stress or strain at the three free surfaces of the body. For the present study the air pressure is insignificant so the normal stress at the free surfaces can be disregarded.

To determine the strain rate and the velocities u and v in the x- and y directions, respectively, the equation valid for Newtonian substances under pure shear is relevant:

$$\dot{\varepsilon}_x = \frac{\sigma_x - \sigma_y}{4\mu}. \tag{9.9}$$

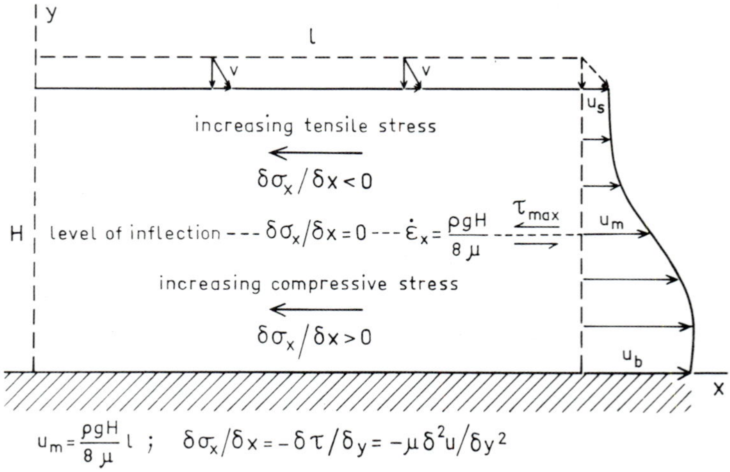

Fig. 9.4. Right-hand half of profile of symmetric spreading rectangular body with free slip along base. Stippled lines indicate initial cross-section; solid line and curve show outline at an early stage of spreading.

Here $\dot{\varepsilon}$ is the strain rate in the direction x, μ the effective viscosity, and σ the normal stress.

Since the normal stress, σ_{x0}, at the sides is zero and furthermore the basal contact is frictionless, we assume tentatively that σ_x at any point in the body vanishes (we shall see below that that is not exactly true). The normal stress in direction y at any point in the body equals the weight of the overburden:

$$\sigma_y = -\rho g(h_0 - y) \tag{9.10}$$

where ρ is density, g acceleration due to gravity, h_0 thickness, and y distance from the base. (Tensile stress is defined as positive, compressive stress as negative.)

The strain rate at a level at level y is accordingly:

$$\dot{\varepsilon}_x = \frac{\rho g(h_0 - y)}{4\mu}. \tag{9.11}$$

The strain is considered homogeneous at any one given level so the velocity in the x-direction is:

$$u = \dot{\varepsilon}_x x = \frac{\rho g(h_0 - y)x}{4\mu} \tag{9.12}$$

where x is the distance from the central cross-section of the body.

Since $v = 0$ at $y = 0$ at all x, the vertical velocity component at level y is:

$$v = \int_0^y -\dot{\varepsilon}_x\, y = -\frac{\rho g}{4\mu}(h_0 y - \tfrac{1}{2}y^2) \tag{9.13}$$

in which the condition of incompressibility: $\dot{\varepsilon}_y = -\dot{\varepsilon}_x$ is applied. The velocity v is thus independent of x in the free-slip model.

According to the above discussion the horizontal spreading of the rectangular body increases in speed the deeper the level, maximum spreading occurring at the bottom, while the uppermost boundary, at which σ_x as well as σ_y vanish, does not spread at all.

As already indicated this simple linear variation of the strains and stresses cannot be strictly correct, because the increase of the velocity u with depth must be associated with a shear strain and a corresponding shear stress, τ_{yx}, in the direction x. However, the existence of a component of simple shear along x in the horizontal plane shows that x is not a principal direction of strain and hence that the intital assumption of a pure-shear model is not exact. The fast-spreading deep strata in the body create a horizontal pull on the slow-spreading strata above which therefore spread somewhat faster than they would otherwise have done. On the other hand the horizontal shear stress diminshes the spreading rate of the layers below. The net result is that the surface spread more rapidly than expected according to the simple model, while the bottom spread more slowly than expected.

The condition that the normal stress σ_y is zero at the surface while $\dot{\varepsilon}_x$ is positive here leads to a significant conclusion: the stress σ_x within the top layer is tensile. Moreover, since the bottom layer spreads less than expected on the basis of the simple model we conclude that σ_x does not vanish in a layer along the frictionless bottom. At the vertical sides at $x = l$, however, σ_x does vanish. We accordingly conclude that the gradient $\partial\sigma_x\partial x$ is not zero in the model.

A consequence of the lack of shear stress at the top surface and at the frictionless base is that the front boundary of the original rectangular body close to the top as well as close to the bottom must be normal to the base and normal to the top surface, respectively, also after the deformation. It follows that the originally vertical and plane sides become sigmoid-shaped as indicated on Fig. 9.4.

The velocity profile for the component u is also sigmoidal. This deviation from linearity of the velocity profile agrees with the conclusion that the stress σ_x has a gradient in the x direction because the term $\partial^2 u/\partial y^2$ in the equation below does not vanish

$$\partial\sigma_x/\partial x \approx -\mu(\partial^2 u/\partial y^2). \tag{9.14}$$

Equation (9.14) follows from the condition of equilibrium between the stresses at two-dimensional strain in the plane x,y. At equilibrium between the normal stress gradient in the x-direction and the shear stress τ_{xy} we have

$$\partial\sigma_x/\partial x = -\partial\tau_{yx}/\partial y \tag{9.15}$$

when the motion is so slow that inertia is negligible, and the body force of gravity does not occur in the equation because x lies in the horizontal plane. For Newtonian materials the relation

$$\tau_{yx} = \mu\dot{\gamma}_{yx} \tag{9.16}$$

is valid. Moreover, $\dot{\gamma}_{yx} = \partial u/\partial y + \partial v/\partial x$. In the model $\partial v/\partial x$ is considered much smaller than $\partial u/\partial y$, hence the approximation $\dot{\gamma}_{yx} \approx \partial u/\partial y$ is permitted and the relationship $\partial\sigma_x/\partial x = -\mu\partial^2 u/\partial y^2$ is obtained.

Below the level of inflection in Fig. 9.4 the term $\partial^2 u/\partial y^2$ assumes negative values which requires that σ_x increases from $x = 0$ in the middle of the body, to the edge at $x = l$. Remembering that compressive stress is negatively defined, this means that the stress becomes less compressive as one goes from $x = 0$ to $x = l$. From the condition $\sigma_x = 0$ at $x = l$ follows that σ_x becomes increasingly compressive from the edge to the central section of the collapsing body. Above the level of inflection the situation is reversed: $\partial^2 u/\partial y^2$ is positive and the stress is tensile at all x, changing from zero at $x = l$ to a maximum value at $x = 0$ and $y = h_0$.

Moreover the strain is essentially pure shear within a zone close to the top surface and within a zone close to the bottom. At the bottom the pure shear is caused by the compressive stress, σ_y, being larger than σ_x, which is also compressive; at the top the strain is caused by the stress σ_x being tensile while the stress σ_y vanishes. In the middle zone of the body, on either side of the level of inflection where $\partial^2 u/\partial y^2 = 0$, there is a combination of pure and simple shear, simple shear reaching maximum value at the level of inflection.

The condition that $\partial^2 u/\partial y^2 = 0$ at a level which is close to the depth $y = \frac{1}{2}h_0$ is significant, since we here see a simple way of determining the velocity of collapse and spreading of the body. At the level of inflection where $\partial^2 u/\partial y^2 = 0$, eqn (9.14) shows that $\partial\sigma_x/\partial x$ vanishes and hence that σ_x is zero at all x. It follows that the longitudinal strain rate at the level of inflection is $\dot{\varepsilon}_x = (\rho g h)/(4\mu)$ where h is the depth of the level of inflection. Experiments with Newtonian substances performed by R. Häll in the Uppsala laboratory show that $h \approx \frac{1}{2}h_0$. (see p. 120). Accordingly, the mean strain rate is $\dot{\varepsilon}_x = (\rho g h_0)/(8\mu)$ and the velocity at the front is $u = (\rho g h_0 l)/(8\mu)$.

The above comments on the ideal free-slip model help to clarify the more realistic model in which there is coherence between the basement and the nappe.

Model coherent to the base

A condition valid for the coherent-base model is that $u = 0$ at $y = 0$ for all values or x. The same holds for v. A condition valid for both models is that the only driving force is gravity; any horizontal shear stress which may exist is created by the gravitational force.

The gravitational collapse and the complementary spreading, combined with the coherence at the bottom, require that u increases from zero at the base to a maximum value at a certain level. Above that level the horizontal

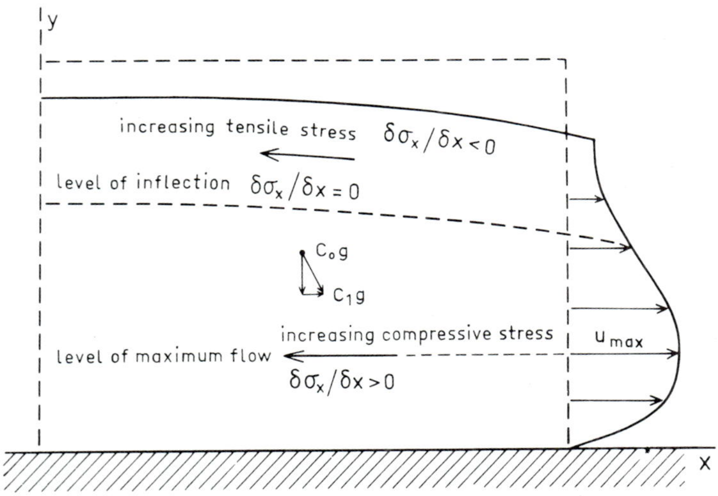

Fig. 9.5. Right-hand half profile of symmetric spreading rectangular body coherent to the base. Stippled lines show initial cross-section; solid curves outline shape at an early stage of spreading.

velocity component must again decrease because of the drag from the superincumbent layers similarly as explained for the free-slip model. Hence, at the front edge the velocity profile is as shown on Fig. 9.5.

Vanishing shear stress at the free surface requires that at the surface $\partial u/\partial y + \partial v/\partial x = 0$. Accordingly—since $\partial v/\partial x$ must be numerically small as compared with $\partial u/\partial y$—there must be an inflection point on the velocity profile somewhere between the top surface and the level of maximum velocity, $u_{\max}$. It is probable that the level at which $\partial^2 u/\partial y^2 = 0$ is located approximately midway between the free surface and the level of maximum horizontal velocity. From the level of maximum velocity to the top surface the coherent-base model is dynamically quite similar to the free-slip model.

It is interesting that the bulge at the level of maximum velocity is well documented by experiments, see below. The discredit which the theory of "extrusion flow" in ice-caps has suffered among glaciologists is therefore not always warranted (see Shumskii, 1964, p. 329). It is true, however, that the level of maximum velocity and the associated bulge is located at higher and higher levels in the collapsing body the greater its length relative to its thickness. In other words, the "extrusion flow" decreases in significance the larger the slenderness ratio. For many ice-caps the process of extrusion may therefore in practice be of little consequence. The reason is simple: the larger the distance between the level of maximum velocity and the sole, the less the shear-strain rate for a given velocity, $u_{\max}$. Furthermore, the longer the nappe relative to its thickness, the more shear-strain energy is dissipated per unit volume and unit time for a given maximal longitudinal strain rate, $\dot{\varepsilon}_{x\max}$. For long and thin nappes and ice-caps, therefore, there is a strong tendency for the level of $\dot{\varepsilon}_{x\max}$ (and $u_{\max}$) to develop close to the surface. This minimizes the shear-strain energy.

Experiments indicate that for slenderness ratios above 4 the level of maximum velocity is very close to the surface.

The velocity profile for the component u suggests that within the whole region from the zone of inflection to the base there is a gradient $\partial\sigma_x/\partial x$ so directed that σ_x at $x = 0$ is compressive all the way from the bottom to the level of inflection. It is safer to say "suggests" because in the coherent-base model $\partial v/\partial x$ is not insignificant, and the correct relation between $\partial\sigma_x/\partial x$ and the velocity variations is

$$\partial\sigma_x/\partial x = -\mu(\partial^2 u/\partial y^2 + \partial^2 v/\partial x\partial y) \tag{9.17}$$

rather than the approximation $\partial\sigma_x/\partial x = \mu\partial^2 u/\partial y^2$ used in the free-slip model. Even in the coherent-base model, however, the term, $\partial^2 v/\partial x\partial y$ is probably small relative to $\partial^2 u/\partial y^2$.

Above the inflection level at $x = 0$ the stress σ_x is tensile in the coherent-base model, just as it is in the free-slip model.

Incidentally, the conclusion that horizontal tensile stress exists at the surface layers of a spreading orogenic nappe and an active ice-cap is in good agreement with the distribution of crevasses in ice caps. For many orogenic nappes the erosion level is too deep to enable this theoretical result to be tested in the field. Numerous experimental model tests, however, verify that the top zone of a spreading body is under tensile stress.

The circumstance that the gradient $\partial\sigma_x/\partial x$ below the level of inflection is steeper in the coherent-base model (the term $\partial^2 u/\partial y^2$ is numerically larger) than in the free-slip model, and moreover that the stress gradient occurs through a thicker zone in the former, have bearing on the contrast in flow behaviour of the two models. The horizontal compressive-stress component is higher and accordingly the rate of vertical sagging is less in the interior of the coherent-base model than in the interior of the free-slip model. At the front at $x = l$, on the other hand, the stress difference $\sigma_x - \sigma_y$ is practically the same in both models. In the frontal part then, both models sag with practically the same strain rate. It follows that the coherent-base model will develop a gentle anticlinal- or dome-shaped profile while the free-slip model spread out as a sheet with practically even thickness. This is well displayed by the experimental models described below (see Figs 9.6–9.10).

Note also the difference in the profiles of ice-caps spreading on land and of shelf ice spreading while floating on the sea. There is a certain friction at the sole of continental ice-caps, hence their typical gentle dome-shaped surface. The lack of sole friction for shelf ice explains why the surface remains straight and the thickness even while the ice sheets spread. See also, e.g. Weertman (1957, 1979).

Application of the stream function to the problem of gravitational viscous collapse

In an attempt to treat the process of gravity collapse quantitatively, i.e. to obtain exact values for the velocities, the strain rates and the stresses, we choose to employ the powerful stream-function method.

A stream function valid for non-inertial two-dimensional slow flow (so-called creeping motion) is defined as a solution of the biharmonic equation

$$\nabla^4 \equiv \partial^4\psi/\partial x^4 + 2\partial^4\psi/\partial x^2\partial y^2 + \partial^4\psi/\partial y^4 = 0 \qquad (9.18)$$

from which the velocity components are derived thus; $u = -\partial\psi/\partial y$ and $v = \partial\psi/\partial x$. Stream functions have the property of automatically satisfying both the condition for continuity—which for incompressible media is $\partial u/\partial x + \partial v/\partial y = 0$ —and the conditions for dynamic equilibrium as expressed in the equations of motion in plane viscous flow, in which inertia is negligible:

$$\begin{aligned} \partial P/\partial x &= \mu(\partial^2 u/\partial x^2 + \partial^2 u/\partial y^2) \\ \partial P/\partial y &= \mu(\partial^2 v/\partial x^2 + \partial^2 v/\partial y^2) - \rho g. \end{aligned} \qquad (9.19)$$

This means that the dynamics in terms of instantaneous velocities, instantaneous strain rates and instantaneous stress distribution is solved for a creeping-motion problem, such as, e.g. a collapsing nappe, if a stream function can be found which satisfies the boundary conditions of the model.

As the first step toward finding a suitable form of a stream function which satisfies the problem at hand we consider the movement pattern in relation to a co-ordinate system whose origin is located in the middle of the base line of the model (Fig. 9.5). Only models coherent to the base are treated. In models of this kind the velocity component u vanishes at all values of x when $y = 0$ and at all values of y when $x = 0$.

At a vertical section at any given distance from origin the velocity profile of the component u is indicated by the deformed shape of the markers in the experimental models (Figs 9.9–9.11). The thus recorded velocity profile can be approximated by a polynomial in y of form

$$u = 2ay + 3by^2 + 4cy^3 + \ldots + (n+1)qy^n \tag{9.20}$$

with suitable coefficients. (The reason for using the integers 2, 3 etc in front of the coefficients will become clear later.)

Since $u = 0$ at $y = 0$ for all values of x there is no place for a constant in the polynomial. Furthermore, the condition that $u = 0$ at $x = 0$ for all values of y implies that the polynomial in y must be multiplied by x and/or by different powers of x.

At this point we note that geometric symmetry of the model allows only odd powers of x to be used (because u is always positive when x is positive and negative when x is negative). We are accordingly led to a function for u of form

$$\begin{aligned} u = {} & (2ay + 3by^2 + 4cy^3 + \ldots + (n+1)qy^n)x \\ & + (2a_1y + 3b_1y^2 + 4c_1y^3 + \ldots + (n+1)q_1y^n)x^3 \\ & + (2a_2y + 3b_2y^2 + 4c_2y^3 + \ldots + (n+1)q_2y^n)x^5 \\ & + \ldots + (2a_my + 3b_my^2 + 4c_my^3 + \ldots + (n+1)q_my^n)x^{(2m+1)}. \end{aligned} \tag{9.21}$$

From the relation $u = -\partial\psi/\partial y$ we obtain the corresponding ψ-function

$$\begin{aligned} \psi = {} & -(ay^2 + by^3 + cy^4 + \ldots + qy^{(n+1)})x \\ & - (a_1y^2 + b_1y^3 + c_1y^4 + \ldots + q_1y^{(n+1)})x^3 \\ & - (a_2y^2 + b_2y^3 + c_2y^4 + \ldots + q_2y^{(n+1)})x^5 \\ & - \ldots - (a_my^2 + b_my^3 + c_my^4 + \ldots + q_my^{(n+1)})x^{(2m+1)}. \end{aligned} \tag{9.22}$$

(Incidentally, the reason for using the integers 2, 3, etc., in front of the coefficients in the velocity polynomial is now evident: it yields simple coefficients in the stream function.)

For the polynomial (9.22) to satisfy the requirement of a stream function, i.e. to be a solution of the biharmonic equation $\nabla^4 = 0$, the coefficients must be related. After much cumbersome arithmetic the following correlations between the coefficients are determined

$$\begin{aligned}\psi = & -(ay^2 + by^3 + cy^4 + dy^5 + ey^6 + fy^7 + gy^8 + hy^9 + iy^{10} + jy^{11})x \\ & + (cy^2 + 1{\cdot}6667dy^3 + 3{\cdot}3333ey^4 + 4{\cdot}6667fy^5 + 7gy^6 + 9hy^7 \\ & + 12iy^8 + 14{\cdot}6667jy^9)x^3 \\ & - (ey^2 + 2{\cdot}333fy^3 + 7gy^4 + 12{\cdot}6hy^5 + 25{\cdot}2iy^6 + 39{\cdot}6jy^7)x^5 \\ & + (gy^2 + 3hy^3 + 12iy^4 + 26{\cdot}4jy^5)x^7 \\ & - (iy^2 + 3{\cdot}66667jy^3)x^9. \end{aligned} \tag{9.23}$$

(The reader may verify that this is a solution of $\partial^4\psi/\partial x^4 + 2\partial^4\psi/\partial x^2\partial y^2 + \partial^4\psi/\partial y^4 = 0$ by repeated partial differentation.)

Here we have arbitrarily terminated the polynomial in y at y^{11} which also implies that the highest permitted power of x is x^9. The relation between the highest power of y employed and the corresponding highest power of x permitted is a consequence of the stream function being a solution of the biharmonic equation.

The coefficients in the principal polynomial—which is the one multiplied by x in the first power—are still mutually unrelated notwithstanding the fact that the whole function (9.23) is a solution of the biharmonic equation. However, all coefficients in the polynomials which are multiplied by x^3, x^5, etc., are expressed in terms of the coefficients in the principal polynomial. Hence, it is only necessary to evaluate the coefficients in the principal polynomial in order to determine completely the stream function.

The as yet undetermined coefficients can be evaluated partly by the boundary conditions of the model—e.g. in our model by the circumstance that the shear stress vanishes at the free surface—and partly by equating the gravitational energy with the strain energy, coupled with a very powerful energy requirement, namely that the *time rate of energy dissipation shall assume maximum value.*

The precision of a polynomial stream function in terms of representing the true dynamics of a model increases with increasing degree of the polynomial. Even so, the lengthy labour of calculating numerous coefficients and the general awkwardness of handling high-degree polynomials make it desirable to reduce the number of terms as much as a reasonable demand of accuracy permits.

It appears that a collapse model of the kind under study is reasonably well represented by a stream function whose maximal power of x is not more than 3. The stream function to use is accordingly

$$\psi = -(ay^2 + by^3 + cy^4 + dy^5)x + (cy^2 + 1{\cdot}66667dy^3)x^3 \tag{9.24}$$

which gives the following expressions for the velocity components

$$u = (2ay + 3by^2 + 4cy^3 + 5dy^4)x - (2cy + 5dy^2)x^3 \tag{9.25}$$

and

$$v = -(ay^2 + by^3 + cy^4 + dy^5) + (3cy^2 + 5dy^3)x^2. \qquad (9.26)$$

To determine the coefficients we first make use of the vanishing shear strain at the free surface at $y = H$ of the initially rectangular collapsing slab. The shear-strain rate follows from the equation

$$\dot{\gamma}_{xy} = \partial u/\partial y + \partial v/\partial x$$

where

$$\partial u/\partial y = (2a + 6by + 12cy^2 + 20dy^3)x - (2c + 10dy)x^3$$

and

$$\partial v/\partial x = (6cy^2 + 10dy^3)x.$$

Accordingly,

$$\dot{\gamma}_{xy} = (2a + 6by + 18cy^2 + 30dy^3)x - (2c + 10dy)x^3 \qquad (9.27)$$

$\dot{\gamma}_{xy} = 0$ at all values of x when $y = H$, therefore:

$$(2a + 6bH + 18cH^2 + 30dH^3)x - (2c + 10dH)x^3 = 0. \qquad (9.28)$$

To satisfy the condition that $\dot{\gamma}_{xy} = 0$ at all values of x the two parentheses must vanish separately. From the second parenthesis we obtain

$$d = -0{\cdot}2cH^{-1},$$

which inserted in the first parenthesis yields

$$c = -\frac{a + 3bH}{6H^2}.$$

Expressed in terms of a and b, d becomes

$$\frac{a + 3bH}{30H^3}.$$

There are accordingly only two unknown coefficients left in the expressions for u and v, which now take the forms

$$u = \left(2ay + 3by^2 - \frac{2}{3}\frac{a + 3bH}{H^2}y^3 + \frac{1}{6}\frac{a + 3bH}{H^3}y^4\right)x + \left(\frac{1}{3}\frac{a + 3bH}{H^2}y - \frac{1}{6}\frac{a + 3bH}{H^3}y^2\right)x^3 \qquad (9.29)$$

and

$$v = -\left(ay^2 + by^3 - \frac{1}{6}\frac{a + 3bH}{H^2}y^4 + \frac{a + 3bH}{30H^3}y^5\right) - \left(\frac{1}{2}\frac{a + 3bH}{H^2}y^2 - \frac{1}{6}\frac{a + 3bH}{H^3}y^3\right)x^2. \qquad (9.30)$$

The absolute and relative values of a and b are evidently related to the velocity distribution and the associated distribution of strains in the collapsing slab. What then controls the velocity distribution? Here the same principles are valid as those implied in the development of the dominant wavelength of buckle folds (Chapter 8) and the preferred spacing of domes (Chapter 7), viz. *the process will take the path characterized by maximum rate of dissipation of energy, coupled with the side condition that the rate of gravity-potential decrement equals the rate of viscous strain-energy dissipation*. The latter condition is valid in the present case since the sole energy input is gravitational.

Let us determine the strain energy first. The viscous strain energy consists of two terms, viz. the shear-strain energy $\mu\dot{\gamma}^2_{xy}$ and the longitudinal-strain energy $4\mu\dot{\varepsilon}^2_x$ (or $4\mu\dot{\varepsilon}^2_y$).

The energy contribution from the shear strain integrated over the whole cross section of the half-slab is

$$\dot{E}_\gamma = \mu\int_0^l\int_0^H \dot{\gamma}^2\partial x\partial y \qquad (9.31)$$

where $\dot{\gamma} = \partial u/\partial y + \partial v/\partial x$ as derived from eqns (9.29) and (9.30).

The longitudinal strain contributes by the amount

$$\dot{E}_2 = 4\mu\int_0^l\int_0^H \dot{\varepsilon}^2_x\partial x\,\partial y. \qquad (9\ 32)$$

Here $\dot{\varepsilon}_x = \partial u/\partial x$ as obtained from eqn (9.29).

Integrated, rearranged and added together, the two strain-energy rates give the following total dissipation rate for a cross-section slice of unit thickness:

$$\begin{aligned}\dot{E}_{\varepsilon\gamma} = \mu[&(3{\cdot}799647r + 1{\cdot}59365r^3 + 0{\cdot}21333r^5 + 0{\cdot}005291r^7)H^4a^2\\ &+ (4{\cdot}025397r + 8{\cdot}342857r^3 + 0{\cdot}32r^5 + 0{\cdot}047619r^7)H^6b^2\\ &+ (4{\cdot}34071r + 6{\cdot}22857r^3 + 1{\cdot}01333r^5 + 0{\cdot}61904761r^7)H^5ab].\end{aligned} \qquad (9.33)$$

Here $r = l/H$ is the slenderness ratio or the aspect ratio of the half-part of the symmetrical model.

For later use it is practical to express eqn. (9.33) more compactly as

$$\dot{E}_{\varepsilon\gamma} = \mu(AH^4a^2 + BH^6b^2 + CH^5ab). \qquad (9.34)$$

Here, A, B and C are defined by the identity of eqns (9.33) and (9.34).

The rate of which the gravity potential changes during the collapse is given by the integral

$$\dot{E}_p = \rho g \int_0^l \int_0^H v \, \partial x \, \partial y \tag{9.35}$$

where v is the vertical velocity component expressed in eqn (9.30). Integration and rearrangement yield

$$\dot{E}_p = -\rho g[(0{\cdot}30556r + 0{\cdot}0416667r^3)H^4 a + (0{\cdot}166667r + 0{\cdot}1250r^3)H^5 b], \tag{9.36}$$

or expressed more compactly for later use:

$$\dot{E}_p = -\rho g(\alpha H^4 a + \beta H^5 b). \tag{9.37}$$

(Note that both $\dot{E}_{\varepsilon\gamma}$ and $\dot{E}_p$ are reckoned per unit thickness of a cross-section slice. To get the correct dimension of the energies, $\dot{E}_{\varepsilon\gamma}$ and $\dot{E}_p$ must be multiplied by the unit of length, cm or m as the case may be.)

The rate of energy change is to be maximized under the subsidiary condition that $\dot{E}_p = \dot{E}_{\varepsilon\gamma}$. In other words, we seek the maximum on the intersection curve between the two surfaces representing the two energies in the co-ordinate system $a,b,\dot{E}$. In an orthogonal co-ordinate system with the axes a and b in the horizontal plane and $\dot{E}$ pointing upward the energy function $\dot{E}_{\varepsilon\gamma} = \mathrm{f}(a,b)$ has the shape of a basin symmetrical about the $\dot{E}$-axis with the deepest point coinciding with the origin, and exhibiting elliptical horizontal cross-sections whose area increases with increasing distance from the base.

The energy function $\dot{E}_p = \mathrm{f}(a,b)$, on the other hand, is represented by an incline plane which contains the origin. Hence the $\dot{E}_p$ plane and the $\dot{E}_{\varepsilon\gamma}$ quadric intersect along a curve which goes through origin and which may or may not be closed (depending chiefly upon the attitude of the $\dot{E}_p$ plane relative to the $\dot{E}$-axis). If the intersection curve is closed it has a maximum which records the maximum rate of energy dissipation under the subsidiary condition that $\dot{E}_p$ and $\dot{E}_{\varepsilon\gamma}$ are equal.

It is this maximum we seek because it occurs at a and b values which are exactly the ones to utilize as coefficients in the stream function and its derivatives.

Ordinarily the energy maximum would have been determined by letting the partial derivatives $\partial\dot{E}/\partial a$ and $\partial\dot{E}/\partial b$ vanish simultaneously. It is known, however, (e.g. Courant, 1956 II, p. 180) that the same maximum can also be determined by the condition

$$\left.\frac{\partial a}{\partial b}\right|_{E_{\gamma\varepsilon}} = \left.\frac{\partial a}{\partial b}\right|_{E_p} \tag{9.38}$$

where $E_{\gamma\varepsilon}$ and E_p signify that the derivatives are derived from eqns (9.33) and (9.36), respectively.

By implicit differentiation of eqns (9.34) and (9.37) one obtains

$$\frac{2BH^2b + CHa}{2Aa + CHb} = \frac{\beta H}{\alpha} \tag{9.39}$$

from which an expression of b in terms of a follows:

$$b = \frac{2\beta/\alpha A - C}{2BH - \beta/\alpha CH} a \equiv \Pi a. \tag{9.40}$$

From the condition $\dot{E}_{\gamma\varepsilon} = \dot{E}_p$ (eqns (9.34) and (9.37)) and the above expression for b, a follows as a function of the properties of the model

$$a = \frac{\rho g(\alpha + \beta \Pi H)}{\mu(A + BH^2\Pi^2 + CH\Pi)} (\tfrac{1}{2} \pm \tfrac{1}{2}). \tag{9.41}$$

Here Π is defined by the identity (9.40).

$a = 0$ evidently coincides with the minimum on the energy function while the non-vanishing value corresponds to the maximum. The definitions of α, β, A, B, C, and Π show that a contains the aspect ratio $l/H = r$ of the model beside its thickness, its density and viscosity coefficient in addition to the acceleration due to gravity. The expression of Π shows that b contains the same model properties.

When the above expressions for a and b are inserted in the stream function and its derivatives, the velocity components and the stresses can be determined quantitatively at any point in a rectangular slab of Newtonian material exposed to the force of gravity. Before doing so, however, we shall estimate how much the selected form of the stream function actually deviates from the physical model.

We have remarked above that the polynomial stream function approaches the true dynamics of a model more closely the higher the degree of the polynomial, but that computation work puts a practical limit on the length of the polynomial. It would, therefore, be desirable to find means of estimating how closely the real physical model is represented by the selected stream function and its derivatives. We are not here referring to the fact that most real systems of geological and geophysical interest are heterogeneous and non-Newtonian, but rather to the circumstance that a polynomial stream function will often only be an approximation even if the model does possess exactly the idealized properties assigned to it. For the model under study in this section, for example, it is possible to demonstrate that the stream function employed is not an exact representative of the model as it is defined. The reason is that the stream function, eqn (9.24), does not result in vanishing shear stress over the whole free surface of the model. The shear stress as derived from the stream function vanishes at the top surface but not at the vertical end faces at $x = \pm l$. To show that, x is put equal to l in eqn (9.28) to yield

$$\dot{\gamma}_{xy} = \left(2a + 6by - 3\frac{a + 3bH}{H^2}y^2 + \frac{a + 3bH}{H^3}y^3\right) l + \left(\frac{a + 3bH}{3H^2} - \frac{a + 3bH}{3H^3}y\right) l^3 \tag{9.42}$$

after the expressions for c and d have been introduced (p. 210).

This may be rearranged thus

$$\dot{\gamma}_{xy} = \left(2al + \frac{a + 3Hb}{3H^2}l^3\right) + \left(6bl - \frac{a + 3bH}{3H^3}l^3\right) y - 3\frac{a + 3bH}{H^2}ly^2 + \frac{a + 3bH}{H^3}l\,y^3. \tag{9.43}$$

If this equation is to vanish at all values of y between 0 and H as required by the physical model, then the two parentheses must vanish individually, as must the terms multiplied by y^2 and y^3. It can be shown that if the equations

$$\begin{aligned} &2al + \tfrac{1}{3}(a + 3Hb)H^{-2}l^3 = 0 \\ &6bl - \tfrac{1}{3}(a + 3Hb)H^{-3}l^3 = 0 \\ &a + 3bH = 0 \end{aligned} \tag{9.44}$$

are to vanish simultaneously then both coefficients a and b must be zero.

Hence the shear strain $\dot{\gamma}_{xy}$ at the vertical end faces at $x = \pm l$ does not vanish in a model represented by the selected stream function. In fact, it is not at all possible to make the said shear strain vanish by the application of a stream function of the form shown in eqn (9.24), unless we have the meaningless condition that all the coefficients are zero.

The question is then: is it necessary to add more members to the polynomial, thereby perhaps succeeding in obtaining a form of the stream function which permits $\dot{\gamma}_{xy}$ to vanish over the whole free surface, or is the discrepancy between the selected stream function and the physical model sufficiently small to neglect higher terms in the polynomial? One way of answering this question is to determine the energy associated with the non-vanishing shear strain at the end faces and compare this energy with the total energy change of the system.

Now, the shear-strain rate $\dot{\gamma}_{xy}$ at $x = l$, $0 \leqslant y \leqslant H$ requires a shear stress of magnitude $\mu\dot{\gamma}_{xy}$ at the end face. (This stress is implied in the stream function but does not exist in the model which the stream function is supposed to represent.) The shear stress corresponds to a force $dF_y =$

$\mu\dot{\gamma}_{xy}\partial y\partial z$ acting on the surface area $\partial y\partial z$. Reckoned per unit of time the force represents an energy change of magnitude $\mathrm{d}E = \mu v\dot{\gamma}_{xy}\,\partial y\partial z$ where v is the vertical velocity component at the point considered on the end face. According to the definition of our model there is no motion in the z direction. The energy associated with the shear strain along a surface band of unit width extending from top to bottom of the end face is then

$$\dot{E}_l = \mu\int_0^H v\dot{\gamma}_{xy}\partial y. \tag{9.45}$$

Introduction of the expressions for v at $x = l$ and $\dot{\gamma}_{xy}$ at $x = l$ (eqns (9.30) and (9.42)), integration and some rearrangement of terms yield

$$\begin{aligned}\dot{E}_l = -\mu[&(0{\cdot}221429r + 0{\cdot}119048r^3 + 0{\cdot}01111r^5)H^4a^2\\ &+ (1{\cdot}107147r + 0{\cdot}2714208r^3 - 0{\cdot}1r^5)H^6b^2\\ &+ (0{\cdot}63333r + 0{\cdot}447597r^3 + 0{\cdot}6667r^5)H^5ab].\end{aligned} \tag{9.46}$$

This is the extra energy implied in the stream function but not actually existing in the model which the stream function is supposed to represent. The discrepancy between the physical model and the stream function applied may now be expressed quantitatively as a percentage of the total energy-dissipation rate of the model. The latter is given in eqn (9.33). Using a good pocket calculator it is possible to find $\dot{E}_l$ in the form of a percentage of $\dot{E}_{\gamma\varepsilon}$ by the relation

$$100\frac{\dot{E}_l}{\dot{E}_{\gamma\varepsilon}}$$

in which the expressions (9.33) and (9.46) for the energies are introduced.

It follows from the energy expressions that the significance of $\dot{E}_l$ in terms of percentage of the total energy balance is chiefly controlled by the aspect ratio r of the model. Table 9.1 shows the relative significance of the extra energy $\dot{E}_l$ as a function of r for a model consisting of a rectangular slab of Newtonian material with viscosity 10^{21} and 10^{20} poise, density $2{\cdot}8\ \mathrm{g\,cm^{-3}}$ and $H = 5\times10^5$ cm. It is seen that the $\dot{E}_l$ is less than 1% of the total energy when $r \geqslant 5$. Note that $\dot{E}_l$, in terms of percentage of $\dot{E}_{\gamma\varepsilon}$, is independent of the viscosity. See also Table 9.2.

This we interpret to mean that the stream function selected is a good approximation to the physical model as long as the aspect ratio of the half-model is not less than 5. We can, therefore, with confidence apply eqn (9.24) and its derivations to determine velocities and stresses in models with large aspect ratios. Incidentally, most nappes have l/H ratios much larger than 5. (Detailed numerical tests reveal minor deviations between

Table 9.1. Velocities at $x = l$, $y = H$ and energy rates of collapsing slabs with different slenderness ratio[a]

μ (poise)	$r = l/H$	u_{lH} (cm year^{-1})	v_{lH} (cm year^{-1})	$\dot{E}_{\gamma\varepsilon}$ (joule year^{-1})	$\dot{E}_l$ (% of $\dot{E}_{\gamma\varepsilon}$)
10^{21}	1	2·16	−1·51	$3{\cdot}35\times10^7$	5·02
10^{21}	2	1·92	−1·18	$2{\cdot}62\times10^7$	6·04
10^{21}	3	1·38	−0·67	$2{\cdot}10\times10^7$	2·133
10^{21}	4	0·95	−0·38	$1{\cdot}50\times10^7$	1·50
10^{21}	5	0·70	−0·24	$1{\cdot}13\times10^7$	0·923
10^{21}	6	0·54	−0·16	$0{\cdot}91\times10^7$	0·632
10^{21}	8	0·38	−0·09	$0{\cdot}64\times10^7$	0·33
10^{21}	10	0·29	−0·056	$0{\cdot}50\times10^7$	0·213
10^{21}	15	0·19	−0·025	$0{\cdot}32\times10^7$	0·093
10^{21}	20	0·14	−0·0137	$0{\cdot}24\times10^7$	0·052
10^{20}	1	21·6	−15·1	$3{\cdot}35\times10^8$	5·02
10^{20}	5	7·0	−2·4	$1{\cdot}13\times10^8$	0·923
10^{20}	20	1·4	−1·137	$0{\cdot}24\times10^8$	0·052

[a] $H = 5\times10^5$ cm, $\rho = 2{\cdot}8$ g cm^{-3}, $g = 981$ cm sec^{-2}, $\mu = 10^{20}$, and 10^{21} poise.

the selected stream function and the physical model. However, these deviations do not concern us here.)

When calculating the velocities at any point in the model it is convenient to introduce the dimensionless ratios $x/l = m$ and $y/H = n$ as well as $r = l/H$. The velocity equations then become

$$u = \left\{2an + 3bHn^2 - \left[(\tfrac{2}{3}n^3 - \tfrac{1}{6}n^4) - (\tfrac{1}{3}n - \tfrac{1}{6}n^2)m^2r^2\right](a + 3bH)\right\}mrH^2 \qquad (9.47)$$

Table 9.2. Velocities, basal shear-strain rate, basal shear stress and rate of energy dissipation of collapsing

r	5		10		20	
μ (poise)	10^{18}	10^{22}	10^{18}	10^{22}	10^{18}	10^{22}
$u_{l,H}$ (cm year^{-1})	$6{\cdot}98\times10^2$	$6{\cdot}98\times10^{-2}$	$2{\cdot}96\times10^2$	$2{\cdot}96\times10^{-2}$	$1{\cdot}4\times10^2$	1·4×10
$u_{\frac{1}{2}l,H}$ (cm year^{-1})	$1{\cdot}36\times10^2$	$1{\cdot}36\times10^{-2}$	43·1	$4{\cdot}3\times10^{-3}$	18·3	1·83×
$u_{l,\frac{1}{2}H}$ (cm year^{-1})	$4{\cdot}96\times10^2$	$4{\cdot}96\times10^{-2}$	$2{\cdot}19\times10^2$	$2{\cdot}19\times10^{-2}$	$1{\cdot}05\times10^2$	1·05×1
v_{lH} (cm year^{-1})	$-2{\cdot}41\times10^2$	$-2{\cdot}4\times10^{-2}$	−56·8	$-5{\cdot}6\times10^{-3}$	−13·9	−1·4×10
$v_{\frac{1}{2}lH}$ (cm year^{-1})	−70·69	$-7{\cdot}069\times10^{-3}$	−14·9	$-1{\cdot}49\times10^{-3}$	−3·52	−3·52×
$v_{O,H}$ (cm year^{-1})	−13·8	$-1{\cdot}38\times10^{-3}$	$-8{\cdot}79\times10^{-1}$	$-8{\cdot}79\times10^{-5}$	$-5{\cdot}5\times10^{-2}$	−5·5×10
$\dot{\gamma}_{l,o}$ (sec^{-1})	$8{\cdot}03\times10^{-11}$	$8{\cdot}03\times10^{-15}$	$3{\cdot}67\times10^{-11}$	$3{\cdot}67\times10^{-15}$	$1{\cdot}78\times10^{-11}$	1·78×
$\dot{\gamma}_{\frac{1}{2}l,o}$ (sec^{-1})	$1{\cdot}3\times10^{-11}$	$1{\cdot}3\times10^{-15}$	$4{\cdot}98\times10^{-12}$	$4{\cdot}98\times10^{-16}$	$2{\cdot}22\times10^{-12}$	2·22×
$\tau_{l,o}$ (bar)	80·3	80·3	36·7	36·7	17·8	17·8
$\tau_{\frac{1}{2}l,o}$ (bar)	13·0	13·0	4·98	4·98	2·2	2·2
$\dot{E}_{\gamma\varepsilon}$ (joule year^{-1})	$1{\cdot}13\times10^{10}$	$1{\cdot}13\times10^6$	$5{\cdot}02\times10^9$	$5{\cdot}02\times10^5$	$2{\cdot}41\times10^9$	2·41×
$\dot{E}_l$ (% of $E_{\gamma\varepsilon}$)	0·924	0·927	0·213	0·213	0·052	0·052

[a] $H = 5\times10^5$ cm, $\rho = 2{\cdot}8$ g cm^{-3}, subscripts l, H; $\frac{1}{2}l$, H, etc., indicate position where velocities, strain, e

and

$$v = -\left\{a + bHn - \left[\left(\tfrac{1}{6}n^2 - \tfrac{1}{30}n^3\right) - \left(\tfrac{1}{2} - \tfrac{1}{6}n\right)m^2r^2\right](a + 3bH)\right\}n^2H^2. \qquad (9.48)$$

An important tectonic question is the magnitude of shear stress at the base of a collapsing nappe. The basal shear stress is given by

$$\tau_{xy} = \mu\dot{\gamma}_{xy} \qquad (9.49)$$

where the expression for $\dot{\gamma}_{xy}$ is applied, putting y equal to zero.

$$\dot{\gamma}_{xy} = \frac{\partial u}{\partial y} + \frac{\partial v}{\partial x} = 2ax + \frac{1}{3}\frac{a + 3bH}{H^2}x^3, \text{ valid when } y = 0. \qquad (9.50)$$

Introducing $m = x/l$ and $r = l/H$ changes the form to

$$\dot{\gamma}_{xy} = \left[2a + \left(\tfrac{1}{3}a + bH\right)r^2m^2\right]rmH. \qquad (9.51)$$

Numerical tests based on eqns (9.33, 9.46–49 and 9.51) have been performed on collapsing slabs, all with height 5×10^5 cm and $\rho = 2{\cdot}8\ \mathrm{g\ cm^{-3}}$. The aspect ratio ranges from 5 to 200, and two different viscosities have been tested, viz. 10^{18} and 10^{22} poise. The data are presented in Table 9.2.

As intuitively expected the velocities as well as the shear-strain rate are inversely proportional to the viscosity at given r-ratio, as is the rate of energy dissipation. The shear stress at the base, however, is independent of the viscosity.

s rectangular slab[a]

50		100		200		r
0^{18}	10^{22}	10^{18}	10^{22}	10^{18}	10^{22}	μ (poise)
	$5{\cdot}53\times10^{-3}$	27·6	$2{\cdot}76\times10^{-3}$	13·8	$1{\cdot}38\times10^{-3}$	$u_{l,H}$ (cm year^{-1})
	$6{\cdot}97\times10^{-4}$	3·45	$3{\cdot}45\times10^{-4}$	1·73	$1{\cdot}73\times10^{-7}$	$u_{\frac{1}{2}l,H}$ (cm year^{-1})
	$4{\cdot}15\times10^{-3}$	20·73	$2{\cdot}07\times10^{-3}$	10·3	$1{\cdot}03\times10^{-3}$	$u_{l,\frac{1}{2}H}$(cm year^{-1})
	$-2{\cdot}2\times10^{-4}$	$-5{\cdot}6\times10^{-1}$	$-5{\cdot}6\times10^{-5}$	$-1{\cdot}4\times10^{-1}$	$-1{\cdot}4\times10^{-5}$	v_{lH}(cm year^{-1})
$\times10^{-1}$	$-5{\cdot}53\times10^{-5}$	$-1{\cdot}38\times10^{-1}$	$-1{\cdot}38\times10^{-5}$	$-3{\cdot}45\times10^{-2}$	$-3{\cdot}45\times10^{-7}$	$v_{\frac{1}{2}lH}$(cm year^{-1})
10^{-3}	$-1{\cdot}4\times10^{-7}$	$-8{\cdot}8\times10^{-5}$	$-8{\cdot}8\times10^{-9}$	$-5{\cdot}5\times10^{-6}$	$-5{\cdot}5\times10^{-10}$	$v_{O,H}$ (cm year^{-1})
$\times10^{-12}$	$7{\cdot}04\times10^{-16}$	$3{\cdot}52\times10^{-12}$	$3{\cdot}52\times10^{-16}$	$1{\cdot}76\times10^{-12}$	$1{\cdot}76\times10^{-16}$	$\gamma_{l,o}$ (sec^{-1})
$\times10^{-13}$	$8{\cdot}84\times10^{-17}$	$4{\cdot}4\times10^{-13}$	$4{\cdot}4\times10^{-17}$	$2{\cdot}2\times10^{-13}$	$2{\cdot}2\times10^{-17}$	$\gamma_{\frac{1}{2}l,o}$ (sec^{-1})
	7·04	3·52	3·52	1·76	1·76	$\tau_{l,o}$ (bar)
	0·89	0·44	0·44	0·22	0·22	$\tau_{\frac{1}{2}l,o}$ (bar)
10^{8}	$9{\cdot}5\times10^{4}$	$4{\cdot}74\times10^{8}$	$4{\cdot}74\times10^{4}$	$2{\cdot}37\times10^{8}$	$2{\cdot}37\times10^{4}$	$\dot{E}_{\gamma\varepsilon}$ (joule year^{-1})
32	0·0082	0·0021	0·0021	0·0005	0·0005	$\dot{E}_l$ (% of $E_{\gamma\varepsilon}$)

ated. Energy is calculated for a 1-cm-thick cross-section slice of the half-model.

Table 9.3. Velocities at top and end faces, and shear-strain rate and shear stress at base of collapsing viscous slabs[a]

n or m	$m = 1$		$n = 1$		$n = 0$	
	u (cm year^{-1})	v (cm year^{-1})	u (cm year^{-1})	v (cm year^{-1})	$\dot{\gamma}_{xy}$ (sec^{-1})	$\tau_{xy} = \dot{\gamma}_{xy}\mu$ (bar)
1	7·01	−2·41	7·01	−2·41	$8{\cdot}03\times10^{-13}$	80·3
0·9	6·85	−2·04	5·31	−1·98	$6{\cdot}00\times10^{-13}$	60·0
0·8	6·55	−1·67	3·95	−1·59	$4{\cdot}34\times10^{-13}$	43·4
0·7	6·15	−1·35	2·86	−1·25	$3{\cdot}04\times10^{-13}$	30·4
0·6	5·59	−1·03	2·01	−0·96	$2{\cdot}04\times10^{-13}$	20·4
0·5	4·96	−0·745	1·36	−0·71	$1{\cdot}30\times10^{-13}$	13·0
0·4	4·18	−0·496	0·88	−0·50	$0{\cdot}78\times10^{-13}$	7·8
0·3	3·33	−0·29	0·55	−0·34	$0{\cdot}433\times10^{-13}$	4·3
0·2	2·32	−0·132	0·31	−0·23	$0{\cdot}217\times10^{-13}$	2·17
0·1	1·21	−0·035	0·136	−0·16	$0{\cdot}866\times10^{-14}$	0·87
0	0·0	−0·0	0·0	−0·14	0·0	0·0

[a] $Or = 5$, $H = 5\times10^5$ cm, $\rho = 2{\cdot}8$ g cm^{-3}, $g = 981$ cm sec^{-2}, $\mu = 10^{20}$ poise

Table 9.4. Shear-strain rate in sec^{-1} and shear stress in bar at bottom corner ($x = l$, $y = 0$) of end of collapsing slabs with different heights[a]

H(cm)	1×10^5	2×10^5	3×10^5	5×10^5	10^6	2×10^6
$\dot{\gamma}_{xy}$(sec^{-1})	$7{\cdot}34\times10^{-14}$	$1{\cdot}47\times10^{-13}$	$2{\cdot}20\times10^{-13}$	$3{\cdot}67\times10^{-13}$	$7{\cdot}34\times10^{-13}$	$1{\cdot}47\times10^{-12}$
τ_{xy}(bar)	7·34	14·7	22	36·7	73·4	147

[a] $r_r = 5$, $H = 5\times10^5$ cm, $\rho = 2{\cdot}8$ g cm^{-3}, $g = 981$ cm sec^{-2}, $\mu = 10^{20}$ poise

The velocities—which are determined at the upper end corner, in the middle of the surface and in the middle of the end face—decrease with increasing aspect ratio of the slab. It is almost an inverse proportionality between r and the velocity component u. The same is true for the stress along the base of the slab.

The velocity v is negative, meaning, as expected, that the vertical component points downward everywhere in the model. At the free surface the downward movement increases rapidly with distance from the centre of the nappe, being four times as rapid at $x = l$ (i.e. $m = 1$) than at $x = \frac{1}{2}l$. Incidentally, in time this is going to generate the domed surface characteristically formed in the experimental models with coherent base (Fig. 9.10).

To show in greater detail the variation of the velocities along the surface and the end face, and the variation of sheer strain and shear stress at the base, a model has been calculated as recorded in Table 9.3, based on eqns (9.47–9.51).

Although it is qualitatively expected that the velocities and the basal shear increase with distance from origin, it is valuable to be able to quantify these values. It is significant for nappe and thrust sheet theories that the gravity collapse model will give maximum basal shear stress at the end of the body while the conventional ice-cap model (p. 199) will give maximum shear wherever the surface slope assumes maximum value.

The factors that affect the basal shear stress are chiefly the density and the aspect ratio for models with same height, and of course the height itself for models with constant aspect ratio. Some data showing the latter relationship are recorded in Table 9.4 based on eqns (9.49 and 9.51).

EXPERIMENTAL TESTS OF SPREADING NAPPES

Experimental tests of spreading models were performed by F. K. Rolf Häll in the Uppsala tectonic laboratory. The models were constructed of a silicone putty with relatively low viscosity and vanishing yield point, so most of the spreading tests could be run without the use of a centrifuge.

To study the flow pattern at various stages during the spreading the models were built of sheets of silicone with different colour, but identical density and viscosity.

Each model was made in duplicate, one with the coloured sheets in vertical orientation, the other with horizontal sheets. This enabled the experimeter to study the flow pattern and strain in great detail during the runs. In some symmetrical models one half was made of horizontal layers, the other of vertical.

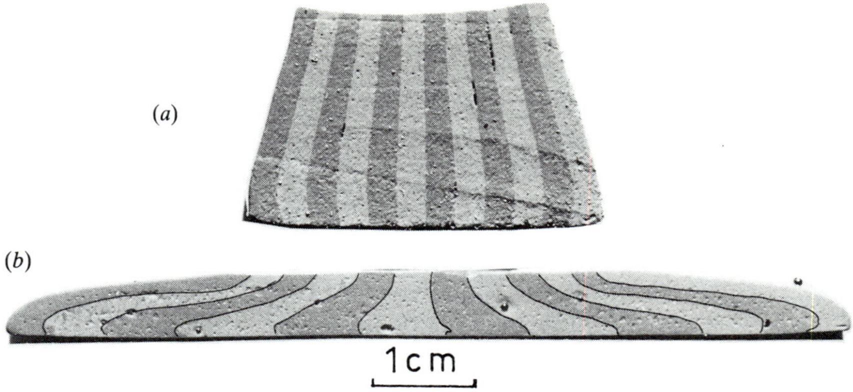

Fig. 9.6. Gravity collapse and spreading of initially rectangular block with quadratic cross-section of silicone putty with free slip at base. Vertical stripes are silicone putty with different colour but identical density and rheological properties. (*a*) early stage of spreading, (*b*) late stage of spreading (unpublished experiment by R. Häll, Uppsala).

Fig. 9.7. Test similar to that in Fig. 9.6, except for the initial cross-section which here is rectangular with width, $2l$, twice as large as height, H. (*a*), (*b*), and (*c*) show three stages of evolution. In (*b*) and (*c*), only the left-hand half of the body shown (unpublished test by R. Häll, Uppsala).

(a)
(b)
(c)
(d)

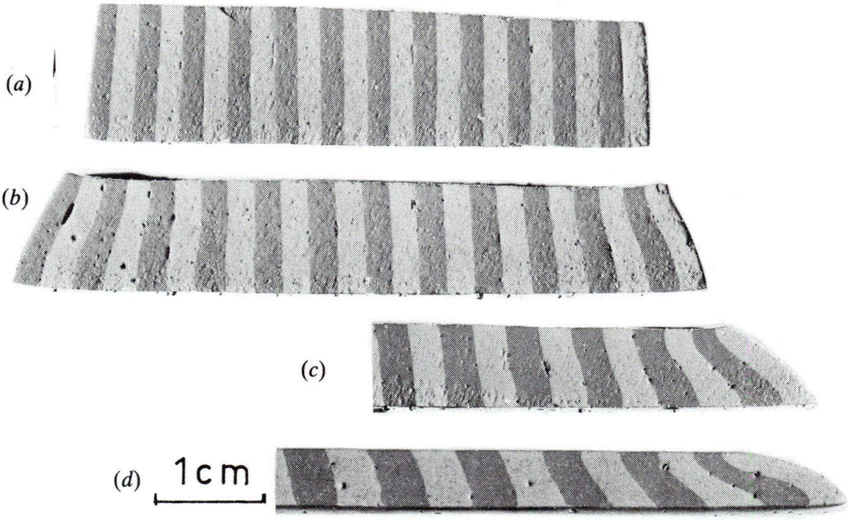

Fig. 9.8. Test similar to those shown in Figs 9.6 and 9.7, except for the initial width/height ratio which here is 4:1. Initial stage and three consecutive stages of evolution shown. Only the right-hand half shown in stages (*c*) and (*d*) (unpublished test by R. Häll, Uppsala).

Two sets of tests were performed, one with the spreading models resting on mercury to simulate free slip along the base, the other with complete coherence between the models and their rigid substratum. Within each set the model started as rectangular blocks which were much longer in one direction than in the two others, the long direction being parallel to the substratum. This geometry made the flow practically two-dimensional in the plane normal to the long axis. The plane of flow is termed x,y with x parallel to the horizontal direction of spreading and y as the vertical direction. Models with different proportions between their width, $2l$, parallel to x, and their height, H_0, parallel to y were tested, the $2l/H$ ratio being 1, 2 and 4 in three different sets of tests.

The tests were not attempts to establish empirical relationships between the velocity of spreading and rheological-geometrical properties of the models but rather studies of the pattern of flow and strain within the models.

Figures 9.6, 9.7 and 9.8 show cross-sections taken at various stages of development of models with free slip along the base. The initial $2l/H$ ratios are unity, 2, and 4, respectively. Models coherent to the base are shown in Figs 9.9, 9.10, and 9.11. The initial $2l/H$ ratios are 1, 2 and 4, respectively.

The difference in flow pattern between the free-slip models and the

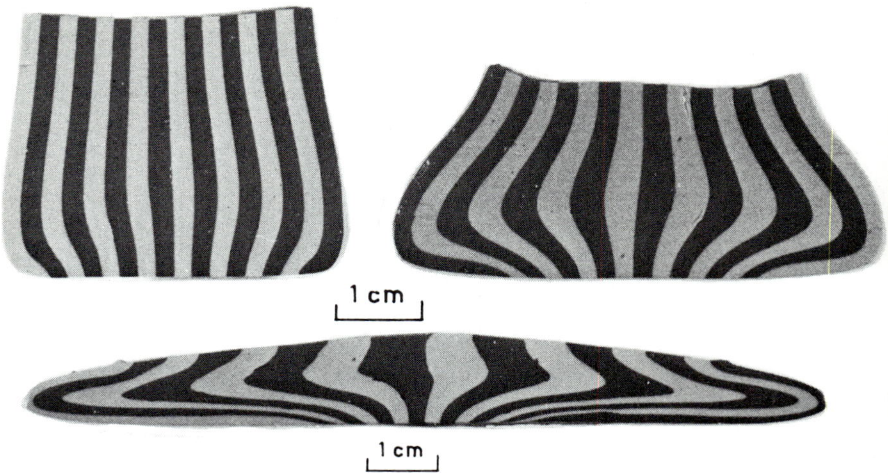

Fig. 9.9. Test of gravity collapse of rectangular parallelepipeds of silicone putty coherent to the rigid base. Initial width/height ratio = 1. Vertical stripes consist of silicone putty with different colour but identical density and rheological properties. Three stages of evolution shown (experiment run in centrifuge by R. Häll, Uppsala).

coherent-base models is striking. As expected, in the free-slip models the maximum spreading is at the base while in the coherent-base models maximum spreading is at some level within the body, the distance from the base chiefly depending upon the $2l/H$-ratio of the model. A second point of contrast in flow pattern of the two groups of models is that the free-slip models remain even-thick during the collapse while the coherent-base models develop a domed surface due to decreasing thickness from the central section to the edges. It is found that the velocity of collapse, under otherwise identical conditions, is much faster in the free-slip models than the coherent-base models.

Let us for a moment focus the attention on the flow pattern of the free-slip models. The fact the the thickness remains uniform along the whole body (except at the ends) demonstrates that the strain is essentially of pure-shear type and that the spreading along the top is almost as fast as the spreading along the bottom. This is especially true within the interior of the models; toward the edges the difference in strain-rate between the top and bottom is more pronounced.

Even though the decreasing thickness remains uniform (except at the ends) during the collapse, the longitudinal finite strain ε_x is not constant at any given level in the model. The detailed pattern of flow as indicated by the colour marker shows that ε_x (and ε_y) varies with x and y in such a

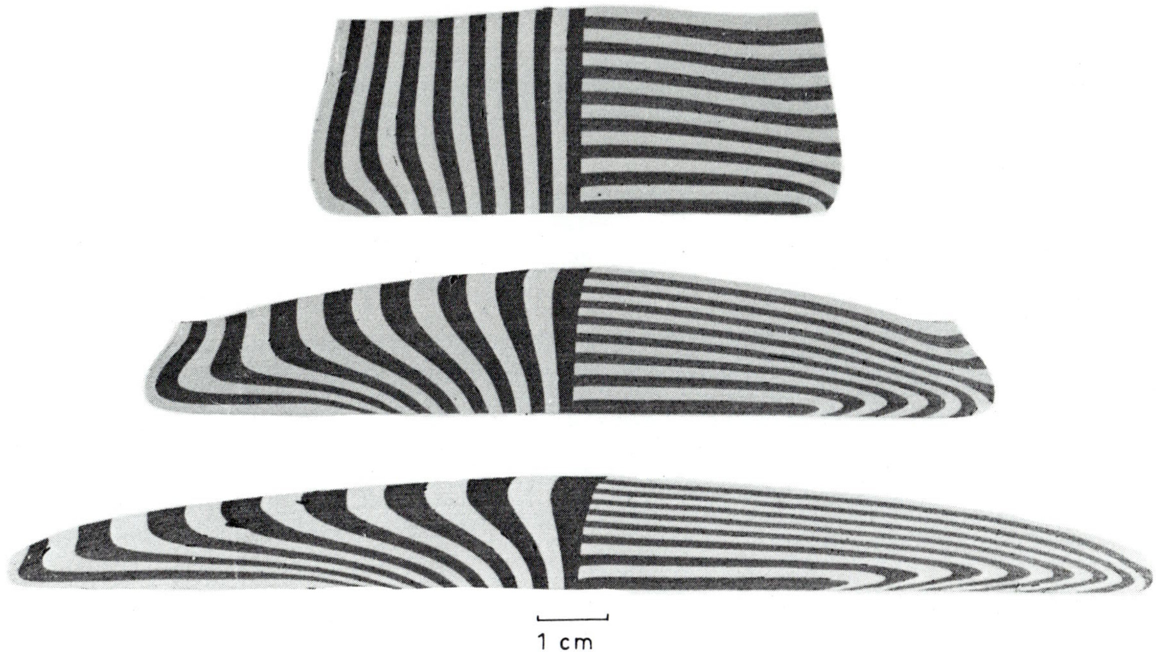

Fig. 9.10. Tests similar to that recorded in Fig. 9.9, except that here the initial width/height ratio is 2:1. Three stages of evolution shown (experiment by R. Häll, Uppsala).

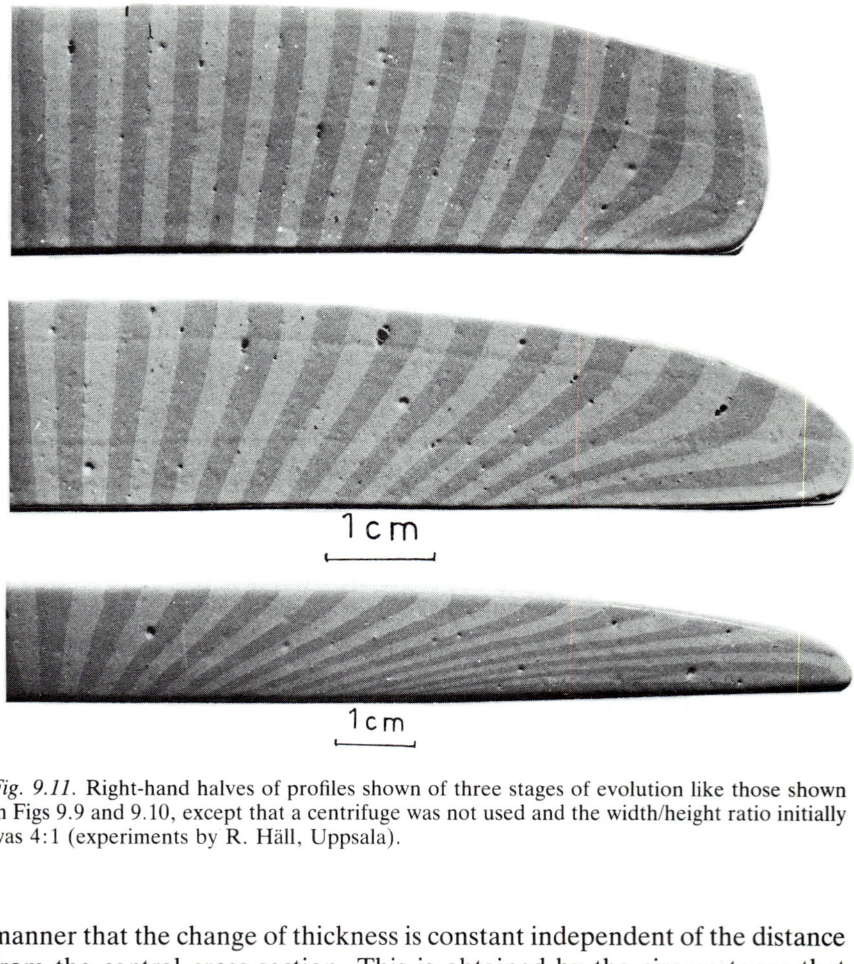

Fig. 9.11. Right-hand halves of profiles shown of three stages of evolution like those shown in Figs 9.9 and 9.10, except that a centrifuge was not used and the width/height ratio initially was 4:1 (experiments by R. Häll, Uppsala).

manner that the change of thickness is constant independent of the distance from the central cross-section. This is obtained by the circumstance that the tensile strain ε_x *decreases* with distance from the centre in the upper part of the models and that ε_x *increases* with distance from the centre in the lower part of the models. This variation of the strain at different levels is demonstrated by the curvature and variation of thickness of the horizontal colour markers in the models (Figs 9.12 and 9.13). There accordingly is experimental evidence that the stress in the upper layers is increasingly tensile as one goes from the edge at $x = l$ to the central cross section. In the lower layers the opposite is true: σ_x becomes more compressive as one approaches the central cross-section.

The sigmoidal shape of the vertical markers as predicted in the theoretical analysis is well developed in the free-slip models.

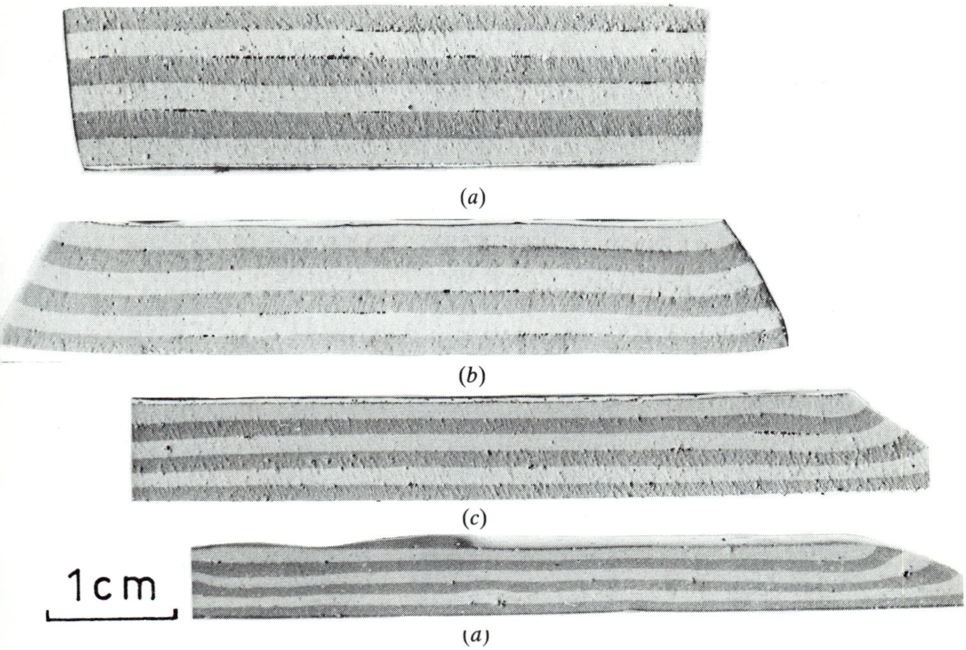

Fig. 9.12. Non-centrifuged collapse and spreading model of silicone putty with free slip along base. Stages (*c*) and (*d*) show right-hand half of model. This model with horizontal neutral colour markers should be compared to that with vertical stripes, Fig. 9.8 (experiments by R. Häll, Uppsala).

Consistent with the lack of shear strain at the top and base, the tendency to maintain the 90° angles between the markers and the free surface, respective the slip-free base, is quite evident in the models. Note, for example, how the surface at the edges is abruptly bent up in order to maintain approximately 90° angle between the deformed vertical marker and the surface.

Turning now to models with coherence along the base, the interesting point as regard the flow is the change of pattern depending upon the length/height ratio of the model. At $2l/H_0 = 1$ it is a pronounced maximum of the function $\varepsilon_x = f(y)$ quite close to the bottom. At the early stage of deformation the level of maximum spreading lies at about $\frac{1}{3}\, H_0$ above the base (see Fig. 9.9). (The level of maximum tensile strain is measured close to the centre of the models because at the ends the bulge of maximum strain has been sagging down and the instantaneous relationships have been disturbed.)

At $2l/H_0 = 2$ the level of maximum tensile strain occurs approximately at $\frac{1}{2}H_0$ (Fig. 9.10) and there is little if any diminishing of the strain above that level. At $2l/H_0 = 4$ it appears that maximum extension is close to the surface. Qualitatively, these are expected relationships based on strain-energy arguments for a given magnitude of extension. The shear-strain parallel to the base diminishes with increasing height of the level of maximum extension. For models which are long relative to their height, the shear strain becomes particularly significant energy wise (because the shear strain increases with distance from the central section). Hence in models with large $2l/H_0$ ratio there is a strong tendency for the level of maximum extension to develop high up in the body. The experimental

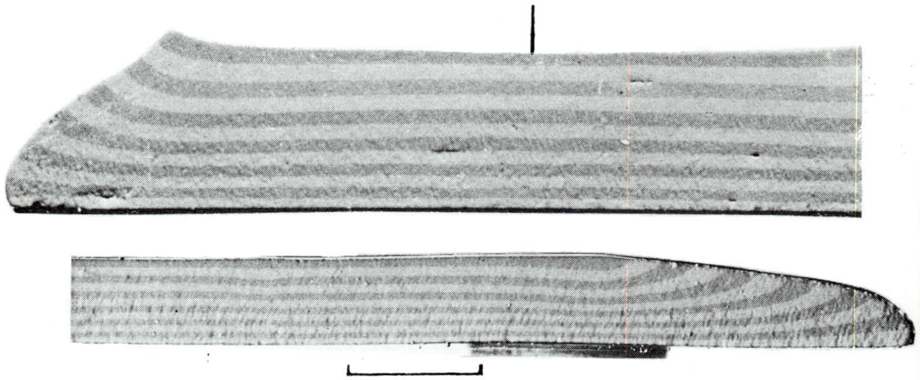

Fig. 9.13. Parts of model similar to the one shown in Fig. 9.12. Note details of deformed shape of the horizontal marker strips at the two stages of evolution shown (experiment by R. Häll, Uppsala).

models indicate that the level of maximum extension occurs very close to the top surface already when the length is not more than four times the height.

These conditions are of interest for the concept of extrusion flow in glaciers. It appears that extrusion flow is physically a realistic concept, but that favourable conditions for such flow are only satisfied in spreading bodies whose length/thickness ratio is less than in most glaciers. Probably the limiting length/thickness ratio for extrusion flow depends upon the rheological properties and the density of the material.

In concluding the remarks on experimental spreading models it is interesting to note the great similarity between the flow pattern in the upper part of the coherent models—above the level of maximum strain—and the flow pattern in the free-slip models from the base and upwards.

10 The Centrifuge Technique

SCALE-MODEL CONSIDERATIONS

A major obstacle in model experiments on gravity-controlled tectonics is the scarcity of suitable natural and manufactured imitation materials. Such materials must be weak enough to sag and spread, and even fracture under their own weight, in order to imitate realistically the evolution of gravity-controlled structures on the Earth. Despite the great crushing and shear strength of rocks, and their high effective viscosity, they yield and flow and fracture under the pull of gravity merely by virtue of the large geometric dimensions of some geological structures and the tremendous length of geological time. For geometrically similar structures of identical materials, the stress caused by gravity at any set of corresponding points increases with increasing linear dimension of the structure, a condition well known from model theory (see Chapter 1). The force ratio $F_g/F_p = \rho lg/\Delta\sigma$, which must be constant for dynamically similar models, shows the same condition (see Table 6.1). It follows that increasing geometric size increases the tendency of gravitational collapse and spreading, and also increases the velocity of viscous flow (cf. the ratio F_g/F_v in Table 6.1) in similar structures of identical substances. It is interesting that this effect which mere size has on the stability of structures was well understood by Galileo, as is shown by the following passage quoted in Hubbert's classic paper (1937) on model theory.

> From what had already been demonstrated, you can plainly see the impossibility of increasing the size of structures to vast dimensions either in art or in nature; likewise the impossibility of building ships, palaces, or temples of enormous size in such a way that their oars, yards, beams, iron bolts, and, in short, all their other parts will hold together; nor can nature produce trees of extraordinary size, because the branches would break down under their own weight; so also it would be impossible to build up the bony structures of men, horses, or other animals so as to hold together and perform normal functions if these animals were to be increased enormously in height; for this increase in height can be accomplished only by employing a material which is harder and stronger than usual, or by enlarging the size of the bones, thus changing their

> shape until the form and appearance of the animals suggest a monstrosity. . . .
> Clearly then, if one wishes to maintain in a great giant the same proportion of limb as found in ordinary man he must either find a harder and stronger material for making the bones, or he must admit a diminution of strength in comparison with men of medium stature; for if his height be increased inordinately he will fall and be crushed under his own weight. Whereas, if the size of a body be diminished, the strength of that body is not diminished in the same proportion; indeed the smaller the body the greater its relative strength. Thus a small dog could probably carry on his back two or three dogs of his own size; but I believe that a horse could not carry even one of his own size.

The reason is that the weight of geometrically similar structures of identical materials increases with the *cube* of the linear dimension whereas their strength, in terms of ability to carry the superincumbent load, increases only with the *square* of the linear dimension.

To achieve spontaneous collapse and sensibly rapid flow in small-scale models driven only by unstable mass distribution in the field of gravity, only mechanically weak and soft substances can be used. Because of this requirement of the model theory, asphalt, heavy oil, soft wax, wet clay, and loose powder have been used as rock-simulating materials in informative model studies by a number of workers (Cloos, 1930; Nettleton, 1934, 1943; Hubbert, 1945; Parker and McDowell, 1955; Bucher, 1956).

However, because the materials must be weak and soft to reproduce with reasonable correctness the natural tectonic events, the choice of materials is limited, and the chance of finding a sufficient variety of substances with the characteristics suited for imitating complex structures is indeed small. Moreover, the softness of the materials makes the models awkward to handle, particularly if complicated structural patterns are under study. Neither the construction of the initial model nor a detailed study of the final pattern is possible in many instances.

Thus much is gained by using experimental techniques which permit the use of the materials with greater strength and firmness without thereby violating significant model laws.

Let us at the outset make it clear that gravity-generated structures can never correctly be imitated by models driven by only surface forces. The high stress created by gravity in large structures cannot therefore be reproduced in a small model merely by increase of the stresses transmitted to the boundary of the model. Thus, for example, the method of loading lead shot over layers to be buckled in a compression box (Willies, 1893) definitely does not give the correct conditions within the buckling layers if they are supposed to imitate large folds over which gravity exercises significant control.

The experimental work on diapir formation by Escher and Kuenen

(1929), interesting as it is, especially as regards the production of folds in enclosed strata, suffers from the same fault. The diapirs were produced by boundary stresses grossly out of proportion to the gravity-induced buoyant stresses in the structures. Such an approach can provide valuable information on problems limited to some local deformation pattern in diapirs such as local buckling, boudinage structures, etc. Experiments of this kind provide no valid information on the evolution of the shape of diapirs as they rise, on their tendency to spread upon approaching the surface of the overburden, or on the evolution of the rim syncline and other typical patterns in the adjacent rocks, all of which are phenomena excellently displayed by properly scaled gravity-driven (or centrifuged) models.

It is the body force per unit mass that must be increased above that of gravity if in our models we want to use materials of greater strength and higher viscocity and still reproduce large-scale tectonic events with reasonable precision.

The effect of increased body force in model theory is indicated by the change occurring in the force ratios gravitational force/pressure force (F_g/F_p), and gravitational force/viscous force (F_g/F_v). When referring to the natural system the acceleration due to gravity is to be used in the force ratios, which then take the forms $\rho lg/\Delta\sigma$ and $\rho l^2 g/\mu v$, respectively, whereas for the model the force ratios of the forms $\rho la/\Delta\sigma$ and $\rho l^2 a/\mu v$ apply, where a is the acceleration (= body force per unit mass) to which the model is exposed. If a can be made much larger than g, say several thousand times the value of g, as is possible in a centrifuge, then the stress in the model and the product μv must be made correspondingly larger relative to non-centrifuged model in order to maintain similarity, because of the conditions

$$(\rho la/\Delta\sigma)_{\text{model}} = (\rho lg/\Delta\sigma)_{\text{prototype}} \tag{10.1}$$

and

$$(\rho l^2 a/\mu v)_{\text{model}} = (\rho l^2 g/\mu v)_{\text{prototype}}. \tag{10.2}$$

There are two obvious experimental methods of achieving increased body force in the model: (1) to make use of the centrifugal force in a large-capacity centrifuge, and (2) to employ a magnetic field on models made ferromagnetic by mixing iron filings, magnetite powder and the like in plastic and visco-elastic materials.

The magnetic method has been tested by the author with some success but all the recorded work discussed below was performed by the centrifuge method, which so far has been found the more fruitful.

The centrifuge technique is very suitable for model studies of many varieties of tectonic phenomena driven by, or merely controlled by, the pull of gravity. The method of whirling models in a centrifuge was, to the best of the writer's knowledge, first utilized by Bucky (1931), who studied static stability problems in mines. Despite the method's suitability for experimental study of tectonic processes of a plastic and visco-elastic nature, no such work appears to have been reported until the present writer (Ramberg, 1963a) published a study of domes, folds, gravity sliding, extrusion of lava, etc., as executed by models of wax, modelling clay, putty and similar substances run in a centrifuge.

The principle of centrifuged dynamic models is simply that the centrifugal force plays the same role in the models as does the force of gravity in geologic structures. However, since the centrifugal force per unit mass, which in magnitude equals the centripetal acceleration but is oppositely directed, may be made several thousand times stronger than the gravitational force per unit mass, model materials can be used that are several thousand times stronger and correspondingly more viscous than materials usable in non-centrifuged models of same size. This is a considerable advantage, as will become evident later.

Besides the centrifugal force acting on models whirled in a centrifuge, there is in theory a small Coriolis force affecting masses moving in a radial direction relative to the centrifuge head, or "up" and "down" in the model. The Coriolis force tends to divert the vertical movements in the model (e.g. ascent of diapirs) in a horizontal direction (eqn 10.3). In the many hundreds of models run, however, we have not noticed any consistent horizontal deviation which could be attributed to the Coriolis force. This is in agreement with its negliglible intensity as shown by its expression (e.g. Hartog, 1948)

$$F_c = 2\,mwv \tag{10.3}$$

where m is the mass of a body that moves with velocity v along the radius of a wheel rotating with an angular velocity w. For most of our model runs w is about 2000 rev min^{-1} or $66{\cdot}6\pi$ rad sec^{-1}, but v is generally much smaller. Ascending model domes, for example, rarely rise at as fast a rate as 1 cm min^{-1} or, say, $0{\cdot}02$ cm sec^{-1}. Inserted in the above expression, these quantities give a value for the Coriolis force per unit mass:

$$\frac{F_c}{m} = 2 \times 66{\cdot}6 \times 0{\cdot}02\,\pi = a_c. \tag{10.4}$$

However, the force per mass equals the acceleration in the body-force field, i.e. the Coriolis acceleration, a_c. In consequence, the ratio between the Coriolis acceleration and the centripetal acceleration is also the ratio between the Coriolis and the centrifugal forces which act on a portion of the model moving with radial component v relative to the frame of the model. The centripetal acceleration is w^2r, where r is the radius of the centrifuge rotor. In the actual case r is about 30 cm in one centrifuge used, and about 50 cm in another. The ratio between the two forces is thus $2v/wr \approx 6\times10^{-6}$ and $3{\cdot}8\times10^{-6}$ respectively. It follows that the component of motion as attributed to Coriolis force is negligibly small compared with the component associated with the centrifugal force.

The above calculations would be stricly correct for a body which was moving through an imaginary fluid with vanishing density but possessing finite viscosity; in other words, a system without buoyant force. In our models, however, the correct radial force to consider is the difference between the centrifugal force per mass and the buoyant force per mass, this difference being $w^2r(\rho_b - \rho_o)/\rho_o$, where subscripts b and o refer to buoyant body and overburden, respectively. However, since $\rho_b - \rho_o$ is rarely less than 10% of ρ_o in the models, the Coriolis force is still negligible relative to the radial force.

We have not found it necessary to consider the small Coriolis effect in our runs so far.

When applying the theory of centrifuged models especially to tectonic experiments, there are two conditions to note: (1) inertial forces may be neglected in most tectonic movements (except in earthquakes), and (2) the body force per unit mass is no longer equal in model and prototype as they would be in non-centrifuged models.

As already noted (p. 76), tectonic movements not involving earthquakes and extrusion of lava are too slow for the inertial force to have a significant effect on the visco-elastic and gravitational forces.

Physically this means that, upon a sudden withdrawing of the motivating gravitational force or pressure force of a mass of rock, the kinetic energy of the mass is generally too small to sustain the movement through a significant distance against the plastico-viscous resistance. In short, the plastic or viscous resistance is an effective brake for the mass.

Similarly, only a small part of the propelling force goes to accelerate a portion of a tectonic system when it starts moving from rest; by far the larger force goes to overcome plastico-viscous drag.

To obtain a more meaningful idea of the relative significance of the inertial force in tectonics, let us attempt to estimate a Reynolds number for one of the more simple gravity-tectonic processes such as the rise of a salt dome. The Reynolds number gives the ratio between the inertial and the viscous forces in a process, as previously discussed.

The viscosity of salt is variable and not well known under the various circumstances encountered by a stratum of salt that develops into domal structures. A value of, say, 10^{17} poise would be reasonable. A rate of rise of 1 cm year^{-1} or $\approx 3\times10^{-8}$ cm sec^{-1} is acceptable, and a radius of 1·5 km = $1{\cdot}5\times10^5$ cm would not be out of order for many salt domes. The density of salt is 2·16 g cm^{-3}.

These figures give a Reynolds number

$$R = \frac{vr\rho}{\mu} \approx 9{\cdot}7 \times 10^{-20}. \tag{10.5}$$

The rise of a circular dome could be compared with flow through a circular pipe. For such flow the critical Reynolds number, which separates laminar flow from turbulent flow, is about 10^3 (see, e.g., Prandtl and Tietjens, 1934). In laminar flow the contribution from inertial forces is small, in keeping with the Reynolds number for such flow being less than the critical value. In turbulent flow inertial forces are significant, for which reason turbulence starts at the critical value of the Reynolds number and increases in intensity with augmented Reynolds number of the systems compared.

The Reynolds number for an ascending salt dome is shown to be no less than 22 orders of magnitude less than the critical value for turbulence. Clearly, then, it is not possible that inertia and turbulence could become noticeable in salt tectonics even if the rate of flow were several orders of magnitude faster than that assumed, and/or the viscosity several orders less. For example, let the salt rise 1000 cm year^{-1}, a quite impossibly fast rate ($\approx 3 \times 10^{-5}$ cm sec^{-1}), and let the viscosity be, say, 10^7 poise (i.e. as soft as asphalt, which obviously is not realistic). Finally, let the radius be 3 km. The corresponding Reynolds number is $R \approx 2 \times 10^{-6}$. The figure is still much too low to permit turbulence and an appreciable contribution from inertial forces in the motion. See also discussion on p. 77.

Similar rough estimations may be made of the Reynolds number of flowage of silicate rocks in tectonic processes. Silicate rocks are more competent (exhibit higher effective viscosity) than salt, and they in all likelihood deform at a slower rate. Even though the dimensions of moving masses in silicate-rock tectonics may often be considerable larger than those of salt domes, it is quite evident that turbulent flow and a corresponding significant contribution from inertial forces can hardly ever occur in tectonics.

It is interesting to note at this point that turbulence did not occur in our centrifuged models of more or less complicated structures, despite a Reynolds number varying between 10^{-6} and 10^{-9}, which must be several orders of magnitude larger than the R-value of similar tectonic systems of silicate rocks. This is experimental evidence that turbulence is not likely to occur in tectonic deformation of crystalline rocks.

Since inertia is insignificant in most tectonic systems—a fact that allows unlike Reynolds numbers in prototype and model as long as inertia is also negligible in the latter—the restriction $l_r t_r^{-2} = 1$ need not apply in this type of model work. The basic model ratios of mass, m_r, of length, l_r and of time, t_r, can be treated as independent. This is extremely important, because strict adherence to the condition $l_r t_r^{-1} = 1$ would render model work of tectonic events impossible as shown by the example discussed on p. 39.

The chief deviation of the centrifuged model theory from that of non-centrifuged models stems from the differerence in body force per unit mass. In non-centrifuged models, the body force per unit mass equals the gravitational acceleration, g (disregarding possible magnetic and electrostatic forces), whereas the body force per unit mass in centrifuged models

Table 10.1. Model ratios used in centrifuged models

Quantity	Model ratios
Linear dimension (e.g. thickness of layers, diameter of domes, etc.)	$\frac{l_m}{l_o} = l_r$
Body force per unit mass (centripetal acceleration a_m and acceleration due to gravity, g)	$\frac{a_m}{g_o} = a_r$
Density	$\frac{\rho_m}{\rho_o} = \rho_r$
Strain	$\frac{\Delta l_m / l_m}{\Delta l_o / l_o} = \varepsilon_r = 1$
Stress	$\frac{\sigma_m}{\sigma_o} = \sigma_r = \rho_r l_r a_r$
Strength of materials	$\frac{s_m}{s_o} = s_r = \sigma_r$
Viscosity	$\frac{\mu_m}{\mu_o} = \mu_r = \sigma_r \varepsilon_r^{-1} t_r = \sigma_r t_r$
Time	$\frac{t_m}{t_o} = t_r = \mu_r \rho_r^{-1} l_r^{-1} a_r^{-1}$

is the resultant of a vector equal to the centripetal acceleration, though oppositely directed, and the vector of the gravitational acceleration. Since most of our models are run at centripetal acceleration greater than 1000 g, the gravitational component in the models may be neglected.

Table 10.1 is a short list of the model ratios most commonly used in the models to be discussed in Chapters 11 to 18. A list of this nature also acts as a guide in selecting model materials with properties consistent with the prototype tectonic system.

For example, a tectonic system consists of two layers of rock 2 and 6 km thick, respectively, with density $3\,\mathrm{g\,cm^{-3}}$ and $2{\cdot}6\,\mathrm{g\,cm^{-3}}$, with strength 2×10^9 and 10^9 dynes $\mathrm{cm^{-2}}$ and effective creep viscosity (at stresses below the strength) of, say, 10^{20} and 10^{19} poise.

For the corresponding model we select two materials with density 1·5 g $\mathrm{cm^{-3}}$ and $1{\cdot}3\,\mathrm{g\,cm^{-3}}$, which gives a model ratio of density of 0·5. The materials are prepared in two layers, say 1 and 3 cm thick, respectively, which give a model ratio of length of 5×10^{-6}. In selecting the strength of the materials we are restricted by the acceleration of the centrifuge, as implied in a_r, since ρ_r and l_r are already fixed.

The maximum value of a_r in our large centrifuge is 4000; hence, the maximum model ratio of strength, s_r, is

$$4000\ \rho_r l_r = 4000\times0{\cdot}5\times5\times10^{-6} = 10^{-2}.$$

If the model is to be run at 4000 g, the strength of the two material has to be

$$2\times10^9 . s_r = 2\times10^7 \text{ dyne cm}^{-2}, \quad \text{and} \quad 10^9 . s_r = 2\times10^6 \text{ dyne cm}^{-2},$$

respectively, which means rather firm model substances; as a matter of fact, 4000 times stronger than the firmness permitted for the materials in corresponding non-centrifuged models.

The chief restriction imposed upon the viscosity of the model materials stems from the condition that the experiments should not run more than a few hours in the centrifuge, nor should the deformations be so rapid that the Reynolds number becomes so large that turbulence occurs. However, since we generally do not know the time involved in the evolution of the tectonic structure under study, we have no guide as to the viscosity of the model materials. Some preliminaries experimentation will have to be performed until we find materials whose viscosity allows the model to mature within a reasonable time. Assume that the two selected materials have viscosities of, say 10^7 and 10^6 poise (e.g. stitching wax and putty), giving a model ratio of viscosity, μ_r, equal to

$$10^7/10^{20} = 10^6/10^{19} = 10^{-13}.$$

The model ratio of time is thereby fixed (see Table 10.1):

$$t_r = \mu_r \rho_r^{-1}\, l_r^{-1}\, a_r^{-1} = 10^{-11}. \tag{10.6}$$

If under these circumstances the model evolves to a certain evolutionary stage in the course of, say, 10^3 sec, the prototype tectonic structure must have evolved through the corresponding evolution in the course of

$$t_o = t_m t_r^{-1} = 10^3 \times 10^{11} = 10^{14}\,\text{sec} \approx 3{\cdot}18\times10^6 \text{ years}.$$

Even for simple structures such as this, difficulties are encountered in obtaining materials whose various properties are so related that they give consistent model ratios. In the above example it may be easy to find two plastic materials with densities of 1·5 and 1·3 to give a defined model ratio of density, but these substances would probably not show strength and/or viscosity so related that a defined model ratio of these properties would follow. Conversely, if reasonable agreement in terms of viscosity or strength is found in two materials, their densities may be out of proportion, etc. For some of the model materials used in our runs, the density can fortunately be changed considerably without having much effect on the rheological characteristics (see p. 240). Clearly, the more complex the structure to be modelled and the greater the number of rheological properties that are significant for the study, the less likelihood that a reasonably consistent scale model can be constructed.

Much of the work reported from the centrifuged model studies is therefore less quantitative than desired. None the less it is hoped that a study of the structural patterns evolved will prove both illuminating and stimulating. At least it has been so to the author, who has gained new understaning of global structures through his study of the experimental results obtained in the Uppsala laboratory.

THE UPPSALA CENTRIFUGE LABORATORY

While watching on television the exciting launching of an early American space capsule, an event which focused attention on gravity as a force of prime significance in nature, the author became inspired to use a large-capacity centrifuge for model study of geotectonic processes. In consequence, a number of domal structures, fold and boudinage structures, and structures simulating oceanic ridges, extrusion of lavas, generation of fold mountains and the rise of batholiths were produced in models run in a centrifuge of moderate acceleration during the winter of 1960–61 at the University of Chicago (Ramberg, 1963a).

Previous published reports described the application of the centrifuge method in the study of mining problems and soil mechanics. Bucky (1931) employed the method for the investigation of the stability of mines, as did Panek (1952). Pokrovsky and Fyodorov (1936) used centrifuges in their soil-mechanics studies.

Since 1963 a number of reports on the practical applications of the centrifuge method—chiefly in soil mechanics—have appeared. For example, Avgherinos and Schofield (1969), Bassett (1973), Malushitsky (1975) and Hoek (1965), the latter being concerned with mining problems.

When it comes to centrifuged models aimed at the understanding of general structural-geological problems of small and large scale, most published reports are based on work done in the Uppsala laboratory (see bibliography).

Professor Beloussov, well-known for his interest in vertical movement and gravity tectonics, has constructed a centrifuge tectonic laboratory in Moscow chiefly aimed at the study of general tectonic problems; see the reports by Sycheva-Mikhailova (1970 and 1973). See also the contribution by Goncharov and Govelov in Beloussov and Vikhert (1975). At Queen's University, Canada, Professor Dixon has recently finished a centrifuge laboratory for the purpose of experimental studies of gravity-controlled geological problems. When the present book goes to press there is no published report available from Dixon's laboratory. See, however, Roeder and Dixon (1977).

The low activity in the field of experimental modelling of geological structures by means of centrifuges is quite surprising in view of the usefulness of the method. As Hubbert stated in his classic work on scale models applied to Earth science in 1937: "The method (*centrifuge method*) is unique, and may be expected ultimately to yield results of great importance, which are otherwise unobtainable" (author's italics).

After moving to the University of Uppsala in 1962, the author established a reasonably well-equipped laboratory for experimental tectonics, and research was continued in several directions. The author wishes at this point to acknowledge the very valuable assistance of Dr Ove Stephansson now Professor at Luleå Technical University, and Mr. G. Haglund in the construction of the laboratory.

Figure 10.1 shows a section through the latest centrifuge arrangement in the improved tectonic laboratory in Uppsala. A stroboscopic light permits inspection of the surface of the model during runs. This is necessary in order to be able to stop the run at appropriate stages of evolution, an important requirement for accurate and quantitative work. A closed-circuit television arrangement is found convenient for demonstration to groups for people, for example, in laboratory courses for students. The latest centrifuge in use is a machine specially built by W. Stock & Co., Marburg, and modified in our own workshop. Its maximum capacity is 10 kg models run at 4000 *g*. The models, which are housed in fixed compartments, are either 19 cm diameter circles or 19 by 19 cm squares. The centrifuge is supplied with a cooling system and thermostat which keep the temperature constant during the runs.

Many of the models described in this book are run in a smaller machine whose maximum capacity is 1·5 kg with model-diameter 10 cm and acceleration up to 3000 *g*. This machine, which incidentally was the only one

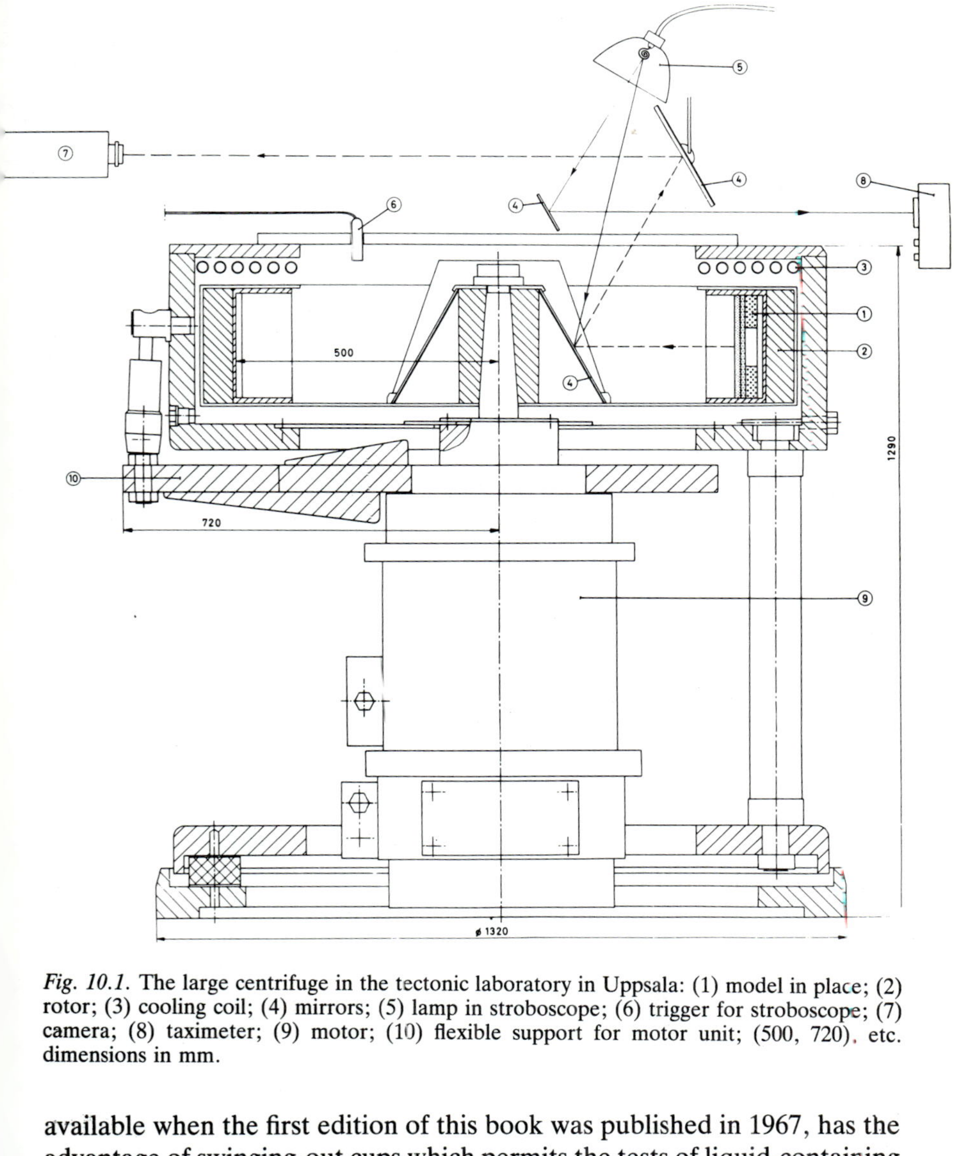

Fig. 10.1. The large centrifuge in the tectonic laboratory in Uppsala: (1) model in place; (2) rotor; (3) cooling coil; (4) mirrors; (5) lamp in stroboscope; (6) trigger for stroboscope; (7) camera; (8) taximeter; (9) motor; (10) flexible support for motor unit; (500, 720), etc. dimensions in mm.

available when the first edition of this book was published in 1967, has the advantage of swinging-out cups which permits the tests of liquid-containing models (e.g. magma-imitation models).

In properly prepared models the cylindrical shape of the equipotential surfaces must be taken into account. For example, the source layer for domal structures, if intended to simulate a flat stratum (or a spherical shell) in the Earth, must be shaped as a part of a cylindrical sheet. If the sagging of surface strata into a basin is being studied, the model basin must be a depression in relation to the equipotential cylindrical surface of the model, not in relation to a plane surface.

If the horizontal layering in models is made straight, lateral flow takes place during the run in the centrifuge until the stable curvature is obtained. This movement may cause unwanted disturbances, which, however, are minimized or prevented by carefully preparing a proper curvature of the layering in the initial model. In the large centrifuge this offers no problem because the permanent bottom of the model compartments is curved exactly parallel to the circumference of the rotor.

In addition to the centrifuge various instruments have been constructed for measuring mechanical and rheological properties of the model materials.

MODEL MATERIALS USED

Modelling clay (i.e. a clay–oil mixture, also called plasticine), various types of silicone putty (= bouncing putty), a plastic-base painter's putty, collophony–ethylene phthalate mixtures (= stitching wax), various wax–oil mixtures, mixtures of heavy powder (e.g. magnetite, tungspar) and silicone putty or stitching wax, burned mixtures of plaster of Paris and sand, plates of concrete, and aqueous solutions were all employed as materials for the simulation of solid rocks with different competency and density, and fluid magmas.

Painter's putty and modelling clay both exhibit yield strength and ultimate strength and are, moreover, relatively dense. These substances are suitable materials for simulating the more competent rocks in various structures and as overburden over a more mobile and buoyant substratum which, e.g., functioned as source layer for domes, batholiths and other buoyant bodies in the models.

Silicone putty, either pure or mixed with a fine powder of magnetite and tungspar to achieve the desired density, and stitching wax by virtue of their low density and viscous behaviour, were used as imitation materials for the least competent rocks and for the buoyant rocks in gravitationally unstable structures.

In a few instances the rise of magma through the crust was simulated by aqueous solutions of $KMnO_4$.

Modelling clay is a mixture of clay and oil whose density and rheological properties depend upon the proportion of solid clay particles to oil. The mixtures employed in the runs described in this book varied in density between $1{\cdot}61\ g\,cm^{-3}$ and $1{\cdot}71\ g\,cm^{-3}$. The substance exhibits a yield strength, depending upon the proportions in the mixture, and a sort of anisotropic hardening effect, i.e. under uniaxial compression the yield strength increases in the compressed direction but decreases for subsequent compression in perpendicular directions. Thus, strain-hardening for compression in one direction is associated with strain-softening for compression in perpendicular directions.

As a measure of the yield strength of the mixture with $\rho = 1{\cdot}71\ \mathrm{g\ cm^{-3}}$ a uniaxial compressive stress (in pure shear) of about 4×10^5 dyne $\mathrm{cm^{-2}}$ or greater was found necessary to produce plastic flow. At stresses above the yield strength a viscosity coefficient varying between $0{\cdot}5\times10^8$ and $7{\cdot}4\times10^8$ poises was determined at room temperature. However, the viscosity depends very much upon the magnitude of the stress, and it is exceedingly sensitive to the deformational history of the sample; it is therefore not possible to decide what value one should assign to the viscosity when the modelling clay is used in the models. Where the conclusions drawn from the runs depended on the quantitative rheological properties of the model material—e.g. conclusions as to the rate of tectonic processes such as rise of salt domes and evolution of fold-mountains—modelling clay was not very suitable. However, in this connection it should be remembered that the rheological properties of the natural rocks are almost unknown, so that even fully known properties of the model materials cannot at the present time provide quantitative information on rate problems in tectonics.

The *painter's putty* employed in the models is of a special plastic-base type with a density of between $1{\cdot}80\ \mathrm{g\ cm^{-3}}$ and $1{\cdot}90\ \mathrm{g\ cm^{-3}}$. In a few runs a type with a density of $2{\cdot}18\ \mathrm{g\ cm^{-3}}$ was used. (Ordinary painter's putty, consisting of mixtures of linseed oil and white pigment, was found unsuitable for work with centrifuged models because of separation in the centrifugal field.) The putty has rather ill-defined rheological properties depending much upon the history of the sample. Fortunately, this kind of plastic-base putty dries very slowly in air.

When stress is applied on a sample of relaxed putty it is at first very soft under an initial rather small strain. The yield strength required to be overcome in order to produce this initial strain is very low. After this initial strain has been produced the sample hardens and exhibits a yield strength which in uniaxial compression lies in the neighbourhood of 3×10^4 dyne $\mathrm{cm^{-2}}$. Under deformations at stresses above the yield strength, the viscosity increased rapidly from 1×10^5 poise which is the lowest value measured on the samples used, to values up to 10^8 poise which is found when the strain has become large and the stress has reached magnitudes of the order of $10^6\ \mathrm{dyne\ cm^{-2}}$. The putty thus exhibits a pronounced strain-hardening effect. Like the modelling clay, the putty hardens anisotropically in the sense that it becomes gradually harder under continually increasing compressive strain in a given direction, while at the same time tending to become softer for compressive strain in perpendicular directions. The rheological character of the putty is, moreover, complicated by the fact that gentle shaking of a hardened sample causes relaxation and disappearance of some of the strain-hardening.

Obviously, with such rheological behaviour it is not easy to know which value of viscosity or strength to employ in scale-model calculations.

Silicone putty or bouncing putty is an artificial visco-elastic substance. Most early models, e.g. those described in the 1967 edition of the book, were made of silicone putty manufactured by Dow Corning Chemical Company. The material was then available in three qualities, C-4551, XC-2-0950, and XC-2-0951, with somewhat different viscosity. At present silicone putty similar to the Dow Corning products is now available from several sources. The various types of silicone putty used in the models described in this book do not differ much in their rheological properties. The chief difference is in the magnitude of their strength. Some types display sufficient strength to support themselves without sagging in pieces up to several cubic centimetres when resting on the table. Other types will flow and spread on the table when the pieces are of the order a few cubic millimetres. In the description of the models silicone putty in often referred to simply as "silicone".

The silcone is quite unique in its rheological behaviour. Rapid impacts cause it to behave as an almost ideal elastic body, bouncing about like a piece of vulcanized rubber when dropped on the floor. Under low and moderate stresses it behaves as a viscous liquid, showing vanishing or very low yield strength. It exhibits an ultimate strength value beyond which fractures develop.

Density, as measured on the batches employed, varied between 1·12 g cm^{-3} and 1·14 g cm^{-3}. To obtain the higher densities needed for scale reasons the silicone was mixed with a fine powder (sieved through a 250 $mesh^{-1}$ in sieve) of magnetite and tungspar. Densities of more than 2 g cm^{-3} were then obtained without destroying the material's ability to behave as an elastic, as well as a viscous, body. However, admixture of

Table 10.2. Viscosity in poise at 24°C

Pure silicone (C-4551), ρ = 1·12 g cm^{-3}	Silicone–magnetite mixture, ρ = 1·30 g cm^{-3}	ρ = 1·81 g cm^{-3}	Silicone–tungspar mixtures, ρ = 1·86 g cm^{-3}
(2·0–2·57)$\times 10^5$ depending on stress	(2·64–3·09)$\times 10^5$ depending on stress	(5·16–7·16)$\times 10^5$ depending on stress	(7·36–12·22)$\times 10^5$ depending on stress

Table 10.3. Viscosity of collophony–ethylene phthalate mixture

Temperature, °C	Viscosity, poise
22·0	$5{\cdot}73\times10^{7}$
24·8	$2{\cdot}46\times10^{7}$
28·2	$9{\cdot}18\times10^{6}$
31·2	$2{\cdot}63\times10^{6}$
35·0	$4{\cdot}83\times10^{5}$
39·5	$1{\cdot}35\times10^{5}$
44·0	$3{\cdot}11\times10^{4}$
47·2	$1{\cdot}52\times10^{4}$

Thanks are due to F. D. O. Stephansson, now professor at Luleå Technical University, for the experimental measurement of the rheological data noted above.

the solid powder increased the viscosity by a factor of up to ten. If the powder is too coarse, separation will occur in the centrifugal field.

Some of the pure silicones are almost Newtonian, but a noticeable, though small, deviation occurs in the mixtures. Table 10.2 gives the viscosity of pure silicone and some silicone powder mixtures.

Mixtures of collophony and ethylene phthalate produce visco-elastic materials with viscosity coefficients varying within very wide limits, depending upon the proportion of the liquid phthalate and the solid and brittle collophony. In the models, mixtures with viscosities in the range 10^6–10^8 poise have been employed. An example is given in Table 10.3

The collophony mixtures behave as almost Newtonian fluids at stresses below their ill-defined ultimate strength, which for some of the mixtures employed lies in the neighbourhood of 10^7 dyne cm^{-2}. Exposed to deviatoric stresses of this magnitude, the material is brittle and shatters. A density of about 1 g cm^{-3} was found for the mixtures.

$KMnO_4$ solutions were found useful as model material for liquid magma when modelling clay and/or painter's putty simulate solid rocks. It is particularly suitable in that manganese precipitates from the "magma" in contact with the putty, thus leaving a black track which shows the movement path of the fluid.

SOME INTRODUCTORY TESTS

To demonstrate qualitatively the significance of the centrifuge technique for model works in the study of gravitational instability, photographs of

Fig. 10.2. Plate of solid granite broken in the centrifugal field in a centrifuge. Distance between supports is 8·15 cm. See text.

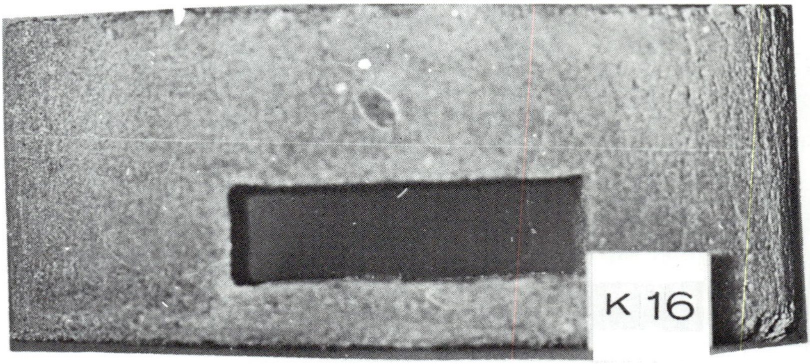

(*a*)

(*b*)

Fig. 10.3 (*a*) Initial stage of simple model of a tunnel; (*b*) collapsed tunnel after run in centrifuge. Width of model 9·8 cm. (Scale bar 1 cm.) See text.

Fig. 10.4. Collapsed roof above tunnel model run in centrifuge. Width of model 9·8 cm. See text.

three simple rock-mechanic tests are presented (Figs 10.2, 10.3 and 10.4). The model in Fig. 10.2 is no more than a plate of solid granite 1·04 cm thick and 10 cm long resting on two supports 8·15 cm apart placed on the bottom of the centrifuge cup. As the velocity of rotation increases, the granite plate becomes increasingly "heavy" in the field of rising centrifugal acceleration until the tensile stress developed at the convex surface of the bending plate exceeds the tensile strength of the material, and collapse occurs.

In all its simplicity this model demonstrates well how the centrifuge method may be useful not only for Earth scientists, but also for construction engineers who plan constructions large enough for gravity to be a problem. For example, what is the maximum span a certain kind of bridge can have without collapse under the heavy traffic of the rush hour? Of course, the stability of a bridge can usually be calculated wihout experimental modelling, but for the sake of explaining the principle let us assume that the theoretical stability calculation is to be tested in a model.

It is necessary that the model is exactly geometrically similar to the prototype, the model ratio of length being l_r. The materials in the model may or may not be the same as in the prototype, but if they are, the interpretation of the test is particularly simple by virture of the model ratio of strength then being unity and the model ratio of density as well, being unity.

The model shown in Fig. 10.2 collapsed at 1700 g in the centrifuge. The maximum span which the prototype can have without collapse can now be determined by the formula for the model ratio of strength:

$$s_r = \rho_r l_r a_r \tag{10.7}$$

where $s_r = 1$, $\rho_r = 1$ and $a_r = 1700$. Accordingly, l_r is 1/1700, so the maximum span of the full-scale bridge is 1700 times the span of the model. A bridge longer than that limit will collapse just as the model collapsed in the centrifuge. In the model the distance between the supports was 8·15 cm and the thickness of the plate 1·04 cm. Accordingly, the tests shows that in nature a granite plate $1{\cdot}04 \times 1700 = 1768$ cm ($= 17{\cdot}68$ m) thick will collapse due to its own weight if the distance between the supports exceeds $8{\cdot}15 \times 1700 = 13\,855$ cm ($= 138{\cdot}55$ m). In practice, however, the thick plate is apt to be mechanically weaker than the small model even if taken from the same granite body. The small plate can be cut out of a fracture-free part of the granite while it would be difficult to escape fractures and other zones of weakness in a plate more than 100 m long.

Suppose that the model was constructed of material with less strength and smaller density than the prototype material, the model ratios being e.g. $s_r = 0{\cdot}1$ and $\rho_r = 0{\cdot}5$. The limiting model ratio of length would then be

$$l_r = s_r \rho_r^{-1} a_n^{-1} = 0{\cdot}1 \times 2 a_r^{-1}. \tag{10.2}$$

However, we will not find that the model, if it is of the same size as above, collapses in the centrifuge already at an acceleration equal to 340 g. Hence the maximum span-length of the bridge which is planned is still 1700 times the span-length of the model. This, of course, we would expect, otherwise experimental modelling would not be very useful!

If the prototype bridge consists of several materials with unlike strength and unlike density, the model must, of course, be similarly constructed and the proportions between the different densities as well as between the different strengths in the model must be the same as in the prototype. This is required by the laws of dynamic similarity.

Models K-16 and K-17 (Figs 10.3 and 10.4) simulate the collpase of a tunnel and are discussed in Ramberg and Stephansson (1965). They consist of a burned mixture of plaster of Paris and magnetite sand, density 1·9 cm^{-3} and tensile strength $5{\cdot}1 \times 10^6$ dyne cm^{-2}.

The total collapse to the state shown in the figures occurred at 610 g for model K-16 and 630 g for model K-17, but fractures had already develped at lower acceleration.

In order that models of this type shall be useful for the planning for underground constructions the detailed structure of the surrounding rocks must be carefully scrutinized for fractures and other zones of weakness. Corresponding structures must be built into the model with great care in order to make it as closely similar to the prototype as possible.

Even if it is difficult to construct models which are dynamically exactly similar to the rock complex within which an underground construction is being planned, it is believed that valuable practical information can be extracted from such models, information which is relevant to the questions of safe roof-span, the spacing of pillars, etc.

It is beyond the scope of this book to discuss models aimed at the solution of technical problems, which are dealt with by Bucky (1931), Hoek (1965), Pokrovsky and Fyodorov (1936), Malushitsky (1975), Panek (1952), and others. See also the discussion of the large-scale centrifuges for geotechnical research planned in the USA (Kornberg, 1979).

11 Dome and Diapir Models

As salt domes or diapirs represent the classic structure generated by gravity, a description and discussion of dome models as generated in the field of a centrifugal force are appropriate. It is to be understood, however, that most of the models not only simulate gravity-generated salt structures, but can also be regarded as replicas of buoyant bodies of granite or other silicate rocks, so-called mantled domes (Eskola, 1948) and batholiths (Grout, 1945) being common examples.

The chief driving power of a magma rising in volcanic vents is probably also the buoyant force of the low-density silicate melt surrounded by more dense plastic rocks. The author, moreover, is of the opinion that the rise of buoyant material plays an important role as a global differentiation process in the deep parts of the Earth, a process which in important ways affects the crust of our globe. The so-called low-velocity layer in the upper mantle, for example, may well be a buoyant layer which tends to rise as ridges and diapirs, and from sources deep in the mantle diapirs or plumes may rise.

The distinction between the terms *dome* and *diapir* is not clear. A dome is a geometric figure with a convex surface facing upward; diapir indicates piercing (forceful penetration) because it stems from the Greek word *diapeirein* which means to pierce (Braunstein and O'Brien, 1968). It seems logical to use the term dome for a body whose upper surface is smooth and convex and does not cut the layering of the overburden. Diapir should then be reserved for bodies which penetrate the layering of the overburden. In practice, however, dome is often used even for salt bodies which have pierced the layering, in fact, bodies which are far from dome-shaped, see e.g. Braunstein and O'Brien (1968). In the following we shall accordingly make no distinction, but use dome and diapir synonymously for bodies which have risen on account of buoyancy.

In geophysical literature the name "plume" is often used when referring to phenomena which structurally seem to be identical to ordinary diapirs. See, e.g., Anderson (1975).

GEOMETRY OF DIAPIRS FORMED IN INITIALLY SYMMETRICAL MODELS OF LAYERED HOMOGENEOUS MATERIALS

The shape of the experimental buoyant bodies depends much upon the mechanical properties of the materials and the initial structure of the system in which the diapirs develop. In systems containing only Newtonian fluids initially arranged in even-thick density-inverted strata parallel to the equipotential surfaces, the diapirs develop as regular symmetrical bodies, first often as even-spaced ridges (anticlines), symmetric about the axial plane, and subsequently as more or less drop-shaped bodies with axial symmetry, rising regularly spaced from the early-formed anticlines. If the overburden fluid is thick, the domes or diapirs usually become quite globular in shape before reaching the surface, connected with the source layer only by a narrow stem which ultimately is pinched off. On reaching the surface the top part of the diapirs spread and may coalesce to a continuous layer. The tests with oil models are good examples on this type of regular symmetrical and often even-spaced ridges and domes (see Figs 7.1–7.8).

In the centrifuged models, on the other hand, some of the materials exhibit a more or less well-defined yield point, some are brittle and may fracture during the run, and often the initial pre-run structures of the models are purposely made irregular in attempts to simulate natural structures and make the tests as comprehensive as possible. In consequence, the buoyant bodies which rise through the model during the centrifugation often give rise to complicated structures quite different from the simple regularity of the diapirs in the oil tests. Even when the initial structure of centrifuged models is in the form of smooth, even-thick layers, the shape of the domes are usually different from those developed in oil-syrup tests. Only when the over-burden as well as the buoyant layer consist of strengthless silicone do centrifuged models develop patterns closely similar to the oil-syrup models. Models S-202, S-204 and S-205 are good examples (Fig. 11.1), as is the model in Fig. 11.2.

When considering models S-202, S-204 and S-205 we recall the dynamic theory which states that the dominant wavelength is proportional to the buoyant-layer thickness in models consisting of the same kinds of materials (cf. p. 72). Even if the spacing of the domes in the models perhaps is not exactly proportional to the thickness of the layers, it clearly is a sympathetic relationship: the thicker the layers, the wider the spacing of the domes. The irregular pattern exhibited in the photographs is partly due to the condition that the profiles are straight sections through a system of diapirs spaced evenly in the horizontal plane. The profiles therefore cut the various domes at different distances from their centres.

The oil-syrup tests discussed on p. 47 and the simple silicone tests

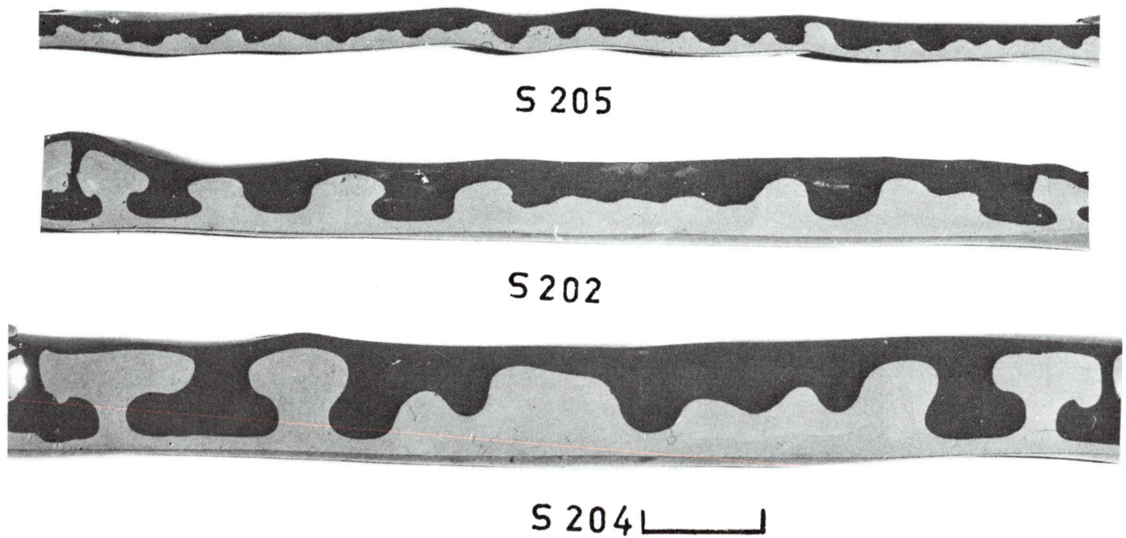

Fig. 11.1. Profiles through three models with inverted density stratification run in centrifuge. S-202 run for 2·5 min at acceleration varying from 800 *g* to 2000 *g*. Density of light-grey bottom layers 1·13 g cm^{-3}, of dark top layer 1·36 g cm^{-3}. Both materials are silicone putties (viscosity of the order of 10^5–10^6 poise). Scale bar 1 cm.

Fig. 11.2. Diapirs of low-density silicone putty having risen through more dense silicone putty (dark grey material) during run in centrifuge. Note down-pointed appendices on circular head of central diapir. Scale bar 1 cm.

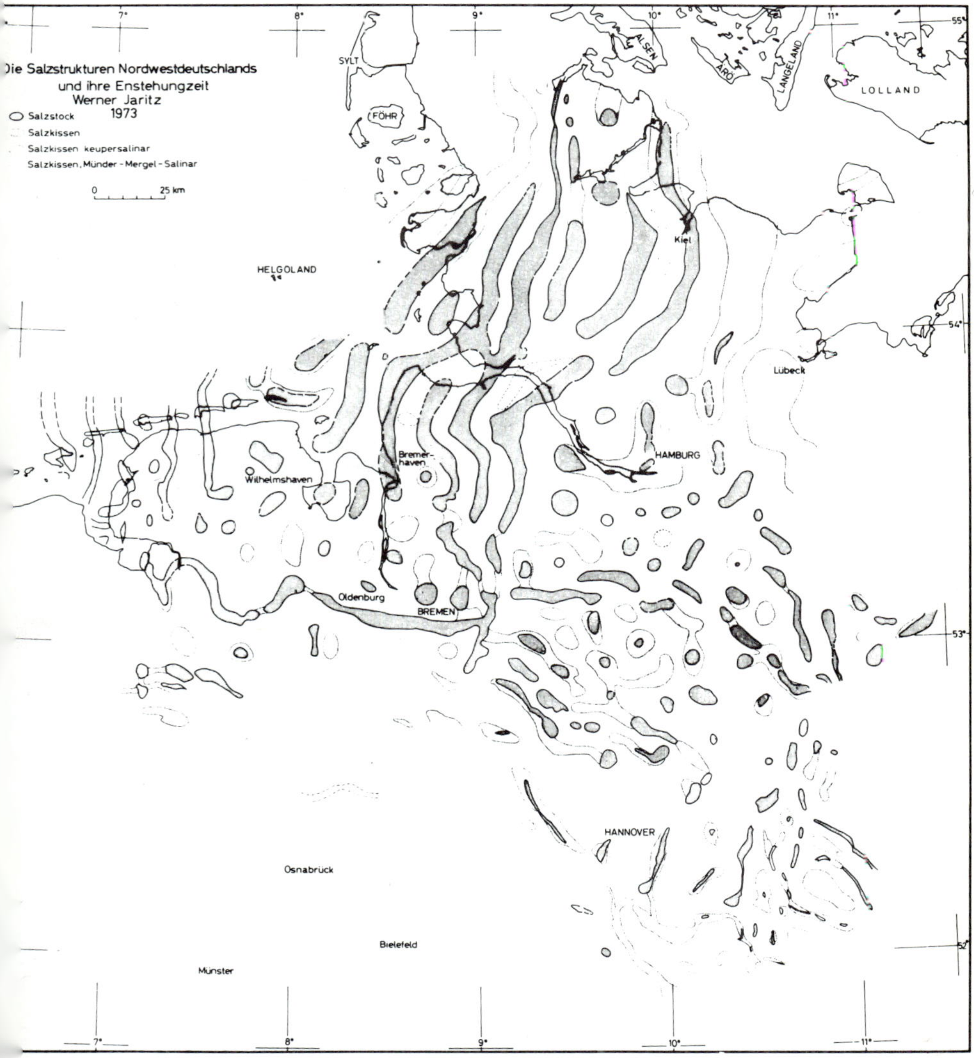

Fig. 11.3. Salt ridges and salt domes in N.W. Germany (from Jaritz, 1973).

S-202, S-204 and S-205 show clearly the theoretically predicted condition that the rise should occur in the form of evenly spaced protuberances, either shaped as elongate ridges or as separate diapirs.

The tendency to develop even spacing is apparent in many areas rich in salt domes and salt ridges. A good example is the N.W. Germany–North Sea region shown in Fig. 11.3. The deep structures of salt pillows, ridges and domes as pictured by Trusheim (1957 and 1960) are also strikingly similar to some of the experimental models. (see Fig. 11.4)

The chief reason for the ascent of magmas in volcanic vents is in all likelihood also the density contrast between silicate melts and the crystalline, but plastic surrounding rocks in the deep parts of the crust and in the mantle.

Fig. 11.4. Salt diapirs rising from salt ridges which in turn have developed from source bed of salt (from Trusheim, 1957).

Mohr and Wood (1976–77), Marsh (1975) and Marsh and Carmichael (1974) have accepted the fluid dynamic theory of buoyant layers as the explanation of the spacing of volcanoes in rift regions and in Island Arcs. It appears that volcanic vents are often quite regularly spaced as demonstrated in Figs 11.5–11.7. Mohr and Wood point out that the spacing depends upon the thickness of the crust in the sense that the average distance between volcanoes increases with the crustal thickness. This is interesting in light of the numerical tests discussed on p. 91 and presented in Table 7.5. If the overburden thickness is of the same order as the wavelength, then theory predicts that decreasing thickness should lead to

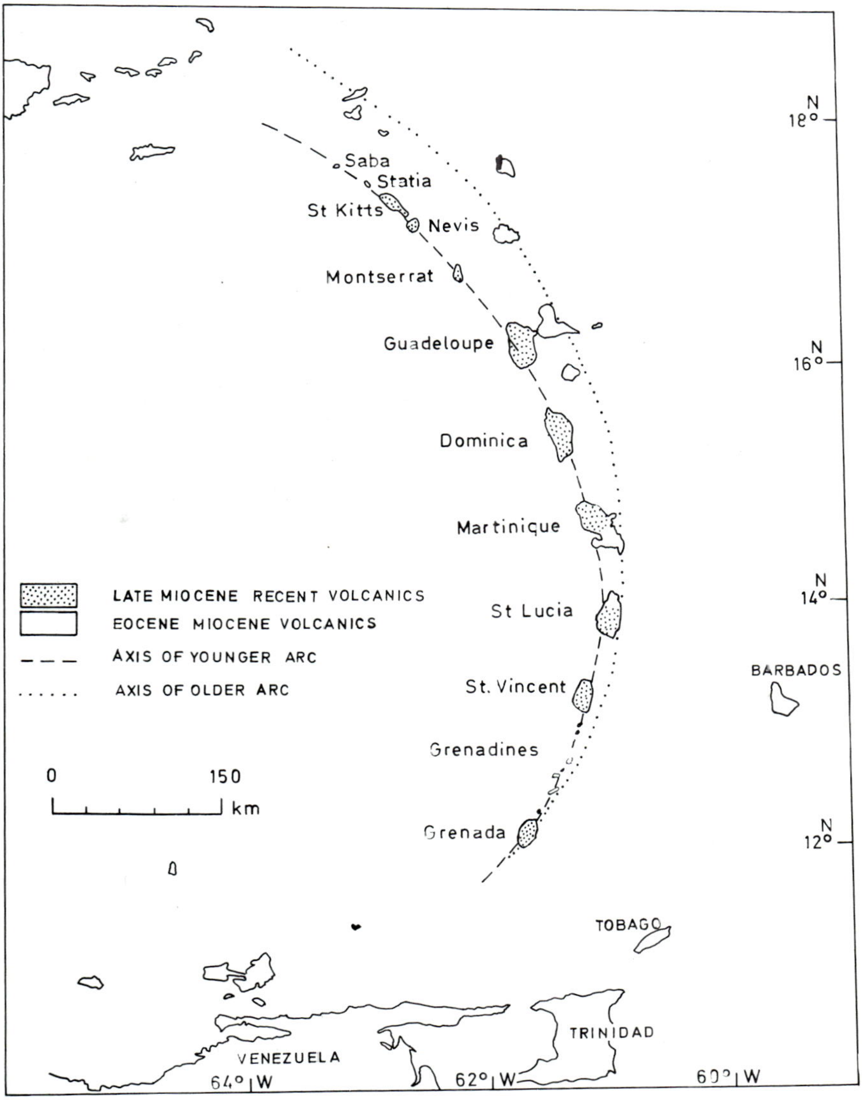

Fig. 11.5. Distribution of volcanic islands along the Lesser Antilles arc (from Powell, 1978).

shortened characteristic wavelength, and accordingly shorter distance between adjacent protuberances rising from an extensive sheet-shaped magma reservoir.

The writer suggests that the rather even spacing in the chain of granitic plutons in S.W. England (Fig. 11.8) can also be interpreted as an example on the dominant-wavelength dynamic model.

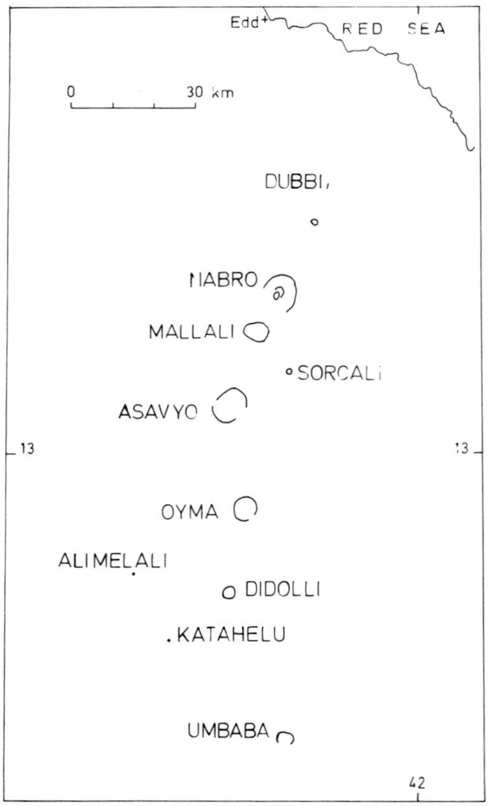

Fig. 11.6. Volcanic centres and calderas of the Dubbi line, N.E. Afar (Mohr and Wood, 1976–77).

Model O-26 (Fig. 11.9) can serve as a representative for mature diapirs generated in centrifuged models consisting of initially even-thick homogeneous layers of silicone as source and window-putty as overburden, arranged parallel to the equipotential surfaces. In this model the position of the diapir was predetermined by making a small bulge in the middle of the model as shown in Fig. 11.10.

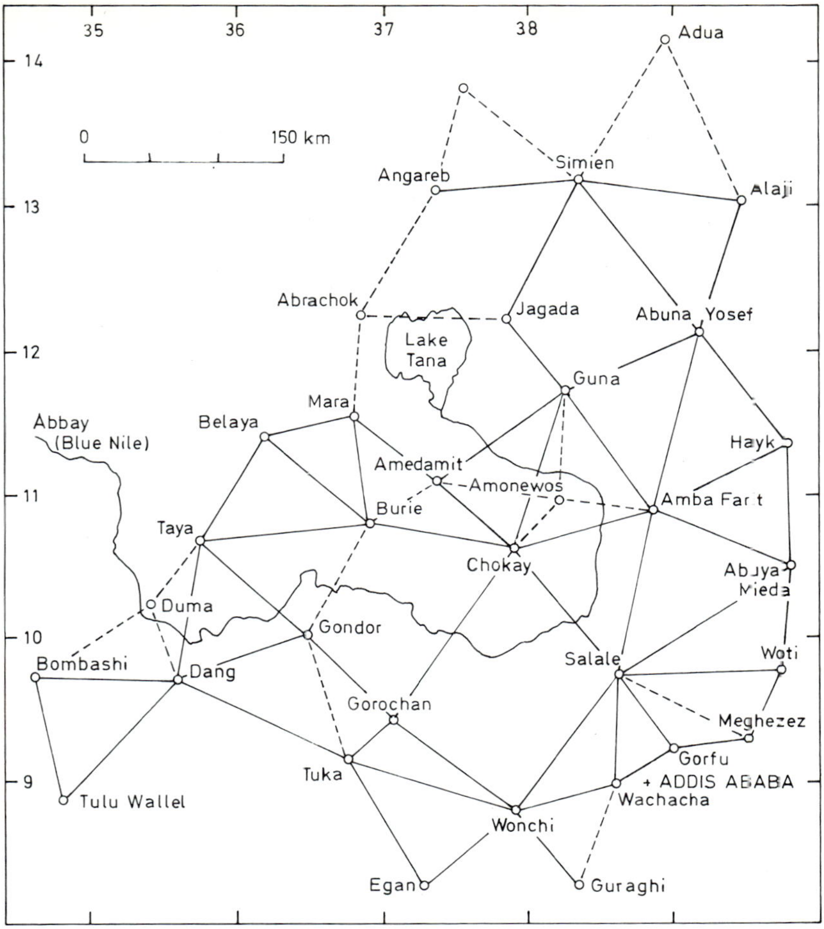

Fig. 11.7. Spacing of volcanic centres in part of N.E. Africa (Mohr and Wood, 1976–77).

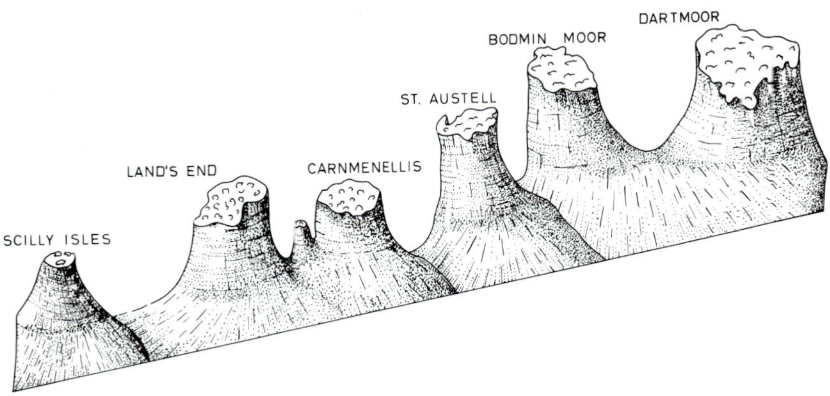

Fig. 11.8. Distribution of the granites between Dartmoor and the Scilly Isles, S.W. England (after Holmes, 1965).

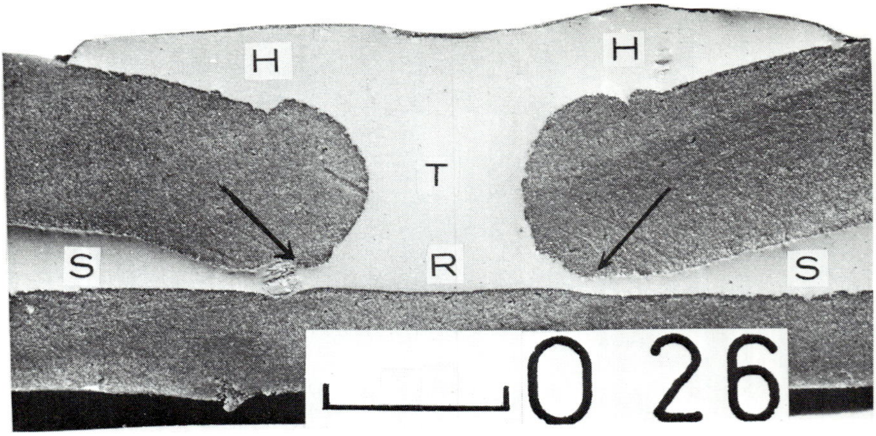

Fig. 11.9. Cross-section through dome of silicone putty (light grey) pierced through overburden of painter's putty (dark grey) with well-developed marginal sink (arrows), trunk (T), hat (H) and root zone (R). S is source layer. Scale bar 1 cm.

(If it makes no difference exactly where in the model one wishes the diapir to develop during the run, it is not necessary to initiate the growth by a bulge or any other irregularity. A diapir (or several) will usually develop somewhere even if the layer is completely even in thickness and parallel to the equipotential surface in the centrifugal field.)

After a run for 4 minutes at 700 *g* in the centrifuge, diapir O-26 developed from the initial state (Fig. 11.10) to the mature mushroom-shape (Fig. 11.9).

The source layer of silicone putty has a density of $1{\cdot}13\ \mathrm{g\,cm^{-3}}$ and a viscosity of about 3×10^5 poise, whereas the density of the overburden of painter's putty is $1{\cdot}87\ \mathrm{g\,cm^{-3}}$ and its viscosity is in the range 10^6–10^7 poise.

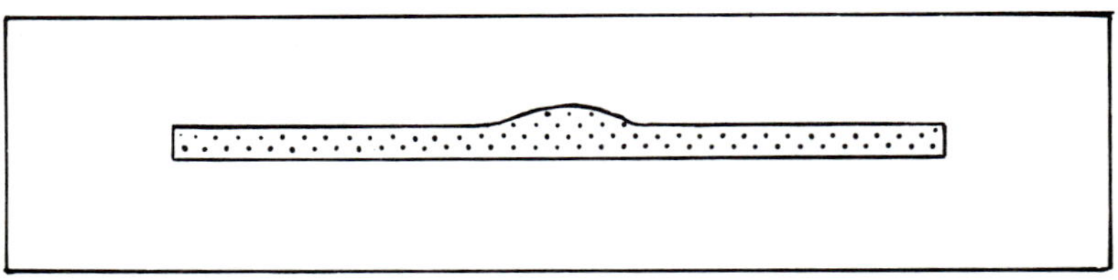

Fig. 11.10. Model O-26 prior to run in centrifuge. Dotted: buoyant layer of silicone putty with initiating bulge embedded in painter's putty.

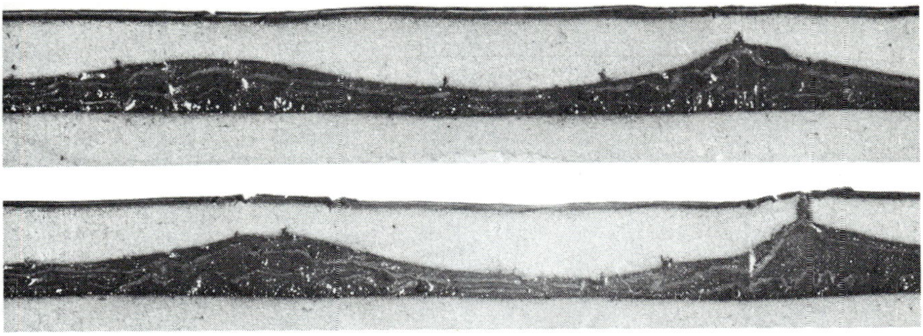

Fig. 11.11. Cross-sections of model S-103 showing early stage of domal growth of black silicone putty mixed with magnetite powder, $\rho = 1{\cdot}34$ g cm^{-3}, in painter's putty, $\rho = 1{\cdot}87$ g cm^{-3}. Run in centrifuge for 30 min at 2000 *g*. Several thin sheets of relative strong modelling clay embedded in the silicone putty have increased the strength of the source layer and hence diminished the velocity of rise.

This model demonstrates the typical geometric shape of diapir structures in general, features which are recognized in most models mentioned in the following discussion though generally less symmetrically and less regularly developed. One recognizes the *source layer* (S), the *marginal sink* (arrows), the *root region*, (R) the *trunk region* (T) and the *hat* (H) of the diapir.

Figures 11.13, 11.14 and 11.38 show other examples of model diapirs with regular and simple shapes.

The development of these geometric features depends on the stage of evolution of the diapir. At a very early stage the body occurs in the form

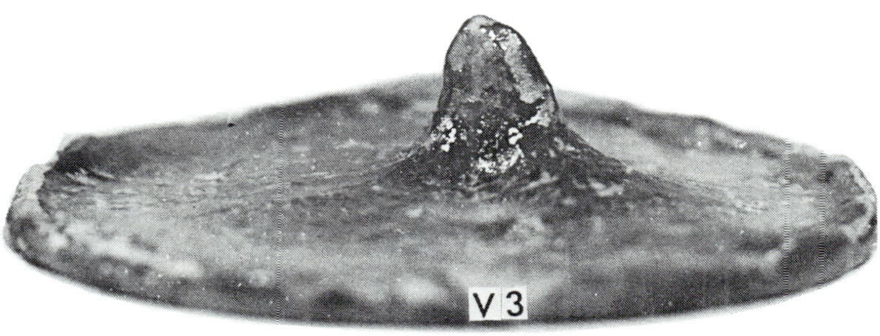

Fig. 11.12. Stage of domal growth somewhat more advanced than shown in Fig. 11.11. Buoyant material is quite stiff wax which has risen through overburden of painter's putty, now removed to show the shape of the dome.

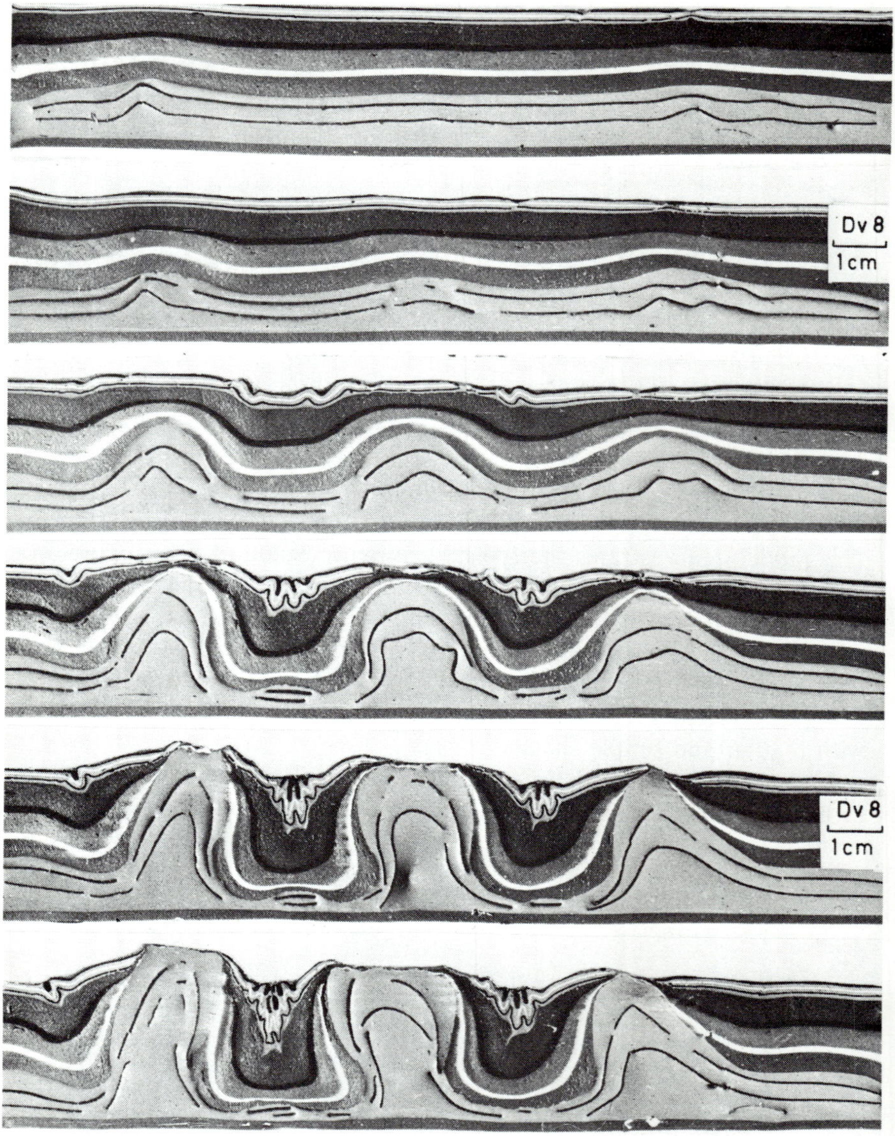

Fig. 11.13. Three ridges of buoyant silicone putty, containing two competent modelling-clay sheets which have ascended into overburden of multilayered painter's putty with top complex consisting of thin sheets of competent modelling clay embedded in silicone putty. Note formation of boudins as well as buckles of the competent sheets in overburden and in buoyant layer. Cross-cuts taken at different distances from culminations on the ridges.

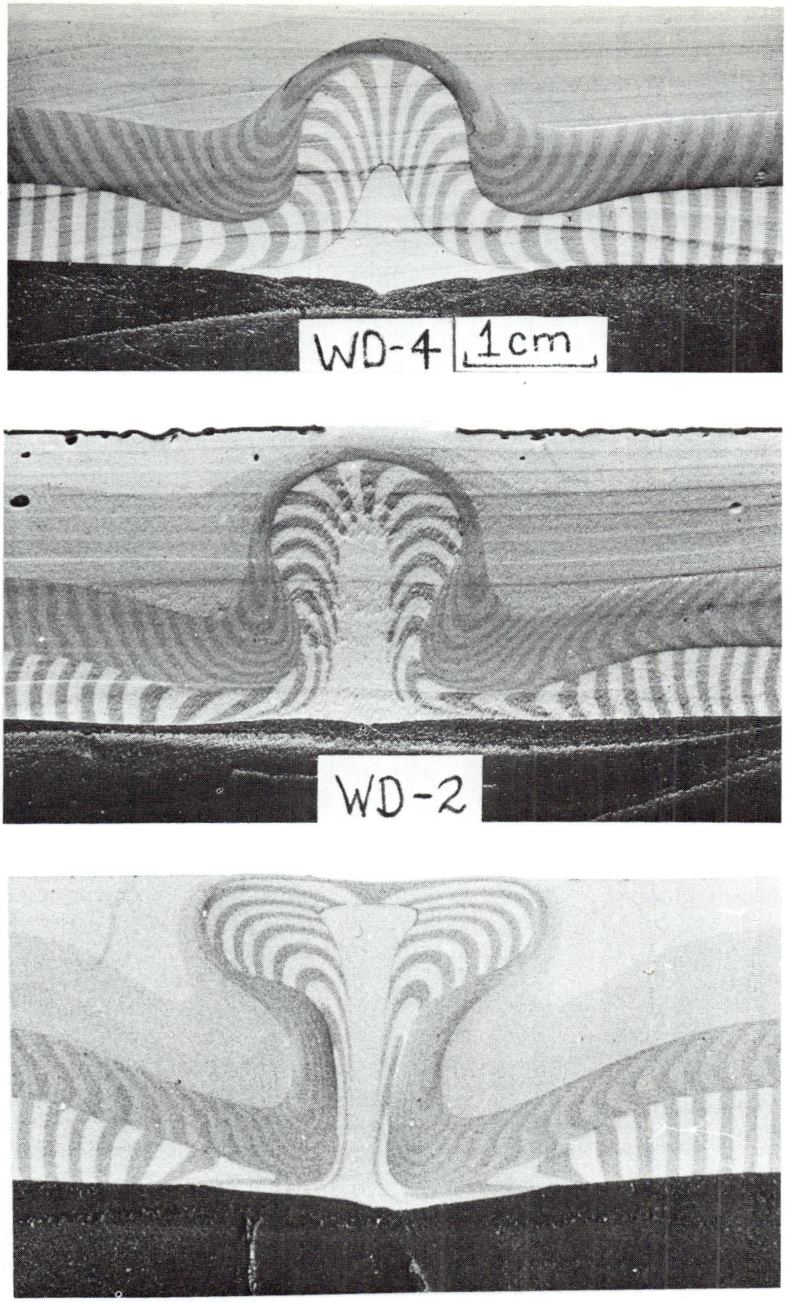

(*a*)

(*b*)

(*c*)

Fig. 11.14. Stages in the evolution of a diapir. Properties of layers, from bottom up: (1) $\rho_1 = 2{\cdot}01$, $h_1 = 10$, $\mu_1 \approx 10^5$; (2) $\rho_2 = 1{\cdot}4$, $h_2 = 6$, $\mu_2 = 1{\cdot}5 \times 10^6$; (3) $\rho_3 = 1{\cdot}56$, $h_3 = 6$, $\mu_3 = 3{\cdot}49 \times 10^5$; (4) $\rho_4 = 1{\cdot}5$, $h_4 = 4$, $\mu_4 = 5{\cdot}61 \times 10^5$; (5) $\rho_5 = 1{\cdot}48$, $h_5 = 4$, $\mu_5 = 5{\cdot}61 \times 10^5$; (6) $\rho_6 = 1{\cdot}49$, $h_6 = 4$, $\mu_6 = 5{\cdot}61 \times 10^5$. Values in g cm^{-3}, mm and poise. Layer 1 consists of painter's putty, layers 2 to 6 of silicone putty. (*a*) run 310 sec, (*b*) 405 sec and (*c*) 505 sec, all at 2000 *g* (from Dixon, 1975).

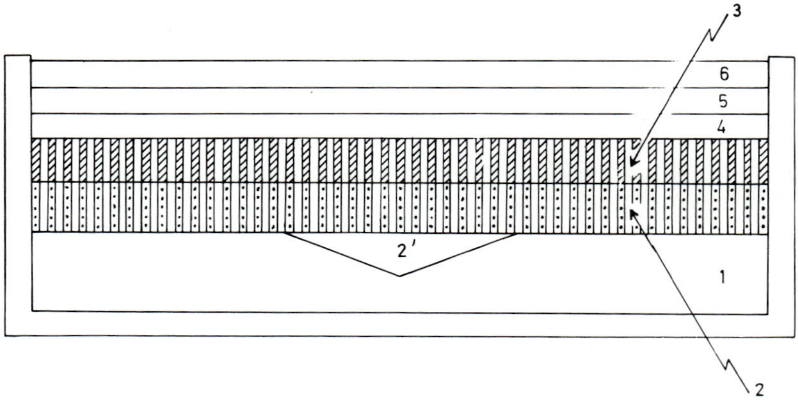

Fig. 11.15. Dixon's model (Fig. 11.14) prior to run in centrifuge. The vertical laminae with different colours in layers 2 and 3 have identical density and identical rheological properties in each layer. Region 2′ shows cross-section of body with same properties as layer 2 which initiated the diapiric rise (after Dixon, 1975).

of a gentle anticline or bulge on the source layer, and the surrounding marginal syncline is correspondingly faintly marked (Fig. 11.11), the marginal syncline becoming deeper as the diapir develops. This is observed in some of the oil-syrup models described above. The bulge or anticline

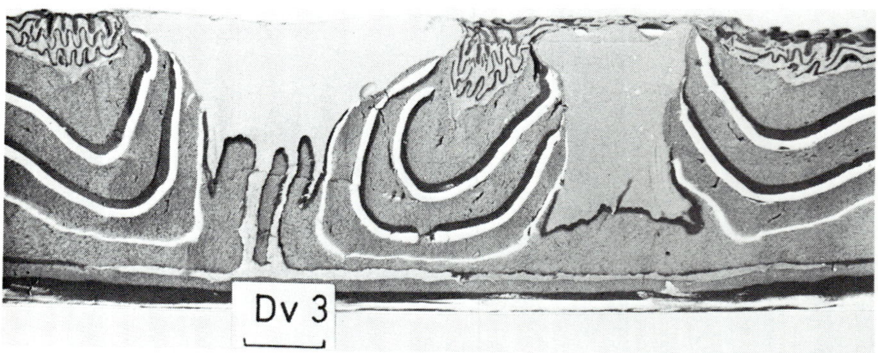

Fig. 11.16. Two diapirs consisting of different silicone putties, light coloured: $\rho = 1{\cdot}36\ \mathrm{g\ cm^{-3}}$; dark coloured: $\rho = 1{\cdot}49\ \mathrm{g\ cm^{-3}}$, having risen through overburden of painter's putty with layers of modelling clay. The buckled and boudinaged surface layer consists of soft silicone putty with thin sheets of modelling clay. Initially the layers were horizontal, the dark and the light coloured buoyant materials occurring as separate sheet-shaped bodies at the bottom of the model. Scale is 1 cm. The apparent rootless angular diapir on the right-hand side is in fact connected with the basal source layer as seen in the section through the model shown in Fig. 11.81.

Fig. 11.17. Dotted material is soft buoyant silicone (density 1·3 g cm^{-3}) which has ascended from the base of the layered overburden and spread on the surface. During this spreading the surface complex (three thin sheets of modelling clay embedded in soft silicone putty) has been buckled, partly broken, and rotated in the stress field caused by the overriding "nappe". The funnel-shaped light-coloured diapir consisting of stiffer silicone putty (density 1·13 g cm^{-3}) has risen later and less rapidly than the "nappe" which it deforms and penetrates. Note that the surface of the funnel-shaped diapir is a little above the surface of the spreading "nappe". This agrees qualitatively with the density differences and isostatic adjustment in the centrifugal field.

formed on the source layer in the early stages is generally wider than the root zone in the later stages. A little later the swell has developed into a steep mound as illustrated by the model shown in Fig. 11.12.

Figure 11.13 shows a model which may serve to illustrate different evolutionary stages of diapir generation, though the stages shown do not actually represent a time sequence, but rather sections across three ridges taken at different distances from culminations rising on the ridges. It is believed that this sequence of cross-sections taken at different distances but at a fixed time, in many respects corresponds to a sequence in time taken at a fixed section across a developing diapir.

For a true time-sequence of diapiric evolution the models made by Dixon (1975) at the Uppsala laboratory are illuminating, particularly for more advanced stages of evolution (Fig. 11.14). Figure 11.15 shows the initial stage of the model.

Some mature diapir models exhibit a curious shape in the region where the trunk changes to the hat. As shown in Fig. 11.16 narrow appendices hang down from the underside of the hat, extended into the inverted overburden. Figures 11.2 and 7.31 (pp. 249 and 124) show other examples on the same phenomenon.

These appendices were hardly noticed in the experimental models until they appeared in numerical finite-element models (Berner *et al.*, 1972). The finite-element result indicates that the structure is a significant dynamic feature which develops in diapirs when the proper conditions are encountered. In the writer's opinion the appendices are drag folds caused by the

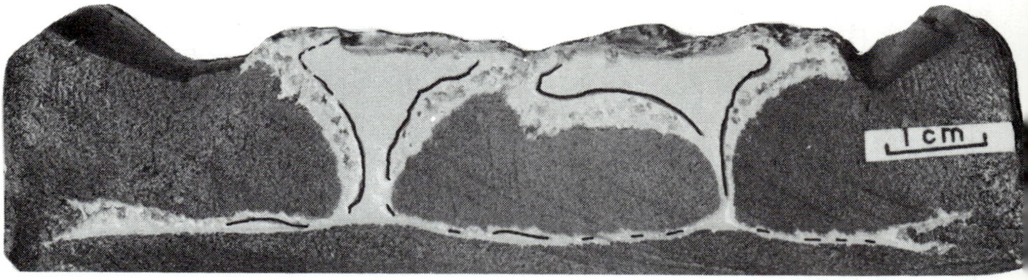

Fig. 11.18. Double dome of silicone with sheets of modelling clay stretched to boudins during rise. Late stage of evolution.

shear stress developed at the contact between diapir and overburden as the former rises and the latter sinks down in the rim synclines.

The spreading hat does not form until the top of the trunk approaches the free surface, hits an impenetrable stratum, or reaches a layer with equal or lower density than the ascending body. One of the diapirs (dotted) in Fig. 11.17 has pierced the surface and the hat has spread in the form of a nappe on top of the model. A later diapir (light-coloured) has penetrated the nappe, spread within it and deformed the structure considerably.

In many models the upper part of the dome starts to spread prior to piercing the surface even if the ascent is not prevented by an impenetrable layer or a low-density stratum in the overburden. This is due to the low confining pressure in the upper part of the overburden. To permit spreading prior to penetrating the surface the overburden must be relatively ductile.

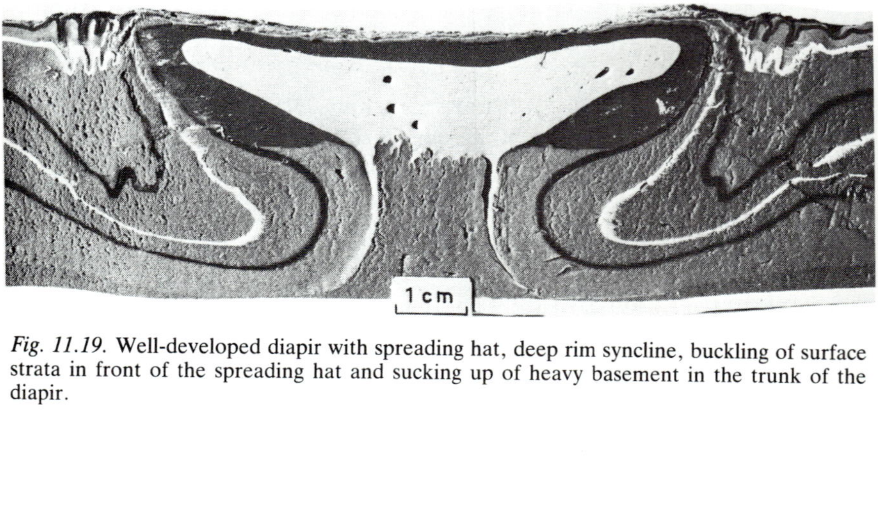

Fig. 11.19. Well-developed diapir with spreading hat, deep rim syncline, buckling of surface strata in front of the spreading hat and sucking up of heavy basement in the trunk of the diapir.

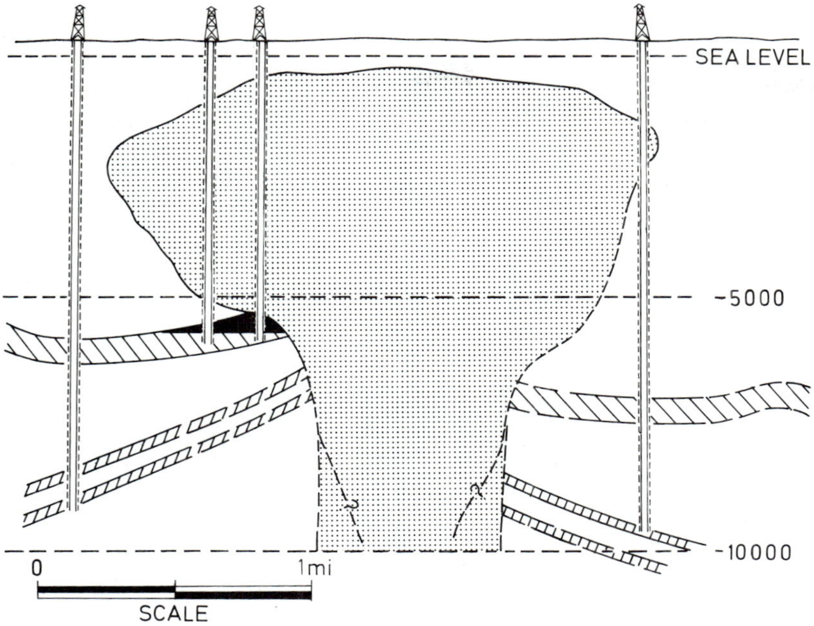

Fig. 11.20. Simplified cross-section of Bethel dome, Andersson County, Texas (from Halbouty, 1967).

Fig. 11.21. Diapir or "stock" of wax having risen through overburden of painter's putty which has been removed to show shape of the diapir. Scale 1 cm.

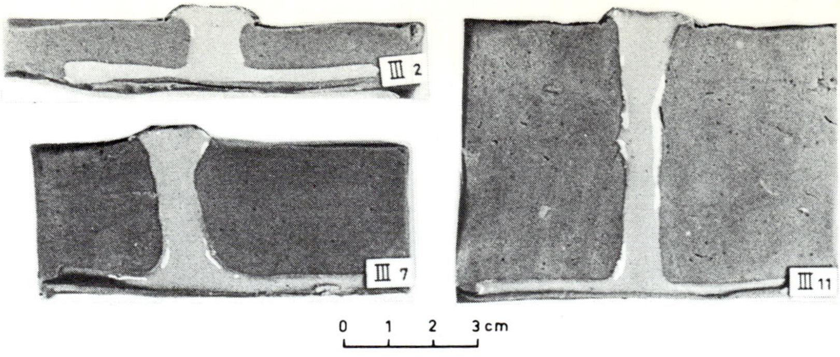

Fig. 11.22. "Stocks" of silicone putty in painter's putty overburden. Examples on different overburden thickness relative to source–layer thickness.

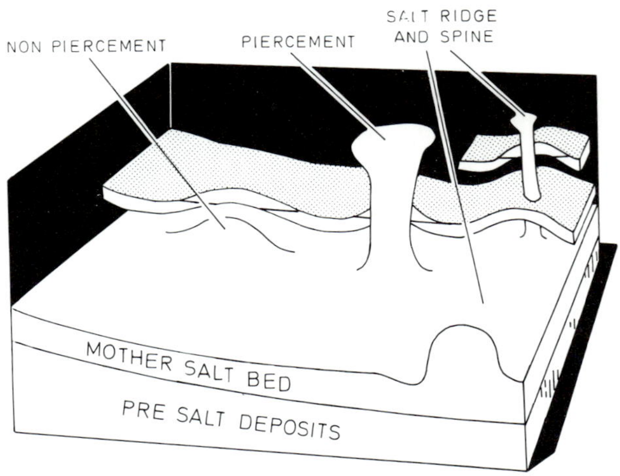

Fig. 11.23. Example on natural salt stocks. Cf. Figs 11.21 and 11.22 (after Halbouty, 1967).

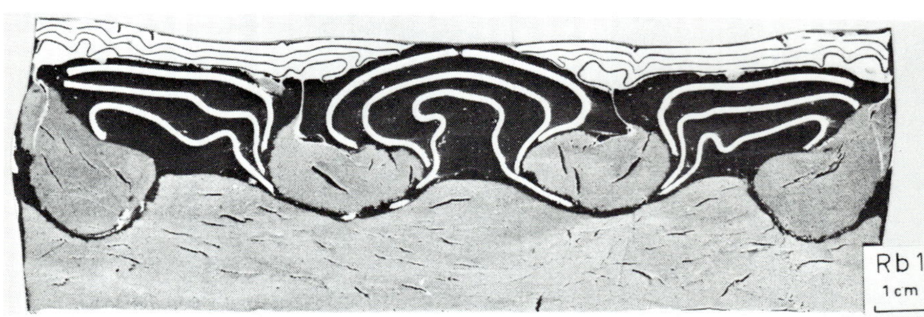

Fig. 11.24. Model simulating e.g. load-casts. Top layer of white low-density silicone putty with 3 thin sheets of black modelling clay rested on a layer of painter's putty, now occurring as four separate lenticular bodies sunk down through the buoyant dark silicone putty layer (containing three sheets of white modelling clay). The buoyant layer rested originally on a basement of heavy painter's putty.

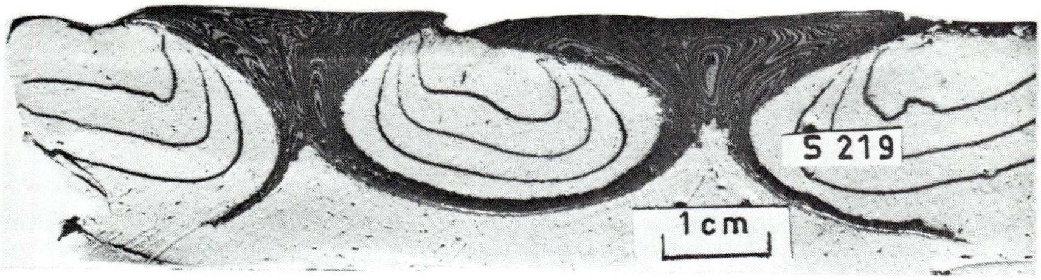

Fig. 11.25. Model showing tendency of flowage during diapiric evolution to produce conformity between initially cross-cutting structures. The model also simulates natural load-cast structures. (For initial structure of model see Fig. 11.26.) Model run 2 min while acceleration increased from 1 *g* to 1500 *g*, 1·5 min while acceleration increased from 1500 *g* to 2600 *g*, and 0·5 min while acceleration decreased to 1 *g*.

Model examples are shown in Figs 11.17–11.19. Natural examples are numerous, and one is seen in Fig. 11.20.

There are also models where the diapirs plunge all the way to the surface without much spreading prior to spilling out on the free surface. This seems to occur when the overburden is somewhat less ductile than in the above cases. Examples on model stocks formed under this condition are shown in Figs 11.21 and 11.22. Salt diapirs often show the same even-thick trunk with but little spreading at the top (Fig. 11.23).

From the point of view of dynamics there is no essential difference between a series of salt diapirs rising into a more dense overburden and segments of a layer of sand sinking down into soft clay to form a so-called load-cast structure. It is essentially a question of different geometric scale and different types of rocks involved. Clearly, the small scale of ordinary load-casts is only possible because of the softness of the unconsolidated

Fig. 11.26. Initial model S-219 (see Fig. 11.25). (1) White painter's putty ρ_1: 1·81 g cm^{-3}; h_1: 8 mm; (2) folded complex of silicone putty mixed with magnetite and tungspar powder and cut to a 4-mm thick plate, ρ_2: 1·35 and 1·34 g cm^{-3}; (3) white painter's putty, ρ_3: 1·81 g cm^{-3}, h_3: 3 mm; (4) modelling clay, 0·3 mm thick.

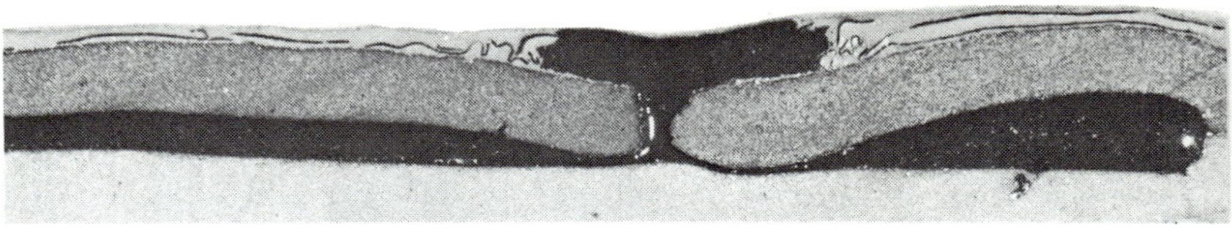

Fig. 11.27. Dark-coloured magnetite-containing silicone putty (ρ_1 = 1·34 g cm^{-3}) having penetrated overlying sheet of painter's putty (ρ = 1.87 g cm^{-3}) and spread within the top layer of pure silicone putty (ρ = 1·14 g cm^{-3}) with thin sheets of modelling clay. Run 1 hour and 10 min at 2000 *g*. Cf. Fig. 11.34.

sediments involved in such structures. By the same token, the compacted and mechanically strong salt and its consolidated overburden are able to flow in response to gravitational forces only because of the huge geometric dimensions. Some of the centrifuged models exhibit geometric patterns directly comparable to many natural load-cast structures (Figs 11.24–11.26).

When the ascending diapir reaches a ductile low-density stratum it does not spread underneath the stratum but ascends into the low-density layer and spreads within it. Model examples are shown in Figs 11.27–11.33.

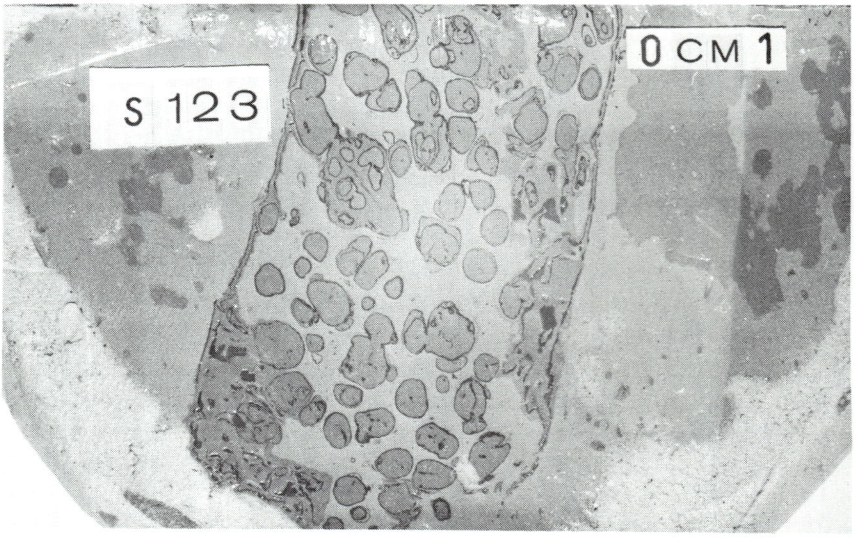

Fig. 11.28. Model (initial structure shown in Fig. 11.29) seen from above after run 10 min at 24000 *g*. Dark grey, red silicone putty (ρ = 1·14 g cm^{-3}). Light grey, overburden of pink silicone putty (ρ = 1·14 g, cm^{-3}). Black mantles around domes are remnants of the heavy black buoyancy-generating sheets shown in Fig. 11.29.

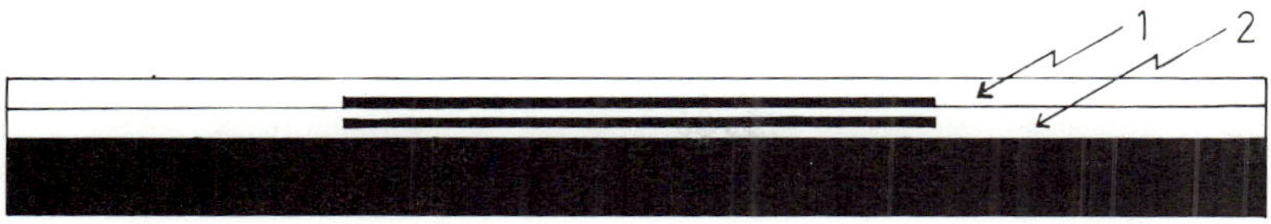

Fig. 11.29. Initial structure of Model S-123. Black: silicone putty with magnetite powder, $\rho = 1{\cdot}35$ g cm^{-3}. (1) Pink silicone putty, $\rho = 1{\cdot}14$ g cm^{-3}; (2) red silicone putty, $\rho = 1{\cdot}14$ g cm^{-3}. Length of section 9·8 cm.

(Note the similarity between the model in Fig. 11.27 and the concept of natural diapirs (Fig. 11.34) by Sanneman (1968).)

Translated to natural geological situations these models demonstrate that a body having risen owing to buoyancy in the Earth may well be surrounded by rocks of lower density. The buoyant force which lifted the ascending mass was generated by dense rocks at lower levels.

Note also the similarity between the famous Rhodesian granite domes (Fig. 11.35) and model S-123 (Fig. 11.28). In model Ny-9 (Fig. 11.36) a cluster of domes has spread in the uppermost strata of low-density silicone, giving a pattern analogous to parts of the Rhodesian region.

The shape of the spreading part of the diapir and its position in the low-density stratum depend greatly on the density contrast between the stratum and the diapir. The less the density of the former relative to the latter, the less does the diapir rise into the low-density stratum. If the density contrast is small, on the other hand, the diapir may reach considerable heights within the low-density stratum even if the latter is less dense than the

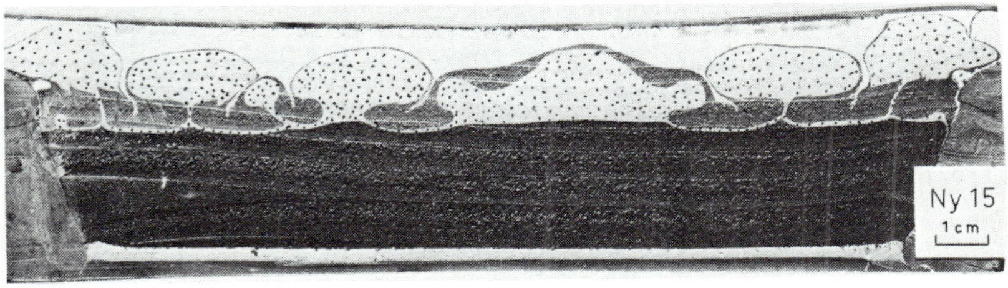

Fig. 11.30. Diapirs developed after centrifugation of model whose initial structure is shown in Fig. 11.31, except that layer 7 does not exist in model Ny-15 and layers 1 and 2 (which are not significant for the dynamic evolution) are not identical in the two models. Dotted: buoyant silicone putty (layer 4 in Fig. 11.31); dark masses sunk down between the domes: heavy silicone putty (layer 5 in Fig. 11.31); second layer from top: (layer 6 in Fig. 11.31). The diapirs have spread before reaching the surface because the diapiring material is more dense than the layer just above the heavy part of the overburden. Scale bar 1 cm.

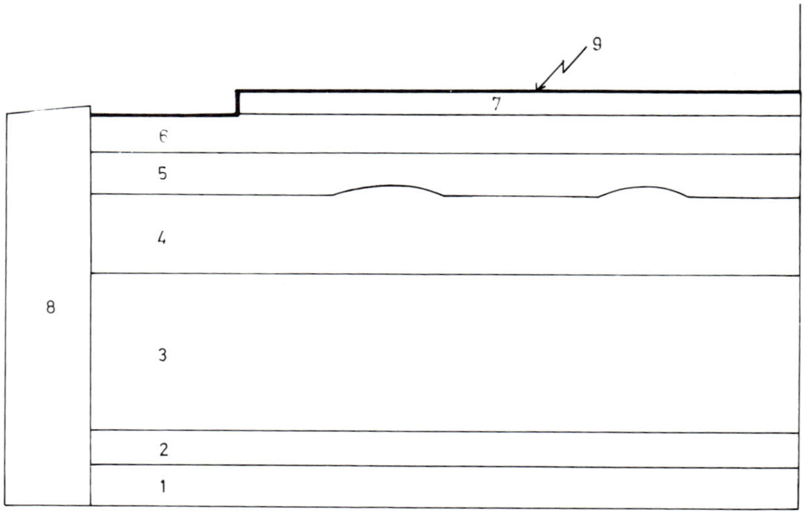

Fig. 11.31. Half-section of initial structure very similar to that shown in Fig. 11.30. (1) white modelling clay, $h_1 = 3$ mm; (2) painter's putty, $h_2 = 2$ mm; (3) silicone putty with magnetite powder, $h_3 = 20$ mm, $\rho_3 = 1{\cdot}32$ g cm^{-3}; (4) silicone putty, $h_4 = 10$ mm, $\rho_4 = 1{\cdot}16$ g cm^{-3}; (5) silicone putty with magnetite powder, $h_5 = 5$ mm, $\rho = 1{\cdot}24$ g cm^{-3}; (6) silicone putty, $h_6 = 5$ mm, $\rho = 1{\cdot}14$ g cm^{-3}; (7) oil-soaked powdered wax; (layer 7 does not exist in Ny-15, Fig. 11.30); (8) modelling clay; (9) sheet of modelling clay, 0·25 mm thick.

diapir. This condition is worth noting when granitoid diapirs in granitoid gneisses are studied: the lack of density contrast, or the reversed contrast, which sometime are found in such occurences does not necessarily mean that the bodies are not true gravity-driven structures. The buoyant force needed to intrude the granitoid plutons may well have been created by density contrast at unexposed deep levels.

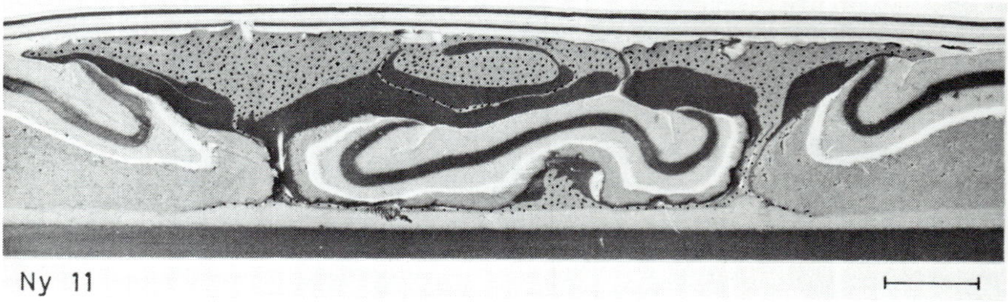

Fig. 11.32. Composite diapirs consisting of light-coloured silicone putty (dotted), $\rho = 1{\cdot}33$ g cm^{-3} and dark silicone putty (black mass along lower part of the spread hats of the diapirs), $\rho = 1{\cdot}45$ g cm^{-3} having risen through a layered overburden of painter's putty and modelling clay and spread underneath a low-density top complex of silicone putty (white) with thin sheets of modelling clay.

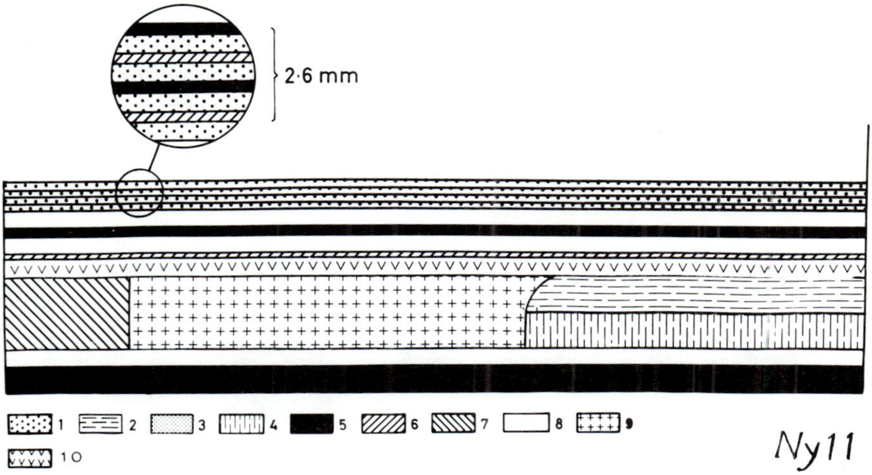

Fig. 11.33. Left-hand half cross-section of initial structure of model shown in Fig. 11.32. (1) Pure silicone putty, $\rho_1 = 1{\cdot}14$ g cm^{-3}; (2) silicone putty with magnetite powder, $\rho_2 = 1{\cdot}45$ g cm^{-3}; (3) not present in this model; (4) silicone putty with tungspar powder, $\rho_4 = 1{\cdot}33$ g cm^{-3}; (5) dark modelling clay, $\rho_5 = 1{\cdot}68$ g cm^{-3}; (6) white modelling clay, $\rho_6 = 1{\cdot}78$ g cm^{-3}; (7) brown modelling clay, $\rho_7 = 1{\cdot}71$ g cm^{-3}; (8) grey painter's putty, $\rho_8 = 1{\cdot}85$ g cm^{-3}; (9) yellow painter's putty, $\rho_9 = 1{\cdot}85$ g cm^{-3}; (10) green putty, $\rho_{10} = 1{\cdot}85$ g cm^{-3}.

It is evident that upon reaching a stiff and strong stratum buoyant bodies try to spread underneath the impenetrable layer. Many models demonstrate this feature, the buoyant material having been forced—by the pressure from below—to spread to a thin, lenticular sill underneath the stiff sheet (Fig. 11.37).

Illustrations in a later chapter show how models of this nature illuminate the evolutionary mechanism of continental rift systems and mid-oceanic ridges, pp. 390 ff.

During the late spreading stages of domal development, inversion of layering in the overburden adjacent to the upper part of the dome is a

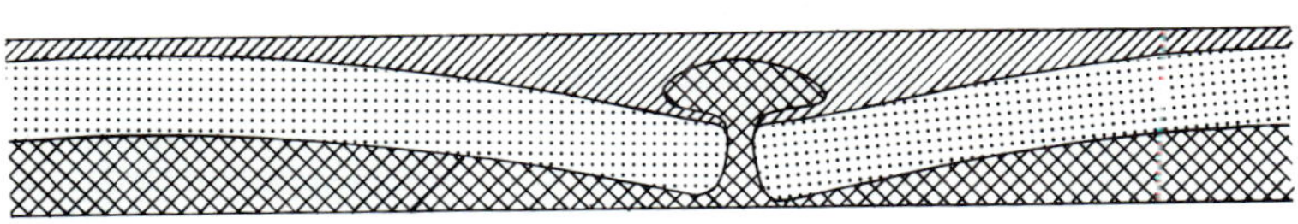

Fig. 11.34. Halokinetic origin of salt diapir according to Sannemann (1968). (Cf. Fig. 11.27.)

typical feature. It is due to drag caused by the rising and horizontally flowing mass. Examples are shown in Figs 11.38 and 11.39.

Complete detachment of the hat from the trunk may occur at a late stage of evolution, the hat portion often becoming a surface layer or a sill below a strong stratum or below a less dense layer (Fig. 11.40).

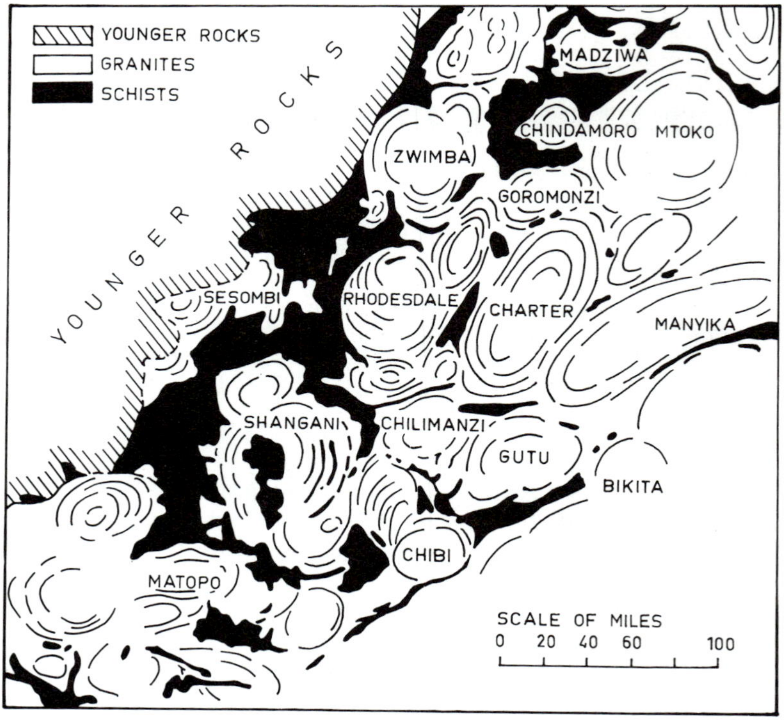

Fig. 11.35. The famous granite domes in Zimbabwe (after Macgregor, 1951).

Some of the model materials are brittle enough to fracture under the deviatoric stresses developed when unstable models are run in the centrifuge. In contrast to viscous flow of Newtonian substances brittle behaviour and fracturing are somewhat erratic phenomena, not possible to predict with precision in relation to the magnitude and geometric character of the

Fig. 11.36. Diapirs of silicone putty having pierced the surface of the overburden and produced buckles in a top sheet of modelling clay while spreading in the centrifugal field. Scale bar 1 cm.

stress field. When fracture occurs in the models, therefore, the movement of both the buoyant bodies and their surroundings is apt to be disordered, resulting in diapirs with irregular and unpredictable shape.

This by no means detracts from the usefulness of the centrifuge method in experimental tectonics: to the contrary, the irregular structures in many instances are more realistic than the smooth, symmetric oil-syrup models.

Examples on irregular diapirs formed because of irregular fracturing in

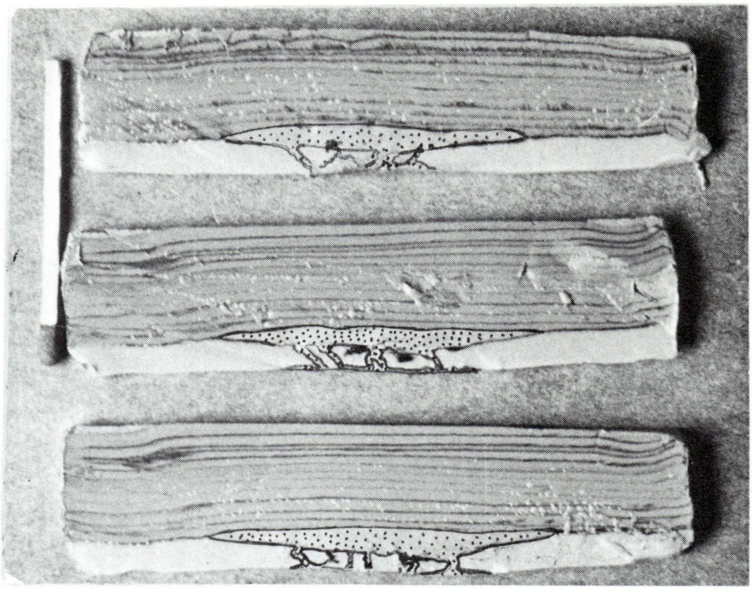

Fig. 11.37. Sections through centrifuged model of subsurface sill of silicone putty (dotted) spread below layered modelling clay. Light grey bottom layer is painter's putty through which the buoyant silicone has ascended. Although the modelling clay is more dense than the silicone putty it is too strong to permit penetration.

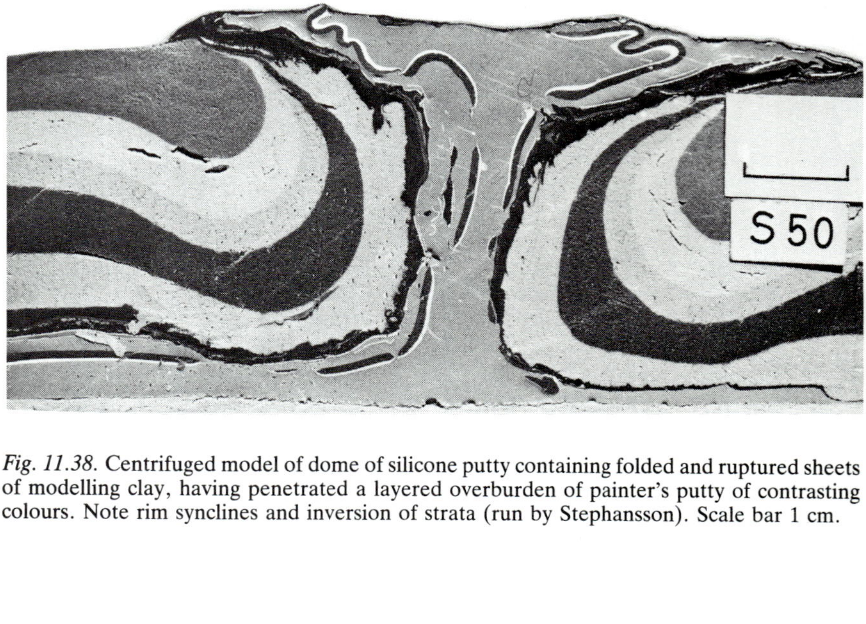

Fig. 11.38. Centrifuged model of dome of silicone putty containing folded and ruptured sheets of modelling clay, having penetrated a layered overburden of painter's putty of contrasting colours. Note rim synclines and inversion of strata (run by Stephansson). Scale bar 1 cm.

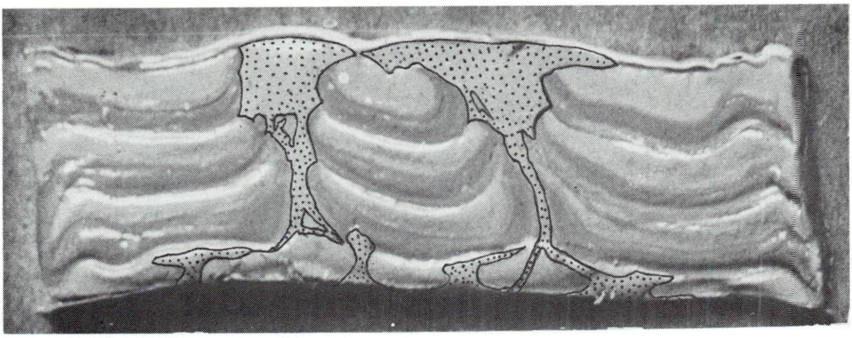

Fig. 11.39. Section through centrifuged model of silicone-putty domes (dotted) having risen through layered overburden of painter's putty (light grey layers and thick dark grey layers) and modelling clay (thin top layer and sheets between the putty layers). Diapirs outlined in ink.

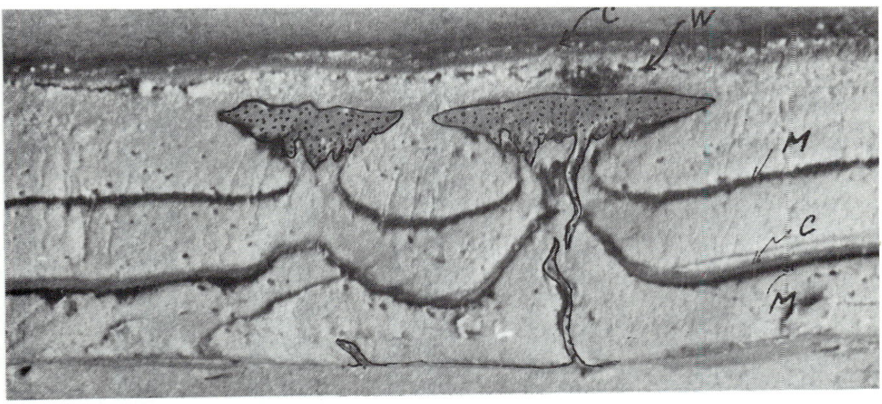

Fig. 11.40. Cut through centrifuged model of buoyant mass of silicone putty (dotted) which has risen through painter's putty and spread underneath powdered wax and modelling clay. (W) layer of powdered wax soaked in oil; (C) thin sheets of modelling clay; (M) originally straight horizontal marker sheets. Length of cut 9·4 cm.

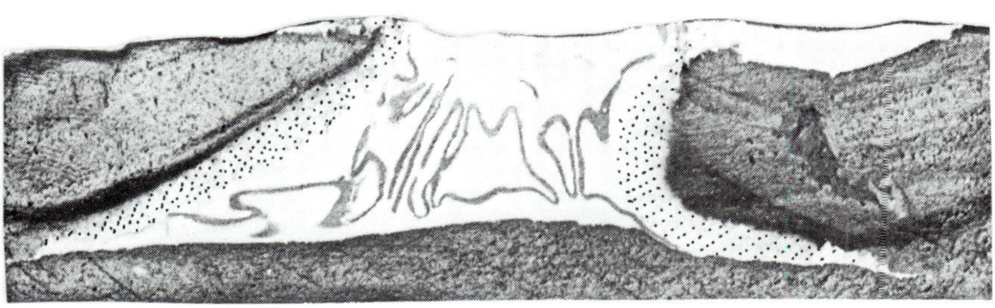

Fig. 11.41. Buoyant layer of silicone putty, with thin, originally straight horizontal sheet of modelling clay, having risen through overburden consisting of basal layer of oil-wetted powder of wax (dotted) below thick layer of painter's putty. cf. Fig. 11.52.

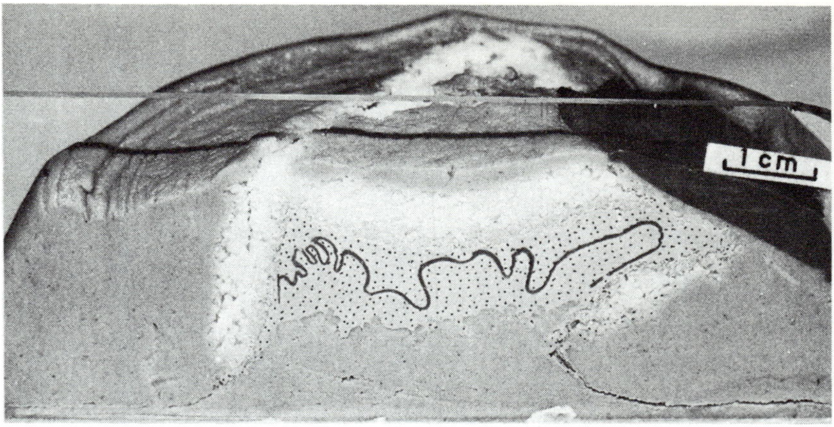

Fig. 11.42. Model quite similar to that shown in Fig. 11.41. Dotted: buoyant silicone putty initially located as even-thick layer along base (note buckled layer of originally straight modelling-clay sheet); white: oil-soaked powdered wax; uniform grey: painter's putty. Cf. natural example shown in Fig. 11.52.

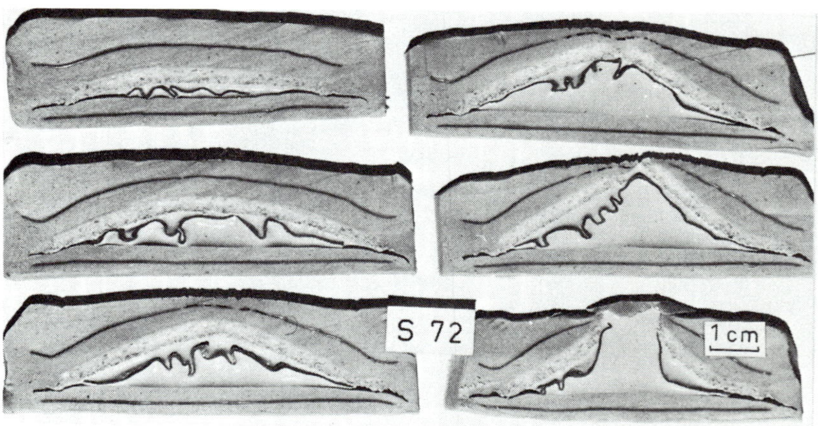

Fig. 11.43. Sections through centrifuged model of silicone putty with powdered-wax layer above, embedded in painter's putty with sheets of modelling clay. Note broad shape of the rise, buckling of the competent dark sheet in the doming body, and boundinage of a similar sheet in the overburden. Run 10 min at 1300 *g*.

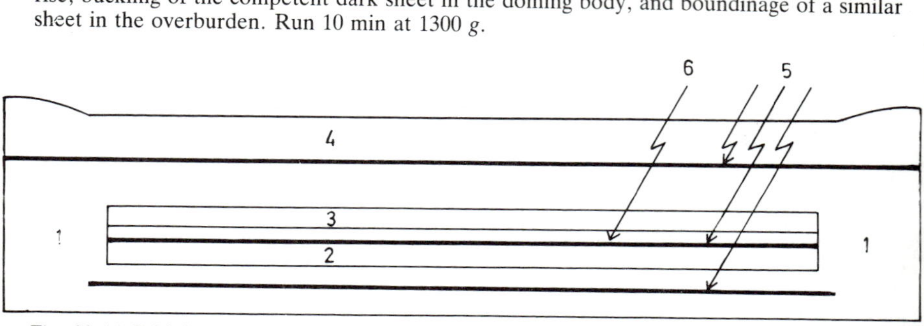

Fig. 11.44. Initial structure of model S-72 (Fig. 11.43). (1) Painter's putty; (2) silicone putty, $h_2 = 2$ mm; (3) wax powder soaked in oil; (4) painter's putty; (5) modelling clay, $h_5 = 0{\cdot}4$ mm; (6) silicone putty, $h_6 = 1$ mm.

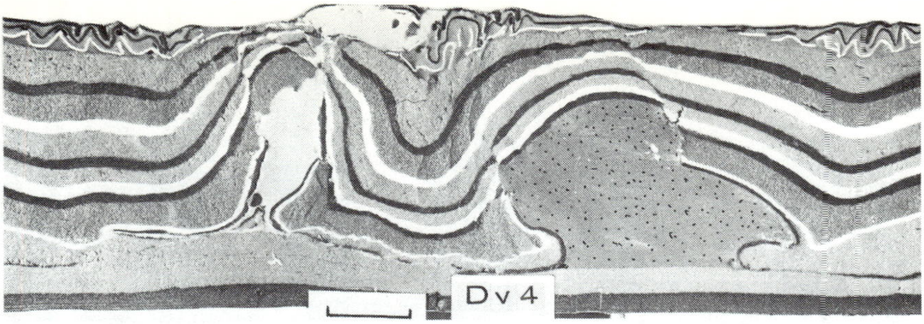

Fig. 11.45. Centrifuged model in which overburden has fractured in the stress caused by two buoyant masses (dark, dotted to the right and white to the left). Displacement has occurred along the faults and the thin sheets of modelling clay at the top have buckled in the rim synclines and broken to boudins above the domes. Scale bar 1 cm

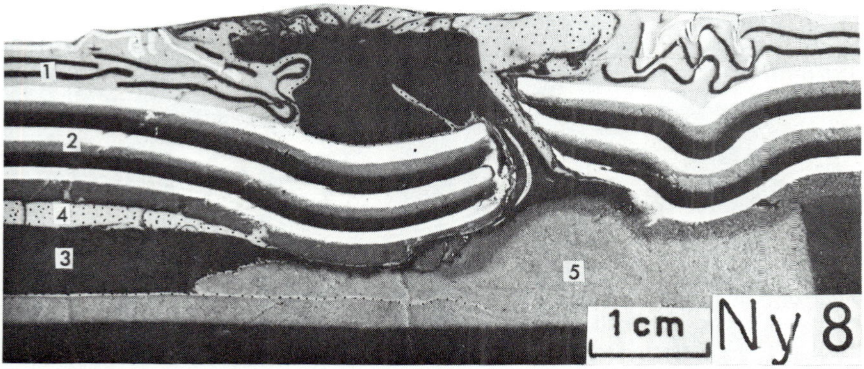

Fig. 11.46. Composite and irregular silicone-putty diapir having broken through relatively strong overburden containing rather thick modelling-clay layers (see Fig. 11.47). While spreading in the surface complex of silicone putty and modelling-clay sheets, the diapir has buckled and torn the modelling-clay sheets in a complicated manner. Dotted part of diapir and (4) are silicone putty, $\rho = 1{\cdot}14$ g cm^{-3}; dark part of diapir and (3) are silicone putty, $\rho = 1{\cdot}45$ g cm^{-3}; (1) is silicone putty, $\rho = 1{\cdot}14$ g cm^{-3}, with embedded sheets of modelling clay (see circle in Fig. 11.47); (2) is interlayered modelling clay and painter's putty; (5) is painter's putty, $\rho = 1{\cdot}85$ g cm^{-3}. Run at 2000 *g* in 10 min and at 3000 *g* in 27 min.

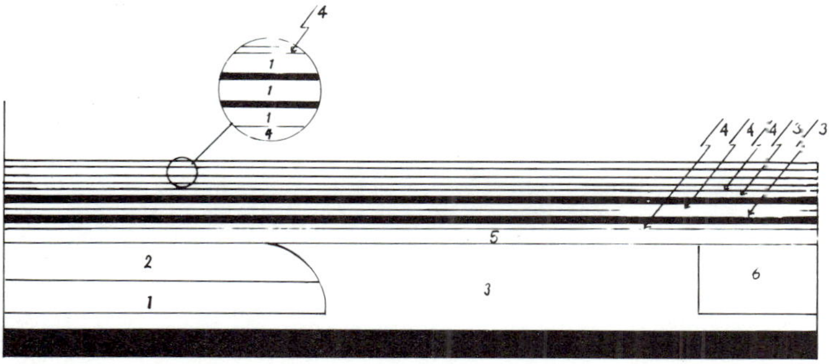

Fig. 11.47. Right-hand half of cross-section showing initial structure of model Ny-8 (Fig. 11.46). (1) Silicone putty, $\rho = 1{\cdot}14$ g cm^{-3}; (2) silicone putty with magnetite powder, $\rho = 1{\cdot}45$ g cm^{-3}; (3) grey painter's putty, $\rho = 1{\cdot}85$ g cm^{-3}; (4) white is modelling clay, $\rho = 1{\cdot}78$ g cm^{-3} and black is dark grey modelling clay, $\rho = 1{\cdot}68$ g cm^{-3}; (5) green painter's putty, $\rho = 1{\cdot}85$ g cm^{-3}; (6) brown modelling clay, $\rho = 1{\cdot}71$ g cm^{-3}.

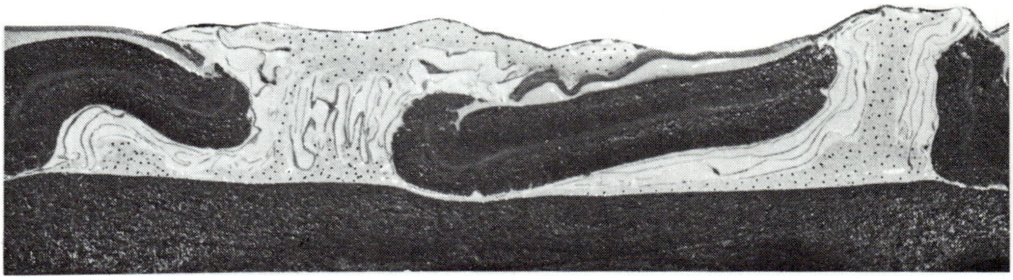

(*a*)

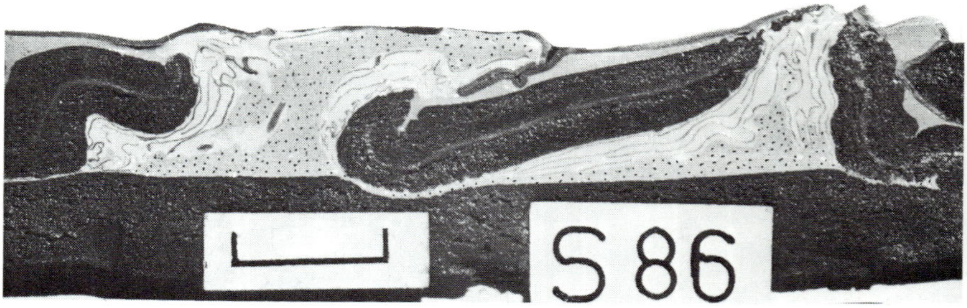

(*b*)

Fig. 11.48. Centrifuged model showing diapirs, one of which has produced a recumbent fold of the surface sheet of modelling clay. Note strong buckling of thin sheets of modelling clay which initially were embedded in the source stratum as horizontal and straight sheets (see Fig. 11.49). Scale bar 1 cm.

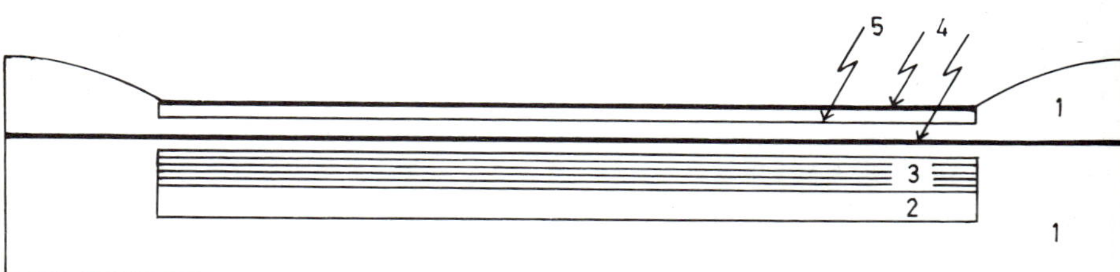

Fig. 11.49. Initial structure of model shown in Fig. 11.48. (1) Painter's putty; (2) silicone putty; (3) silicone putty with four thin sheets of modelling clay; (4) sheets of modelling clay; (5) sheet of silicone putty.

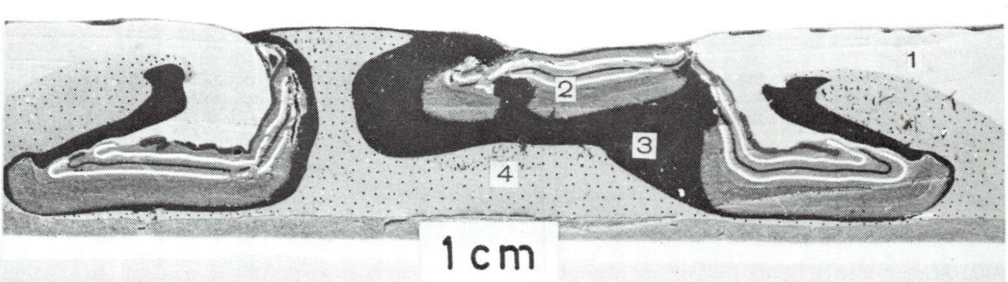

Fig. 11.50 Irregular pattern produced when a buoyant double layer of silicone putty (3) and (4) rises through relatively brittle overburden (2).

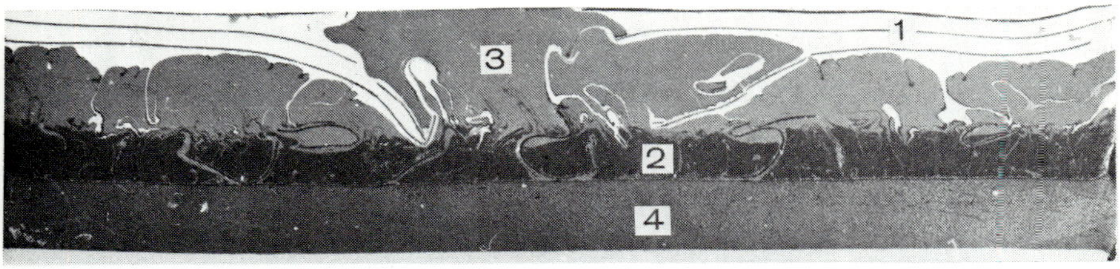

Fig. 11.51. Centrifuged model in which the diapirs have assumed very irregular shapes. (3) Buoyant layer initially situated below the more dense overburden (2). Layer (3) contained a thin sheet of relatively stiff modelling clay. (1) Complex of low-density white silicone putty containing three thin sheets of stiff modelling clay. The irregular pattern of the diapirs is due to the stiffness of the thin embedded sheets of modelling clay. Note that the channels through which the diapirs must have passed across the heavy overburden (2) are almost completely collapsed after penetration. (4) Substratum of painter's putty.

the overburden are presented in Figs 11.41–11.51. Figure 11.52 is a classic cross-section of a German salt stock (from Cloos, 1936); it compares well with the models exhibited in Figs 11.41 and 11.42.

INITIATION OF DIAPIRS

If both source layer and overburden are uniform with respect to thickness, composition and structure, and, moreover, the layering is parallel to the equipotential surfaces in the body-force field, then diapirs should theoretically not start growing from the metastable source layer. In theory a small irregularity in geometry or material property is necessary to generate doming. However, if neither source nor overburden materials do possess a finite yield strength the required irregularities become infinitely small, and doming starts in the metastable system even without detectable irregularities. This is what happens in the oil-syrup models described above (pp. 44–55). The substances are true liquids with vanishing yield strength; consequently, diapirs start to grow sooner or later from a seemingly straight horizontal layer of oil underlying heavy syrup.

Although the source-layer imitation materials—some of the silicone putties and the collophony mixture—used in many of the centrifuged models do not show a measurable yield strength, the overburden materials—the painter's putty and modelling clay—do possess yield strength. Theoretically, therefore, the unstably stratified centrifuged models require finite geometric perturbations or heterogeneous material property in order to generate diapirs from the source layer.

Diapir-generating irregularities are of various kinds, the simplest being

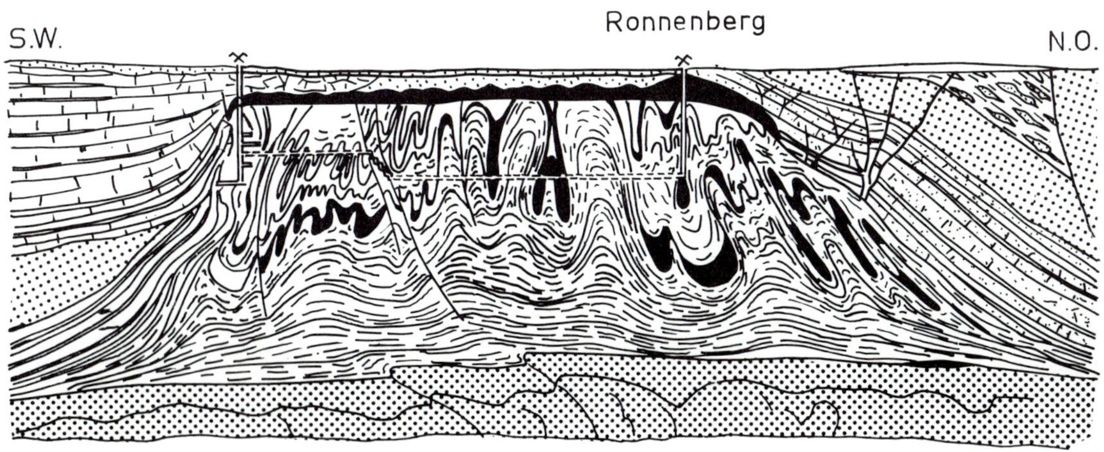

Fig. 11.52. Flow folds in a salt diapir close to Hanover, Germany (after a reproduction by Cloos (1930) originally by Seidel).

an initiating bulge built on the buoyant layer in the initial model. In order that such an initiating bulge shall effectively generate doming, its height in a given model must exceed a critical value depending upon the yield strength of the overburden, its thickness, the density contrast and the centrifugal force per unit mass. If the buoyant material also exhibits finite yield strength, this factor adds to the critical height of the initiating bulge.

When analysing the situation more closely, it is convenient to introduce the term "plunging strength" of the overburden. This we shall define as the minimum force—not stress—needed to push a cylindrical solid plunger of given cross-section into the plastic body.

For our purpose the body to be penetrated is sheet-shaped (Fig. 11.53). The plunging strength of such a body is related to (1) the yield strength

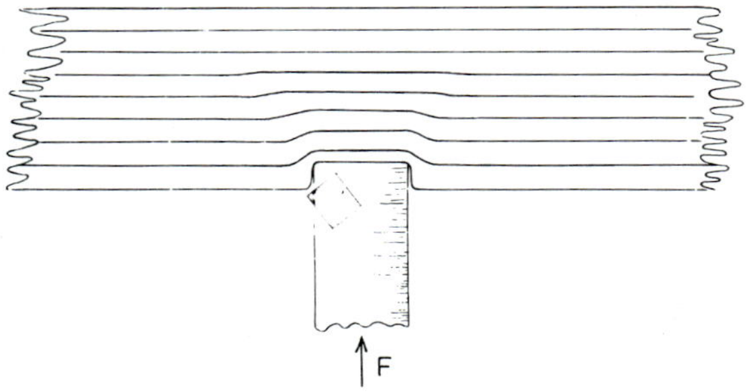

Fig. 11.53. Deformation of passive marker sheets in a viscous body in which a solid plunger is forced.

of the material, (2) the diameter of the plunger and (3) the thickness of the sheet. The plunging strength increases with increasing values of all these quantities. With reference to the thickness of the sheet there is a maximum limit above which further increase in thickness has no effect. This is because the strain produced by the plunger penetrates only a limited distance into the material in front of the plunger. Addition of material beyond this distance does not affect the force which resists the thrust of the plunger.

Let S_p be the plunging strength of a given overburden when the circular cross section of the plunger is A cm^2. In order that an initiating circular dome of cross-section A cm^2 should generate a vertical force sufficient to penetrate the overburden and thus be capable to grow, the height, h, of the dome must exceed the limit

$$h_{\text{crit}} \geqslant \frac{S_p}{\Delta\rho a A} \tag{11.1}$$

(see Fig. 11.54). Here $\Delta\rho$ is the difference in density between overburden and source material, and a is the centrifugal acceleration.

The plunging strength of a strain-hardened sample of a given kind of painter's putty was determined to be $1{\cdot}4\times10^5$ for a circular plunger of 1 cm^2 cross-section. Let the difference in density between painter's putty and silicone putty be $0{\cdot}7\,g\,cm^{-3}$. On this basis the minimum height of an initiating cylindrical bulge of 1 cm^2 cross-section is

$$h_{crit} = \frac{1{\cdot}4\times10}{0{\cdot}7\times10^3\times981} \approx 0{\cdot}2\ \text{cm},$$

for models centrifuged at $a = 1000\,g$. Theoretically, then, only interface irregularities more than 2 mm high should be able to develop into ascending diapirs.

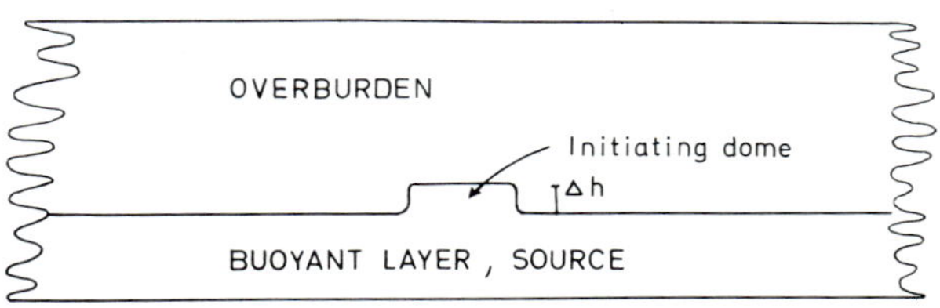

Fig. 11.54. Initiating dome of buoyant material in a model.

It appeared that in our runs with pure silicone as source and painter's putty as overburden initial perturbations much less than 2 mm proved capable of generating diapirism. In fact, it was virtually impossible to prevent diapiric growth at centrifugal acceleration exceeding 1000 g, irrespective of how carefully one attempted to prepare perfect layering parallel to the equipotential surface. A possible reason is the almost vanishingly low yield strength shown by the painter's putty during the first part of a strain and stress test (see p. 239). Perhaps this easily produced initial deformation prior to strain-hardening of the putty allows perturbation to grow to heights sufficient to generate the plunging force that is necessary for further penetration into the now strain-hardened overburden putty.

Although diapirs readily formed from uniform source layers of silicone with putty overburden when centrifuged at 1000 g or more, the geometric distribution of the domes would generally not be as regular as that found in the oil-syrup tests. This is probably due to the yield strength of the putty overburden. It may be imagined that the rise of a diapir which by chance happened to penetrate the "strength barrier" of the overburden would be so favourable that the relatively rapid withdrawal of material from the source layer through this diapir prevented initiation of diapirs in the close neighbourhood, thereby invalidating the theoretical characteristic wavelength concept.

In this connection reference is again made to the models shown in Fig. 11.1. The overburden in these models was magnetite-containing silicone with a density of 1·36 g cm^{-3}, a mixture with negligibly small yield strength. The initial models consisted of uniformly thick, homogeneous layers parallel to the equipotential surface. The result after centrifugation is a dome

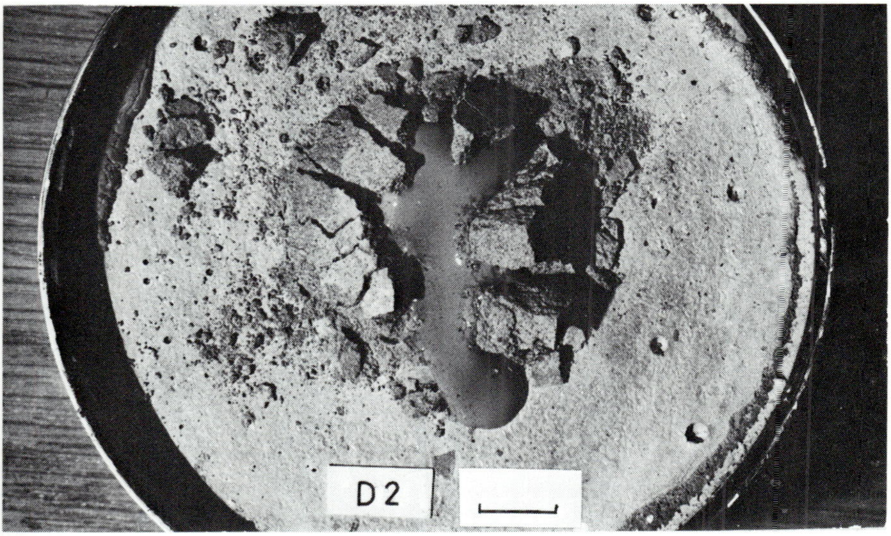

Fig. 11.55. Silicone-putty diapir having penetrated overburden of concrete. Run 6 min at 2900 *g*. Scale bar 1 cm.

structure with quite even spacing similar to the pattern developed in the oil-syrup tests.

With overburden material possessing considerable strength—such as modelling clay, plaster of Paris and concrete—initiating domes of considerable height are necessary to start ascent in the centrifuged models. A case is shown in Fig. 11.55. In this model the concrete overburden is so strong that an initial silicone dome not less than 2 cm high was needed to create enough buoyant force to fracture the overburden and make the silicone rise to the surface at 2900 *g* after 6 min.

In some models a sheet of soft painter's putty was placed between the

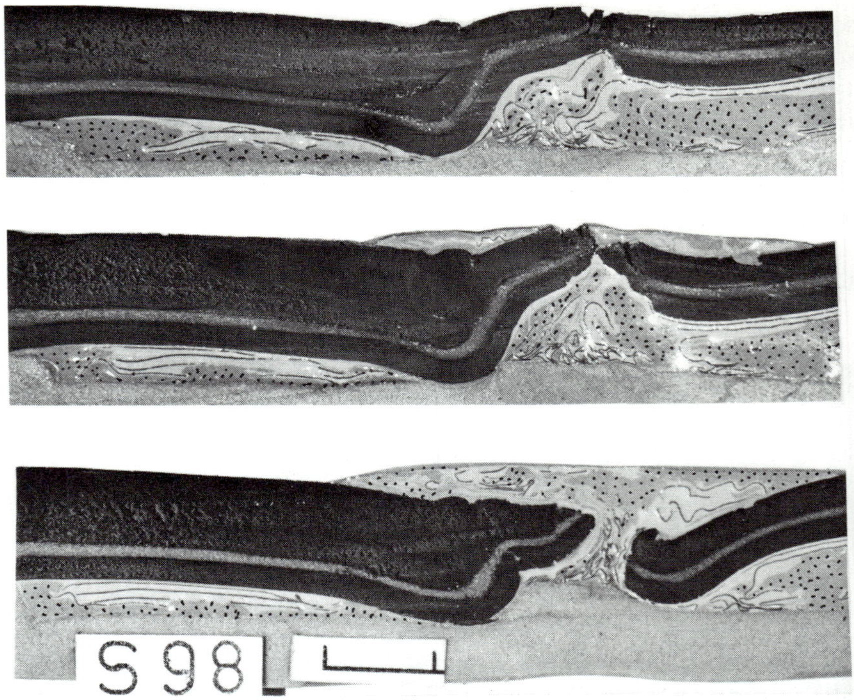

Fig. 11.56. Silicone-putty diapirs containing thin sheets of competent modelling clay having risen along the edge of the thick part of overburden of painter's putty and modelling clay. The stress above the rising silicone putty has been sufficient to break the relatively strong overburden. Run 7 min at 2400 *g* and 60 min at 2900 *g*. Scale bar 1 cm.

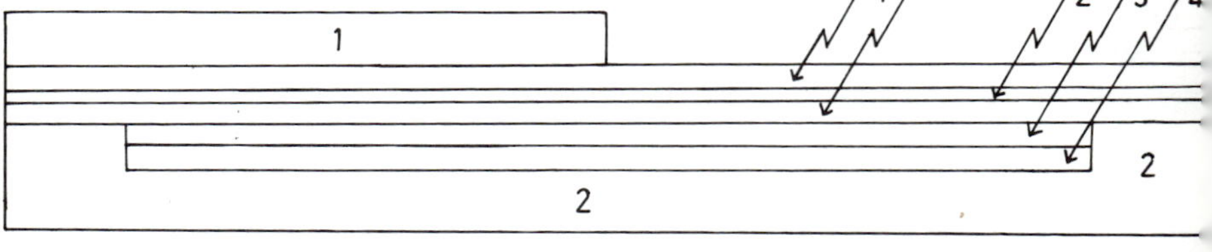

Fig. 11.57. Initial structure of model shown in Fig. 11.56. (1) Modelling clay; (2) painter's putty; (3) silicone putty with foru thin sheets of modelling clay; (4) silicone putty, shown dotted in Fig. 11.56. Length of section 9.4 cm.

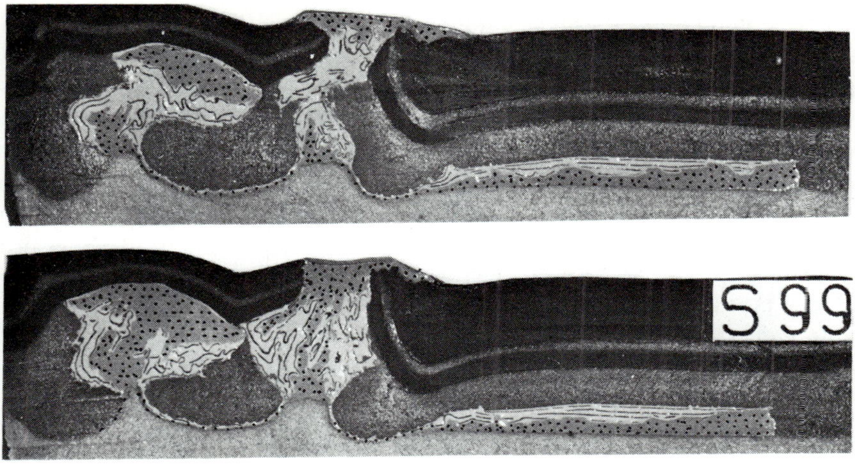

Fig. 11.58. Silicone-putty diapirs with thin sheets of modelling clay having risen along the edge of the thick part of modelling clay-putty overburden. Run 15 min at 2900 *g*. Note that much less time was needed for the diapirs to penetrate the overburden in this model than in the similar model S-98 because in S-99 (unlike S-98) a layer of soft painter's putty existed between the buoyant layer and the strong modelling-clay overburden. Cf. Figs 11.57 and 11.59.

strong overburden and the buoyant layer as shown e.g. in Fig. 11.37. Diapirs would then readily rise through the soft putty, and on reaching the strong part of the overburden the plunging force of the buoyant bodies may become sufficient for further penetration into the strong layer. This has been the case in models S-98 and S-99 (Figs 11.56–11.59), whereas the upper part of the overburden was too strong in the model shown in Fig. 11.37. In this model the rising silicone was forced to spread to a conformable sill underneath the strong layers of the overburden.

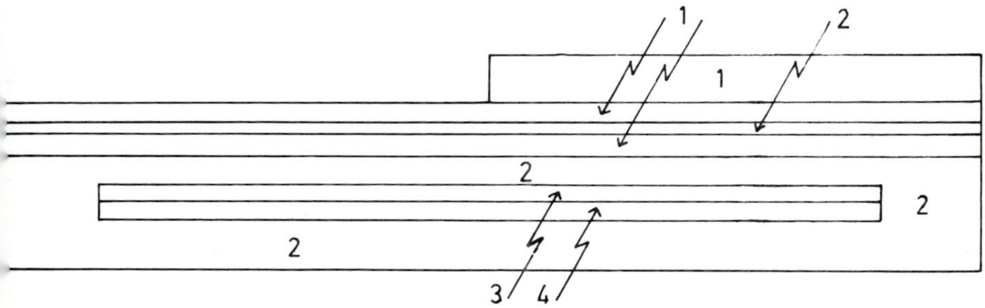

Fig. 11.59 Initial structure of model shown in Fig. 11.58. (1) Modelling clay; (2) painter's putty; (3) silicone putty with four thin sheets of modelling clay; (4) silicone putty.

The building up of a high plunging force by an easy initial rise through the soft deep zone of the overburden is believed to be of geological consequence in certain situations. The common upheavals of crustal blocks by movement along boundary faults (e.g. a number of marginal basement windows in the Scandinavian Caledonides, cf. p. 378, and numerous uplifts in the Rocky Mountains; Eardley, 1963) may well have been pushed up by deep-seated buoyant bodies that have gained high plunging force from their rise through the less rigid deep part of their overburden. See also models M-20, M-21, M-22 and M-28, Figs 11.70–11.74.

Besides bulges on the buoyant layer in the initial models, any discontinuity of sufficient magnitude in source layer or overburden would start off diapirs. For example, an abrupt change in thickness or density of either

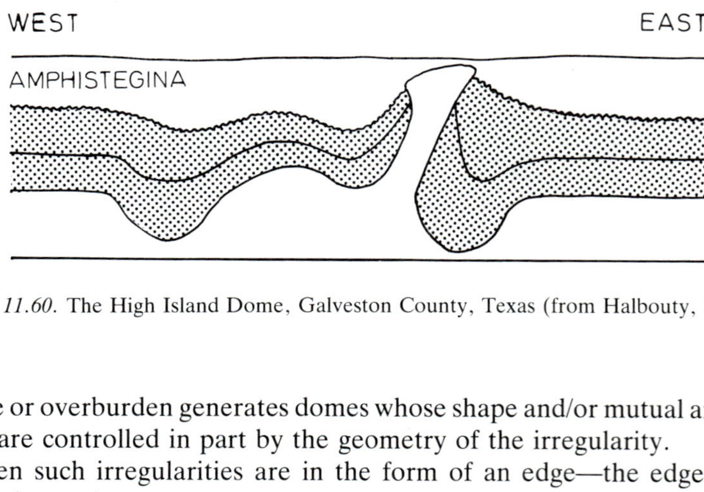

Fig. 11.60. The High Island Dome, Galveston County, Texas (from Halbouty, 1967).

source or overburden generates domes whose shape and/or mutual arrangement are controlled in part by the geometry of the irregularity.

Often such irregularities are in the form of an edge—the edge of the source layer, for example, or the edge of a layer with contrasted density or strength or thickness, in the overburden. Edge discontinuities of this kind tend to produce anticlines elongate parallel to the edge similar to those shown in Figs 11.56 and 11.58. (Incidentally, note the geometric similarity between model S-99 and the High Island Dome, Fig. 11.60.) These buoyant anticlines may sometimes rise to the surface as a solid wall, but more often the anticline is rendered unstable before reaching the surface. More or less regularly spaced culminations form on the ridge, which thus splits up into a row of diapirs (Figs 11.61–11.63). Compare the evolution of the oil-syrup models (Figs 7.1–7.8).

Salt anticlines or salt walls, such as are found in the saline deposits of northern Germany (Trusheim, 1957, 1960), and in Colorado and Utah

(Cater and Elston, 1963), are generally considered to be the products of unidirectional horizontal compression. We have shown above that such structures may well form in response to pure buoyancy forces in systems with edge discontinuities. The same is true for some gneiss-filled anticlines in orogens.

The dome-generating effect of an edge discontinuity is not difficult to understand in terms of dynamics. In Fig. 11.64 the thickness and weight of the overburden change rapidly across the edge. This creates a pressure

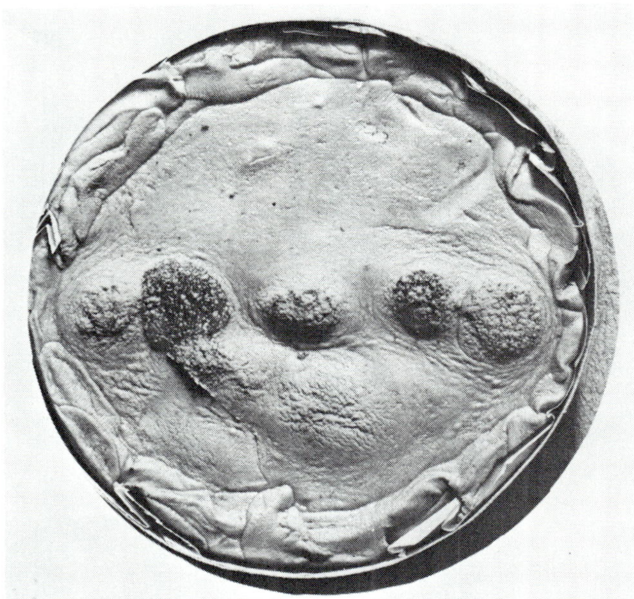

Fig. 11.61. Row of domes of wax formed along an edge in the overburden.

difference in the source layer on either side of the edge discontinuity, and horizontal flow starts in the source, its thickness decreasing under the heavy side of the overburden and increasing below the thin and lighter side. An initiating bulge is thus produced along the edge discontinuity, and an anticline will rise if not prevented by the strength of the materials.

At the boundary of the source layer a similar situation exists. At point X_o in Fig. 11.65 the pressure is higher, because of the weight of the overburden, than at the same level within the source layer. Flow from the edge into the source follows, and a ridge is produced which may continue to develop into a buoyant anticline and subsequent isolated diapirs.

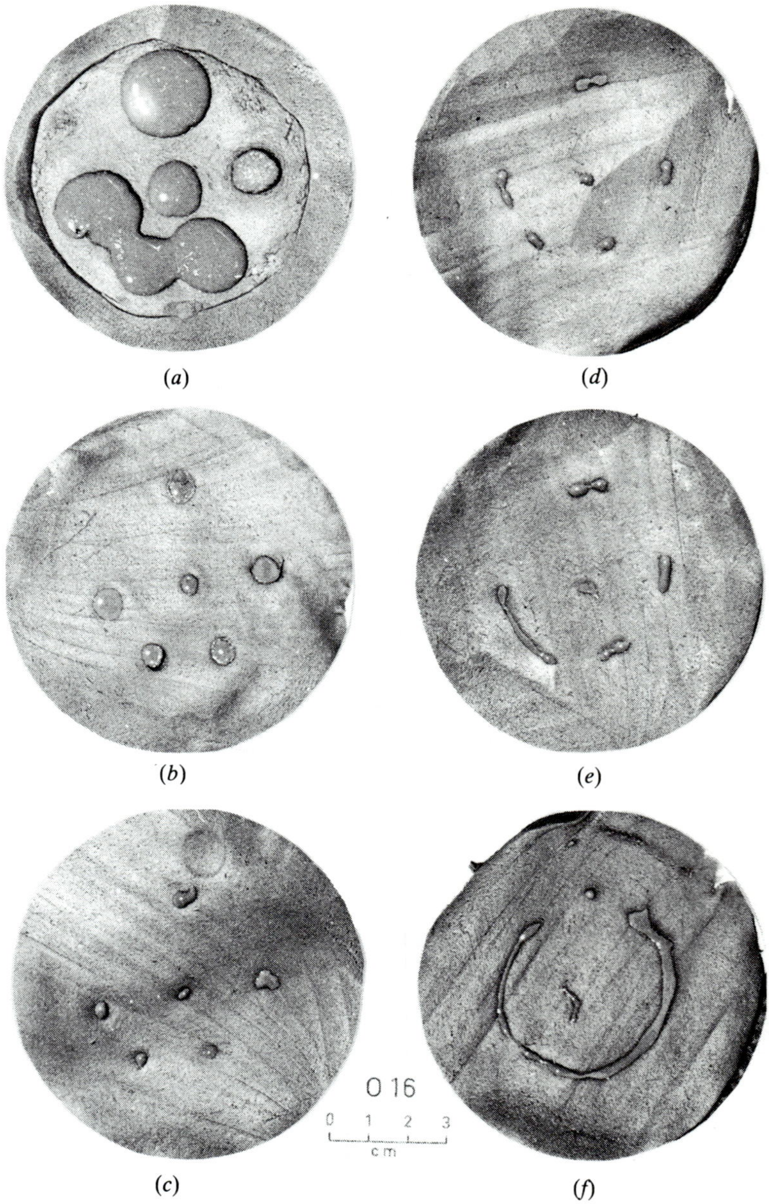

Fig. 11.62. Horizontal cuts at different levels through model containing five diapirs protruding from a circular ridge rising from a sheet of buoyant silicone putty. A sixth diapir occurs in the centre. The cuts are about 2 mm apart, (*a*) being the surface and (*f*) the deepest cut. The circular basal ridge is controlled by the circular outline of the model. The overburden consists of painter's putty.

The change of thickness or density in source or overburden need not be abrupt. Even a gradual change generates lateral flow along the thickness gradient, which also represents a pressure gradient, and dome-initiating ridges tend to develop normal to the thickness gradient.

The dynamics of the process of generation and rise of domes from a buoyant layer is such that the velocity increases rapidly with increasing geometric dimension of geometrically similar structures of a given set of

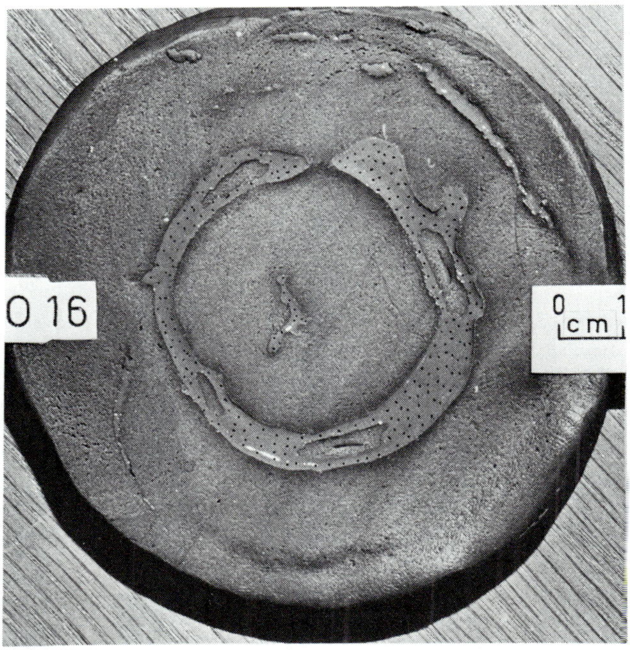

Fig. 11.63. A deep section of model O-16, showing the silicone putty ridge (dotted) and the basement underneath the buoyant sheet, filling the cores of anticlines situated directly below the diapirs on the ridge. See Fig. 11.62.

substances. This theoretical results was qualitatively verified over and over again during our experiments. The thicker the source layer of silicone or collophony mixture in the models, the shorter the time needed and/or the smaller the centrifugal force needed in general for production of diapir structures. Figure 7.5 demonstrates this particular condition in a non-centrifuged oil test. The oil layer was initially wedge-shaped and varied in thickness from about 2 cm at the one end of the container to just a thin

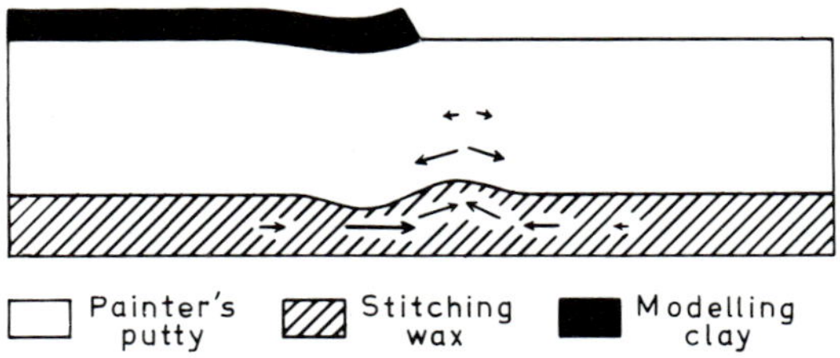

Fig. 11.64. Initiation of a dome or ridge under the edge of excess overburden. Flow indicated by arrows.

film at the other end. The photographs show how the initiation and growth of the domes start at the thick end of the oil sheet and gradually expand towards the thin end.

FORMATION OF MARGINAL SYNCLINES

It is well known from studies of salt domes in their natural environment as well as in experimental models that the material in the domes is not supplied evenly from the entire source layer, but is chiefly taken from a region close to the root of the domes. The flow of salt from the region around a dome results in a thinning of the source layer and a collapse of the superincumbent sediments. Such depressions around salt domes are called marginal or peripheral sinks. (Incidentally, the term "sink" is an unfortunate choice from the point of view of fluid dynamics, because the marginal sink is the source of the flow of salt; the dome itself is the sink.)

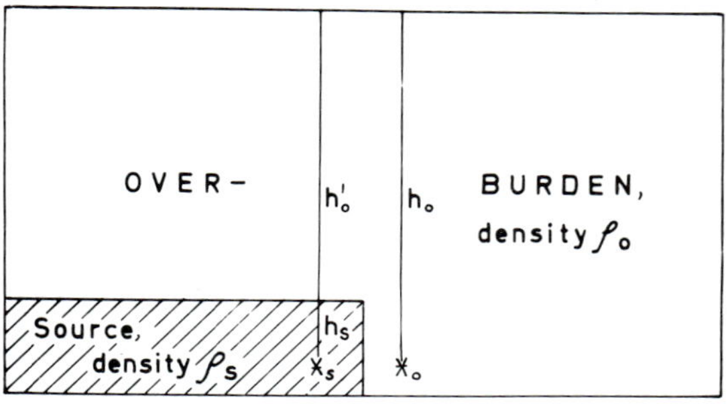

Fig. 11.65. Buoyant layer with sharp edge below heavy overburden.

Marginal sinks or synclines of this kind are prominent features of most dome models studied in the Uppsala laboratory, and the writer regards them of great tectonic importance on both a local and regional scale. For example, some submarine trenches adjacent to island arcs may be explained as marginal synclines located above the source for the intruded and extruded material which builds up the islands. The "extremely deep" syncline around the Vredefort dome in South Africa (Du Toit, 1954) is a striking example of a marginal sink from which the dome-building material probably has come (Fig. 11.94).

With respect to the mechanism of formation of the marginal synclines it is immaterial whether or not the Vredefort dome was initiated by the impact of a meteor.

To explain in principle how a marginal syncline develops around a dome, consider a stiff circular plate which is loaded down on a layer of viscous

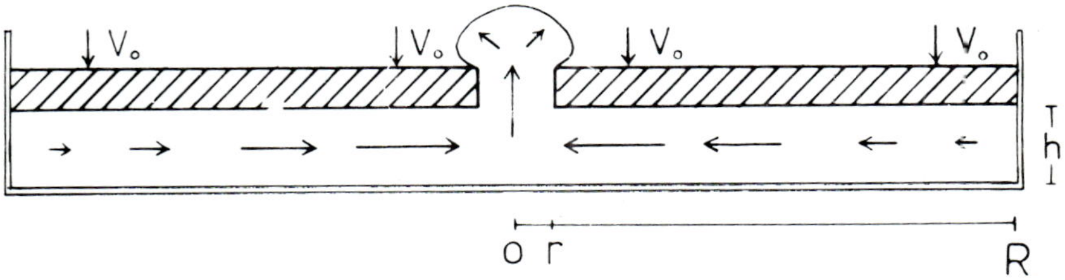

Fig. 11.66. Cross-section of plate with central hole being pressed down on a viscous mass. V_0 is velocity of motion of the plate. Flow indicated by arrows.

material resting on the bottom of a cylindrical vessel whose inner diameter matches the outer diameter of the stiff plate. The plate has a circular hole in the centre, through which the viscous material extrudes in a manner similar to that in which natural domes are pushed up through weak spots in the overburden (Fig. 11.66).

The distribution of pressure in the viscous layer is of prime importance in the study of the development of marginal synclines. The pressure distribution is closely related to the flow pattern in the viscous layer. Since the flow is directed radially towards the central hole, it follows that the horizontal component of the pressure gradient points radially away from the centre of the plate.

Provided that the thickness of the viscous layer remains uniform during the vertical compression, the horizontal pressure gradient will vary along the radius approximately in proportion to the variation of the rate of flow as measured in volume per unit time per unit cross-section of the centripetal flow. Now, the volume of viscous material which, because of the vertical compression, in unit time passes through a cylindrical surface with radius r is readily found to be

$$V = \pi(R^2 - r^2)\, \partial h/\partial t \tag{11.2}$$

where R is the outer radius of the layer and $\partial h/\partial t$ the rate of change of the thickness of the viscous layer, i.e. the rate at which the stiff plate moves downward. The area of the cylindrical surface with radius r, i.e. the cross-section of the centripetal flow at distance r from the centre, is $2\pi rh$ where h is the thickness of the viscous layer at a given instant.

Hence, in terms of volume per unit time per area of cross-section, the rate of centripetal flow through the cylindrical surface with radius r is determined by

$$v = \frac{V}{2\pi rh} = \frac{R^2 - r^2}{2rh}\frac{\partial h}{\partial t}. \tag{11.3}$$

Provided that the thickness of the layer is small relative to its horizontal dimensions, it is sufficiently accurate for our purpose to assume that the horizontal pressure gradient is proportional to the rate of centripetal flow,

$$\frac{\mathrm{d}P}{\mathrm{d}r} = c\frac{R^2 - r^2}{r}\frac{\partial h}{\partial t} \tag{11.4}$$

where the constant c depends upon the viscosity of the material, the thickness of the layer, and is possibly also a function of the horizontal level in the layer. (Since the pressure varies along the vertical direction in the layer, it is possible that the horizontal pressure gradient also varies with the vertical distance from the base of the layer.) Integration of eqn (11.4) gives pressure as a function of distance from the central hole,

$$P_r = c(R^2 \ln r - \tfrac{1}{2}r^2)\frac{\partial h}{\partial t} + c'. \tag{11.5}$$

The constant of integration, c', may be determined either in relation to the pressure, P_R, at the outer edge of the layer, or in relation to the pressure, P_0, at the edge of the central hole. Selecting P_0 as a known reference pressure, the constant of integration follows by putting $r = r_0$ and $P_r = P_0$ in eqn (11.5), r_0 being the radius of the hole, thus

$$c' = P_0 - c(R^2 \ln r_0 - \tfrac{1}{2}r_0^2)\frac{\partial h}{\partial t} \tag{11.6}$$

which, introduced into equation (79), gives

$$P_r = c(R^2 \ln r - \tfrac{1}{2}r^2 - R^2 \ln r_0 + \tfrac{1}{2}r_0^2)\frac{\partial h}{\partial t} + P_0. \tag{11.7}$$

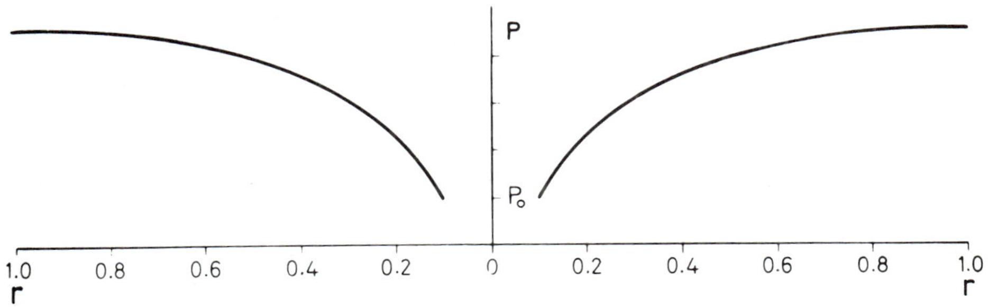

Fig. 11.67. Pressure distribution in the viscous mass shown in Fig. 11.66, along a radius around the central hole.

Equation 11.7 gives the pressure profile along a radius as shown in Fig. 11.67.

Since the coefficient c varies with the thickness, h, the pressure profile in Fig. 11.67 must be regarded as an instantaneous variation along a radius. If the rate of vertical compression, $\partial h/\partial t$, remains uniform throughout the layer, the pressure profile changes with time, but remains similar in general shape to that shown in the figure.

We shall now assume that the stiff plate above the viscous layer is replaced by an overburden which is ductile enough to yield under its own weight, and so heavy that the total load on the viscous layer is the same as that caused by the stiff plate. As the new overburden is not rigid, it sags

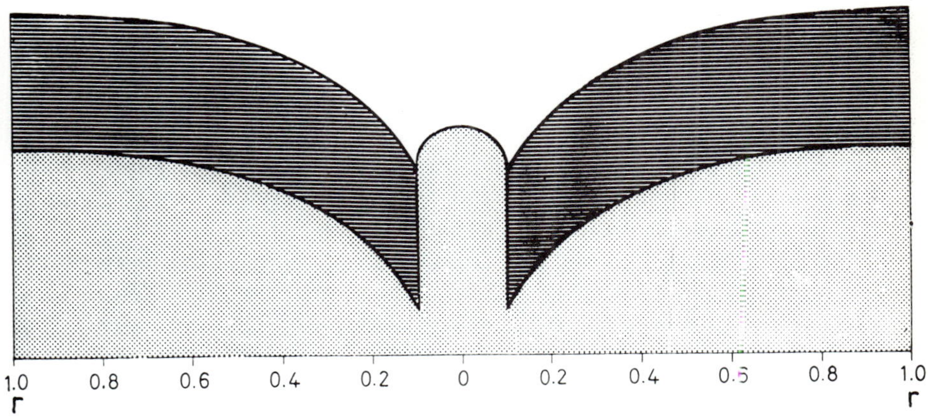

Fig. 11.68. Idealized profile of viscous mass extruded through a central hole owing to the weight of overlying plastic plate.

down unevenly at a rate which varies with the local pressure in the subjacent viscous layer. In other words, each small local portion of the overburden must be supported by the local pressure in the subjacent layer. (The horizontal dimensions of the smallest portion of the overburden which is supported by the local pressure below depend upon the strength of the overburden.) Thus, the geometric profile of the overburden and of the upper boundary of the viscous layer tends to become parallel to the pressure profile in the viscous layer. For a hypothetical case with negligible drag along the vertical wall of the dome, the profile will be as shown in Fig. 11.68.

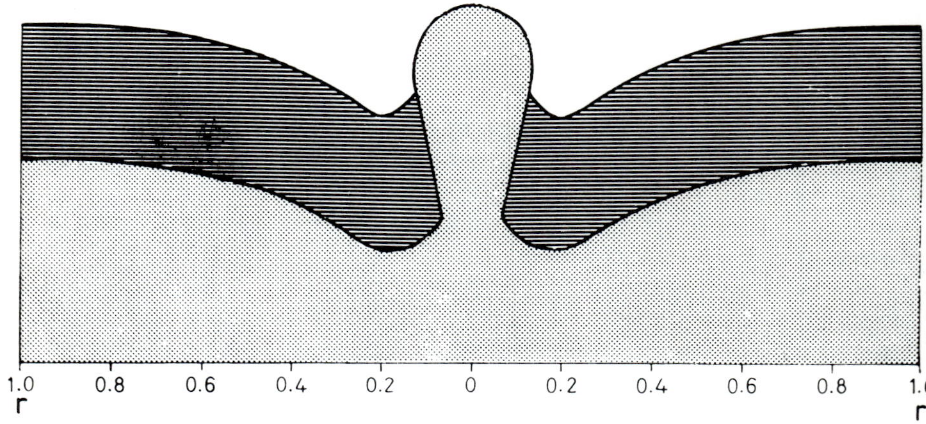

Fig. 11.69. Same as Fig. 11.68 but here drag along the sides of the rising diapir is indicated.

In model experiments and in nature, however, vertical drag along the boundary of the domes is always present. The drag tends to lift the immediately adjacent portion of the overburden, thereby displacing the depression in the overburden some distance away from direct contact with the dome. Consequently, a marginal syncline is formed a short distance away from the root of the dome as shown in Fig. 11.69.

It has been suggested that flow from distant portions of the source layer cannot pass the marginal syncline and, hence, that the active source of a dome is limited to the region inside the marginal sink. This is not necessarily true, because the pressure in the viscous layer underneath the lowest point of the marginal syncline may or may not be higher than the pressure at the same level outside the syncline. If the pressure below the sink is less than that outside, flow can pass the sink and material from distant points may contribute to the growth of the domes. The pressure underneath the depression may be great enough to make it function as an impassable threshold only in case the overburden directly above is sufficiently thickened by infilling or by gravity collapse. Incidentally, if the overburden in the low region is thickened by sediments laid down on the surface, the pressure in the buoyant layer underneath the sink may become great enough to prevent crossing of flow only if the density of the sediments is greater than that of the source stratum.

An interesting and geologically significant deduction follows, viz. that the depressed region must show negative gravity anomaly indicating mass deficiency relative to the surroundings despite the fact that the surface continues to founder relative to the immediate adjacent regions. This point is particularly pertinent in connection with the negative gravity anomaly found over many deep submarine trenches.

BUOYANT UPLIFT OF CRUSTAL BLOCKS ALONG STEEP FAULTS

So far it has not been possible to find model materials which under the conditions of dynamic similarity with tectonic structures would develop the kind of smooth, extensive tension and/or shear fractures which are common in regions of crystalline schists, gneisses and granites. The problem of generating such fractures has not therefore been studied in our centrifuged experiments. The experimental study must be limited to the effect which such fractures may have—once they have formed—upon the geometry of flow and strain of structures such as diapirs, nappes, etc.

Zones of weakness can be built into the initial state of the model as cuts which may or may not be "welded" by materials which are weaker than the rest of the model materials. This arbitrary predetermined localization and orientation of fractures is of course not useful as a guide to the understanding of how fractures are generated in natural structures, but their effects on the evolution of subsequent structures in the models may still be informative.

To study how the evolution of a buoyant plastico-viscous body may be influenced by extensive fractures in a mechanically otherwise strong overburden, models M-20, M-21, M-22 and M-28 were run in the centrifuge (Figs 11.70–11.75). In an overburden of modelling clay and painter's putty overlying buoyant silicone, two parallel vertical cuts were made. The cuts were "welded" by weak painter's putty.

The surface of the overburden block between the two cuts was made a little lower than the surface of the rest of the overburden, but the bottom of the block was level with the adjacent overburden. This made the pressure

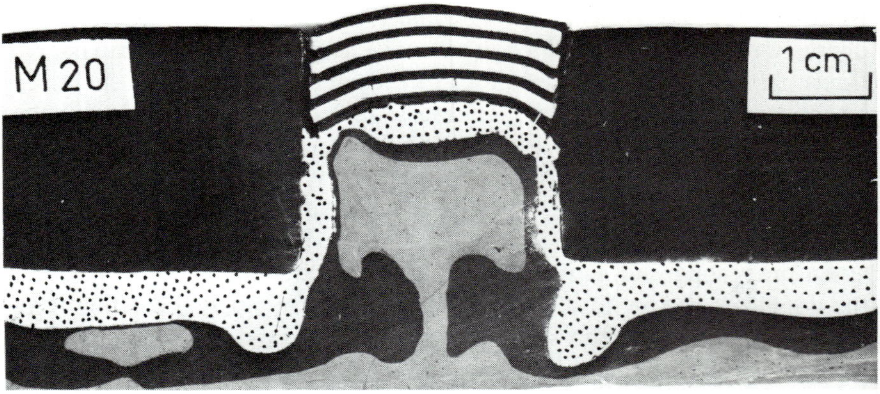

Fig. 11.70. The model whose initial structure is shown in Fig. 11.71, run 7 min at acceleration varying between 1000 *g* and 1500 *g*. Basal layer not shown.

in the buoyant layer relatively low underneath the block, and flow would be directed toward the bottom of the block which then would be lifted.

After a short run in the centrifuge the silicone forced the cut-out block upward until its surface was elevated a few millimetres above the surface of the adjacent overburden. Continued run in the centrifuge did not elevate the block further—apparently a state of isostatic equilibrium had been reached.

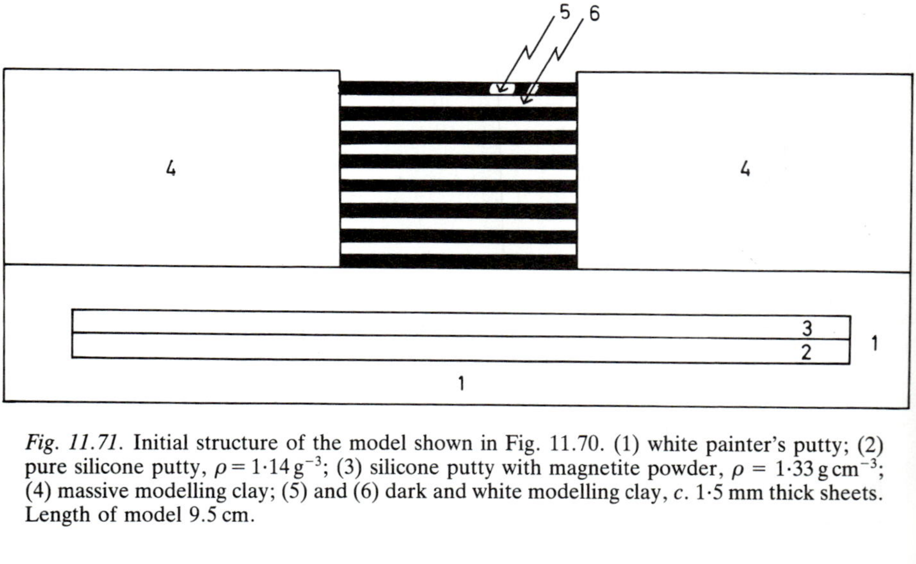

Fig. 11.71. Initial structure of the model shown in Fig. 11.70. (1) white painter's putty; (2) pure silicone putty, $\rho = 1{\cdot}14\,g^{-3}$; (3) silicone putty with magnetite powder, $\rho = 1{\cdot}33\,g\,cm^{-3}$; (4) massive modelling clay; (5) and (6) dark and white modelling clay, *c.* 1·5 mm thick sheets. Length of model 9.5 cm.

Fig. 11.72. Model similar to M-20 (Fig. 11.71) but differing in details, run 22·5 min at acceleration changing from 1000 *g* to 3000 *g*. In M-21 the initial sheet of dark- and light-coloured buoyant silicone putty was much thinner than in M-20.

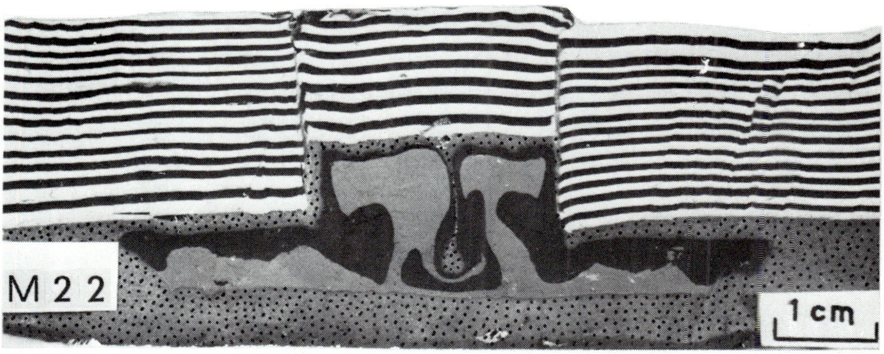

Fig. 11.73. Model of type initially similar to M-20 (Fig. 11.71) but differing in details, run 11·5 min at acceleration varying between 1000 *g* and 2000 *g*.

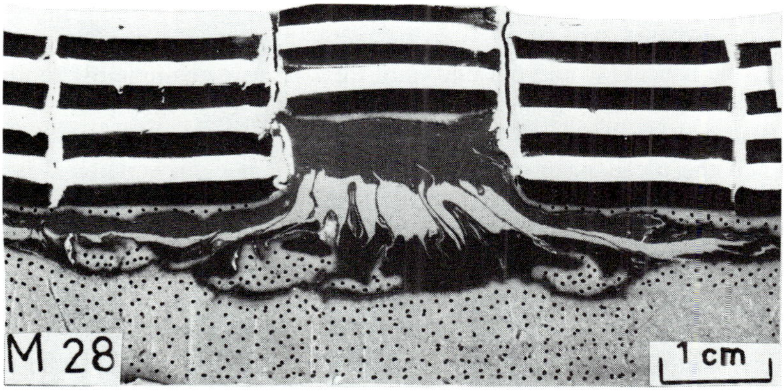

Fig. 11.74. Part of model whose initial structure is shown in Fig. 11·75 run ∠ min 10 sec at acceleration varying between 1000 *g* and 3000 *g*.

In nature the locally elevated surface would be exposed to erosion with consequent disturbance of the isostatic equilibrium. To simulate erosion in the model the centrifuge was stopped and a top sheet of the elevated block removed. Continued centrifugation then forced the cut-out block upward—owing to the buoyant silicone—until the "eroded" surface reached a level somewhat higher than the pre-erosion stage. After removal of a second top layer from the elevated block continued centrifugation elevated the block another step. This stepwise process: elevation → "erosion" → elevation → "erosion", etc., would continue until all the buoyant silicone had moved into the space below the elevated block. If the amount of

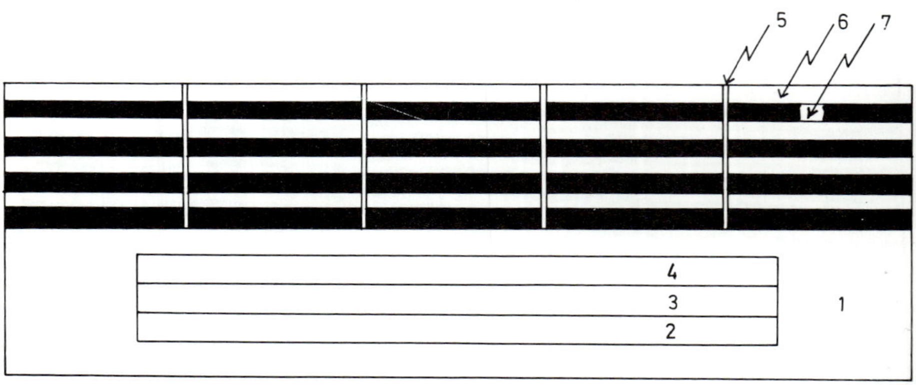

Fig. 11.75. Initial structure of model M-28 (Fig. 11.74). (1) Painter's putty; (2) silicone putty, $\rho = 1{\cdot}26\ \mathrm{g\,cm^{-3}}$; (3) silicone putty, $\rho = 1{\cdot}44\ \mathrm{g\,cm^{-3}}$; (4) silicone putty, $\rho = 1{\cdot}58\ \mathrm{g\,cm^{-3}}$; (5) painter's putty; (6) white modelling clay; (7) dark modelling clay.

silicone available was larger than the volume of the cut-out overburden block, the process would not end until the block was completely removed by erosion and the silicone was spilling over the edge of the well left by the removed block.

When the mechanism of granitoid batholiths is contemplated it is also necessary to take into account the dynamic consequence of the contrast in rheological properties of a plastic but crystalline body and a liquid magma. It has been demonstrated experimentally that buoyant materials which is very fluid relative to the model-Earth crust produce quite different final structures even if the initial structure is identical to the above models. The buoyant rise of model magmas is considered in Chapter 13, and models relevant to the studies in the present section are shown in Figs 11.14, 11.15 and 11.16.

In areas containing granitoid gneisses and other regional metamorphic rocks it is conceivable that the heavy cover is ductile while the buoyant granitoid basement is brittle. The cover may for example consist of biotite schist and marble, both rocks being appreciably more dense than an ordinary granitoid gneiss. At moderate grade of regional metamorphism the mica schist is also less competent than the gneiss. To test the possibility that under such conditions the buoyant but refractory granitoid basement would rise as horsts into the ductile overburden the following experiments were run. Blocks of wood, density 0·9 g cm^{-3} (model I) or Plexiglass, density 1·14 g cm^{-3} (model II) imitated the fractured upper part of a buoyant basement, and silicone, density 1·14 g cm^{-3} represented the deep ductile part of the basement while silicone, density 1·60 g cm^{-3}, imitated the heavy but ductile cover. To initiate the rise of horsts the ductile part of the basement was made thicker in the middle of model II. After centrifugation, low-density blocks of the rigid basement had risen into the ductile cover (Figs 11.76–11.79). The ductile deep part of the basement had moved horizontally to fill the space left below the rising blocks while small portions of the cover had found their way down along the fractures between some of the rigid buoyant blocks.

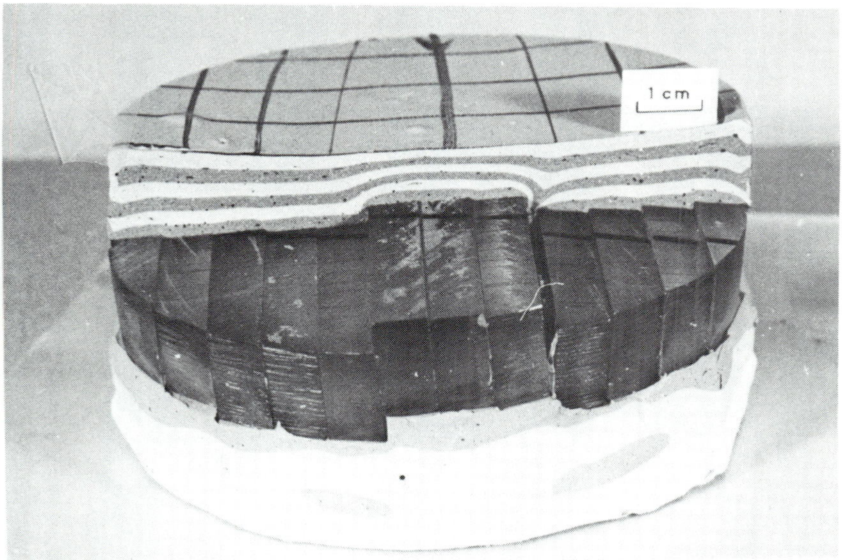

Fig. 11.76. Model simulating the buoyant rise of blocks of a rigid basement into a plastic overburden. Run 2 min at 1600 *g*. See initial structure (Fig. 11.77). Half the overburden cut away (run by R. Häll, Uppsala).

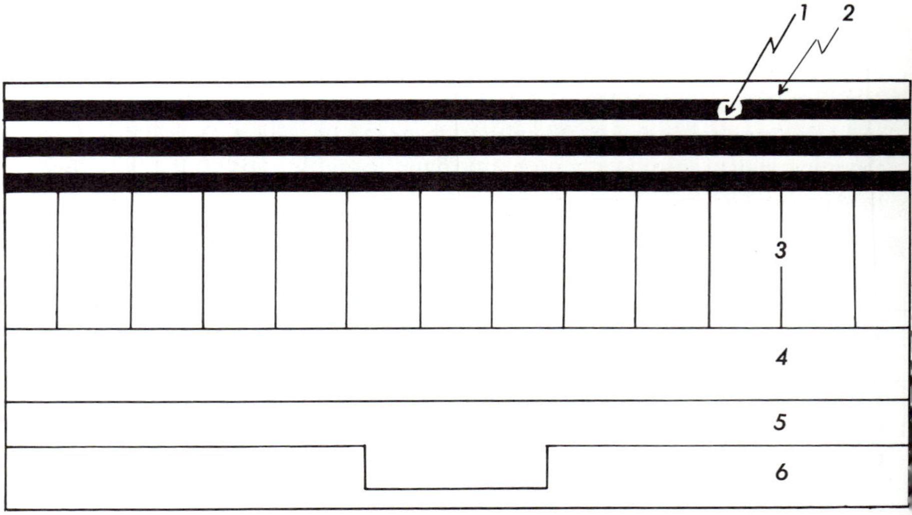

Fig. 11.77. Initial structure of model shown in Fig. 11.76. (1) Red silicone putty, $\rho = 1{\cdot}60$ g cm^{-3}; (2) pink silicone putty, $\rho = 1{\cdot}60$ g cm^{-3}; (3) blocks of Plexiglass, $\rho = 1{\cdot}14$ g cm^{-3}, separated by thin oil films; (4) silicone putty, $\rho = 1{\cdot}60$ g cm^{-3}; (5) silicone putty, $\rho = 1{\cdot}14$ g cm^{-3}; (6) painter's putty (prepared by R. Häll, Uppsala). Length of model 9·5 cm.

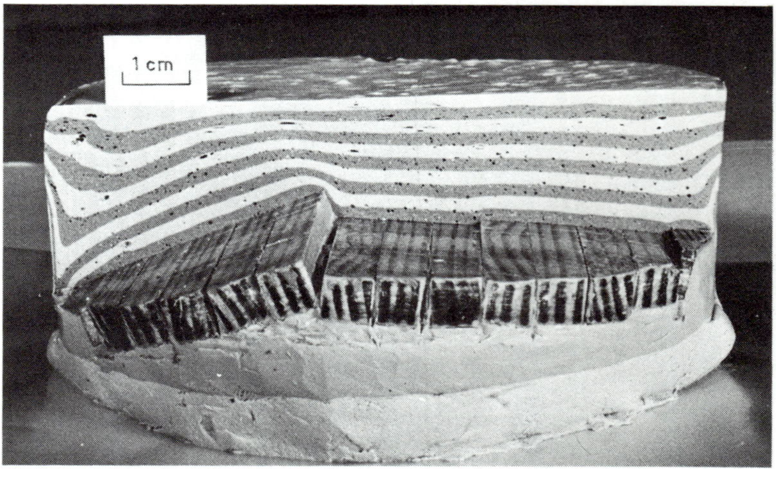

Fig. 11.78. Model simulating uneven rise of fractured rigid, but buoyant, basement into a plastic overburden. Run 8 min at 1600 *g*. See Fig. 11.79. Half the overburden cut away (run by R. Häll, Uppsala).

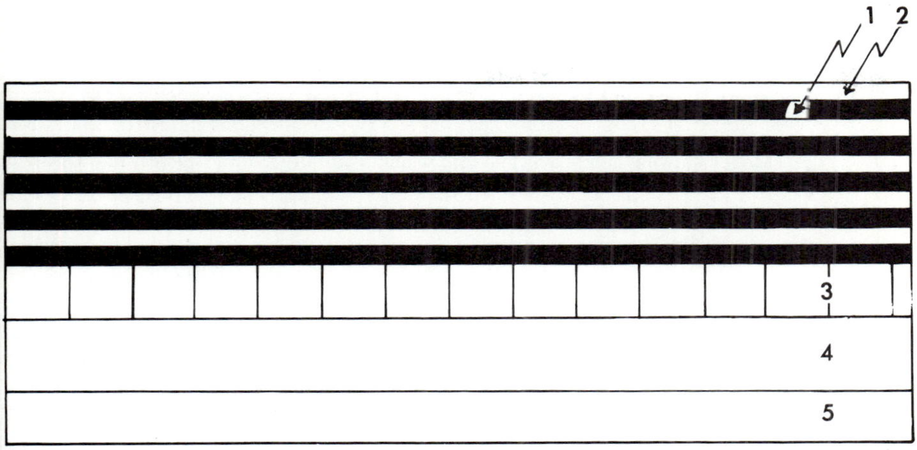

Fig. 11.79. Initial structure of model shown in Fig. 11.78 (1) Red silicone putty, $\rho = 1{\cdot}60$ g cm^{-3}, (2) pink silicone putty, $\rho = 1{\cdot}60$ g cm^{-3}; (3) blocks of wood, $\rho = 0{\cdot}9$ g cm^{-3}; (4) silicone putty, $\rho = 1{\cdot}14$ g cm^{-3}; (5) painter's putty (prepared by R. Häll, Uppsala).

STRUCTURES GENERATED IN OVERBURDEN AND SUBSTRATUM TO A BUOYANT LAYER

In an originally planar layered overburden the first structures produced by the rise of domes from a source layer and the complementary subsidence of the rim synclines are simple bending folds. Examples are numerous: model DV-8 (Fig. 11.13); S-219 (Fig. 11.107); S-50 (Fig. 11.38); S-86 (Fig. 11.48), etc. The horizontal tensile stress generated above the domes or buoyant ridges produces various types of stretched structures in the crests of the folds (supertenuous folds), whereas the sagging down in the rim synclines produce second-order buckle folds within these regions. There is thus a tendency to form synclinoria, but not anticlinoria, in multilayered overburden above a buoyant substratum.

Model S-109 (Fig. 11.80) shows an example on the formation of gentle buckles in a shallow rim syncline at the early stage of diapiric growth. Figures 11.81 and 11.82 provide additional examples on the development of synclinoria by buckling of surficial strata in rim synclines.

As the hat of diapirs spreads in the form of lobes over the surface the uppermost non-brittle strata of the overburden may be thrown into recumbent folds—fold nappes—with lobes of the diapir as cores. Model S-86 (Fig. 11.48) is an example.

Usually this recumbent folding of the topmost layers of the overburden

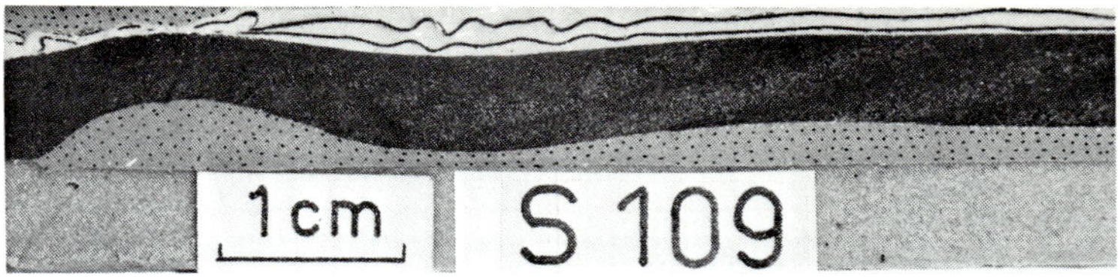

Fig. 11.80. Model showing gentle syncline into which a surface complex of two thin sheets of modelling clay embedded in silicone putty have slid, producing gentle buckles in the process.

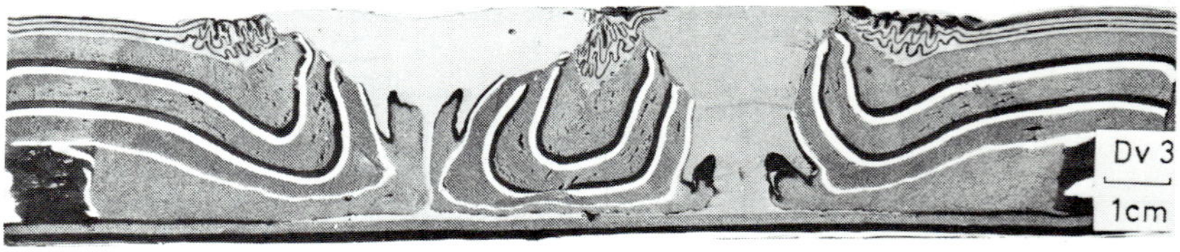

Fig. 11.81. Section through model shown in Fig. 11.16, showing more clearly the buckling of the layered surface complex in the rim synclines as well as that both diapirs are rooted in the basal source layers.

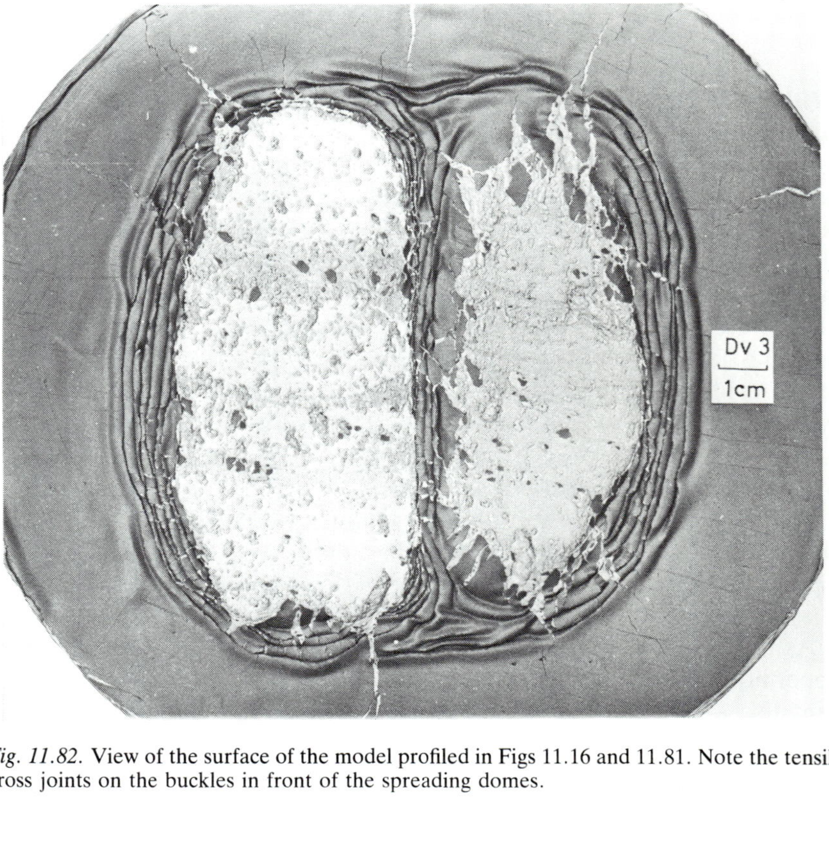

Fig. 11.82. View of the surface of the model profiled in Figs 11.16 and 11.81. Note the tensile cross joints on the buckles in front of the spreading domes.

Fig. 11.83. Section through model DV-7, also shown in Fig. 11.17. In this section the late diapir (the angular white body in the middle of the figure) has just started to rise, thereby gently deforming the surface lobe (nappe) which has spread out from the early-formed diapir (dotted).

is associated with fracturing and the formation of boudins or pinch-and-swell structure of the uppermost fold limb.

The advancing front of the spreading nappe often compresses the surficial strata into buckle-folds of the type exhibited by model DV-3 (Fig. 11.81) or model S-116 (Fig. 15.14) and many other models shown in this book.

When the spreading hat of a diapir overruns the frontal buckles the buckles become rotated into recumbent position and even broken to fragments stacked on top of one another like shingles on a roof, e.g. Fig. 11.83.

Spreading of diapirs just underneath the uppermost layers of the overburden may produce fracture systems quite similar to some natural salt domes. Compare the model shown in Fig. 11.84 with the structure above the Clay Creek Dome (Fig. 11.85).

Model S-90 (Fig. 11.86) shows the overburden surface at the beginning of spreading of a diapir underneath. A network of fractures has developed at the surface—much like desiccation fractures in clay—and small portions of the buoyant material are just beginning to pierce the surface. Greater spreading and more widely separated fragments are generated in e.g. model DV-3 (Fig. 11.82) and the models shown in Figs 11.87 and 11.88.

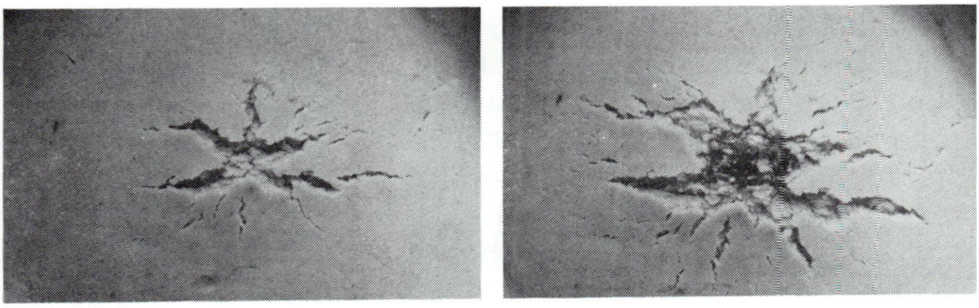

(*a*) (*b*)

Fig. 11.84. Dome of wax in the process of piercing the surface of painter's putty overburden, (*b*) taken a little later than (*a*). Cf. the fracture pattern above the Clay Creek Dome (Fig. 11.85).

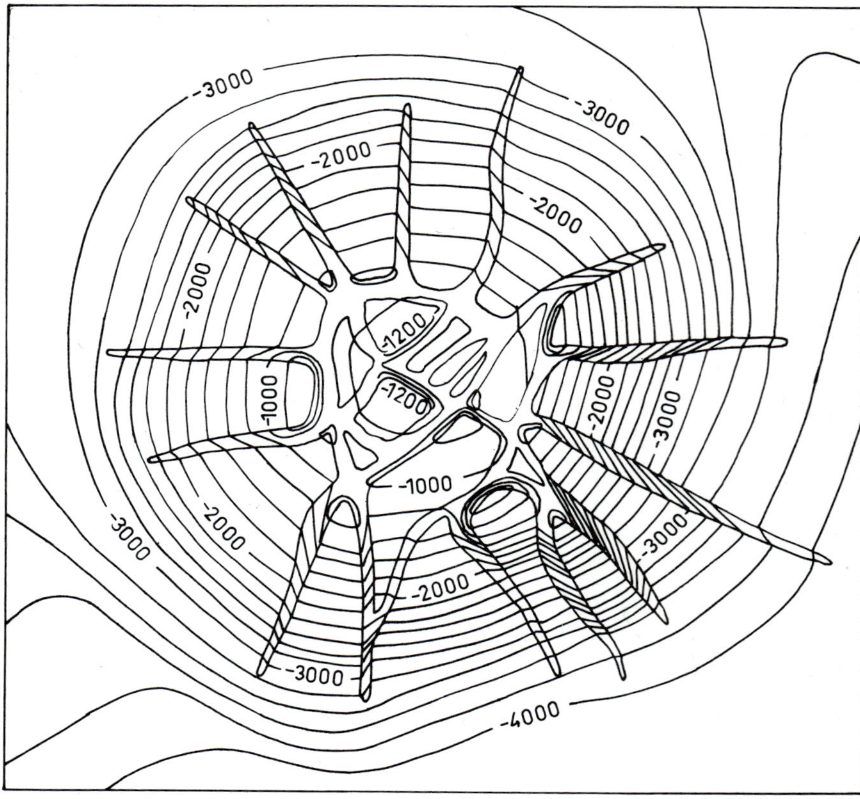

Fig. 11.85. Fractures above the Clay Creek Dome, Washington County, Texas (from Gussow in Braunstein and O'Brien, 1968).

If the surface stratum is somewhat brittle, breaks often occur in the sharp bends at the crests of the folds, see e.g. model DV-4 (Fig. 11.89).

Where the diapiric lobes are curved with the convex side facing outward the spreading creates a tensile stress parallel to the circumference. This may produce tension cracks normal to the fold axes in the buckles formed in front of the lobe. Examples of such cracks are seen in models S-104 (Fig. 11.90) and DV-3 (Fig. 11.82). The folds in the surface layer are formed normal to the divergent lines in the spreading lobe underneath. The upper parts of domes are often funnel-shaped. Around such domes inversion of the layering in the overburden is commonly developed in the models (e.g. Figs 11.38 and 11.14).

If the substratum below a layer of buoyant material is not rigid, it tends to participate in the flow and develop a bulge or ridge with the crest directly

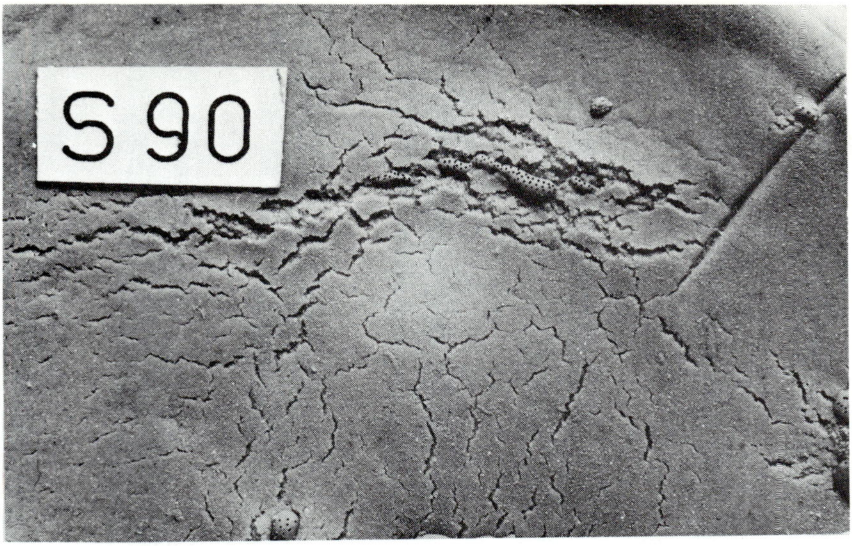

Fig. 11.86. Fractures produced in painter's putty surface by a silicone-putty dome that rises from below, gently elevates the surface, and at places pierces the surface (small dotted body).

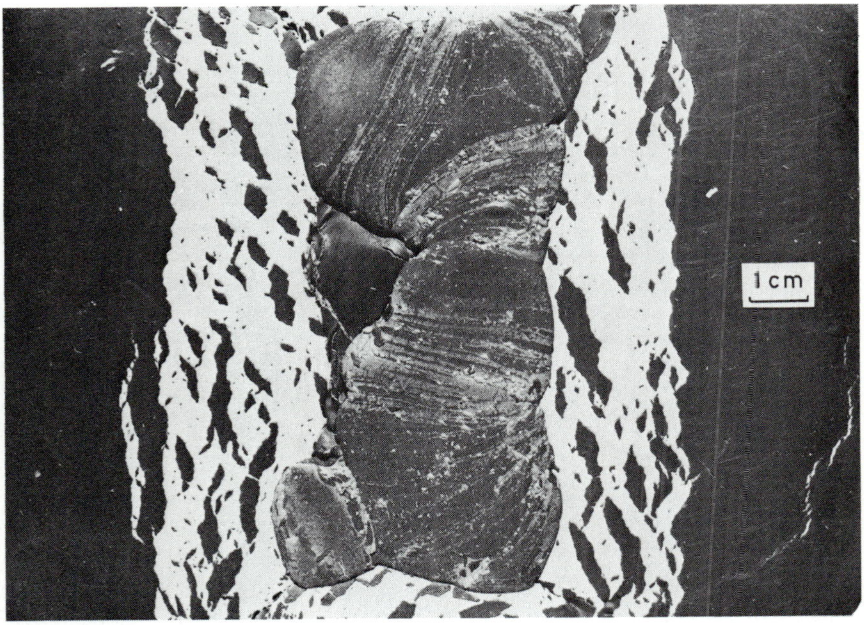

Fig. 11.87. Fragmentation to angular pieces of a thin surface sheet of modelling clay caused by the rise and spreading of a silicone-putty diapir. The light-coloured material is the surface of the diapir in the middle of which a secondary diapir has formed. Note the fragment shape which suggests split along diagonal fractures. Cf. fragmentation of the surface sheet in Fig. 11.88.

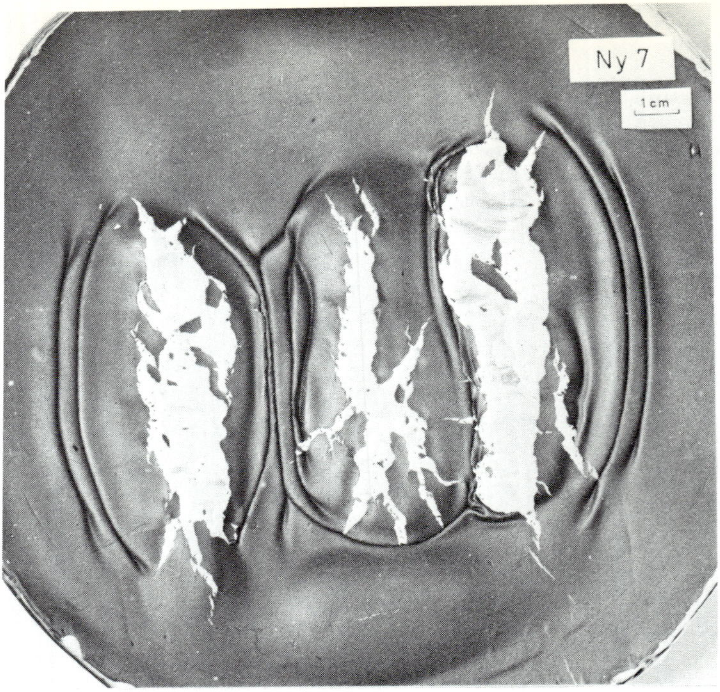

Fig. 11.88. Fracturing and fragmentation of surface sheet of modelling clay above three ridges of buoyant silicone putty. See cross-section of same model (Fig. 15.8).

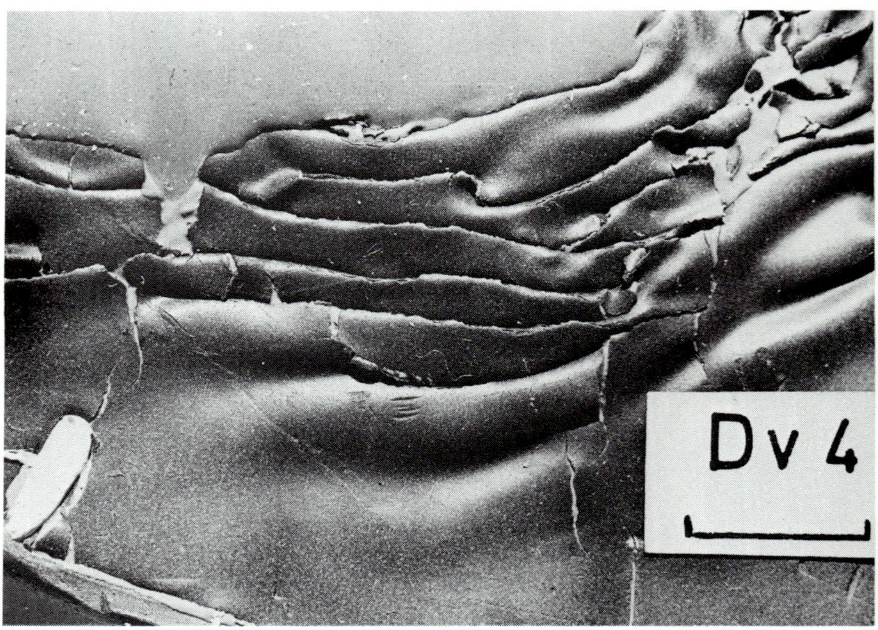

Fig. 11.89. Buckling and fracturing of surface sheet of modelling clay in front of spreading lobe of a diapir (smooth-looking mass at upper left and center). See cross-section of same model (Fig. 11.45). Scale 1 cm.

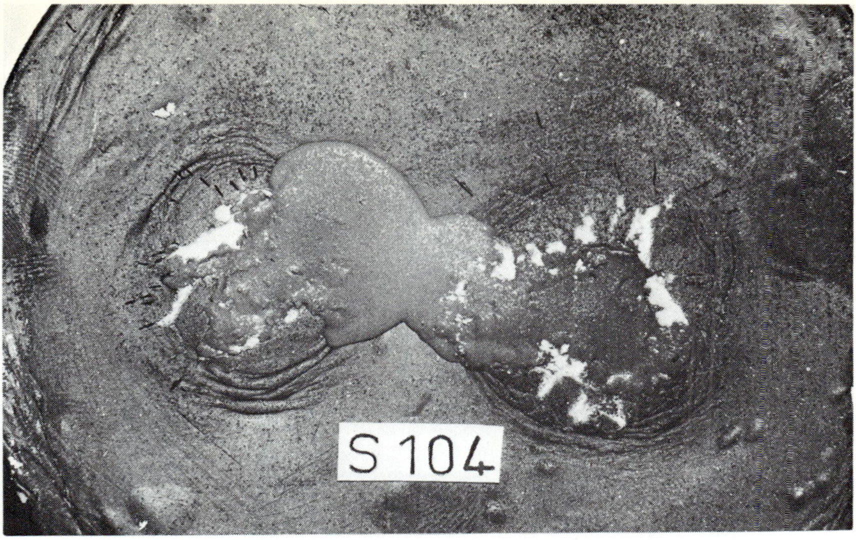

Fig. 11.90. Buckles and tension fractures in surface sheet of modelling clay above two domes. Width of original about 8 cm.

Fig. 11.91. Source layer and dome of silicone putty (light grey) with embedded thin sheet of modelling clay. Overburden of painter's putty and substratum of heavy silicone putty with magnetite powder. Note gentle rise of heavy substratum below dome. Run 8 min at acceleration changing from 700 *g* to 2100 *g*.

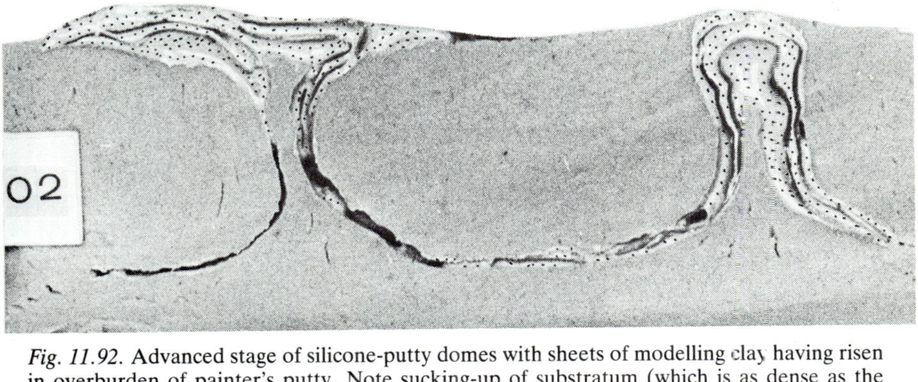

Fig. 11.92. Advanced stage of silicone-putty domes with sheets of modelling clay having risen in overburden of painter's putty. Note sucking-up of substratum (which is as dense as the overburden) high into the trunk of the domes. Run 9 min at 1350 *g*. Cf. Figs 11.93 and 16.11. Length of original 9 cm.

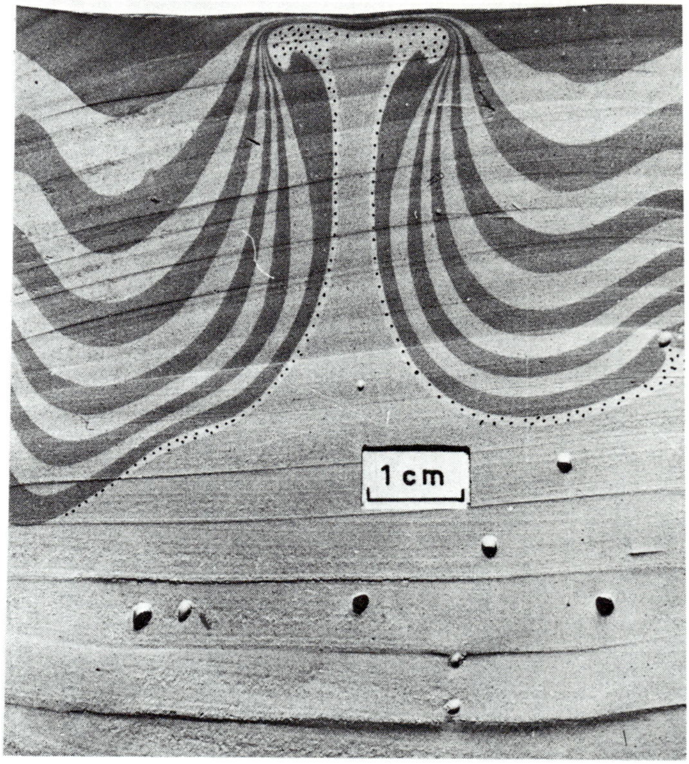

Fig. 11.93. From a thin buoyant sheet (dotted) of silicone putty ($\rho_1 = 1{\cdot}14\ \mathrm{g\ cm^{-3}}$) originally even-thick and located in the middle of the model, a single diapir has developed in the centre of the sheet and ascended to the surface. The heavy substratum has been brought almost to the surface in the trunk of the diapir. Overburden consists of horizontal sheets of silicone putty with contrasted colour but identical density ($\rho_2 = 1{\cdot}3\ \mathrm{g\ cm^{-3}}$) and identical viscosity. Substratum made of the same silicone putty ($\rho_4 = \rho_2 = 1{\cdot}3\ \mathrm{g\ cm^{-3}}$), but uniformly coloured. See text.

below the trunk of the dome or directly below a buoyant anticline, as shown in Fig. 11.91. In some cases the substratum is pulled far up into the trunk, as happened in the models shown in Figs 11.92, 11.93 and e.g. 16.11. This tendency of non-rigid basement to be sucked up into the root zone of diapirs, and even all the way to their hat, is predicted theoretically as discussed on p. 107. The feature may be of considerable interest for the study of natural diapirs. A rise of a relatively heavy substratum would give a gravity high in the central region of a natural diapir, quite contrary to expectation according to the conventional picture of domal structures. In

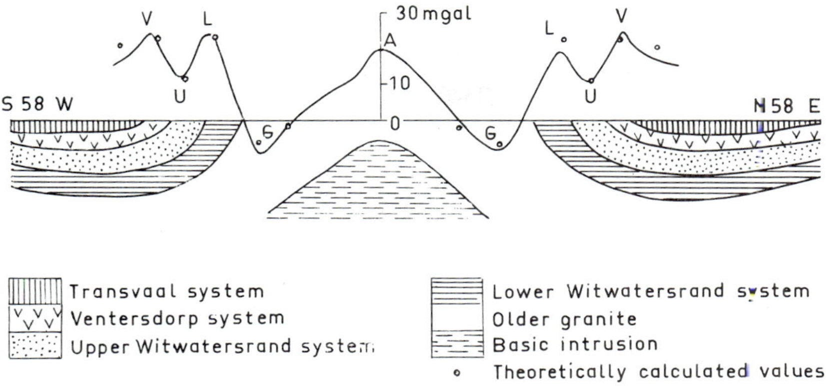

Fig. 11.94. Profile of structure and gravity variation across the Vredefort dome, South Africa (after Daly, 1947).

this connection, one may refer to the gravity field measured across the famous Vredefort dome in South Africa (see, e.g., Daly, 1947; Fig. 11.94).

The centripetal flow in the source layer towards the root of a dome generates, by friction coupling, a corresponding flow in the contact zone of the substratum. There is therefore a certain tendency for competent layers enclosed in the substratum to develop buckle folds more or less parallel to those formed in the source layer itself. Model Ny-9 (Fig. 11.95) furnishes a well-developed set of cusp-and-arc type buckle fold in the soft basement underneath the root zone of a diapir.

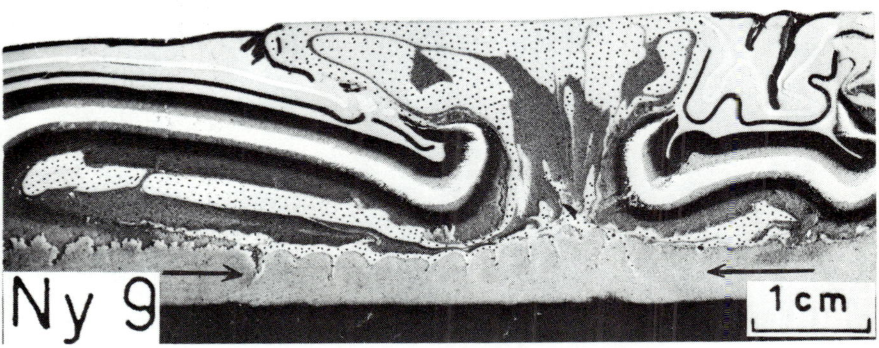

Fig. 11.95. In the source layer (black and dotted silicone putty) horizontal flow was directed toward the root of the diapir thus transmitting shear stress to the basal layer of painter's putty which was therefore compressed (arrows), producing cusp-and-arc type folds with the convex side of the arc facing the softer medium, the silicone putty. See text.

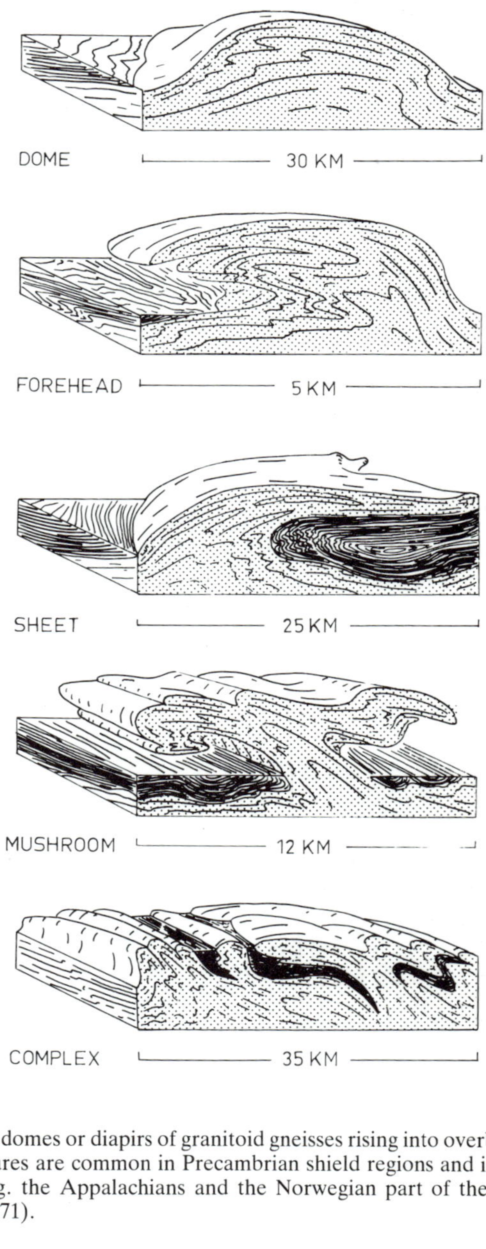

Fig. 11.96. Types of domes or diapirs of granitoid gneisses rising into overburden of crystalline schists. Such structures are common in Precambrian shield regions and in the core of deeply eroded orogens, e.g. the Appalachians and the Norwegian part of the Caledonides (from Haller, 1955 and 1971).

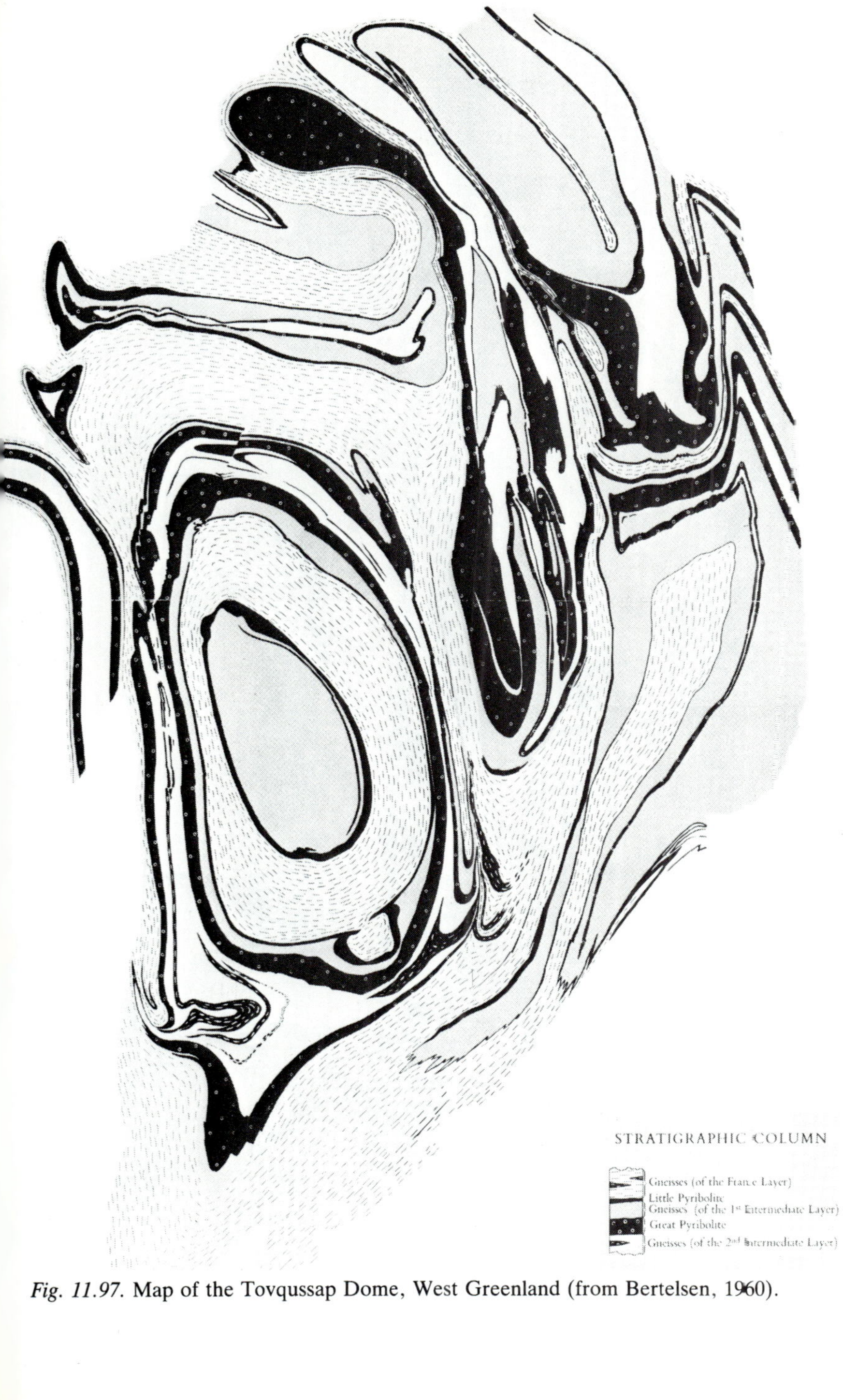

Fig. 11.97. Map of the Tovqussap Dome, West Greenland (from Bertelsen, 1960).

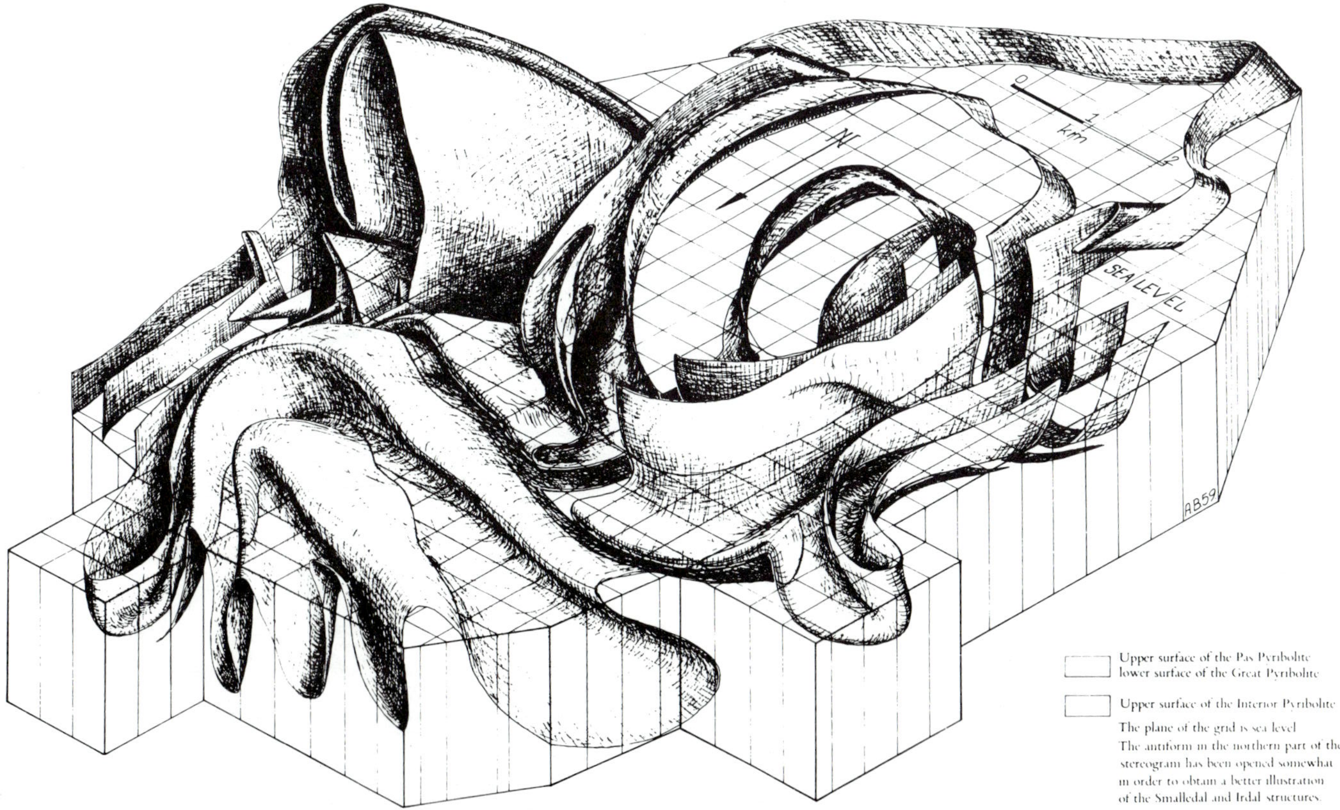

Fig. 11.98. Structural stereogram of the Tovqussap Dome, West Greenland (from Bertelsen, 1960).

Many of the model domes illustrated in this book are strikingly similar to the natural gneiss-granite domes such as those from East Greenland described by Haller (1956) (Fig. 11.96), from West Greenland described by Berthelsen (1960) (Figs 11.97 and 11.98), and the classic ones of Eskola (1948) from Maryland, U.S.A. More recent descriptions of Appalachian gneiss domes are found in Zen *et al.*, (1968) (see Figs 15.18–15.21).

INTERNAL STRUCTURES DEVELOPED IN SOURCE LAYER AND RISING DIAPIRS

The internal flow pattern developed during rise of more or less circular diapirs is generally as follows. In the source layer flow is centripetal and convergent towards the root of the diapir. At least in the early history of domal growth, the flow is able to cross under the rim syncline, thus feeding the diapir from an area also outside the rim syncline. This is demonstrated in models with strain markers, e.g. models O-7 and O-22 (Figs 11.99–

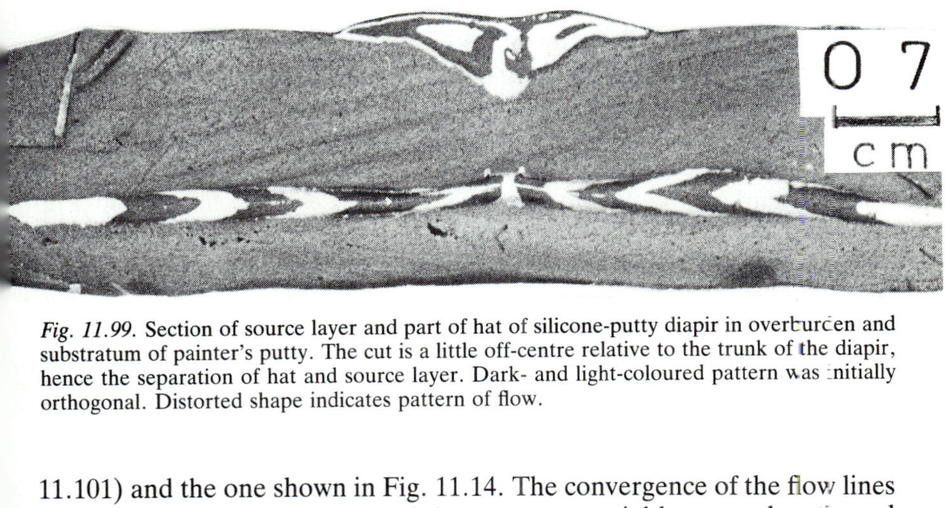

Fig. 11.99. Section of source layer and part of hat of silicone-putty diapir in overburden and substratum of painter's putty. The cut is a little off-centre relative to the trunk of the diapir, hence the separation of hat and source layer. Dark- and light-coloured pattern was initially orthogonal. Distorted shape indicates pattern of flow.

11.101) and the one shown in Fig. 11.14. The convergence of the flow lines means that a volume element of the source material becomes lengthened parallel to the flow lines and shortened normal to the flow lines in the horizontal plane, i.e. there is a radial lengthenening and a tangential shortening of the "fluid" element. When the flow passes underneath the rim syncline there is even a shortening in the vertical dimension.

In addition to the convergent pattern of the flow lines in the source layer outside the root zone, drag along the boundaries against the overburden and the substratum adds a component of horizontal shear-strain to the flow

Fig. 11.100. Initial structure of model O-22 (Fig. 11. 101). (1) Painter's putty; (2) light-coloured silicone putty; (3) dark-coloured silicone putty. Length of section 9.7 cm.

geometry, as is readily demonstrated by neutral markers (Fig. 11.100). As the flow lines bend upward in the root zone and become vertical in the trunk of the diapir, the convergent pattern changes to an almost parallel one. On crossing from the trunk to the spreading hat, the flow lines change from almost parallel to divergent in the horizontal plane. This flow pattern is readily made visible by employing both passive and active markers.

Passive or neutral markers may consist of strips, layers, spheres and other known shapes, whose colour contrasts with that of the main substance, but their rheological properties must be identical with those of the main substance. Such markers follow passively the flow whose finite pattern is displayed in more or less detail by the deformed markers. Attention is directed to the carefully prepared model (see Fig. 11.14) run by Dixon (1975).

Active markers, on the other hand, do not possess the same rheological properties as the bulk of the flowing mass. Such markers have higher yield strength and/or higher viscosity coefficient than the enclosing material. In our runs thin sheets of modelling clay were usually employed as active markers in silicone or stitching wax. Active markers respond to the strain

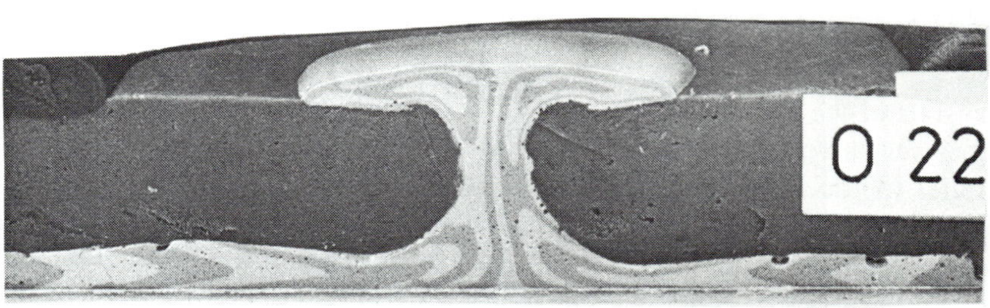

Fig. 11.101. Section through centre of silicone-putty diapir in overburden of painter's putty. Distorted shape of initially orthogonal pattern of dark- and light-coloured source layer indicates geometry of flow. Length of section 9 cm.

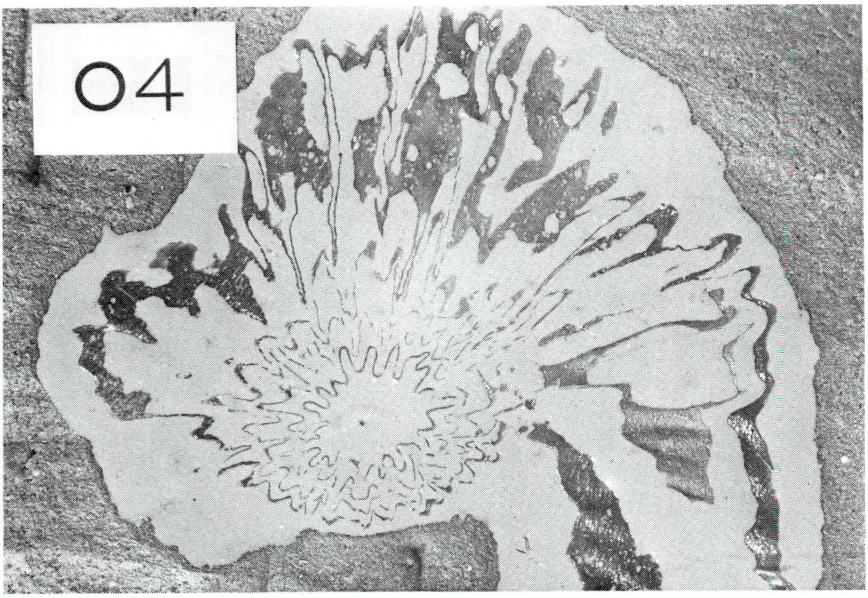

Fig. 11.102. Horizontal cut through the root zone of silicone-putty dome and source layer with four thin sheets of modelling clay strongly buckled in the field of convergent flow around the root of the dome. Horizontal radial fold axes assume vertical position in the trunk of the dome. Run 10 min at 850 *g*.

in the flow by buckling or by stretching to pinch-and-swell or boudinage structures. Because of their contrasted rheological character the markers disturb the strain and flow pattern in their neighbourhood—they become surrounded by a zone of contact strain with a geometry different from that of the undisturbed enclosing material. See also the discussion on the theory of folding, Chapter 8.

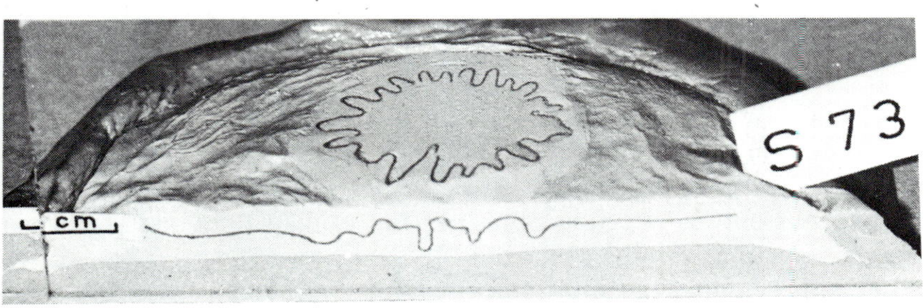

Fig. 11.103. Buckled sheet of modelling clay originally embedded as a straight sheet in a buoyant layer of silicone putty. Model cut vertically and horizontally and overburden removed to show how embedded sheet has been distorted during centrifugation.

The behaviour of active markers in the form of relative competent sheets is of particular relevance to structural geology and geotectonics. Competent layers of modelling clay embedded in the source layer of silicone or stitching wax develop buckle folds with radially arranged axes in the region outside the root of domes, shown clearly in Figs 11.102 and 11.103. If weak enough, the embedded sheets also break under tension with cracks developing at an obtuse angle to the principal direction of stretch such as demonstrated

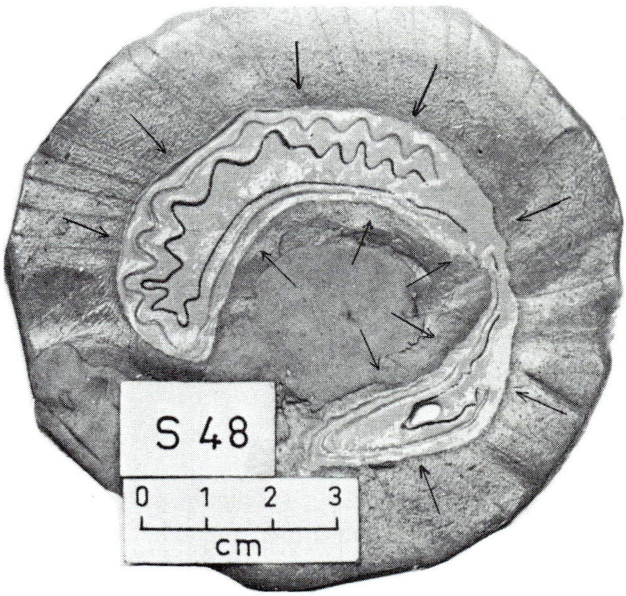

Fig. 11.104. Horizontal cut through buoyant, circular ridge of silicone putty with two embedded sheets of modelling clay. Run at 750 *g* for 4 min. During rise of buoyant ridge, flow in source layer towards ridge was centripetal from edge of model and centrifugal from centre of model, i.e. the former flow was convergent and the latter divergent (arrows). In consequence, stretching (boudins) of modelling-clay sheets occurred in concave part of the ridge, and compression (buckling) in its convex part.

in Figs 11.13, 11.18 and others. Thus, when the fold axis is also an axis of stretch, not only a direction obtuse to the principal compressive strain, cracks develop across the fold axis.

In the terminology of geotectonics in use among students of petrofabrics, the elongation associated with these folds would be called *a*-elongation, i.e. an elongation parallel to the flow direction, the flow direction coinciding with the axis of the folds. In the root region the fold axes bend upward

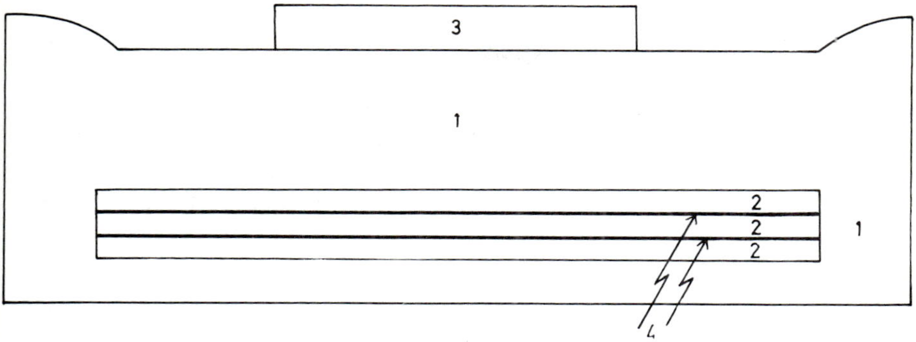

Fig. 11.105. Model S-48 prior to run in centrifuge (see Fig. 11.104). (1) Painter's putty; (2) buoyant silicone; (3) extra load on overburden; black: sheets of modelling clay.

more or less parallel to the flow lines, and as the cross-section of the flow decreases markedly in the transition from the root zone to the trunk, the buckling of the embedded sheets is intensified in this region, so also is the stretching parallel to the fold axis. Buckel folds and boudinage structure are therefore especially prominent in the trunk of domes, e.g. Fig. 11.103.

In the strongly divergent flow pattern in the spreading hat the enclosed modelling clay sheets become fragmented and the small fragments may become widely separated (see Fig. 11.87).

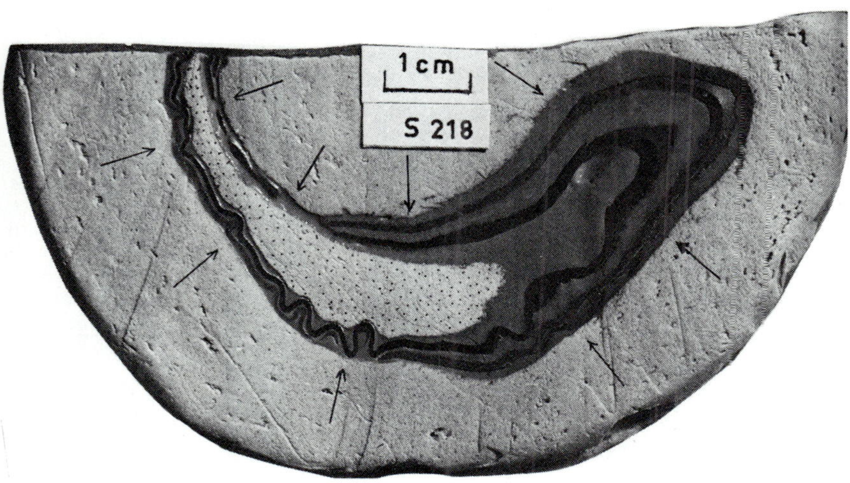

Fig. 11.106. Model similar to S-48, Fig. 11.104. In S-218 the substratum has been sucked up into the rising buoyant circular ridge, forming the core (dotted) of the anticline which the ridge constitutes. The model is cut horizontally and vertically. Buckling and stretching of the embedded sheets of modelling clay follow same pattern as in model S-48 (Fig. 11.104).

Competent sheets embedded in a source layer that rises as a straight anticline along, e.g. a line discontinuity are not exposed to significant compression, neither in the horizontal flow in the source layer nor in the rising anticline itself. The flow lines connected with strongly elongate anticlines are essentially parallel except around the end of the ridges where convergence occurs. However, in many tests a buoyant ridge formed parallel to the circular boundary of the model, in which case the flow lines in the source layer would converge as they point toward the convex side of the ridge, and diverge as they point toward the concave side. Enclosed competent sheets of modelling clay accordingly develop buckle folds as they rise in the convex half of the buoyant ridge, but become stretched to boudins in the concave half of the ridge (see, for example, Figs 11.104–11.106).

CONFORMATION AND REORIENTATION OF STRUCTURES DURING BUOYANT RISE

Culminations of basal gneisses in orogenic belts often show a flow structure conformable to the overlying folded strata. This is true, for example, of

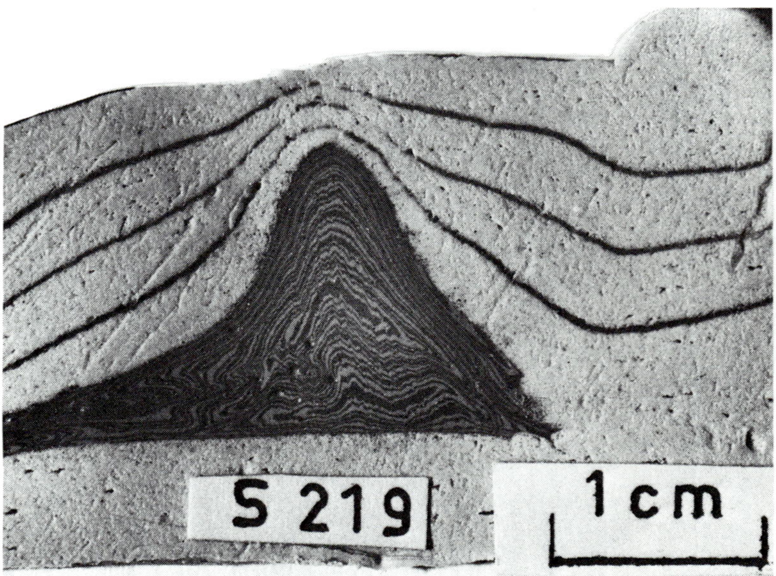

Fig. 11.107. Domal rise of buoyant silicone-putty layer with originally discordant structure against layered overburden. Secondary conformity produced during flow in ascending dome, and boudins of competent modelling-clay sheets generated in overburden above crest of dome. See text and Fig. 11.25.

Fig. 11.108. Diapir model of silicone putty showing foliation (colour layering) parallel to the contact toward painter's-putty overburden. The foliation is produced by laminar flow caused by contact drag when the buoyant material flows towards and into the diapir.

many central-gneiss culminations in the Scandinavian Caledonides (see p. 376). There is little reason to assume that the conformable structure is primary—one must rather think of the geosynclinal column of sediments and lavas as having rested upon a discordant basal structure. In this connection models shown in Figs 11.107 and 11.108 are of interest. The buoyant body was a layer cut out of a complexly folded mixture of dark- and light-coloured silicone (see Fig. 11.26). Accordingly it was initially a sharp discordance between the overburden and the structure in the source layer. The situation is similar to the disconformity existing between a basement of gneisses and crystalline schists with more or less complicated fold structure cut by erosion, and the overlaying column of sediments and lavas prior to the orogenic deformation. The rise of the dome and the associated flow in the source layer dragged and rotated the old structure in the basement into quasi-parallelism with the contact. In this way conformity was generated between the layering in the overburden and the rotated layering and gneissosity in the basement.

INTERFERENCE OF FLOW PATTERN IN SOURCE LAYER BETWEEN ADJACENT DIAPIRS

Two simultaneously rising diapirs interfere with the flow fields of one another. Figure 11.109 shows the particle paths in the source layer between two diapirs equal in size and rising with equal rate. The significant difference between this flow pattern and the flow around a single dome is the strong stretching in the region between the domes, and the field of divergent flow lines occurring within this region.

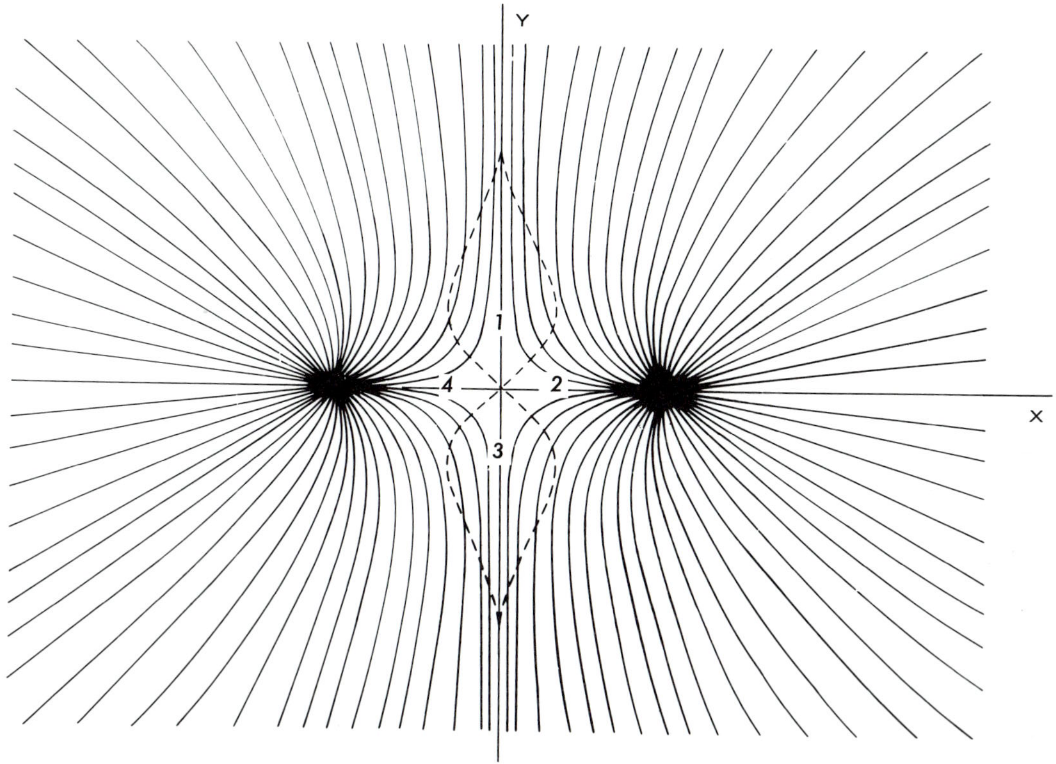

Fig. 11.109. Flow lines in buoyant layer from which two diapirs have risen: (1) and (3) are regions of divergent flow (when projected on the horizontal plant); (2) and (4) are regions of especially strong convergent flow (when projected on the horizontaal plane.) See text.

To obtain a general understanding of the flow geometry in the source layer within the region of interference between two diapirs, assume that the rate of thinning of the source layer (due to the material extruded into the two diapirs) is uniform throughout the region which acts as source for the diapirs. In other words, we assume that a rim syncline does not develop and that the diapirs are fed from the whole source area by uniform rate of sinking of the overburden.

Let under these conditions the velocity of flow up through the trunk of each diapir be Q as measured in terms of volume per unit time; for example, $Q\,\mathrm{cm}^3\,\mathrm{sec}^{-1}$. At a radial distance r_1 from diapir 1 the average horizontal velocity component pointing toward diapir 1 is then

$$v_1 = \frac{Q}{2\pi h r_1}\ \mathrm{cm\ sec}^{-1}. \tag{11.8}$$

Here h is the instantaneous thickness of the source layer.

Because of the withdrawal of source material through diapir 2 there will also be an average horizontal velocity component directed toward this diapir at radial distance r_2, defined by the expression

$$v_2 = \frac{Q}{2\pi h r_2}\ \mathrm{cm\ sec}^{-1}. \tag{11.9}$$

At any point x,y in the source layer the two horizontal vectors add up to give the resultant average horizontal flow velocity. From Fig. 11.110 we

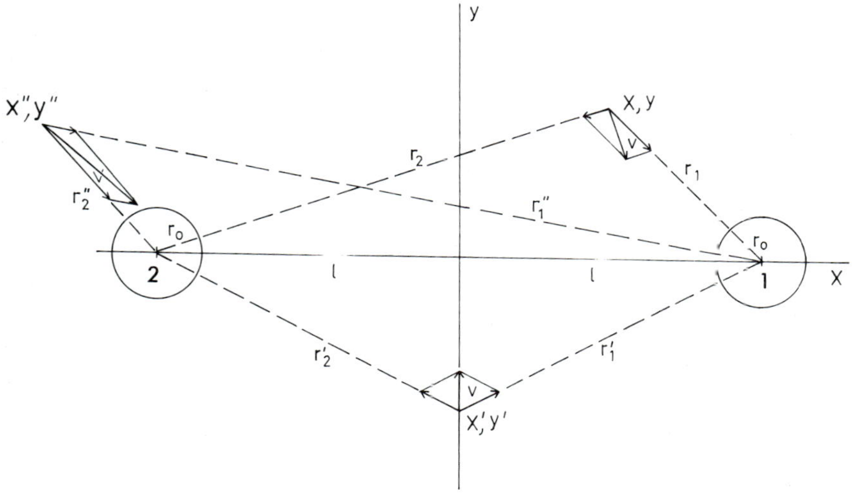

Fig. 11.110. Velocity vectors at selected points in a source layer feeding two simultaneously rising domes whose roots are marked (1) and (2). See text.

see that the two vectors are

$$v_1 = \frac{Q}{2\pi h\sqrt{(l-x)^2 + y^2}} \text{ cm sec}^{-1} \tag{11.10}$$

and

$$v_2 = \frac{Q}{2\pi h\sqrt{(l+x)^2 + y^2}} \text{ cm sec}^{-1} \tag{11.11}$$

respectively. Here l is the half-distance between the two diapirs. These equations give the particle paths shown in Fig. 11.109.

Consider a small volume of the source material around a point lying on the y-axis which is equidistant from the two diapirs (Fig. 11.110). The particle path through the point is oriented normal to the line joining the diapirs, and points toward this line which in our scheme coincides with the x-axis. The shape of the particle paths adjacent to the point in the small volume considered, shows that the flow is divergent in the sense that the particle paths diverge in the horizontal plane. This conclusion follows from the above expressions for the velocity components v_1 and v_2. The small volume lying on the y-axis will accordingly widen in the x-direction and shorten in the y-direction as it moves along the y-axis. If it contains horizontal sheets of competent materials, buckles will form whose fold axis as well as stretch axis are normal to the direction of flow. That is, *b*-tectonites are formed.

A small volume of source material lying on the x-axis which joins the two diapirs will move toward the closest diapir and at the same time become lengthen in the direction of flow and compressed normal to the flow. Within regions close to the x-axis the buckle-folds will be similar to those generated around an isolated diapir, i.e. the fold axis as well as the axis of stretching coincide with the direction of flow. *a*-tectonites are produced in the flow field, see p. 312.

An infinitesimal small volume around the point exactly in the middle between the two diapirs (the origin of our co-ordinate system) does not move at all. It becomes stretched parallel to x and shortened parallel to y. Here buckling will occur without sensible displacement of the small region in relation to the surroundings.

Outside the zones close to the y- and x-axis the flow pattern is a little

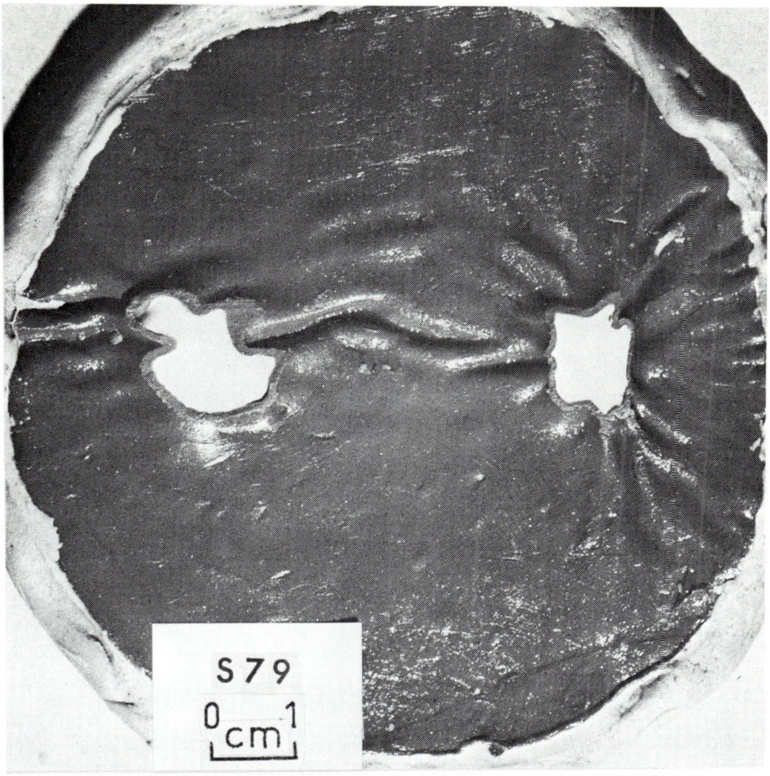

Fig. 11.111. Model of two diapirs of silicone putty (white) with thin top sheet of modelling clay (dark grey) rising through painter's overburden (removed in photograph). Note buckling of modelling-clay sheet around and between domes.

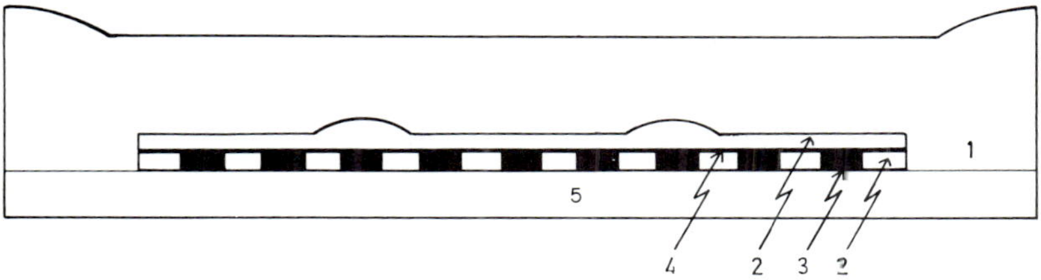

Fig. 11.112. Initial structure of the model shown in Fig. 11.111 and 11.113. (1) Painter's putty; (2) pink silicone putty; (3) dark-coloured silicone putty; (4) thin sheet of modelling clay; (5) mercury. Length of section 9.5 cm.

Fig. 11.113. Deformation of source layer around two diapirs. Figure shows the underside of source layer of silicone putty, consisting initially of straight strips of alternating light and dark colours. To prevent drag along bottom, the source layer was floating on mercury during run in centrifuge. The deformation pattern shows how the material was sucked towards the roots of the two domes resulting in strong stretching in the region between the domes.

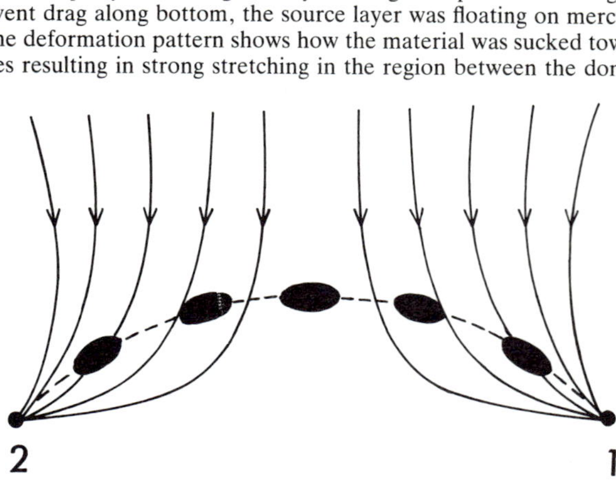

Fig. 11.114. Shape of horizontal projection of selected flow lines in source layer within the region between two diapirs rising at localities (1) and (2). Strain ellipse also schematically indicated along one of the arcs which coincides with the axes of of buckle folds in the region between the diapirs.

more complicated in the sense that neither the direction of maximum stretching nor the direction of maximum shortening (i.e. neither of the principal directions of strain) coincides with the direction of flow. That is, the flow produces rotational strain. Hence the fold axis and/or mineral lineation do not develop either normal to or parallel to the direction of movement but assume inclined angular orientation which varies from place to place. This is faintly indicated in model S-79 (Figs 11.111–11.113). In Fig. 11.114 the variation of the orientation of the strain ellipsoid is schematically indicated in the region between the two diapirs. The axes of stretch and the fold axis constitute curves extending from the one diapir to the other, except for the straight axis of the fold which coincides with the line joining the two diapirs. Only at three points is a curved fold axis either normal or parallel to the particle paths: at the intersection between the fold axis and the *y*-axis (here the fold axis is normal to the particle path) and at the two places where the fold enters the trunks of the diapirs (here the fold axis coincides with the particle path). At all other places the fold axis is inclined to the particle path, the angle of intersection changing continuously from being normal to becoming parallel along the fold as one goes from the *y*-axis to either of the diapirs.

This experimental and theoretical result is believed to be of considerable consequence for the interpretation of fold geometry whether generated by gravity or not. There seems to be no physical constraint which demands that fold axis and direction of movement in terms of particle path shall only exist in a normal or a parallel relationship. Indeed, the more general situation is that the two structures are neither parallel nor normal to one another.

The folding close to the line connecting two diapirs is well developed in the model shown in Figs 11.115 and 11.116, where horizontal cuts at several levels are exhibited.

The strong stretching along the line joining two diapirs often gives rise to boudinage or pinch-and-swell structure of competent sheets embedded in the source layer such as exemplified in model S-43 (Fig. 11.117).

DOME-IN-DOME STRUCTURE

The theory of dome generation predicts that at the proper relationship between thicknesses, densities and viscosities of the layers in stratified models two or more orders of diapirs may form. A theoretical model is discussed on p. 89.

We have not attempted especially to produce dome-in-dome structure experimentally, but such structures have occasionally developed in models

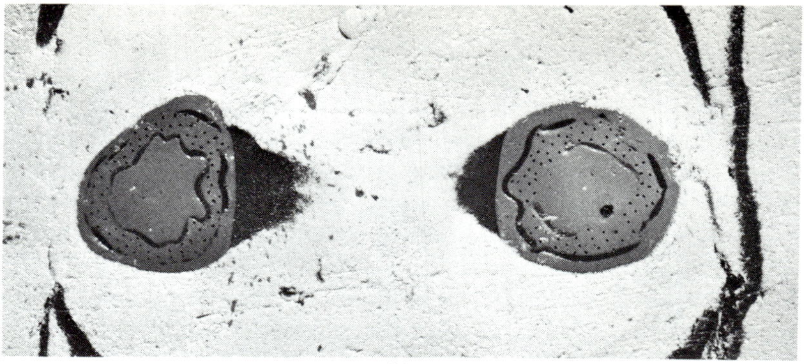

(*a*)

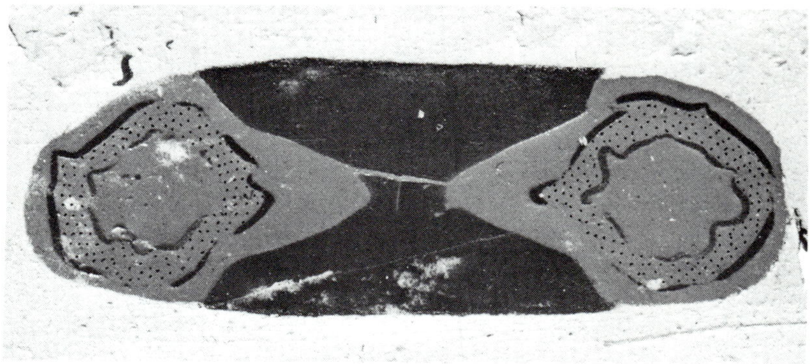

(*b*)

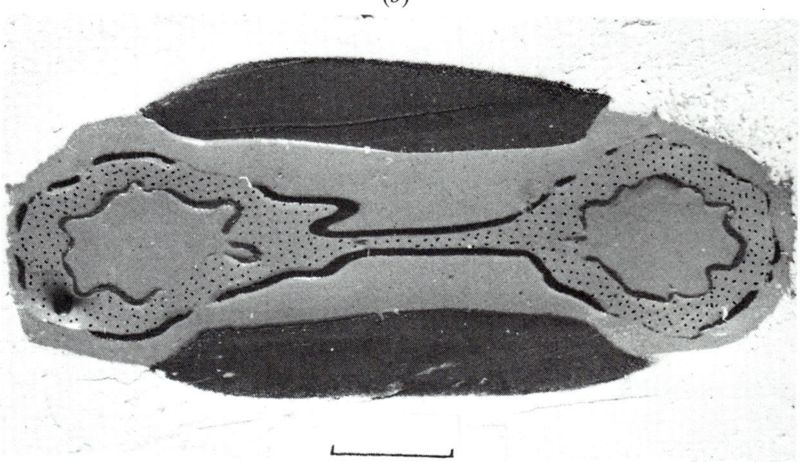

(*c*)

Fig. 11.115. Horizontal cuts at different levels in model containing two simultaneously rising diapirs of silicone putty with two embedded sheets of modelling clay. The sequence (*a*) to (*f*) goes from a cut close to the surface down to the root zone of the diapirs. White grainy material in (*a*), (*b*) and (*c*) is painter's putty overburden (1 in Fig. 11.116). Dark-coloured mass in (*b*) to (*f*) is silicone putty containing magnetite powder (3 in Fig. 11.116). Light-coloured and dotted material is buoyant silicone putty (2 in Fig. 11.116). Dark folded bands are cuts through the deformed sheets of modelling clay initially embedded in the layer of buoyant silicone putty as indicated in Fig. 11.116. Silicone putty between the two sheets of modelling clay is dotted in the photographs. The dotted narrow ellipse-shaped field in the centre of (*f*) is a synform (model run by R. Häll in the Uppsala laboratory). Scale bar 1 cm.

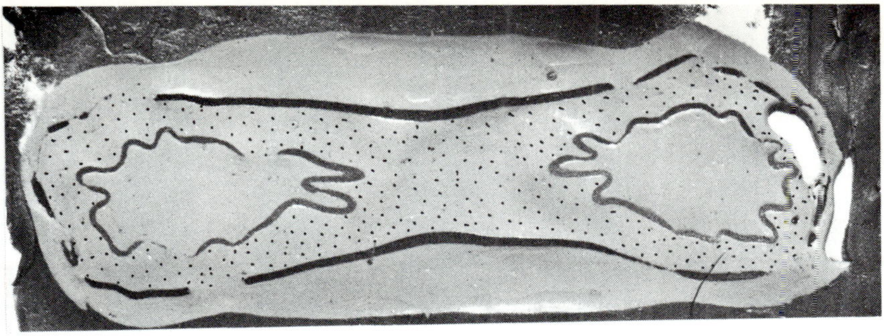

(*d*)

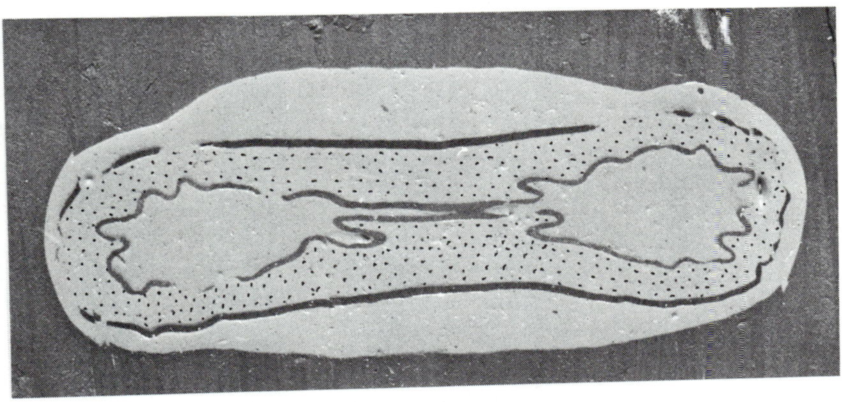

(*e*)

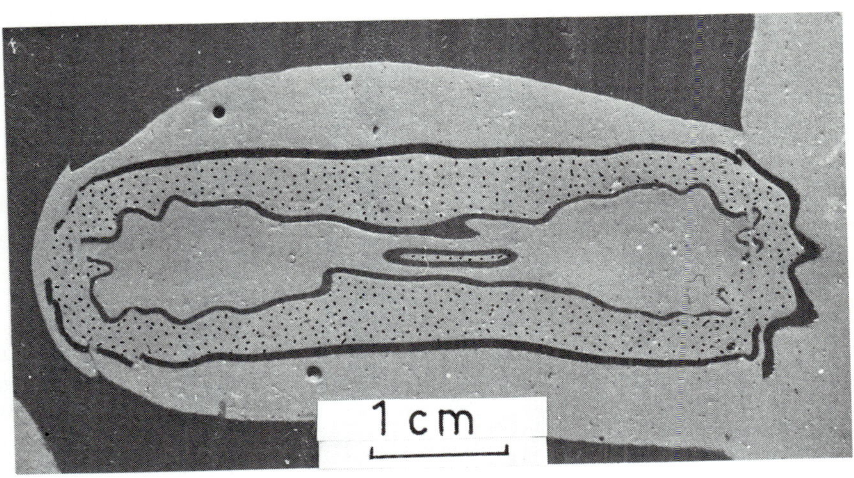

(*f*)

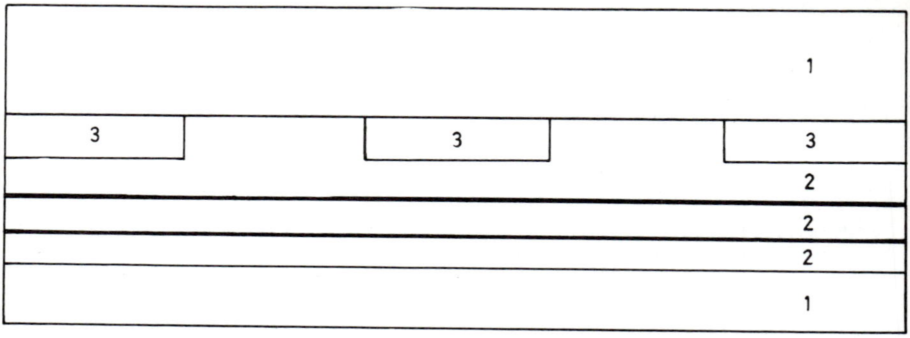

Fig. 11.116. Initial structure of model shown in Fig. 11.115 (1) Painter's putty, $\rho = 1{\cdot}8$ g cm^{-3}; (2) silicone putty, $\rho = 1{\cdot}14$ g cm^{-3}; (3) modelling clay, $\rho = 1{\cdot}7$ g cm^{-3}. Solid lines in (2) are thin sheets of modelling clay. Length of section 9·5 cm.

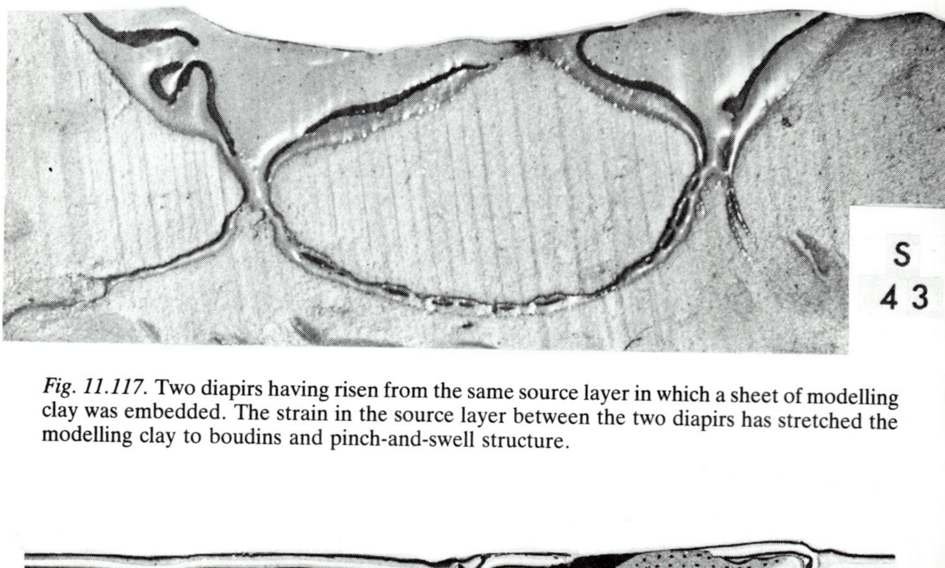

Fig. 11.117. Two diapirs having risen from the same source layer in which a sheet of modelling clay was embedded. The strain in the source layer between the two diapirs has stretched the modelling clay to boudins and pinch-and-swell structure.

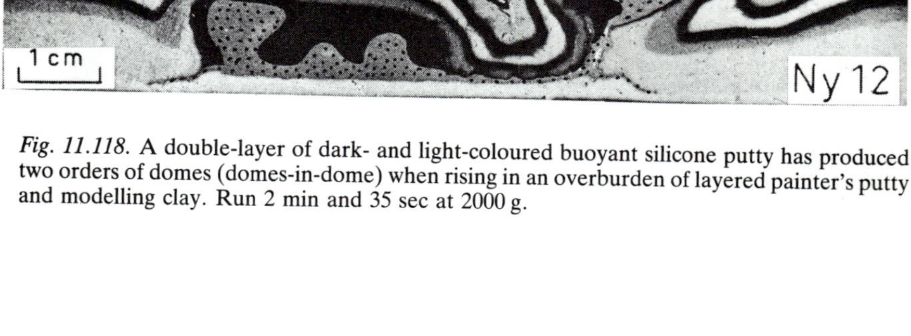

Fig. 11.118. A double-layer of dark- and light-coloured buoyant silicone putty has produced two orders of domes (domes-in-dome) when rising in an overburden of layered painter's putty and modelling clay. Run 2 min and 35 sec at 2000 g.

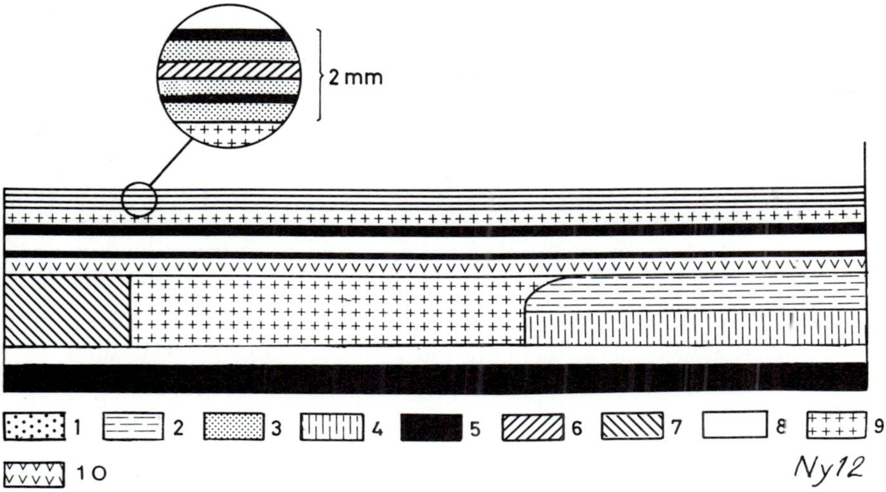

Fig. 11.119. Left-hand half section of initial structure of model shown in Fig. 11.118. (1) Not present in this model; (2) silicone putty with magnetite powder, $\rho = 1{\cdot}45\ \mathrm{g\,cm^{-3}}$; (3) silicone putty with tungspar powder, $\rho = 1{\cdot}28\ \mathrm{g\,cm^{-3}}$; (4) silicone putty with tungspar powder, $\rho = 1{\cdot}36\ \mathrm{g\,cm^{-3}}$; (5) dark modelling clay, $\rho = 1{\cdot}68\ \mathrm{g\,cm^{-3}}$; (6) white modelling clay; (7) brown modelling clay, $\rho = 1{\cdot}71\ \mathrm{g\,cm^{-3}}$; (8) grey painter's putty, $\rho = 1{\cdot}85\ \mathrm{g\,cm^{-3}}$; (9) yellow painter's putty, $\rho = 1{\cdot}85\ \mathrm{g\,cm^{-3}}$; (10) green painter's putty, $\rho = 1{\cdot}85\ \mathrm{g\,cm^{-3}}$.

studied for other purposes. Examples are shown in Figs 11.118–120. Here a buoyant double-layer of silicone is overlaid by an interlayered complex of modelling clay and painter's putty, both being more dense than either of the two buoyant layers. The relatively dense ($\rho = 1{\cdot}45\ \mathrm{g\,cm^{-3}}$) buoyant layer (dark shade in photograph) was initially placed above the less dense

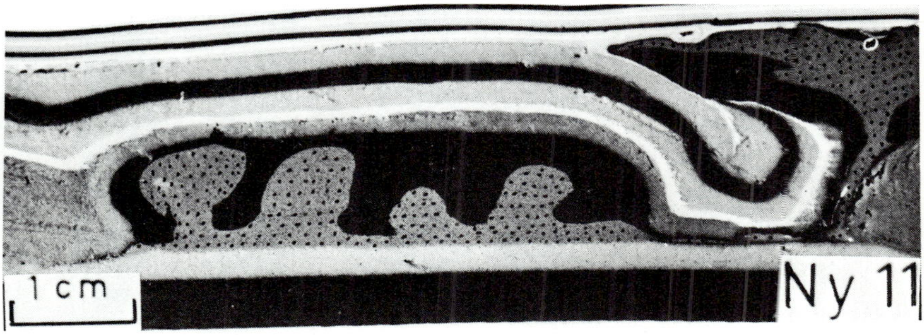

Fig. 11.120. Model with double layer of dark- and light-coloured silicone putty having risen in the form of domes-in-dome into overburden of painter's putty interlayered with modelling-clay sheets. Run 4 min at 2000 *g* (by H. Sjöström, Uppsala). See also Figs 11.32 and 11.33.

buoyant layer ($\rho = 1{\cdot}36\ \mathrm{g\,cm^{-3}}$, light shade in photograph). On the top of the dense overburden—varying between $\rho = 1{\cdot}68\ \mathrm{g\,cm^{-3}}$ and $1{\cdot}90\ \mathrm{g\,cm^{-3}}$ for the putty and modelling clay—a complex of low-density silicone and thin modelling clay sheets was placed. The purpose of the thin-layered and low-density top complex was twofold: to facilitate buckling of the surface in response to a spreading diapir, and to make the lobes of the diapirs spread below the surface.

As usual the buoyant material would rise during centrifugation, but what we are interested in at present is the two types of rise: the rise of the light silicone in the more dense silicone, and the rise of both the light and the more dense silicone into the heavy putty–modelling clay overburden.

Figure 11.120 shows a double-diapir consisting of the two types of silicone which have penetrated the heavy part of the overburden and spread underneath the low-density top of the overburden. A wide dome which has not yet pierced the layering of the overburden is also seen. The latter is a well-developed composite dome consisting of four small diapirs of the low-density silicone risen into the more dense silicone, the two buoyant layers having risen together as sort of a composite superdome.

In other models the tendency to form dome-in-dome is also evident, even if the development is rarely as clear as in models Ny-11 and Ny-12. Model S-112 (see Fig. 15.1) is a case in point. Here the stage of evolution is well advanced for which reason the dome-in-dome pattern is less clear.

12 Examples of Scale-Model Calculation of Rate of Rise of Salt Diapir

The somewhat uncontrollable rheological characteristics of the painter's putty and the modelling clay used as overburden in dome models are not a very good basis for quantitative scale-model calculations. Nevertheless, we provide an example of how the rate of rise of a natural domal structure may be estimated from an experimental dynamic model. Model S-216 (Fig. 12.1) is suitable. The initial structure is shown in Fig. 12.2.

Significant geometric dimensions and physical properties of the model are listed in Table 12.1, which also gives the dimensions of a similar natural dome of salt and reasonable physical properties of the constituent rocks. Model ratios of the various properties are also given.

We wish to find the time needed for the evolution of a natural salt dome dynamically similar to the model. To find the rate of rise of the natural dome, it is necessary to know the time of evolution of the model and the centripetal acceleration to which it was subjected. Since the centrifuge needs some time to reach a given angular velocity, the model was not subjected to a constant acceleration throughout the run, which lasted 26 min 35 sec. It took 3 min 45 sec to raise the velocity to the 2200 rev min^{-1} (= 1520 g) at which the model was run for 18 min 15 sec. The velocity was then increased rapidly to 2800 rev min^{-1} (= 2500 g) and maintained at this level for 4 min 35 sec, until the run was stopped.

Because of low average acceleration during the relatively short time before the 2200 rev min^{-1} level was reached, we do not introduce a large error by disregarding the first 3 min 45 sec in our calculations. (After all, lack of exact knowledge of the physical properties of the rocks would introduce much larger error). The model is therefore regarded as being subjected to only two periods of acceleration: 1095 sec at 1520 g and 275 sec at 2500 g. The model was run in the smaller centrifuge described in Ramberg (1967b).

Because of the dissimilar accelerations the two periods have different ratios of time, as given in Table 12.1.

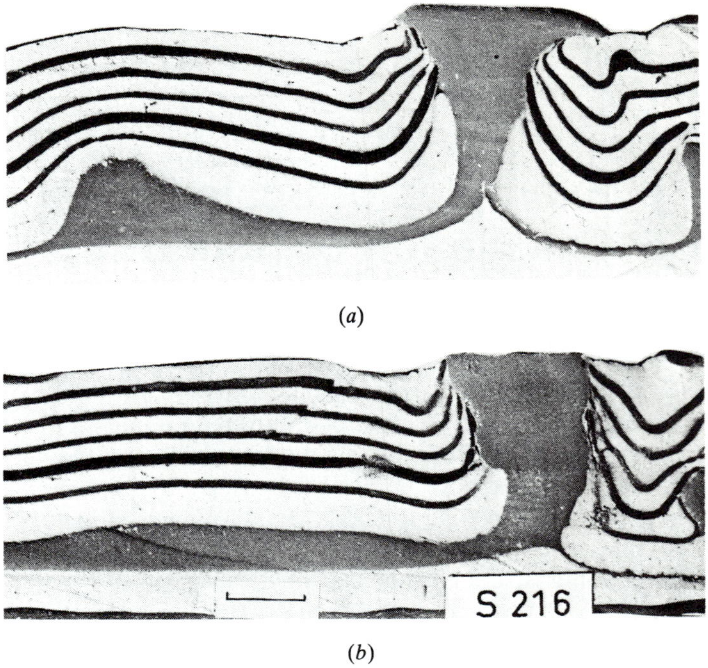

Fig. 12.1. Cross-sections of two models of silicone-putty diapirs which have risen through alternating layers of putty (light colour) and modelling clay (darker colour). Difference in shape of the two models is partly due to difference in density of the buoyant silicone, $\rho = 1{\cdot}13\ \mathrm{g\ cm^{-3}}$ in model S-215 (*a*), $\rho = 1{\cdot}48\ \mathrm{g\ cm^{-3}}$ in S-216. The surface of the dome in S-215 is therefore a little higher than in S-216, the spreading more pronounced, and the substratum sucked a little higher up in the dome in S-215 than in S-216. See Table 12.1.

Fig. 12.2. Initial structure of model S-216 (Fig. 12.1). (1) Painter's putty, $\rho = 1{\cdot}80\ \mathrm{g\ cm^{-3}}$; (2) silicone putty, $\rho = 1{\cdot}48\ \mathrm{g\ cm^{-3}}$; (3) sheets of modelling clay; $\rho = 1{\cdot}72\ \mathrm{g\ cm^{-3}}$.

Table 12.1. Example of scale-model calculation of time of rise of salt dome

Quantity	Model S–216	Natural salt dome	Model ratios
Thickness of buoyant layer	5 mm	500 m	$l_r = 10^{-5}$
Total thickness of overburden	20 mm	2000 m	$l_r = 10^{-5}$
Thickness of each competent layer	1 mm	100 m	$l_r = 10^{-5}$
Thickness of incompetent layer directly above source	5 mm	500 m	$l_r = 10^{-5}$
Thickness of each incompetent layer	2 mm	200 m	$l_r = 10^{-5}$
Diameter of dome	Variable (Fig. 12.1)	Variable	
Density of buoyant material	$1{\cdot}48$ g cm^{-3}	$2{\cdot}22$ g cm^{-3}	$\rho_r = 0{\cdot}67$
Density of competent overburden	$1{\cdot}72$ g cm^{-3}	$2{\cdot}55$ g cm^{-3}	$\rho_r = 0{\cdot}67$
Density of incompetent overburden	$1{\cdot}80$ g cm^{-3}	$2{\cdot}70$ g cm^{-3}	$\rho_r = 0{\cdot}67$
Viscosity of buoyant layer	5×10^{5} P	10^{18} P	$\mu_r = 5\times10^{-13}$
Viscosity of incompetent overburden	10^{6}–10^{7} P	2×10^{18}–2×10^{19} P	$\mu_r = 5\times10^{-13}$
Viscosity of competent overburden	3×10^{8} P	6×10^{20} P	$\mu_r = 5\times10^{-13}$
Body force per unit mass, first period	1520 *g* cm sec^{-2}	1 *g* cm sec^{-2}	$a'_r = 1520$
Body force per unit mass, second period	2500 *g* cm sec^{-2}	1 *g* cm sec^{-2}	$a''_r = 2500$
Stress, first period	Variable	Variable	$\sigma'_r = \rho_r l_r a'_r = 10{\cdot}2\times 10^{-3}$
Stress, second period	Variable	Variable	$\sigma''_r = \rho_r l_r a''_r = 1{\cdot}68\times10^{-2}$
Time, first period	1095 sec	t'_o	$t'_r = \mu_r \sigma_r^{-1} \approx 5\times10^{-11}$
Time, second period	275 sec	t''_o	$t''_r = \mu_r \sigma_r^{-1} \approx 3\times10^{-11}$
Total time of rise	1370 sec	t_o	

$t'_o = t'_m/t'_r = 1095/5\times10^{-11} = 2{\cdot}19\times10^{13}$ sec; $t''_o = t''_m/t''_r = 275/3\times10^{-11} = 0{\cdot}92\times10^{13}$ sec. Total time of rise: $3{\cdot}11\times10^{13}$ sec $\approx 10^{6}$ years.

The period of 1095 sec in the model evolution gives the following time for the natural structure:

$$t'_o = t'_m/t'_r = 1095 \text{ sec} \times \tfrac{1}{5} \times 10^{11} = 2{\cdot}19 \times 10^{13} \text{ sec}. \qquad (12.1)$$

The second period of 275 sec corresponds to

$$t''_o = t''_m/t''_r = 275 \text{ sec} \times \tfrac{1}{3} \times 10^{11} = 0{\cdot}92 \times 10^{13} \text{ sec} \qquad (12.2)$$

for the natural structure.

The result is then that a natural structure whose dimensions and constituent rocks are as given in the table needs about 10^6 years to evolve to the same stage as the model has evolved.

One million years for a salt diapir to rise two thousand metres from a salt stratum is not an unreasonable span of time. It means an average velocity of 0·2 cm per year.

13 The Rise of Magma due to Buoyant Forces†

INTRODUCTION

In the models discussed up to now the contrast in viscosity and/or strength of the buoyant materials (the silicone putty) and the overburden (the mixtures of modelling clay and painter's putty) has been moderate, as one would expect to find among crystalline rocks of contrasted mineralogical compositions and among silicate melts of unlike chemical composition. The ratio of the effective viscosities of crystalline rocks probably varies within the range 1 to 10^3 (as indicated by e.g. the wavelength/thickness ratio of buckle folds in crystalline schists and gneisses). The viscosity contrast between crystalline rocks and a crystal mush with but a small amount of fluid may also be of the same moderate order.

On the other hand, the viscosity contrast between a fully liquid magma and crystalline rocks in nature is generally much larger than the contrast employed in the models so far. Table 13.1 shows examples of the viscosity of some silicate magmas. We have unfortunately little information on the magnitude of the effective viscosity of solid silicate rocks except that it is generally greater than say 10^{17} poise (which is the measured viscosity of one of the most ductile rocks, namely rock salt). The measured average viscosity of the deep crust and upper mantle, namely about 10^{22} poise, also points toward a very high effective viscosity of crystalline silicate rocks. For the sake of argument let us assume that the solid Earth's crust through which a magma rises shows a viscosity of say 10^{22} poise. In contrast the viscosity of even the most viscous granitic melt (obsidian) shown in Table 13.1 is only $4{\cdot}4 \times 10^6$ poise. The ratio between the viscosities of the obsidian and the solid crust is consequently

$$\frac{\mu_{\text{obs}}}{\mu_c} = \frac{4{\cdot}4 \times 10^6}{10^{22}} = 4{\cdot}4 \times 10^{-16}. \qquad (13.1)$$

Now, one scale-model requirement is that the model must show the same ratio between the viscosities of the "obsidian" and the "crust" respec-

† This chapter is based on the author's contribution to Newall and Rast (1970).

Table 13.1. Viscosity of melted rocks and natural lavas (from Clark, 1966)

Rock type	Temperature(°C)	Viscosity (poise)
Obsidian		
Oki	1400	10^6
Niijima	1400	$4{\cdot}4\times10^6$
Arita	1400	$4{\cdot}4\times10^6$
Erevan	1400	$1{\cdot}7\times10^6$
Obsidian "O"	800	10^{11}–10^{12}
Obsidian "O"	1000	10^9–10^{10}
Obsidian "O"	1200	10^7–10^8
Hornblende granite	1400	2×10^6
Hornblende-mica andesite	1400	$1{\cdot}6\times10^4$
Andesutes	1400	150 to 1500
Diabases	1400	15 to 400
Alika flow, Hawaii 1919	1400	$4{\cdot}3\times10^4$
Kau flow, Hawaii 1887	1400	$4{\cdot}77\times10^4$
Mihara, Japan 1951	1038	$2{\cdot}3\times10^5$
———	1083	$7{\cdot}1\times10^4$
———	1108	$1{\cdot}8\times10^4$
———	1125	$5{\cdot}6\times10^3$

tively, if dynamic similarity and similar movement pattern in model and nature shall prevail. This particular condition is obviously not met in our models because the stiffest material that it is practical to use in the present case has a viscosity of some 10^8 poise. Taking this as the model crust the model obsidian must show an impossibly low viscosity, viz.

$$\mu_{\text{model obs}} = 4{\cdot}4 \times 10^{-16} \times 10^8 = 4{\cdot}4 \times 10^{-8}\,\text{poise}$$

which means a substance much more fluid than the ordinary gases. It would be still worse if we were to study dynamic models of say the Mihara flow (Table 13.1).

The ratio between Mihara flow at 1125°C and the Earth's crust is

$$\frac{\mu_{\text{Mihara}}}{\mu_{\text{crust}}} = \frac{5{\cdot}6 \times 10^3}{10^{22}} = 5{\cdot}6 \times 10^{-19} \tag{13.2}$$

which translated to our models with the same "crust" as above would mean a substance whose viscosity is vanishingly small, viz.

$$\mu_{\text{model Mihara}} = 5{\cdot}6 \times 10^{-19} \times 10^8 = 5{\cdot}6 \times 10^{-11}\,\text{poise.}$$

For comparison we note that the viscosity of liquid water is of the order 10^{-2} poise and that of steam at 1 atmosphere pressure and 100°C is $1{\cdot}2\times 10^{-4}$ poise.

Realizing that dynamic similarity of magmatic intrusion in crystalline rocks with respect to the viscosity contrast cannot be obtained in our runs with the present set-up, we have made a number of qualitative tests in which the viscosity contrast between the model crust and the model magma has been as large as available model materials permit. Aqueous solutions of $KMnO_4$ have been used to simulate magmas, and modelling clay and painter's putty to simulate the solid Earth's crust. This gives a maximum viscosity ratio

$$\frac{\mu_{\text{magma}}}{\mu_{\text{crust}}} = \frac{10^{-2}}{10^{8}} = 10^{-10} \tag{13.3}$$

Incidentally, this would not be far from the ratio between a viscous acid magma and a rock-salt deposit. If we consider obsidian glass as a magma the ratio employed in the models would probably be reasonable for the study of obsidian glass flowing amongst such rocks as marble and the most mobile silicate rocks (e.g. chlorite schist?).

The following models are of interest when magma tectonics is considered.

MODEL M

A ring of painter's putty 3·2 mm thick was placed on the bottom of the centrifuge cup; the central hole was 40 mm in diameter and the outer diameter of the ring was 94 mm, which coincides with the diameter of the cup. The hole was filled with a concentrated solution of $KMnO_4$. Such a solution gives a brown stain when it comes in contact with putty or modelling clay and thus leaves a mark where it passes the overburden, even if the liquid itself is completely squeezed out.

The overburden consisted of six sheets of modelling clay about 1·2 mm thick interlayed with seven sheets of painter's putty 3 mm thick, the complete column being 28·2 mm high.

The model was run at 700 rev min^{-1}, which corresponds to an acceleration of 210 g, for 4·5 min in the centrifuge. The $KMnO_4$ "magma" penetrated the surface between 3 and 4·5 min after the run was started. (This test was made prior to adapting a stroboscope to the centrifuge, so the exact time could not be determined.) The fact that much fluid remained in lenticular pockets (secondary "magma" chambers) below several of the modelling-clay layers when the run was discontinued after 4·5 min (Fig. 13.1) indicates that the liquid penetrated the surface only a few seconds before the run was discontinued. (Experience from several other runs with fluids enclosed in putty and modelling clay shows that all the enclosed fluid will be squeezed out in the course of a few seconds after penetrating the surface.) After

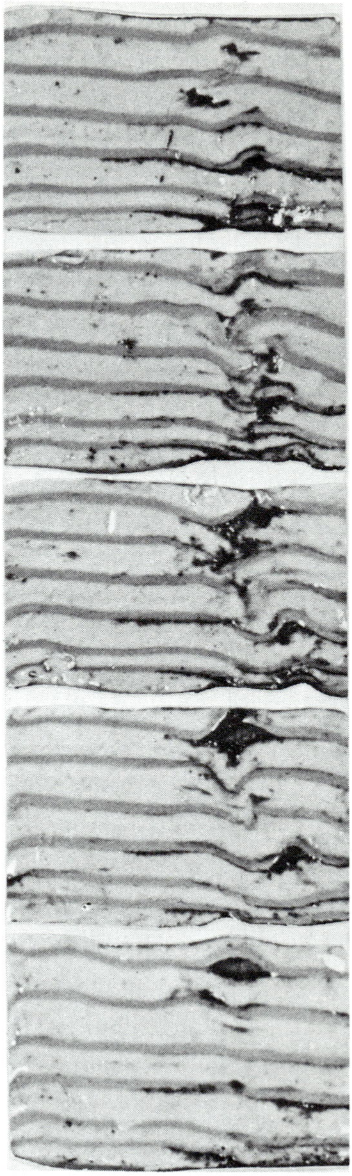

Figs. 13.1. Slices about 2 mm thick cut through centrifuged model of putty (light grey) and modelling clay (dark grey) through which a solution of $KMnO_4$ has risen. The black areas are openings in the overburden lined with a stain of manganese oxide from the rising solution. Sections 2 and 3 from the top are cuts through the main vent for the solution. Note strong deformation and fracturing of the competent modelling-clay sheets. Length of cuts covers about half the diameter of the model. See text.

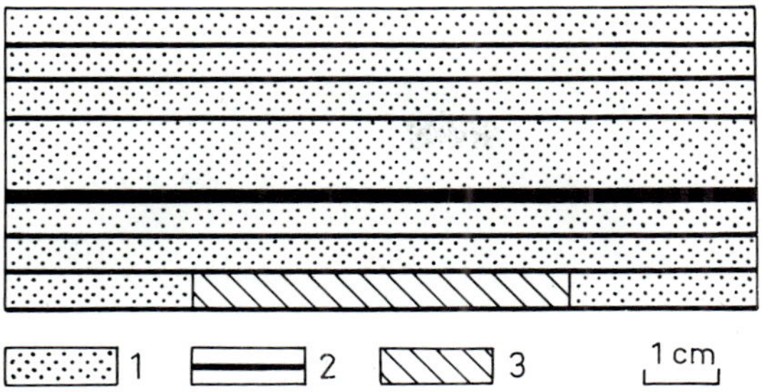

Fig. 13.2. Initial structure of model shown in Figs 13.3 and 13.4.

3 min run there was no penetration of the surface, though a dome-shaped elevation had developed showing that the "magma" had started to rise through the layered overburden and was probably temporarily trapped below some of the relatively strong sheets of modelling clay. The time needed for the "magma" to rise through the 28 mm thick overburden is then 3·75 ± 0·75 min.

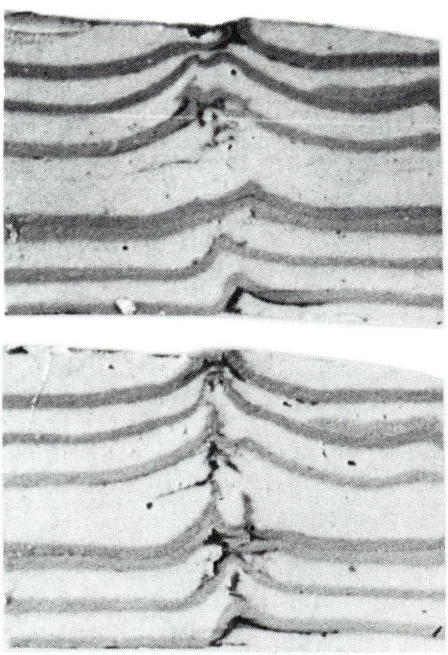

Fig. 13.3. Cross-sections through vent of ascent of $KMnO_4$ solution through overburden in model shown in Fig. 13.2. The vent extends vertically from one edge of the collapsed original "magma" chamber. Run for less than 1 min at 1000 *g*.

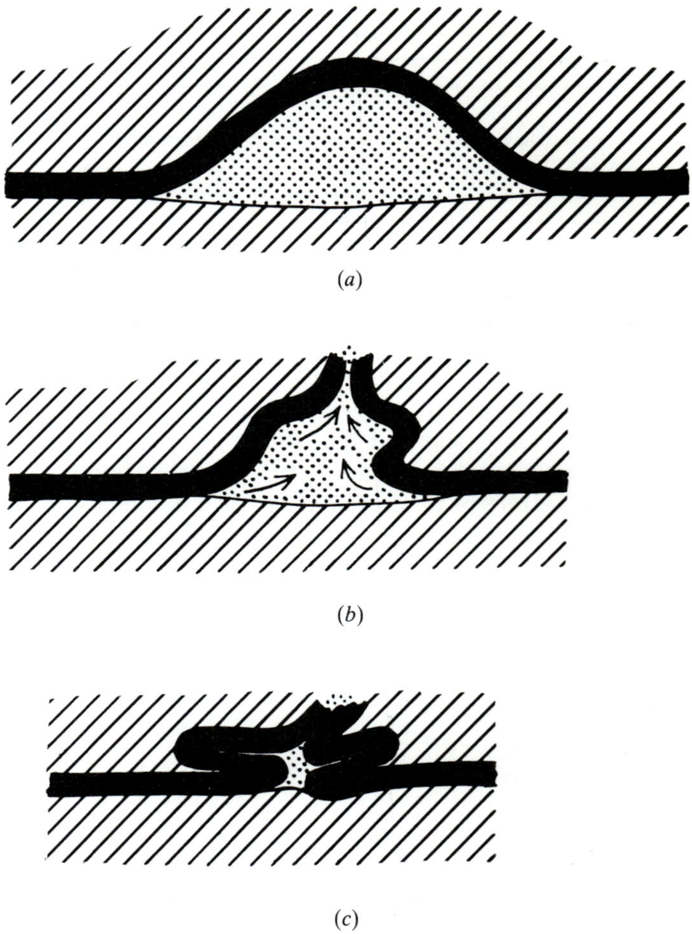

Fig. 13.4. Sketch showing development of collapse structure along the vent in Fig. 13.3.

MODEL M-0

The overburden consisted of six layers of painter's putty, one about 7 mm thick and the rest about 3·5 mm thick intercalated with five single sheets about 1 mm thick, and a double sheet some 2 mm thick of modelling clay. The whole column was about 33 mm high. A 3·5 mm central circular chamber at the bottom was filled with an aqueous solution of $KMnO_4$. The construction is shown in Fig. 13.2.

As noted above, the viscosity ratio between a dilute aqueous solution and a mixture of putty and modelling clay in a model of this size corresponds rather well to the viscosity ratio between a very viscous obsidian glass and the most mobile crystalline rocks in nature.

The model was run for less than 1 min at an acceleration corresponding to some 1000 *g*. At the end of the run practically every trace of the "magma"

Fig. 13.5. Sections through model (M-1) of magma-imitation fluid ($KMnO_4$ solution) rising through imitation crust (layers of putty with contrasting colours), run for 15 min at 210 *g* in centrifuge with centrifugal force pointing downward in the figure. Dark irregular areas show the uneven path of the rising "melt".

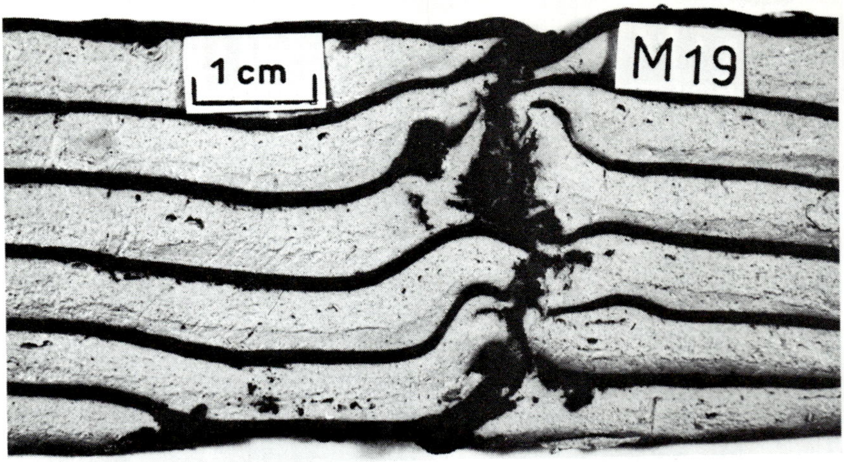

Fig. 13.6. Magma-simulation model of $KMnO_4$ solution having risen through layered overburden of putty (thick light-coloured layers) and modelling clay (dark sheets). Collapsed "magma" chamber (left of centre, bottom) and vent of ascent shown. Initial stage shown in Fig. 13.7 (from Ramberg in Newall and Rast, 1970).

had been extruded on the surface, but the $KMnO_4$ solution had left a distinct brownish-black coating along the path of the ascending "magma", as shown in Fig. 13.3.

At several points along the path the layering has been strongly disturbed. Not only have the sheets of modelling clay been dragged along the path of ascent, but at some points strong folding has occurred (e.g. at the double layer of modelling clay in the middle of the overburden). The particular geometry of this folding is probably best explained by a collapse of a "magma" chamber immediately below the double layer of modelling clay.

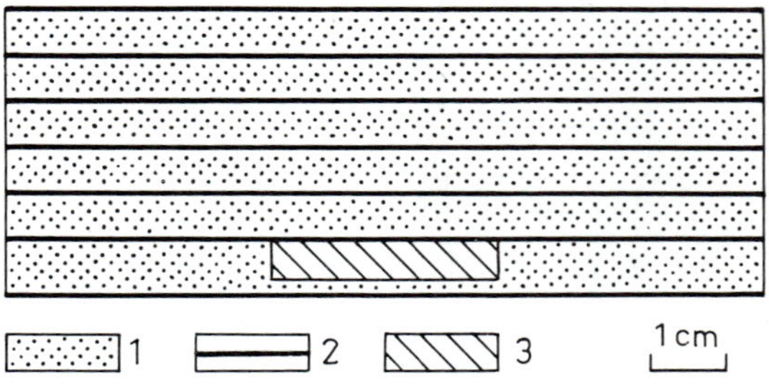

Fig. 13.7. Initial structure of model shown in Fig. 13.6. (1) Painter's putty; (2) Layers of modelling clay; (3) $KMnO_4$ solution (from Ramberg in Newall and Rast, 1970).

The double layer of modelling clay constituted a dome-shaped roof of the chamber which collapsed when the "magma" burst through the roof. It is easy to imagine how the relatively competent sheet of modelling clay would buckle during the collapse (see Fig. 13.4).

Uncollapsed chambers still filled with "magma" were preserved in model M which was run at a lower centrifugal acceleration.

MODEL M-1

A 5 cm thick overburden of painter's putty consisting of sheets of different colour (Fig. 13.5) was placed above a chamber containing a $KMnO_4$ solution. After 15 min run at 210 g, the imitation magma had penetrated the overburden along a very irregular path with many horizontal branches. Though the overburden putty was homogeneous, its structure was probably somewhat anisotropic with a certain flakiness parallel to the horizontal plane resulting from the preparation of the layers of different colour. It appears that the $KMnO_4$ solution tends to follow this flaky structure much as a natural magma tends to spread laterally between sedimentary strata.

MODEL M-19

Model M-19 (Figs 13.6 and 13.7) is quite similar to models M-1 and M-0. It shows the irregular and narrow magma channel characteristic for the models with a low-viscosity fluid as the buoyant medium. However, the details of the irregular channel deviate sufficiently from the other models to warrant publication of the photograph. Figure 13.7 shows the model prior to run.

The above models of the buoyant rise of a fluid magma show geometric patterns quite unlike those characteristic for diapiric rise in models with less contrasted viscosity and strength of overburden and buoyant layer. Owing to the high fluidity of the magma, in contrast to the rigidity of the crystalline crust, the melt can rise with considerable speed through the crust, even through a narrow and irregular path. The subsidence of the crust as fair-sized blocks and the rise of the magma through narrow channels are therefore energetically favoured processes. With more nearly equal viscosity of overburden and buoyant layer, however, the subsidence and the rise are less contrasted processes—the sinking parts are relatively more mobile and therefore decrease in lateral extent, whereas the rising bodies are relatively less ductile and increase in lateral dimension, the subsiding and the ascending masses becoming more equal in dimensions. This is significant in discussions of the emplacement of plutonic bodies.

MODELS 1, 2, 3, 5 AND 7

In models 1, 2, 3, 5, and 7, the country rock was imitated by layered and somewhat folded modelling clay in which the "magma" (aqueous solution of $KMnO_4$) was initially housed in a spherical compartment as shown in Fig. 13.8.

Except for the bottom layer in model 3 the material surrounding the magma chamber was stronger and less ductile than in models M to M-19 described above. The results after centrifugation are recorded in Figs 13.9–13.13.

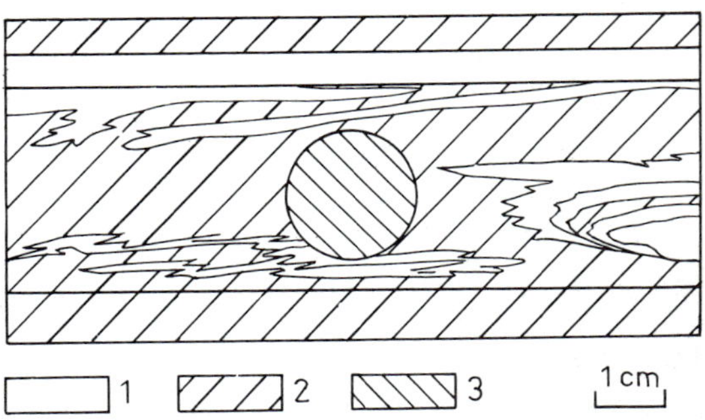

Fig. 13.8. Initial structure of model 5 (Fig. 13.12). (1) White modelling clay; (2) black modelling clay; (3) $KMnO_4$ solution (from Ramberg in Newall and Rast, 1970).

The empty space seen in the photographs is the partially collapsed magma chamber which was actually filled with the remaining portion of the magma when the model was sectioned after the runs. The rest of the magma had been squeezed out and found its way up to the surface via fractures which often followed the folded banding in the surrounding and overlying model crust.

It is interesting to note that in some of these models, especially 1, 2 and 7, the magma chamber collapsed by a cataclastic mechanism rather than by ductile plastic flowage as found in most models described in Chapter 11. Pieces have broken off from the walls and ceiling of the magma chambers and were found on the bottom of the chambers after the runs, or the ceiling had fallen in, due to the development of extensive fractures

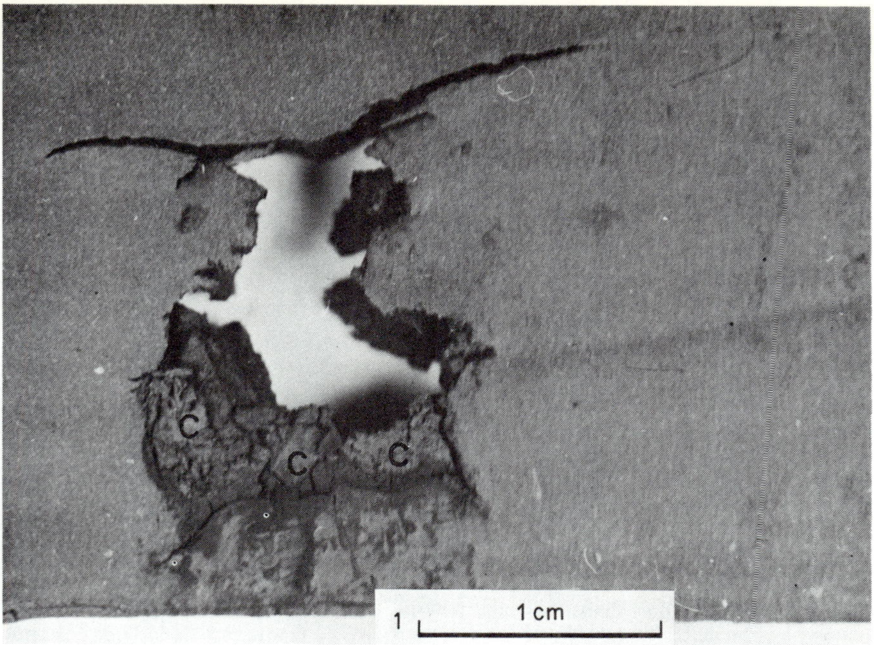

Fig. 13.9. Details of partly collapsed "magma" chamber of model 1 with initial structure similar to that shown in Fig. 13.8. Note fragments (*c*) from sides of chamber lying on floor (from Rambert in Newall and Rast, 1970).

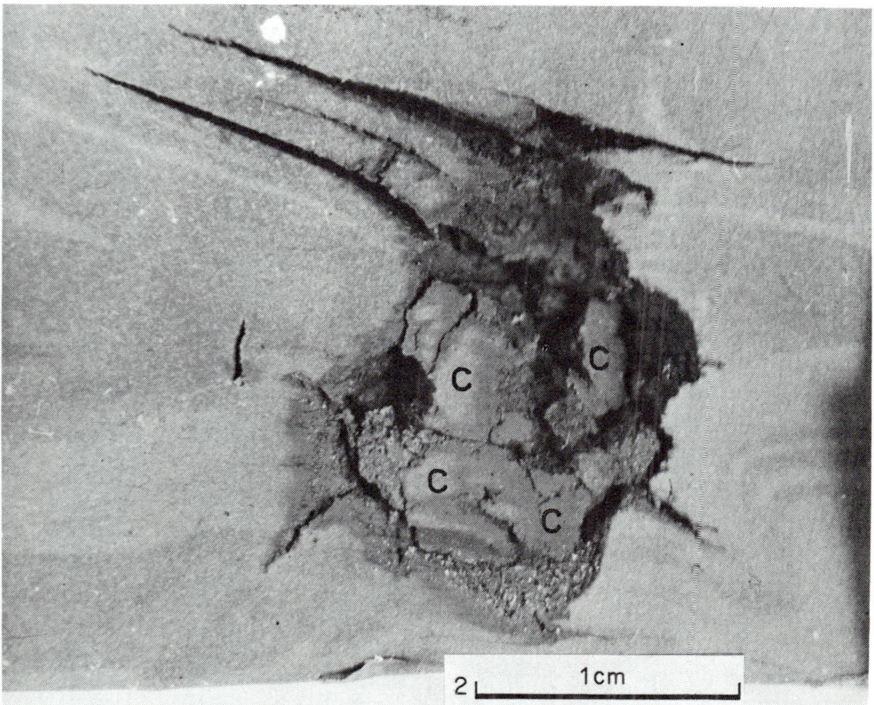

Fig. 13.10. Details of partly collapsed "magma" chamber in model 2 with initial structure similar to that shown in Fig. 13.8. Note loose fragments of country "rock" in chamber, and cracks along faint layering in surrounding "rock" (from Ramberg in Newall and Rast, 1970).

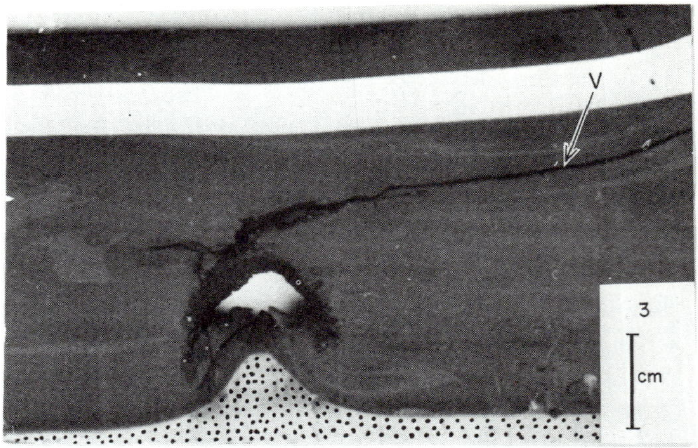

Fig. 13.11. Model rather similar to 1 and 2. Note upheaval of floor in "magma" chamber (caused by flow in the basal layer of soft painter's putty, dotted), and the inclined fracture along which the "magma" has escaped (V) (from Ramberg in Newall and Rast, 1970).

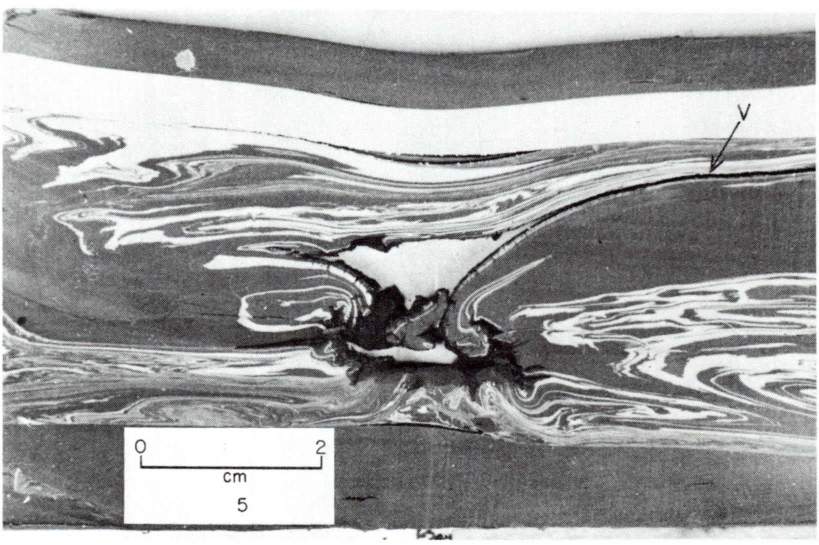

Fig. 13.12. Model showing partly collapse of "magma" chamber and fracture parallel to layering along which the "magma" has escaped (V). Initial structure shown in Fig. 13.8.

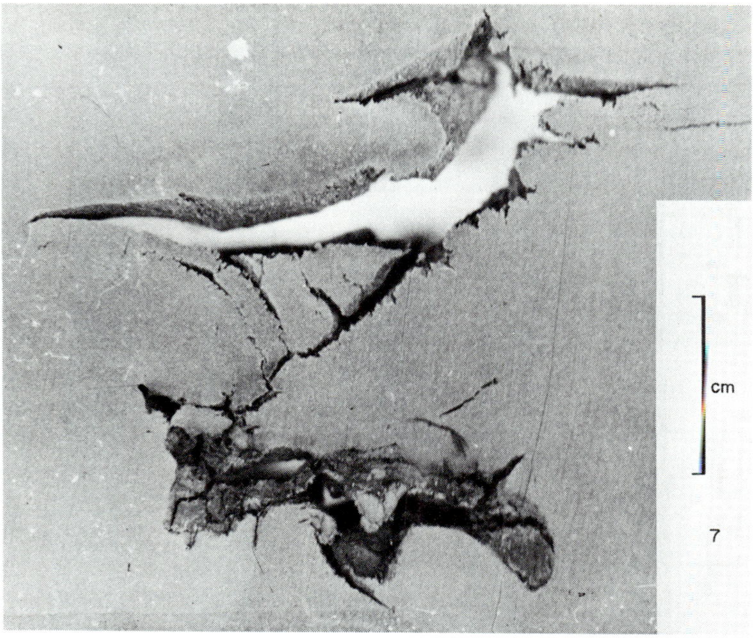

Fig. 13.13. Detail of model 7 exhibiting fractures above collapsing "magma" chamber (lowest structure in middle). Initial structure quite similar to model 5, Fig. 13.8.

in the overburden above the magma chamber, model 7. In model 3 the collapse of the chamber has occurred by an upheaval of the floor rather than an overhead stoping mechanism. This is so because of the greater ductility of the floor in this model whose basal layer was a sheet of painter's putty.

Some granitoid batholiths in orogens have steep walls which coincide with reverse faults. Examples are especially plentiful in the Andes cordillera, see, e.g. Childs and Beebe (1963). In models described in the chapter on domes it was demonstrated that a buoyant body with sufficient high viscosity could lift the overburden *en bloc* along the fractures thus creating a horst delimited by reverse faults. The ductile but very viscous ascending mass would fill the space left by the lifted overburden block, the vertical displacement of the overburden being dynamically permissible because of continuous erosion of the elevated top.

To study whether a silicate melt could produce the same structure several models were tested in which the magma in the form of potash permanganate solution was housed in a compartment surrounded by painter's putty underneath a fractured Earth's crust. The crust consisted of modelling clay

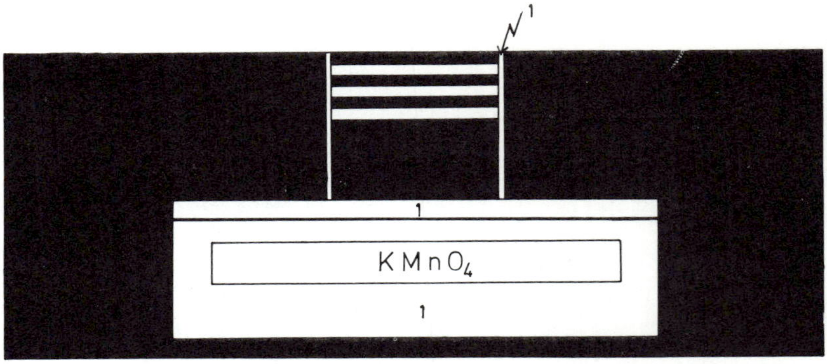

Fig. 13.14. Model M-24 prior to run in centrifuge. (1) Painter's putty, green above the "magma" chamber, grey below; black is dark modelling clay; white layers at top are white modelling clay; $KMnO_4$ is compartment filled with $KMnO_4$ solution. Length of section 9·5 cm.

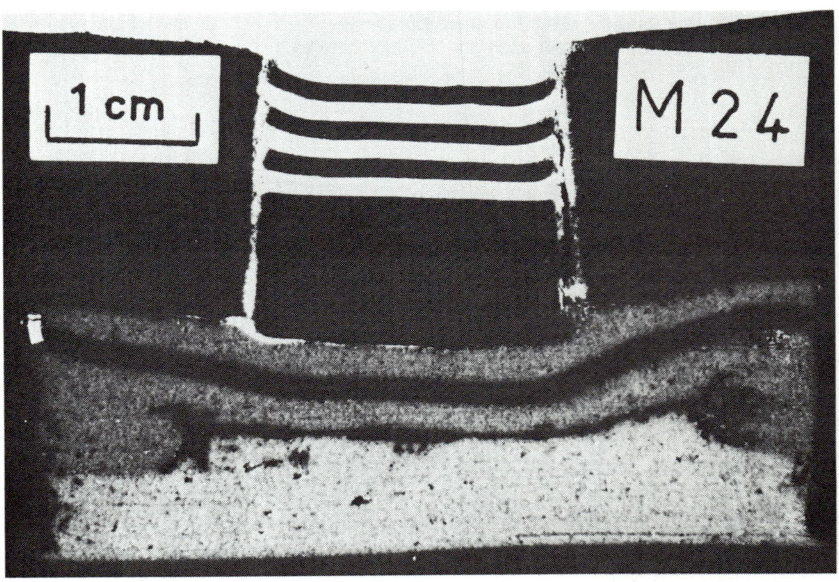

Fig. 13.15. Model M-24 after run 5 min at acceleration increasing from 1000 *g* to 2500 *g*. $KMnO_4$ solution has escaped along the vertical sides of the subsided block (from Ramberg in Newall and Rast, 1970).

Fig. 13.16. Model similar to M-24, run 5 min at acceleration increasing from 1000 *g* to 2500 *g* during which the $KMnO_4$ solution, squeezed out from the "magma" chamber, escaped along the sides of the subsiding block (from Ramberg in Newall and Rast, 1970).

in which two vertical cuts were made, extending from the surface to the roof of the magma chamber. The cuts were cemented together with putty (Fig. 13.14). Thus the initial stages of the models were similar to the models tested in connection with the generation of plastic diapirs, Chapter 11, except that the buoyant silicone was replaced by an aqueous solution. It is therefore particularly interesting to note the great contrast in final structure after the models were exposed to the centrifugal force, compare Figs 13.14–13.16 with 11.70–11.73. Whereas the silicone, whose viscosity is comparable to that of the putty which was used to seal the fractures in the overburden, lifts the overburden block between the fractures, the model magma simply penetrates the putty-sealed fractures and extrudes at the surface. Instead of rising to a horst, the block between the fractures sinks down into the collapsing magma chamber. Indeed, if the fractures were circular instead of planar, and the overburden was made of a model material which would fracture more easily than the modelling clay, then the models would have simulated the process of cauldron subsidence.

The above models of the buoyant rise of a fluid magma show geometric patterns quite unlike those characterizing diapiric upheavals in models with less contrasted viscosity of overburden and buoyant layer. Owing to the high fluidity of the magma, in contrast to the rigidity of the crystalline crust, the melt can rise with considerable speed through the crust, even through a narrow and irregular path. The subsidence of the crust as fair-sized blocks and the rise of the magma through narrow channels are energetically efficient processes. With more nearly equal viscosity of overburden and buoyant layer the subsidence and the rise are less contrasted processes—the sinking part is relatively more mobile and therefore decreases in lateral extent, whereas the rising body is relatively less ductile and increases in lateral cross-section. This is significant in discussions of the emplacement of plutonic bodies.

14 Subsiding Sheets

INTRODUCTION

The rise of a buoyant mass of solid rock or fluid magma in the Earth is coupled with the subsidence of material of approximately equal volume. It is the displacement of matter towards levels of low gravitational potential which in fact furnishes the energy necessary to elevate the rising mass. The rise of specific light material and the subsidence of specific heavy material are complementary processes of the same major circle of mass displacement. Domes and diapirs, batholiths, buoyant geanticlinal ridges, and extruding lavas are forced upward to make room for the subsiding masses.

Despite the fact that the subsidence is the more primary process—and the more significant in terms of energy—the rise or upheaval is by far the more conspicuous in terms of geological structures produced. There are several reasons for this. Not only are the rising masses more accessible for study than the foundering masses, but the contrasting movement pattern also makes the rise more conspicuous than the subsidence. The buoyant materials, such as salt relative to limestone or sandstone, or sial relative to sima, or the low-velocity layer in the mantle relative to its overburden, are usually more mobile (less competent) than the subsiding overburden. This is due partly to the different mineralogical composition of the materials and partly to the higher temperature which generally can be attributed to the rising material compared with the sinking material.

The most striking example is hot and light lava rising through a stiff crustal overburden, but even a mantled granitic dome or a batholith retains its high temperature from depth on its way upward. The geothermal isotherms are carried upward by the buoyant masses.

This greater mobility of the buoyant masses means that the rising movement tends to concentrate within relatively small areas, with considerable deep-seated lateral flow towards the centres of rise. Here again a volcano is an extreme example of concentrated vertical movement because of strong contrast in fluidity; a salt dome or a granitic plug represents somewhat less extreme movement concentration, a batholith even less so, and the movement concentration in an epeirogenically rising block need not be greater than the movement concentration in the compensating subsidence of the surroundings.

Because of the generally greater stiffness of the overburden, the subsidence is distributed over a wider area than the rise. As the two oppositely moving masses are of almost equal volume, it follows that the average vertical displacement, and the velocity of vertical motion as well, are considerably less for the subsiding masses than for the rising ones. Were it not for the striking shape of a salt plug which makes the prevailing buoyant-rise theory inevitable, the much less conspicuous marginal syncline would hardly have initiated the theory of vertical forces. We would probably have attributed the gentle depression to lateral compressive forces or to primary irregularities related to the sedimentation processes.

In the model studies of subsidence structures we have chiefly tested the structures developed in connection with horizontal layers sinking through a plastic mass. Such models ought to be of significance for the "secondary" behaviour of huge lopoliths such as the Bushveld in Africa and the Moskox in Canada, the latter described by Smith and Kopp (1963).

THEORETICAL CONSIDERATIONS

A plate of heavy material lying on the top of a light, plastic material tends to sink through the light material. The fluid dynamics of this process is not simple, and we shall not attempt a comprehensive analysis but rather point out a few significant facts. In so doing we assume that the heavy plate is uniformly thick and consists of a viscous material with high viscosity, or a visco-elastic material with a low yield point.

For the plate to sink, the subjacent material has to flow laterally from under the centre of the plate towards, and past, its edges (Fig. 14.1). To maintain this lateral flow, there must be a pressure gradient in the substratum material pointing from the edge of the plate towards its central region, as indicated on Fig. 14.2. Because of this pressure distribution in the substratum, the foundering plate is not evenly supported, but more strongly supported in its central region than towards its edges. It follows that the edges will sink more rapidly than the central part, and in theory a bell-shaped profile is likely to develop.

The flow in the subjacent layer across the edge of the extra load may in these cases be treated as two-dimensional, provided that the load has the shape of an elongated rectangular strip. Let the horizontal x-axis be normal to the long edge of the load and situated in the middle of the subjacent layer, the y-axis, which is normal to the layering, being placed

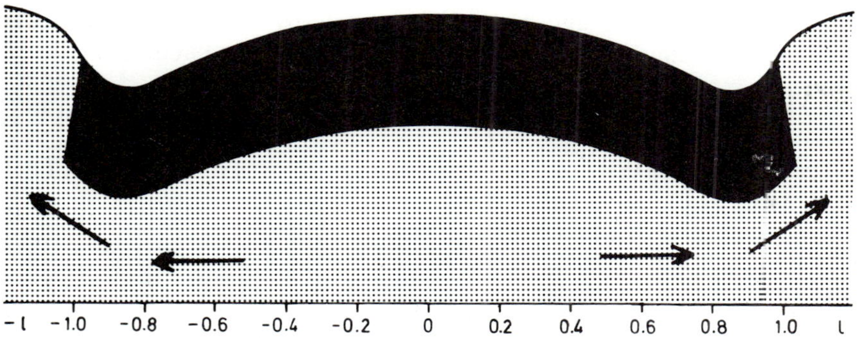

Fig. 14.1. Profile of viscous substratum flowing out from underneath a sheet-shaped load of plastic or viscous material.

in the centre of the system. The z-axis is parallel to the long axis of the loaded strip. The flow is approximately planar in x, y. In the first approximation we shall assume that the heavy plate remains planar during the flow. This simplification enables us to determine the variation of the pressure and the stresses in the (x,y)-plane in the substratum.

The case is analogous to the solved problem of a viscous fluid being squeezed out from between two plates which approach one another (Jaeger, 1956). To solve this problem a co-ordinate system is assumed whose origin is in the midpoint of the fluid mass, the x-axis being parallel to the plates and the y-axis normal to the plates. If the plates are moved in a direction

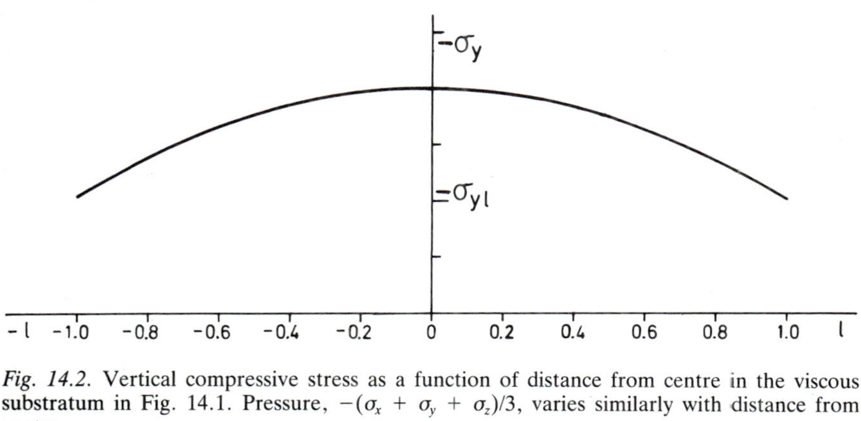

Fig. 14.2. Vertical compressive stress as a function of distance from centre in the viscous substratum in Fig. 14.1. Pressure, $-(\sigma_x + \sigma_y + \sigma_z)/3$, varies similarly with distance from centre.

parallel to the y-axis and towards the x-axis by a velocity v_0 then the pressure in the fluid varies with x and y according to the equation

$$P \frac{3\mu}{4h^3} v_0 (y^2 - h^2 - x^2 + l^2) + P_l \tag{14.1}$$

where μ is the viscosity of the fluid, h half the thickness of the layer, l half its length parallel to x and P_l pressure at $x = \pm l$.

The normal stresses parallel to x and y are

$$\sigma_x = \frac{3\mu}{4h^3} v_0 [3(h^2 - y^2) + x^2 - l^2] + \sigma_{xl} \tag{14.2}$$

and

$$\sigma_y = \frac{3\mu}{4h^3} v_0 [y^2 - h^2 + x^2 - l^2] + \sigma_{yl}. \tag{14.3}$$

When y is put equal to h the equations give the pressure and stresses at the contact between the plates and the fluid mass. The force which resists the movement of the plates is found by integrating the expression for the stress σ_y between $+l$ and $-l$ when y is kept equal to $\pm h$.

Similarly, the force which resists the sinking of a heavy plate which rests on the top of a fluid layer follows from the integration of a somewhat simplified version of eqn. (14.3), viz.

$$\sigma_y = cv_0(x^2 - l^2) + \sigma_{yl} \tag{14.4}$$

Here c is a constant depending on the thickness of the fluid, its viscosity, the extent to which the plate has sunk down in the fluid, etc. To calculate the magnitude of σ_y at different x we put the terms cv_0 and l equal to unity and obtain the information shown in Fig. 14.2

We shall now assume that the overburden is ductile and able to sag locally, depending on the magnitude of the supporting stress, σ_y, at $y = h$. When a steady state is attained, the profile of the overburden and the upper boundary of the viscous layer will be adjusted to the pressure profile as shown in Fig. 14.1 where the effect of the drag at the edges is also taken into consideration.

The particular flow pattern and stress distribution described here result in an undulation of a strip of stratified overburden when placed on a less dense plastic or viscous layer. Two synclines (marginal sinks) separated by a gentle anticline are likely to develop.

The pattern is readily produced experimentally in centrifuged models containing a strip of modelling clay or painter's putty resting on a layer of stitching wax or bouncing putty. Indeed, if such models are run at sufficient acceleration the tensile stress in the central anticlinal bulge is strong enough to cause fracturing.

Consideration of the distribution of the stress that supports the overburden leads to the geologically significant conclusion that the more fluid the buoyant material relative to the overburden, the flatter the profile of the marginal synclines. The more fluid the buoyant material, the less steep the pressure gradient needed to maintain a certain rate of horizontal flow in the source layer. A gentle pressure gradient gives a gentle geometric profile according to the discussion above.

The accentuated subsidence of the edges of the heavy plate is analogous to the formation of a rim syncline adjacent to many domes. Both are dynamic features which remain as a sort of dynamic equilibrium structure as long as the flow process takes place. The edge depression and the rim syncline are not statically stable, however, and both must disappear—flatten out—when, and if, static stability is established.

It is particularly significant to note that there will be a gravity low above the edge depression of a foundering heavy layer, just as a negative anomaly may exist above the rim syncline adjacent to a domal structure (see p. 255). Despite the presence of the gravity low, the local depressions will not tend to rise, as long as the dome is in a state of active rise and the layer has not yet sunk to its stable position. Instead, the marginal depressions may even show a tendency to continue to subside in relation to their immediate surroundings. This is opposite to the conditions more commonly found, viz. that areas showing a local gravity low tend to rise relative to their surroundings.

The substratum material tends not only to flow out from underneath the foundering sheet, but also to pierce through it. This buoyant tendency is at a maximum in the centre of the sheet, at the crest of the anticline which it forms. Whether or not the substratum material does actually burst through the overlying sheet depends upon the rheological and mechanical properties of the two materials, their density contrast and their geometric dimensions. In a given field of gravity or centrifugal field the tendency to burst through the central anticline increases with increasing size of geometrically similar structures, if they consist of materials with the same rheological and mechanical properties. Application of increasing centrifugal acceleration to identical models raises the bursting force of the central bulge of the substratum.

It is plain that the central domes develop more easily when the foundering heavy sheet is wide and thin, whereas the substratum material is more likely to flow around a sheet which is thick in relation to its lateral dimensions.

The flow connected with a subsiding layer gives rise to a backwash pattern in the superincumbent strata which the author believes is of great tectonic and geophysical significance.

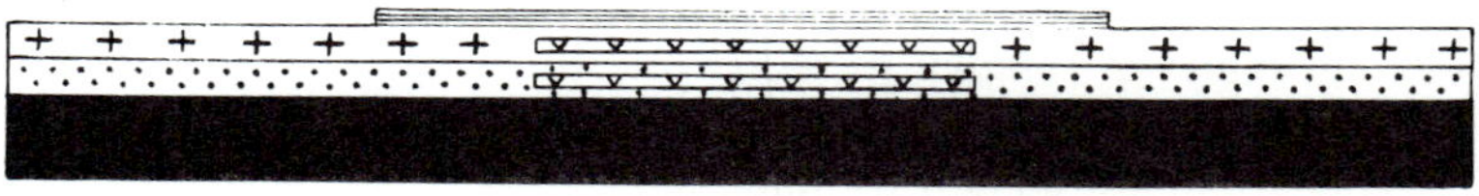

S 126

Fig. 14.3. Initial structure of model S-126 (Fig. 14.4). Angular marks, painter's putty ($\rho = 1{\cdot}95\ \text{g cm}^{-3}$); dotted, silicone putty ($\rho = 1{\cdot}14\ \text{g cm}^{-3}$); thin horizontal lines, silicone putty with sheets of modelling clay; black, silicone putty with magnetite powder ($\rho = 1{\cdot}35\ \text{g cm}^{-3}$); crosses, silicone putty ($\rho = 1{\cdot}14\ \text{g cm}^{-3}$) (a little darker in colour than the dotted silicone). Length of section 9·5 cm.

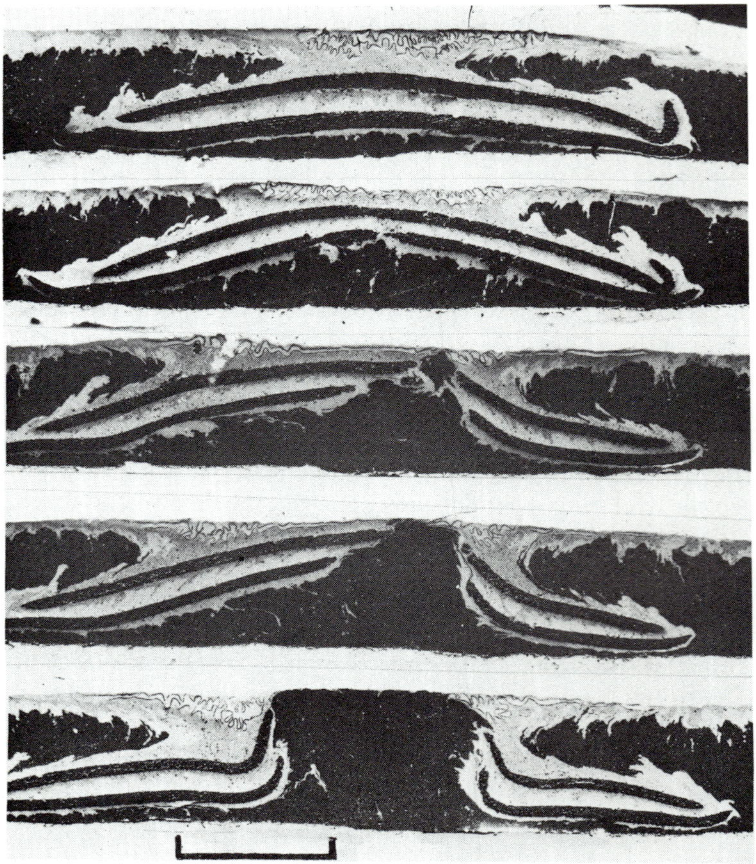

Fig. 14.4. Sequence of cross-sections of model S-126 (Fig. 14.3) after run 3 min at 1000 *g*. Scale bar 1 cm.

EXPERIMENTAL TESTS

Figure 14.3 shows the initial stage of model S-126 before a few minutes' run at 1000 *g* in the centrifuge, a process which produced the structure shown in Fig. 14.4.

The deformation structure indicates the pattern of flow generated during the sinking of the double layer of heavy painter's putty through the sub-

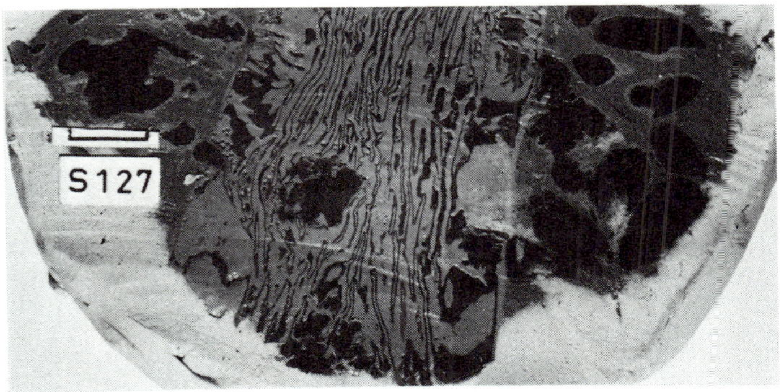

(*a*)

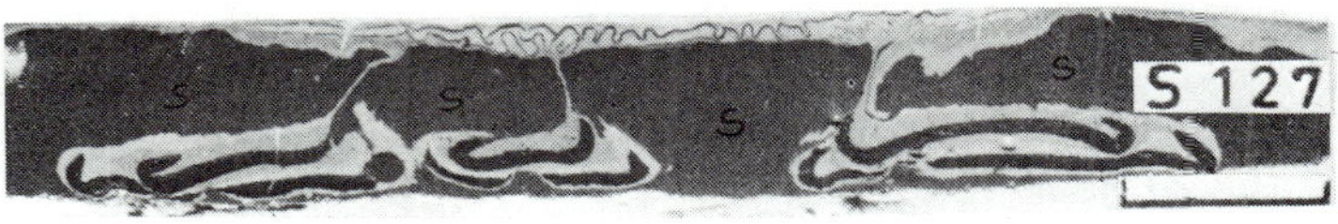

(*b*)

Fig. 14.5. Horizontal section cut 2 mm below the surface of half the model (*a*), and cross-section (*b*) of model initially similar to S-126, Figs 14.3 and 14.4. S-127 is run longer and at higher *g* in centrifuge, 1·5 min at 600 *g*, 4 min at 1700 *g*, and 2·5 min at 2200 *g*. This explains that the two heavy sheets have sunk to the bottom in S-127. Note strong buckling of the surface sheets of modelling clay in the backwash of the subsiding layers. Scale bar 1 cm.

stratum of dark and light silicone. The theoretically predicted anticlinal shape of the sinking layers is well developed. It is also interesting to note the upward and backward bend of some of the edges of the sheets, caused by the drag against the enclosing medium which has overridden the subsiding sheet from both sides. This centripetal flow above the subsiding layers is particularly significant from a geological point of view, in as much as it creates a local compression of the thin surface sheets, which are thrown into compression buckles.

Model S-127 (Fig. 14.5) initially similar to S-126, has been run at somewhat higher acceleration than the previous model, namely for 1·5 min at

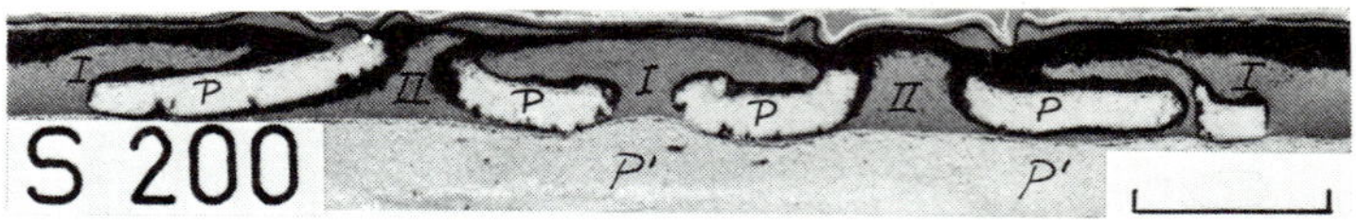

Fig. 14.6. Model in which a layer of painter's putty (P), density 1·80 g cm^{-3}, has sunk through a double-layer substratum of black and grey silicone putty, ρ = 1·39 g cm^{-3} and 1·41 g cm^{-3} respectively. During subsidence the heavy layers ruptured first in the middle and dome I rose, then the two halves ruptured to make room for domes II. Two thin surface sheets of modelling clay (black and white) were forced to buckle in the backwash of the sinking sheet. Run 12 min at 2900 *g*. Initial structure shown in Fig. 14.7. Scale bar 1 cm.

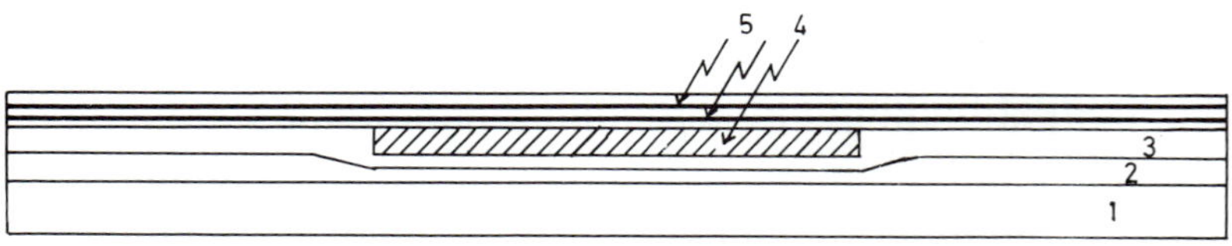

Fig. 14.7. Initial structure of model S-200. (1) Blue painter's putty, ρ = 1·87 g cm^{-3}; (2) silicone putty with tungspar powder, ρ = 1·41 g cm^{-3}; (3) silicone putty with magnetite powder, ρ = 1·39 g cm^{-3}; (4) white painter's putty, ρ = 1·80 g cm^{-3}; (5) silicone putty with 2 sheets of modelling clay, ρ = 1·15 g cm^{-3}. Length of profile 9·5 cm.

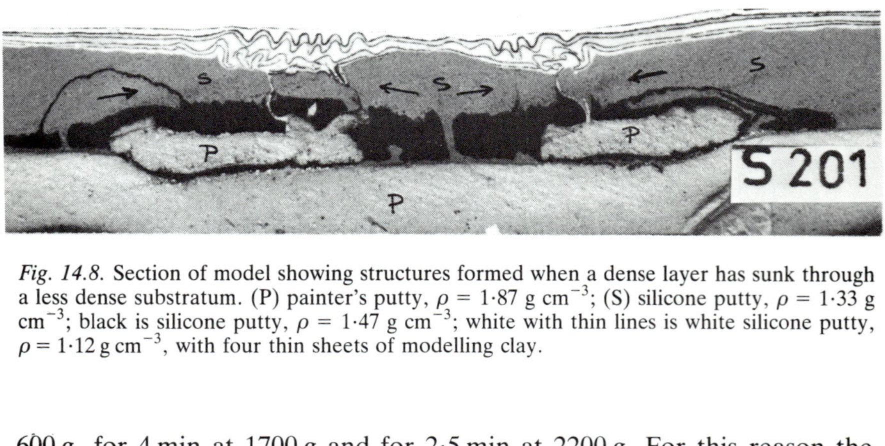

Fig. 14.8. Section of model showing structures formed when a dense layer has sunk through a less dense substratum. (P) painter's putty, ρ = 1·87 g cm^{-3}; (S) silicone putty, ρ = 1·33 g cm^{-3}; black is silicone putty, ρ = 1·47 g cm^{-3}; white with thin lines is white silicone putty, ρ = 1·12 g cm^{-3}, with four thin sheets of modelling clay.

600 *g*, for 4 min at 1700 *g* and for 2·5 min at 2200 *g*. For this reason the heavy sheets have sunk almost to the bottom, they have broken into several fragments as a result of penetration of domal protuberances from the substratum, and their edges are strongly bent into an overturned position. Otherwise the essential pattern is similar to that shown in model S-126. One notes in S-127 a strong buckling of the originally flat surface "sedimentary" complex, a buckling which has developed in the backwash of the sinking bodies.

Model S-200 (Figs 14.6 and 14.7) is interesting for the following reasons.

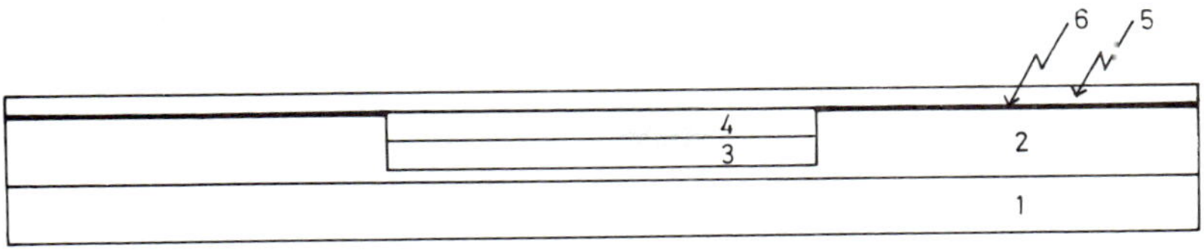

Fig. 14.9. Initial structure of S-201, Fig. 14.8.(1) and (4) painter's putty, $\rho = 1{\cdot}87\ g\ cm^{-3}$; (2) silicone putty, $\rho = 1{\cdot}33\ g\ cm^{-3}$; (3) dark silicone putty, $\rho = 1{\cdot}47\ g\ cm^{-3}$; (5) and (6) silicone putty, $\rho = 1{\cdot}12\ g\ cm^{-3}$, with sheets of modelling clay embedded. Length of section 9·5 cm.

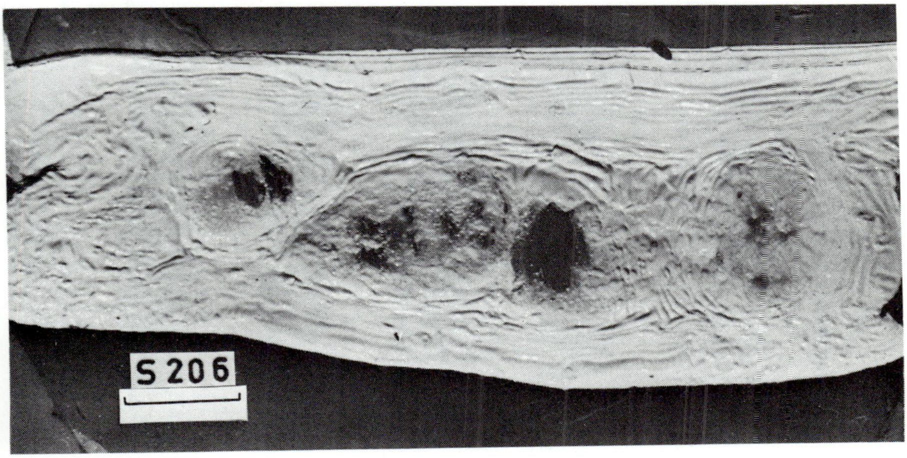

Fig. 14.10. Surface of model whose initial structure is exhibited in Fig. 14.13. Four domes have reached the surface and crumpled the top layer of silicone putty with embedded modelling-clay sheets. Run for 2 min at acceleration increasing from 130 *g* to 2200 *g*, and for 3 min at 2200 *g*.

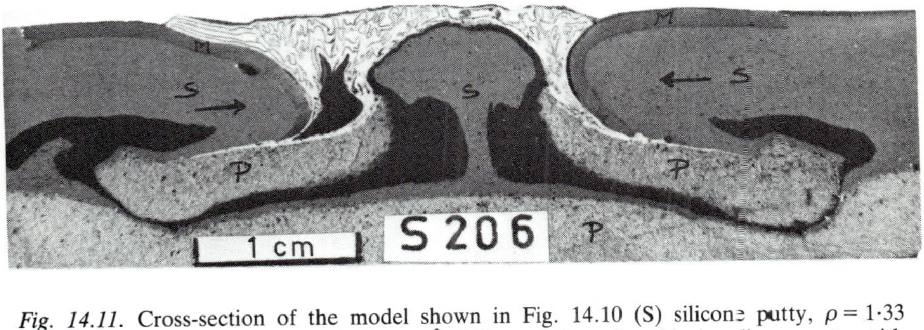

Fig. 14.11. Cross-section of the model shown in Fig. 14.10 (S) silicone putty, $\rho = 1{\cdot}33\ g\ cm^{-3}$; (P) painter's putty, $\rho = 1{\cdot}87\ g\ cm^{-3}$; (M) modelling clay; black: silicone putty with magnetite powder, $\rho = 1{\cdot}47\ g\ cm^{-3}$.

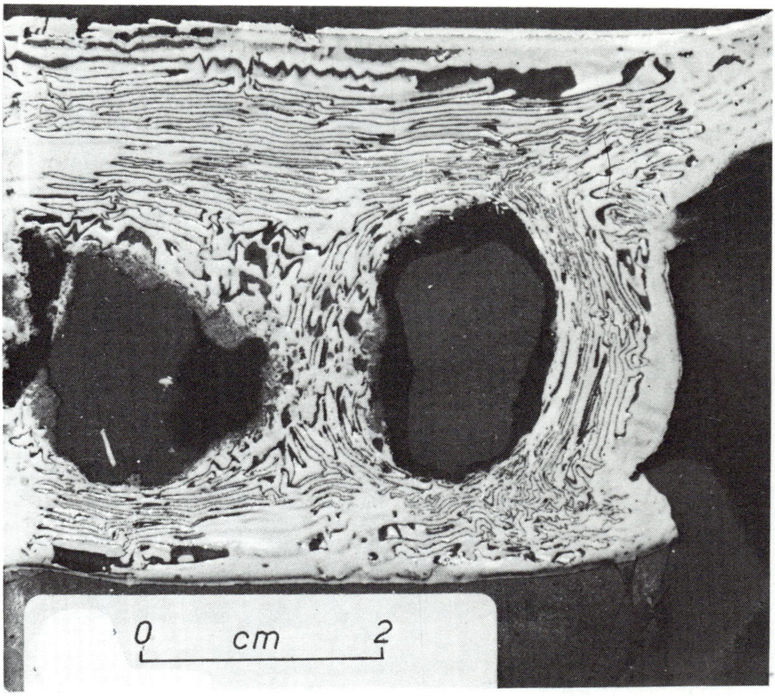

Fig. 14.12. Model S- 206 (viewed from above after cut in two and a thin top sheet sheared off) shows complex folds produced around the diapirs.

First, it demonstrates the significance of scaling of mechanical properties of the model materials. In S-200 the surface layers of modelling clay were accidentally made a little thicker, and thus stronger, than in S-126, S-127 and some of the models referred to below; as a consequence the backwash flow connected with the subsidence was not able to throw the surface layers into the tight folds characteristc of some of the other models.

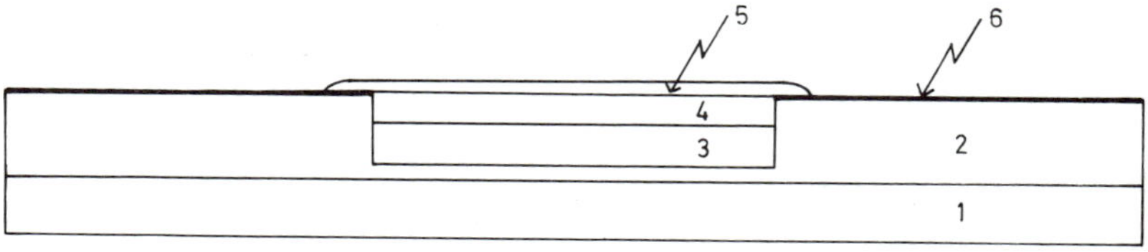

Fig. 14.13. Initial structure of model shown in Figs 14.10, 14.11 and 14.12. (1) and (4) painter's putty, $\rho = 1{\cdot}87$ g cm^{-3}; (2) silicone putty, $\rho = 1{\cdot}33$ g cm^{-3}; (3) silicone putty with magnetite powder, $\rho = 1{\cdot}47$ g cm^{-3}; (5) silicone putty with four thin sheets of modelling clay. Length of section 9·5 cm.

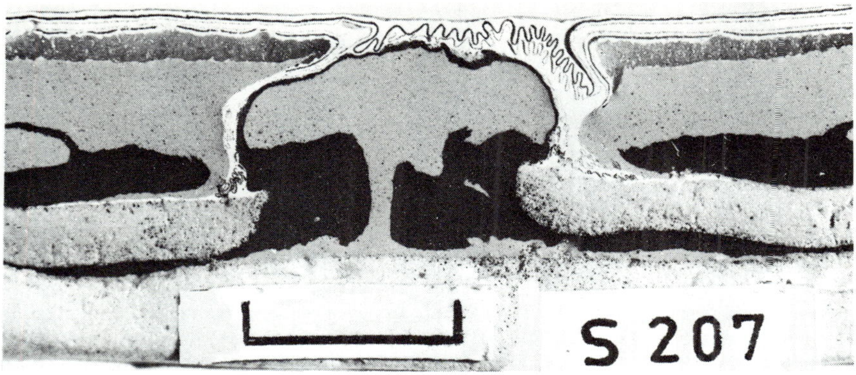

Fig. 14.14. Cross-section of model whose initial structure is shown in Fig. 14.15, showing isoclinal buckling as well as stretching of surface multilayer in connection with the domal rise of the buoyant silicone, and the subsidence of the sheet of heavy putty. Run for 1·5 min at acceleration increasing from 730 *g* to 2200 *g* and for 10 min at 2200 *g*.

Secondly, model S-200 exhibits very clearly two generations of domes. In the middle of the subsiding sheet a dome occurs which shows considerable spreading of the hat above the heavy sheet. This dome is older—developed at an earlier stage of evolution—than either of the smaller domes closer to the edge of the sheet. This is perhaps trivial. While the heavy sheet sunk it was moulded into a wide anticlinal shape. The strongest penetrating force created by the buoyant substratum was directed towards the crest of the anticline at the centre of the sheet. Here, then, the substratum would first burst through. A part of the sheet was thus split in two halves, each of which would sink independently. They would form two smaller anticlines, which in turn were ruptured in the middle of new domal protuberances.

Various geologically realistic-looking structures were generated in surface strata of models with heavy plastic layers sinking through a less dense substratum. Though the basic pattern is the same, the details are different

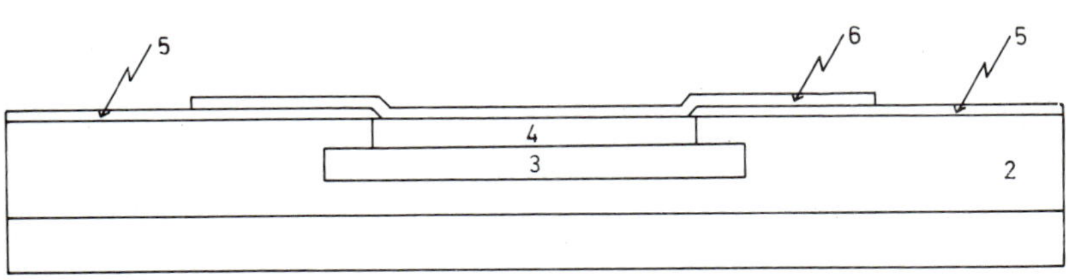

Fig. 14.15. Initial structure of model S-207. (1) and (4) painter's putty, $\rho = 1{\cdot}87\ \mathrm{g\ cm^{-3}}$; (2) silicone putty, $\rho = 1{\cdot}33\ \mathrm{g\ cm^{-3}}$; (3) silicone putty with magnetite powder, $\rho = 1{\cdot}47\ \mathrm{g\ cm^{-3}}$; (5) sheet of powdered wax soaked in oil; (6) silicone putty, $\rho = 1{\cdot}12\ \mathrm{g\ cm^{-3}}$, in which four thin sheets of modelling clay are embedded. Length of section 9·5 cm.

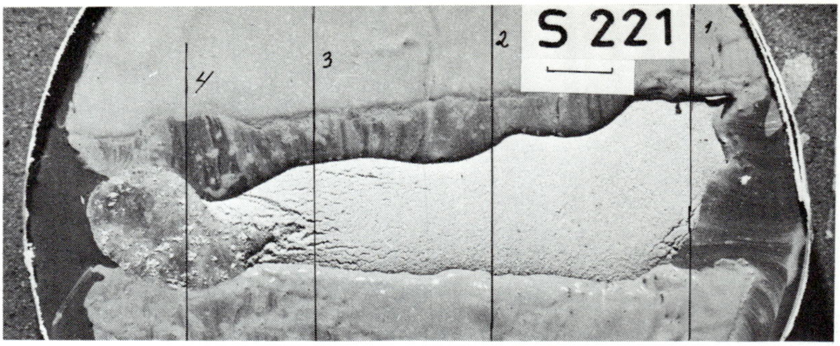

Fig. 14.16. View from above of model with buoyant ridge rising in the middle. White grainy structure: painter's putty overburden covering the silicon-putty ridge which has pierced the overburden at the left side of ridge. Lines 1 to 4 show positions of profiles in Fig. 14.17. Run 10 min while acceleration increased continuously from 0 to 2000 *g*.

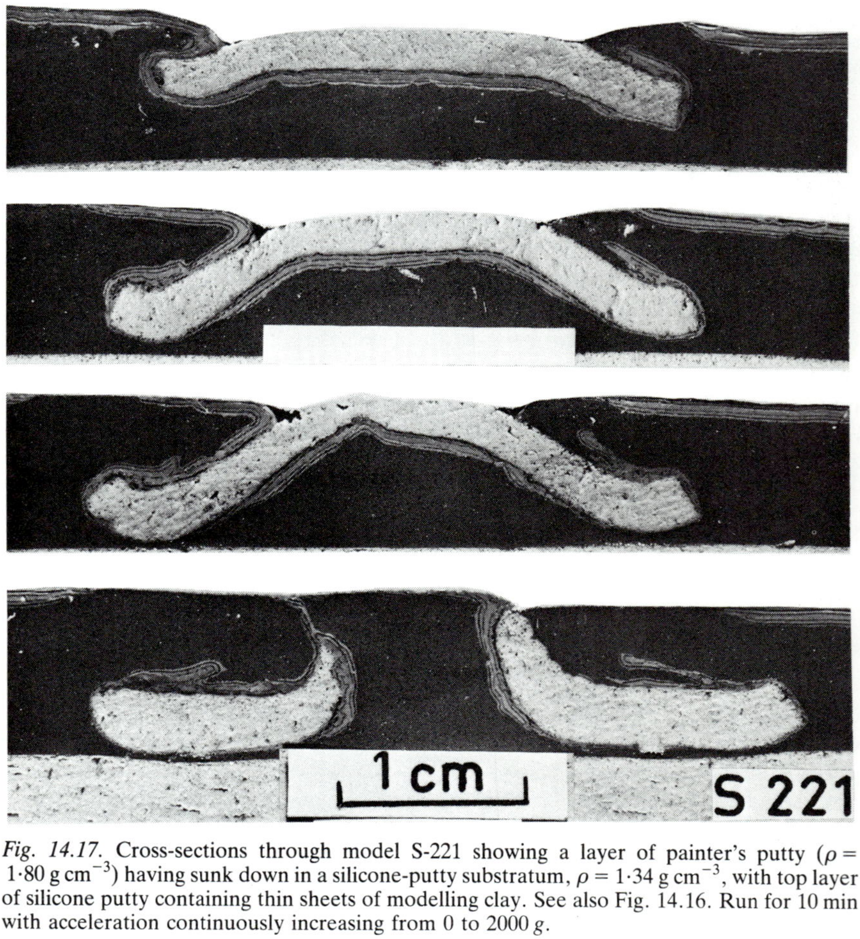

Fig. 14.17. Cross-sections through model S-221 showing a layer of painter's putty ($\rho = 1{\cdot}80\ \mathrm{g\ cm^{-3}}$) having sunk down in a silicone-putty substratum, $\rho = 1{\cdot}34\ \mathrm{g\ cm^{-3}}$, with top layer of silicone putty containing thin sheets of modelling clay. See also Fig. 14.16. Run for 10 min with acceleration continuously increasing from 0 to 2000 *g*.

enough to warrant presentation of a number of photographs of these structures (Figs 14.8–14.15).

Whereas the models so far discussed have shown what happens to surface strata in the backwash above a subsiding body, models S-221 (Figs 14.15 to 14.17) and S-222 (Fig. 14.18) show structures produced in strata below a subsiding body. We note in particular the strong thinning and stretching of the strata below the sinking edges of the subsiding layer in model S-221, and the drag-folding of some of the strata as they are pulled along the upper side of the layer.

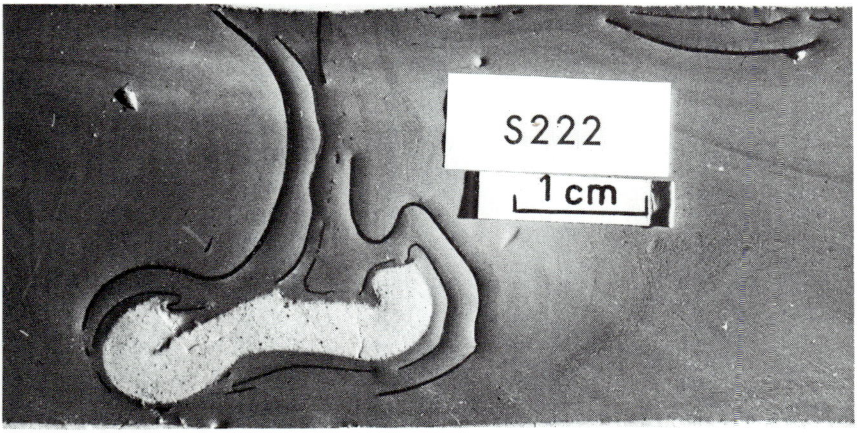

Fig. 14.18. Layer of painter's putty (light grey), $\rho = 1{\cdot}80$ g cm^{3}, sunk through silicone ($\rho = 1{\cdot}13$ g cm^{-3}). Two horizontal sheets of modelling clay were embedded in the substratum at 1 mm and 3 mm, respectively, below the putty layer, which originally rested on the surface. Note fracturing of modelling-clay sheets in the divergent flow below the sinking layer. Note, moreover, that the shape of the layer agrees qualitatively with theoretical prediction (see text). Run 1·5 min at 550–2200 *g*.

In model S-222 the lateral centrifugal flow in the substratum below the subsiding layer has been strong enough to pull the two embedded competent sheets into separated boudins. In this model the bending of the edges of the subsiding layer is especially conspicuous, and it is in good agreement with theoretical expectation as based on the flow pattern predicted by theoretical analysis.

In summary, the principal trends of the models of subsiding heavy layers are as follows:

(1) The subsiding layer assumes an anticlinal shape because of the rapid subsidence of the edges relative to the central part.

(2) Lateral centrifugal flow occurs below the subsiding sheet; this means lateral extension and consequent stretching and boudin formation in competent layers in the substratum.
(3) Lateral centripetal flow occurs in the region above the subsiding sheet; this produces buckling of surface layers immediately above the sheet, but lateral extension and tension fractures with boudinage in the surface regions outside the subsiding sheet.
(4) Surface strata, though much too light to sink by themselves, are often pulled down to great depths with the sinking heavy sheet.
(5) Domes may burst through the subsiding sheet; spreading of surface strata above the hat of the domes, and compression and crumpling of the strata in the synclines outside the domes follow.

15 Models Simulating the Role of Gravity in Orogenic Belts

GENERAL CONSIDERATIONS

The numerous hypotheses proposed to explain the generation of orogens may be grouped in two classes; *lateral-compression hypotheses* and *vertical-movement hypotheses.*

The lateral-compression hypotheses assume a large lateral shortening across geosynclines and orogenic belts, and consider lateral compressive stresses in the crust as the primary forces that generate the characteristic orogenic structures. This is the "giant vise" (or vice) model of orogenesis: the fold belts have been "squeezed between the jaws of a giant vise" as Rodgers (1970) so vividly pictures the evolution of the Appalchians. By this reasoning the vertical movements that unquestionably have occurred during orogenesis are secondary effects of the compressive forces, due partly to buckling of the crust and partly, perhaps, to uneven compressive strain and uneven crustal thickening, or the stacking of slices of the crust on top of one another.

The vertical-movement hypotheses take vertical forces and vertical motion as primary, the folding and thrusting being the results of down-slope gliding of surface strata, of spreading of the upper portion of rising masses, and of local squeezing of strata between rising batholiths and the like. The idea of primary vertical movements in fold-mountain formation was discussed by Naumann (1849, 1858), and developed and modified by Haarmann (1930), van Bemmelen (1933, 1960), Ramberg (1945, 1966), Beloussov (1960), Ramberg and Sjöström (1973) and others. Even in Playfair's *Illustration of the Huttonian Theory of the Earth* (1802, reprinted 1964) vertical forces and vertical movements were regarded as dominant in tectonic deformation. To quote:

> "Though the primary direction of the force which thus elevated them must have been from below upwards, yet it has been so combined with the gravity and resistance of the mass to which it was applied, as to create a lateral and oblique thrust, and to produce those contortions of the strata, which, when on the great scale, are among the most striking and instructive phenomena of geology."

Evolution from the geosynclinal stage to the uplifted finished orogen comprises a vertical rise of some 10–15 km of the crust. This is true at least for the eugeosynclinal furrow within the geosyncline zone (Stille, 1940; Aubouin, 1965). Since geophysical measurements show that isostatic equilibrium is closely approached across old and young orogenic belts, vertical displacement of this magnitude cannot take place without a large change in the average density of the rocks within—or below—the orogens.

One can think of a number of processes which may cause a change of the average density of crustal belts caught up in orogenic upheavals. Granitization in the form of regional-scale introduction of low-density sialic material into the heavy simatic basement of an eugeosynclinal basin is one possibility. In response to the decrement of the average density adjustment in the form of isostatic rise would follow. That granitization is one way of changing the average density of the crust in orogenic belts is strongly suggested by the experience that deeply eroded core zones of orogens are exceptionally rich in granitic material. Here aggregates of potash feldspar, acidic plagioclase and quartz—the low-density minerals charateristic for granites—occur in the form of plutons, pegmatite veins and various types of impregnations in crystalline schists. Another way of decreasing the average density of the geosynclinal floor is to compress, and correspondingly thicken, the sialic crust selectively in the orogenic belt.

On the other hand, geosynclinal depressions themselves form by subsidence of a part of the crust. This process must be coupled with increased average density in order to keep the subsided crust in isostatic balance. One readily can prove that it is physically impossible to buckle the crust down in the form of geosynclines that are several hundred kilometres wide. Such a process would not only create a large negative gravity anomaly above the depression, but would also require compressive forces greatly in excess of the strength of crustal rocks as discussed in detail on p. 175.

Stretching of the sialic crustal layer followed by isostatic adjustment is equivalent to an increment of the average density of the region, and submergence of the surface below sea level will follow. We have seen how in the model tests a light crust becomes stretched and fractures in tension above a spreading pillow whose density is somewhat greater than that of the model crust. The surface of the tenuated sialic crust will then form a basin whose depth depends upon the degree of tenuation and the density of the spreading pillow of anomalous mantle material. Maximum depth will occur where the fractured crust has been pulled apart to expose the simatic pillow below.

Another process that leads to increased average crustal density and hence to possible development of a geosyncline is the invasion of basic and ultrabasic magmas in the form of huge quantities of lava pouring out on the top of the crust and injected as sills, dikes and plutons within the sialic part of the crust as suggested by the author some time ago (Ramberg, 1945).

The centrifuged models demonstrate that, if large enough for gravity to be effective, these heavy sheet-shaped bodies will sink through the sialic crust, squeezing the subjacent material sideways as they subside. Only little sialic material may remain within a region of a subsiding complex of basic lavas and sills; much of the sialic material will be pushed to the sides where it may rise as batholiths.

If the region invaded by the simatic masses is wide relative to the composite thickness of the lavas and sills, however, both experiments and characteristic-wavelength theory indicate that the sialic material will ascend as diapirs, etc., also within the interior of the subsiding basin, not only along its edges.

It has been suggested (Gorman *et al.*, 1978) that processes of this nature are chiefly responsible for "greenstone belts" encountered within many Precambrian shields.

The writer believes that much of the structures we see in orogenic belts also is formed because of gravitational instability, not only due to the emplacement of heavy simatic material on top of sial, but also because of the density contrast between a column of metamorphosed average sediments and their granitoid basement. The latter conditions have been employed in the numerical model discussed on p. 102.

In this connection it is interesting that many orogenic belts appear to be ensialic, i.e. the basement below the cover of deformed and metamorphosed supracrustals is granitic to granodioritic. This is, for example, the situation in the Scandinavian Caledonides, as will be discussed in the next section of the book. In the Caledonides we also find that most of the present cover—whether consisting of the original geosynclinal column of supracrustals or of thrust sheets—has a higher density than the substratum. The latter consists chiefly of quartzo-felsphatic rocks, usually with a gneissoid texture, while much amphibolites, garnet–mica schists, lime silicate schists, marble, etc.—all more dense than an average granite—occur in the cover.

When in the geosynclinal period of orogenesis the basement was covered by a 10–15 or even 20 km thick pile of sediments and lavas, the basement was particularly mobile because of the relatively high temperature, and the stage was set for gravity to work efficiently on the instable density distribution. To study what may happen dynamically subsequent to an initial synclinal stage of the kind described above, a number of models were run in the centrifuge, some of which are presented in Figs 15.1–15.14.

Besides the tectonic effect of buoyant-basement diapirism, gravity collapse and spreading are important phenomena in orogenic belts. The

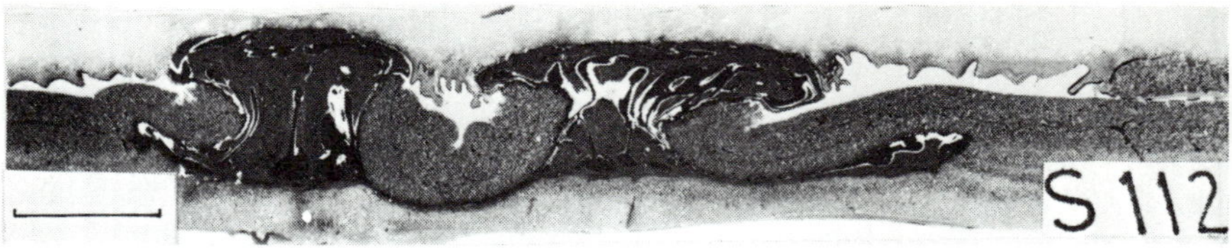

Fig. 15.1 Section through model of domes of black, white and dark-grey silicone putty penetrating painter's putty overburden (inclined hatching) and top layer of white and red silicone with thin sheet of modelling clay. Run 11 min at 2000 *g*. See also Fig. 15.4.

spreading is both in the form of mushrooming hats of diapirs and in the form of collapsing piles of the sedimentary-volcanic cover as visualized by Price in De Jong and Scholton (1973) and by Elliot (1976). Based on experience from folds and other deformation structures generated in glaciers, Huddleston (1977) considered much of the recumbent-type folds in the Alps and the Appalachians to be gravity-produced structures.

A semiquantitative theory, supported by mostly non-centrifuged experiments, is presented in Chapter 9. Spreading lobes of basement diapirs have developed in many centrifuged models, (see, e.g. Figs 11.83, 15.13 and 15.14). Some of the models are believed to simulate quite realistically the movement of pennine-type nappes.

Recent descriptions of orogenic belts are for example found in the following publications: the Alps as described by Gwinner (1971); the Himalayas by Gansser (1964); the Canadian Rockies by Price in De Jong and Scholten (1973); the Appalachians by Zen *et al.* (1968), Fisher *et al.* (1970) and Rodgers (1970); and the Caledonides by Haller (1971, the Greenland Caledonides), by Johnson and Stewart (1963, the British Caledonides) by Strand and Kulling (1972, the Scandinavian Caledonides). See also Closs *et al.* (1978) in which traverses through the Alps, the Apennines and the Hellenides are described. Gansser (1973) has given an interesting analysis of the Andes, and Spencer (1977) has edited a comprehensive volume of Mesozoic and Cenozoic belts. For 1980 works see Bibliography.

Along the continental margins where thick piles of sediments generally form, gravitational instability is created if heavy submarine lavas pour over the sedimentary pile. In models believed to simulate this situation the diapirism, folding and other deformations which characterize the rearrangement to a more stable state end up with structures which have much in common with natural island-arc systems. Examples are presented in Figs 15.15 and 15.16.

For comparison with the experimental orogen models simplified maps and profiles from the Appalachians and the Caledonides are reproduced in Figs 15.17–15.24.

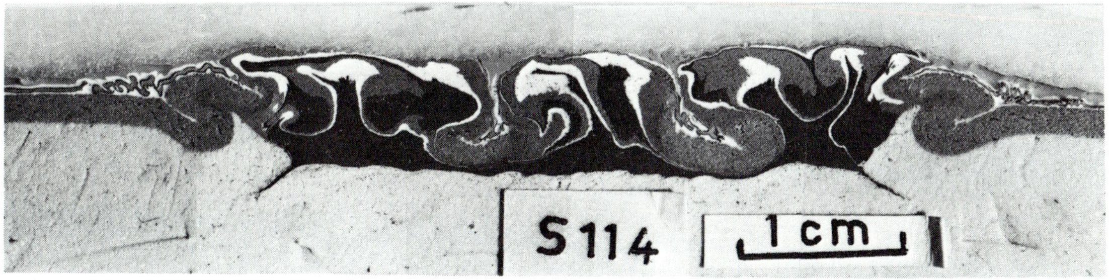

Fig. 15.2. Section through model S-114 of domes and nappes of grey, white and black silicone risen through putty overburden (dark layer shown at each side of model), and thin "sedimentary" layers of silicone and modelling clay under cover of soft wax. Run 8 min at 1300–2200 *g*. Note the deeply pulled down surface layers in the synclines between the domes and nappes. See also Fig. 15.4.

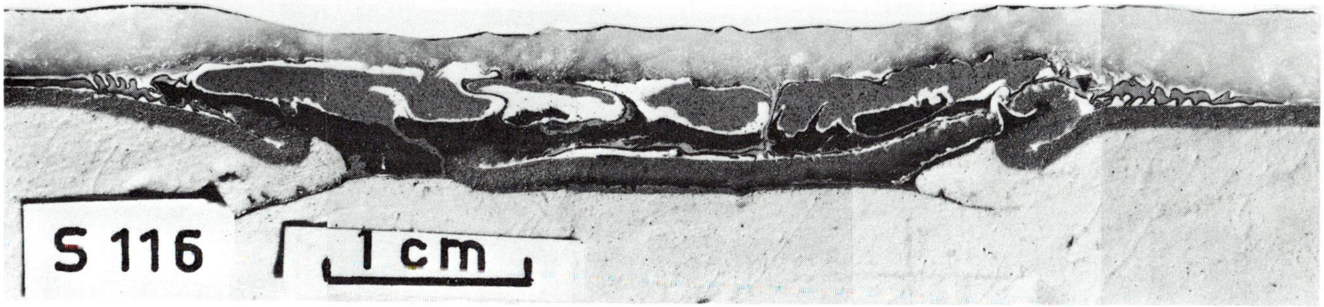

Fig. 15.3. Section through model S-116, which is quite similar to model S-114 (Fig. 15.2). In S-116, however, the dark-grey and light-grey silicone is inverted, and the overburden of putty is less thick. Run 15 min at 2400–2900 *g*. Note well developed buckle folds in front of nappe on left side of model.

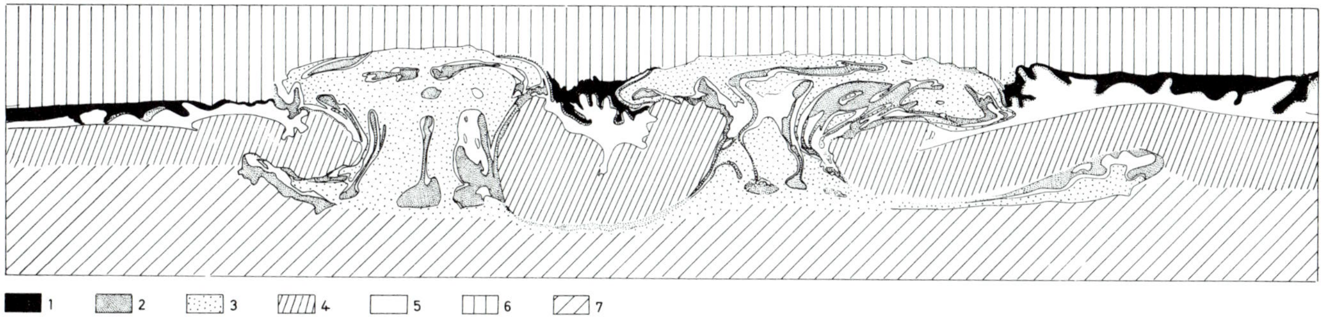

1
2
3
4
5
6
7

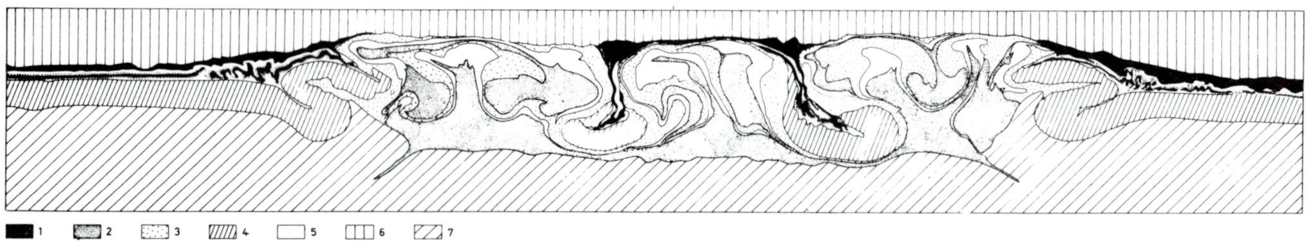

1
2
3
4
5
6
7

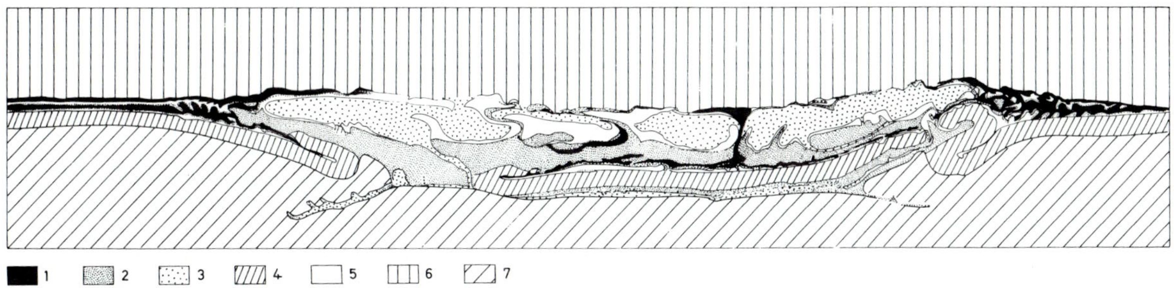

Fig. 15.4. Trace drawings of profiles of S-112 and S-114 as shown in Figs 15.1 and 15.2. Drawings of S-116 refers to profile slightly different from that shown in Fig. 15.3. (1) Red silicone, $\rho = 1{\cdot}14\ \mathrm{g\ cm^{-3}}$; (2) black silicone, $\rho = 1{\cdot}35\ \mathrm{g\ cm^{-3}}$; (3) grey silicone, $\rho = 1{\cdot}25\ \mathrm{g\ cm^{-3}}$; (4) putty, $\rho = 1{\cdot}87\ \mathrm{g\ cm^{-3}}$; (5) silicone, $\rho = 1{\cdot}14\ \mathrm{g\ cm^{-3}}$; (6) oil–powdered wax mixture; (7) painter's putty, $\rho = 1{\cdot}87\ \mathrm{g\ cm^{-3}}$.

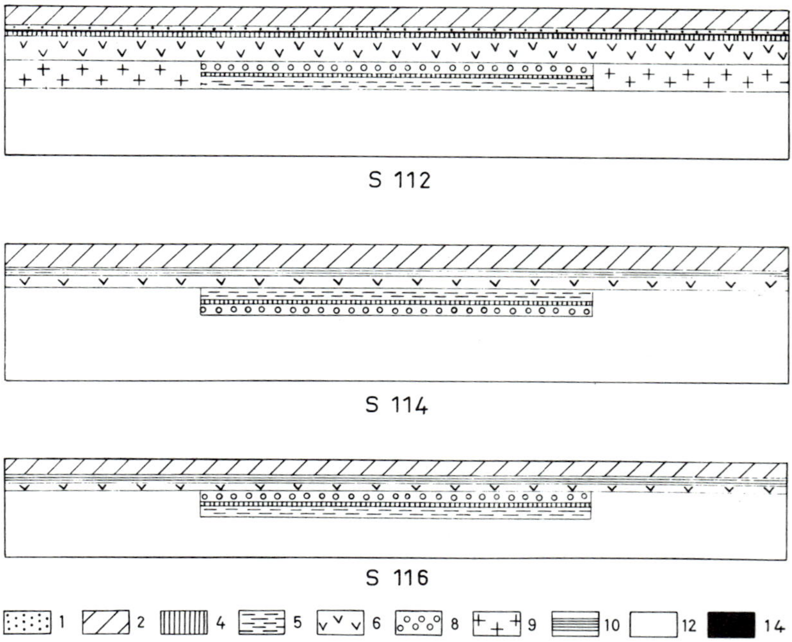

Fig. 15.5. Cross-sections through models S-112, S-114, and S-116 prior to run in centrifuge. See Figs 15.1–4. (1) Red silicone, $\rho = 1{\cdot}14\ \mathrm{g\ cm^{-3}}$; (2) oil–wax mixture, $\rho = 0{\cdot}9\ \mathrm{g\ cm^{-3}}$; (4) white silicone, $1{\cdot}14\ \mathrm{g\ cm^{-3}}$; (5) grey silicone, $\rho = 1{\cdot}25\ \mathrm{g\ cm^{-3}}$; (6) painter's putty, $\rho = 1{\cdot}87\ \mathrm{g\ cm^{-3}}$; (8) black silicone, $\rho = 1{\cdot}35\ \mathrm{g\ cm^{-3}}$; (9) painter's putty, $\rho = 1{\cdot}87\ \mathrm{g\ cm^{-3}}$; (10) red silicone, $\rho = 1{\cdot}14\ \mathrm{g\ cm^{-3}}$ with thin layers of modelling clay; (12) painter's putty, $\rho = 1{\cdot}87\ \mathrm{g\ cm^{-3}}$; (14) modelling-clay sheet (in S-112).

Fig. 15.6. Composite diapirs of two types of silicone putty (dotted, $\rho = 1{\cdot}14\ \mathrm{g\ cm^{-3}}$, and black, $\rho = 1{\cdot}33\ \mathrm{g\ cm^{-3}}$) having pierced a layered overburden of painter's putty and modelling clay with a top complex of silicone with embedded thin sheets of modelling clay. Note complicated folding of the initially straight strata. See also Fig. 15.7. Run 4·5 min at 2200 *g*, 30 sec between 2200 *g* and 3000 *g*, and 10 min 32 sec at 3000 *g*. Cf. for example, the left part of the model with the Appalachian structure shown in Fig. 15.24.

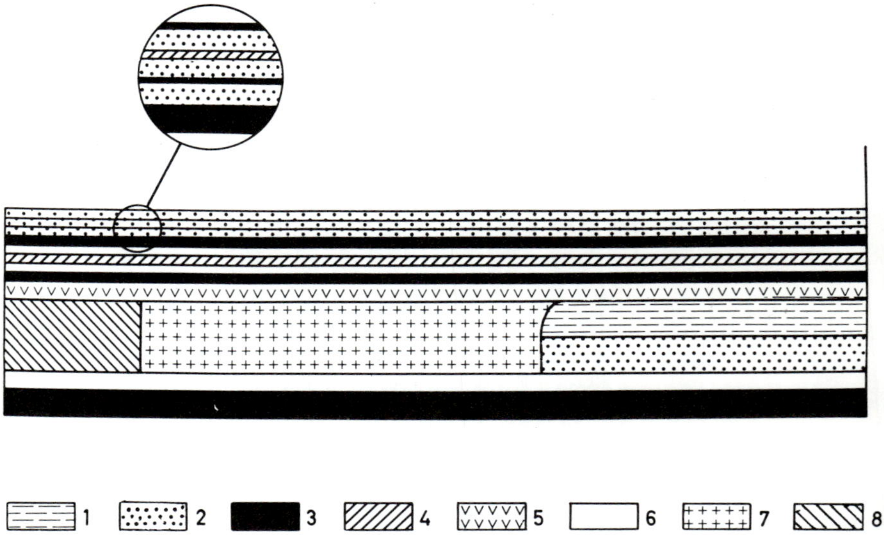

Fig. 15.7. Left-half section of initial structure of model Ny-9, Fig. 15.6. (1) Silicone putty, with magnetite powder, $\rho = 1{\cdot}33\ \mathrm{g\ cm^{-3}}$; (2) silicone putty, $\rho = 1{\cdot}14\ \mathrm{g\ cm^{-3}}$; (3) dark green modelling clay, $\rho = 1{\cdot}68\ \mathrm{g\ cm^{-3}}$; (4) white modelling clay, $\rho = 1{\cdot}78\ \mathrm{g\ cm^{-3}}$; (5) green painter's putty, $\rho = 1{\cdot}85\ \mathrm{g\ cm^{-3}}$; (6) grey painter's putty, $\rho = 1{\cdot}85\ \mathrm{g\ cm^{-3}}$; (7) painter's putty; (8) brown modelling clay, $\rho = 1{\cdot}71\ \mathrm{g\ cm^{-3}}$. Length of half-section 9·7 cm.

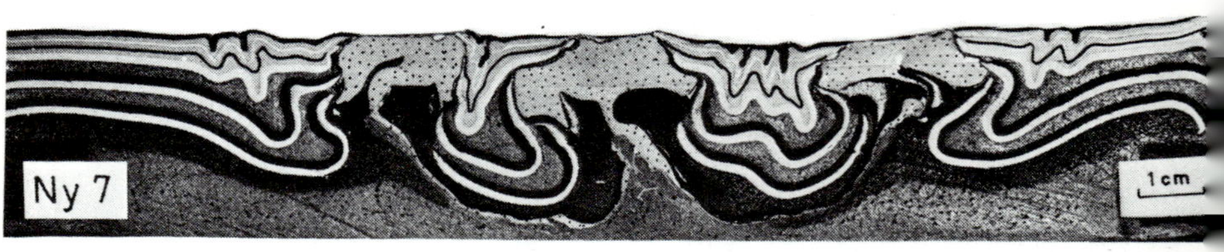

Fig. 15.8. Three composite diapirs of two types of silicone putty, one with $\rho = 1{\cdot}14\ \mathrm{g\ cm^{-3}}$ (dotted) and one with $\rho = 1{\cdot}58\ \mathrm{g\ cm^{-3}}$ (dark) having risen through a layered overburden of painter's putty and modelling clay with a surface complex of silicone putty containing thin sheets of modelling clay (run by H. Sjöström, Uppsala) 1 min 35 sec at 2000 *g*. Although the initial stage of the model was quite similar to model Ny-9 (Figs 15.6 and 15.7) Ny-7 has developed much less–is less mature–than Ny-9, whose diapirs have spread to form recumbent-fold nappes. This is chiefly because of the difference in running time.

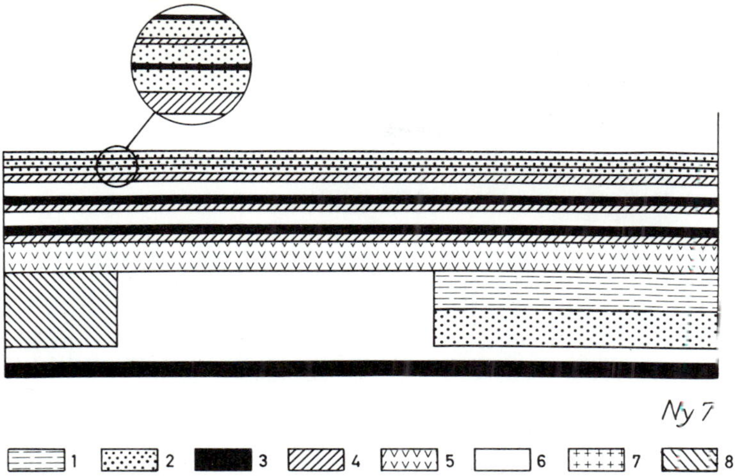

Fig. 15.9. Left-half section of initial structure of model Ny-7, Fig. 15.8 (1) Silicone putty with magnetite powder, $\rho = 1{\cdot}58$ g cm^{-3}; (2) silicone putty, $\rho = 1{\cdot}14$ g cm^{-3}; (3) dark grey modelling clay, $\rho = 1{\cdot}68$ g cm^{-3}; (4) white modelling clay, $\rho = 1{\cdot}78$ g cm^{-3}; (5) green painter's putty, $\rho = 1{\cdot}85$ g cm^{-3}; (6) grey painter's putty, $\rho = 1{\cdot}85$ g cm^{-3}; (7) does not exist in this model; (8) brown painter's putty, $\rho = 1{\cdot}71$ g cm^{-3}. Length of half-section 9·7 cm).

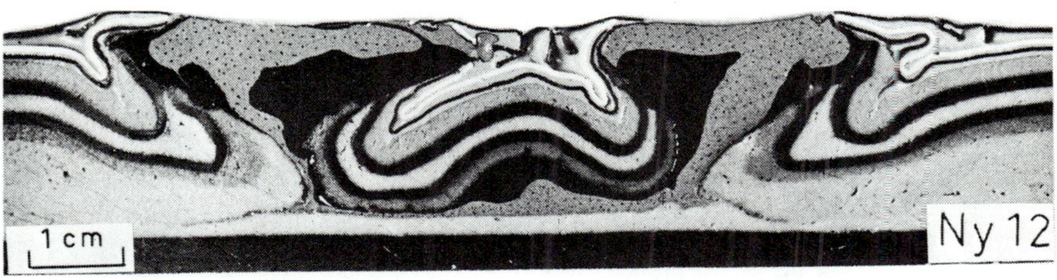

Fig. 15.10. Two composite diapirs of light- and dark-coloured silicone putty having ascended to the surface through a layered overburden. See also Figs 11.118 and 11.119.

Fig. 15.11. Model quite similar to Ny-12 (Fig. 15.10). Note the deformation of the competent sheets of modelling clay below the lobe of the composite diapir. Run for a BBC TV programme "The Restless Earth".

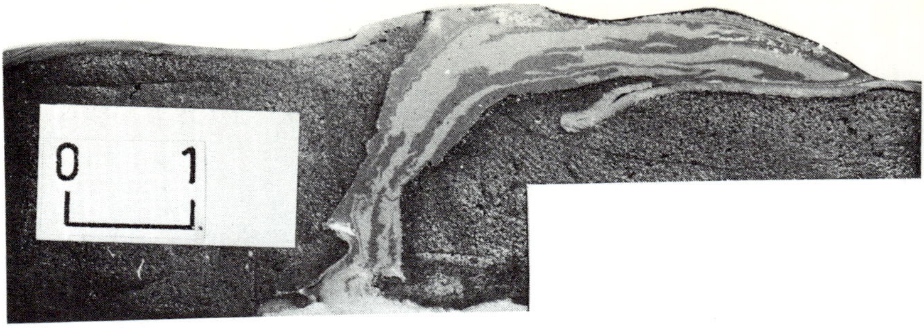

Fig. 15.12. Diapir with one-sided spreading lobe.

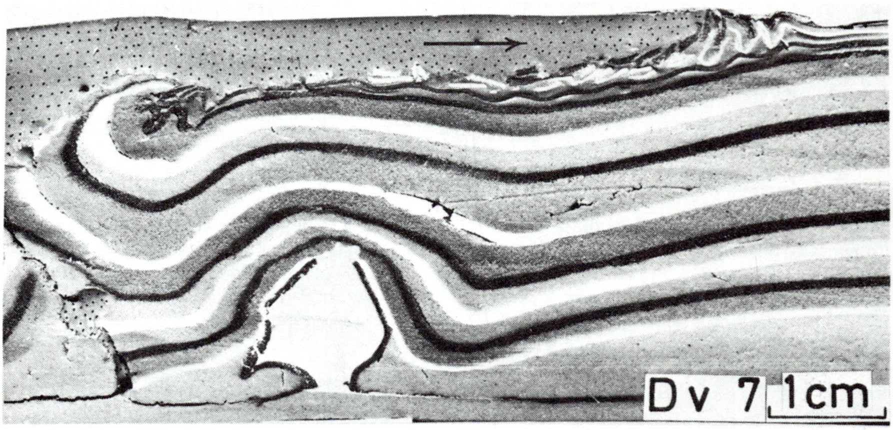

Fig. 15.13. Detail of model DV-7 showing folding and fracturing of surface layered complex overrun by a spreading lobe of a silicone-putty nappe. See also Figs 11.17 and 11.83.

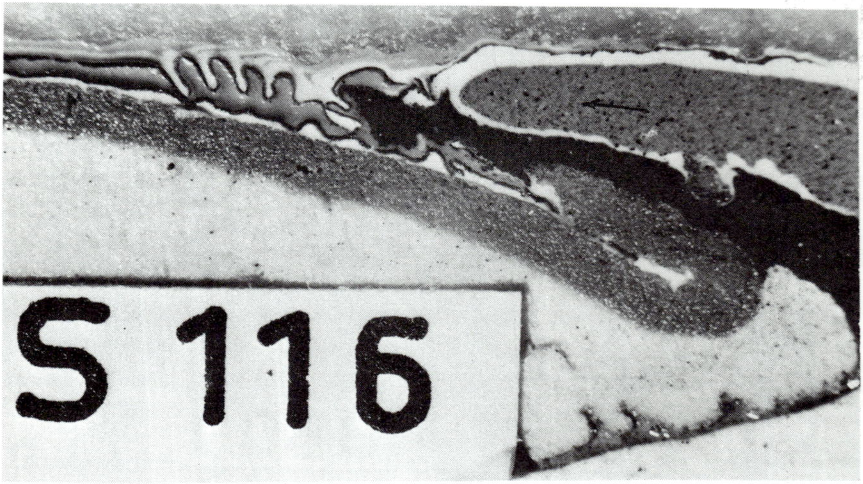

Fig. 15-14. Detail of model S-116 showing recumbent fold of thick layer (dark grey) underneath the spreading lobe of a diapir, and isoclinal buckles formed in front of the lobe. See also Fig. 15.3.

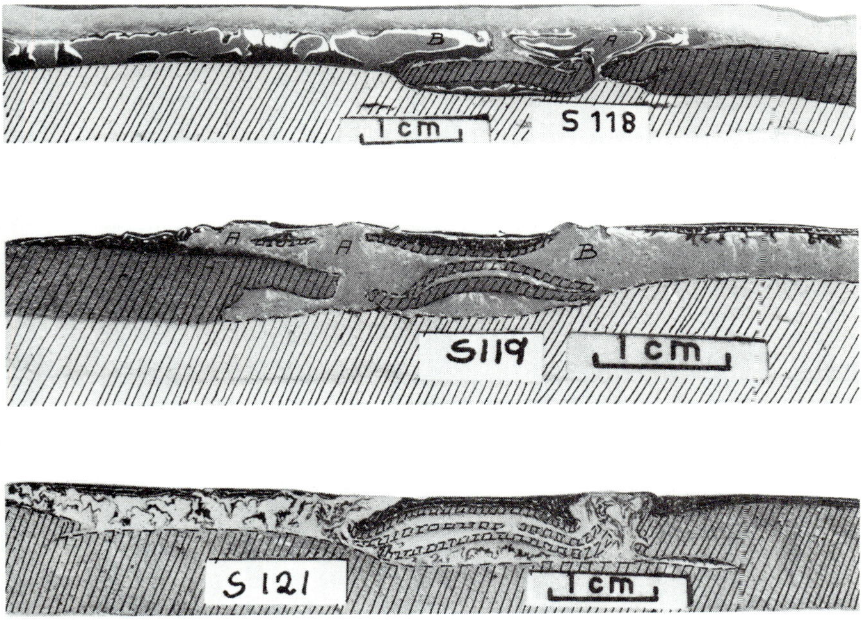

Fig. 15.15. Profiles through three centrifuged models of possible tectonic evolution within the border region between continent and ocean of circum-Pacific type. Ocean on the right-hand side in S-118 and S-121, on the left-hand side in S-119. S-118 run 2 min at 700 *g* and 14 min at 2000 *g*; S-119 run 15 min at 2500 *g* and 10 min at 2900 *g*; S-121 run 3 min at 700 *g* and 7 min at 2650 *g*. *A*, island arcs, *B*, batholiths at continental edge. Inclined hatching, putty (of contrasting colours), $\rho = 1{\cdot}87\ \mathrm{g\,cm^{-3}}$. In S-118: grey: silicone putty, $\rho = 1{\cdot}25\ \mathrm{g\,cm^{-3}}$; black: silicone putty, $\rho = 1{\cdot}34\ \mathrm{g\,cm^{-3}}$; white: silicone putty, $\rho = 1{\cdot}14\ \mathrm{g\,cm^{-3}}$; greyish-white layer on top: oil–wax mixture. In S-119: grey with white irregular spots: silicone putty, $\rho = 1{\cdot}41\ \mathrm{g\,cm^{-3}}$, and $1{\cdot}14\ \mathrm{g\,cm^{-3}}$ (white); black with light-coloured folded sheets on top of model: silicone putty with sheets of modelling clay. In S-121: light grey with dark irregular "streaks": silicone putty, $\rho = 1{\cdot}34\ \mathrm{g\,cm^{-3}}$; black with light-coloured buckled sheets on top of model: silicone putty with sheets of modelling clay. For information on initial structures see Fig. 15.16.

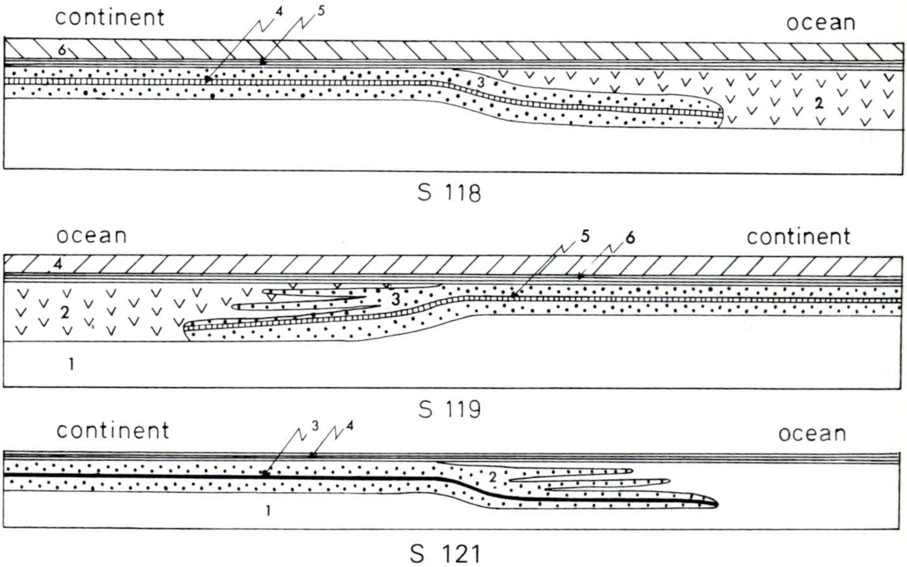

Fig. 15.16. Sections through models S-118, S-119 and S-121 prior to run in centrifuge. White: painter's putty, $\rho = 1{\cdot}87\ \mathrm{g\ cm^{-3}}$; angular marks: painter's putty (dark colour), $\rho = 1{\cdot}87\ \mathrm{g\ cm^{-3}}$; dotted: silicone, $\rho = 1{\cdot}25\ \mathrm{g\ cm^{-3}}$, $1.34\ \mathrm{g\ cm^{-3}}$, and $1.41\ \mathrm{g\ cm^{-3}}$ (see Fig. 15.15); vertical lines: white silicone putty, $\rho = 1{\cdot}14\ \mathrm{g\ cm^{-3}}$; black sheets in S-121: dark silicone putty; thin horizontal lines: sheets of modelling clay in silicone putty; inclined hatching: oil–wax mixture.

SCALE-MODEL TEST OF AN OROGENIC STRUCTURE

Although the rheological properties of rocks in their natural environment are practically unknown, one may endeavour to scale up one of the centrifuged orogen models to simulate a reasonable proportioned natural mountain chain.

Model S-116 has been selected as an example. The result is shown in Table 15.1, which also lists the assumptions made concerning density, geometric dimensions and effective viscosity of the natural rocks. The resulting time of evolution from the geosynclinal stage to the folded and elevated orogen—122 million years—seems to be in surprisingly good agreement with generally accepted values for such evolutions.

A brief discussion of the implication of gravity-tectonic models for the Scandinavian Caledonides follows below.

Table 15.1. Example of scale-model calculation of time of evolution of fold mountain

Quantity	Model S–116	Natural orogen	Model ratios
Thickness of basalt cover	1 mm	4 km	$l_r = 2{\cdot}5\times10^{-7}$
Total thickness of subjacent sial layers	4 mm	16 km	$l_r = 2{\cdot}5\times10^{-7}$
Width of final orogen, from nappe front to nappe front (Figs 15.4; 15.5)	50 mm	200 km	$l_r = 2{\cdot}5\times10^{-7}$
Density of basalt	$1{\cdot}87$ g cm^{-3}	$3{\cdot}0$ g cm^{-3}	$\rho_r = 0{\cdot}62$[a]
Density of sial, average of layers 4, 5 and 8 (Fig. 15.5)	$1{\cdot}24$ g cm^{-3}	$2{\cdot}7$ g cm^{-3}	$\rho_r = 0{\cdot}46$[a]
Density contrast	$0{\cdot}63$ g cm^{-3}	$0{\cdot}3$ g cm^{-3}	$\Delta\rho_r = 2{\cdot}1$[b]
Viscosity of sial	3×10^{5} poise	10^{21} poise	$\mu_r = 3\times10^{-16}$
Viscosity of sima	10^{6}–10^{7} poise	$3{\cdot}3\times10^{21}$–$3{\cdot}3\times10^{22}$ poise	$\mu_r = 3\times10^{-16}$
Body force per unit mass, first period	$2{\cdot}4\times10^{3}\ g$	$1\ g$	$a'_r = 2{\cdot}4\times10^{3}$
Body force per unit mass, second period	$2{\cdot}9\times10^{3}\ g$	$1\ g$	$a''_r = 2{\cdot}9\times10^{3}$
Buoyant stress, first period	Variable	Variable	$\sigma'_r = \Delta\rho_r l_r a'_r = 1.26\times10^{-3}$
Buoyant stress, second period	Variable	Variable	$\sigma''_r = \Delta\rho_r l_r a''_r = 1{\cdot}52\times10^{-3}$
Time, first period	660 sec	t'_o	$t'_r = \mu_r\sigma_r'^{-1} = 2{\cdot}4\times10^{-13}$
Time, second period	240 sec	t''_o	$t''_r = \mu_r\sigma_r''^{-1} = 2{\cdot}0\times10^{-13}$
Total time of evolution	900 sec	t_o	

$t'_o = t'_m/t'_r = 660\ \text{sec}/2{\cdot}4\times10^{-13} = 2{\cdot}7\times10^{15}$ sec; $t''_o = t''_m/t''_r = 240\ \text{sec}/2{\cdot}0\times10^{-13} = 1{\cdot}2\times10^{15}$ sec. Total time of evolution: $3{\cdot}9\times10^{15}$ sec $= 122\times10^{6}$ years.

[a] The model ratio of density is not constant, showing that the stress cannot be correctly scaled.

[b] By applying the ratio of the density contrast in the expression for the ratio of stress and ratio of time, however, the most important force in the systems, viz. the buoyancy, is reasonably well scaled.

DIAPIRISIM AND GRAVITY SPREADING IN THE SCANDINAVIAN CALEDONIDES

Basement diapirism

Along the eastern and south-eastern border of the Scandinavian Caledonides (Fig. 15.17) the foreland basement dips gently below the autochthonous Cambrian sediments, volcanics and the overriding Caledonian nappes. Little or no sign of Caledonian metamorphism is visible in the granitoid basement here, though low-temperature cataclasis and mylonitization show that the basement was not completely undisturbed during the Caledonian orogenesis. This is in marked contrast to the conditions in the more or less isolated windows of basement rocks exposed within the Cambro-Silurian strata and the nappes of the Caledonides, especially the windows within the core of the orogen along the coast of Norway. The latter region consists chiefly of quartzo-felspathic gneisses whose grade of metamorphism, much of which is of Caledonian age, reaches high amphibolite facies and even granulite facies. Among the gneisses are highly metamorphosed and metasomatosed supracrustals, some of which demonstrably are of Eocambrian to Silurian age (e.g. Eocambrian sparagmite and Cambro-Silurian so-called Trondheim schists including pillow lava metamorphosed to green schists). Anorthosite and eclogite are found, especially in the huge Möre-gneiss culmination. Metadiabases are abundant in some regions within the gneisses.

Field studies by many geologists (see references in e.g. Strand and Kulling, 1972) have established that, though the basal-gneiss complexes are generally positioned stratigraphically—but not always tectonically—below the Cambro-Silurian strata, the gneisses often show a young, somewhat penetrative relation to the Cambro-Silurian supracrustals.

Pegmatitization and migmatitization are common in the schists within the core region of the orogen. Here the Cambro-Silurian schists in the neighbourhood of the basal gneiss exhibit a grade of regional metamorphism corresponding to that of the gneisses, and the Caledonian deformation structure—i.e. schistosity and mineral lineation—is parallel in basal complex and adjacent Cambro-Silurian schists. The planar structure is often steep, and the positions are frequently inverted in recumbent folds with gneiss cores. These are structures similar to many of the experimental models of buoyant anticlines and overturned lobes of domes.

If some of the gneisses are altered schists and lavas of Eocambrian-Silurian age—recrystallized sparagmite, for example, constitutes a considerable part of the Möre and Vestranden regions—the major portion of the basal complexes appears to be the Precambrian basement exposed to

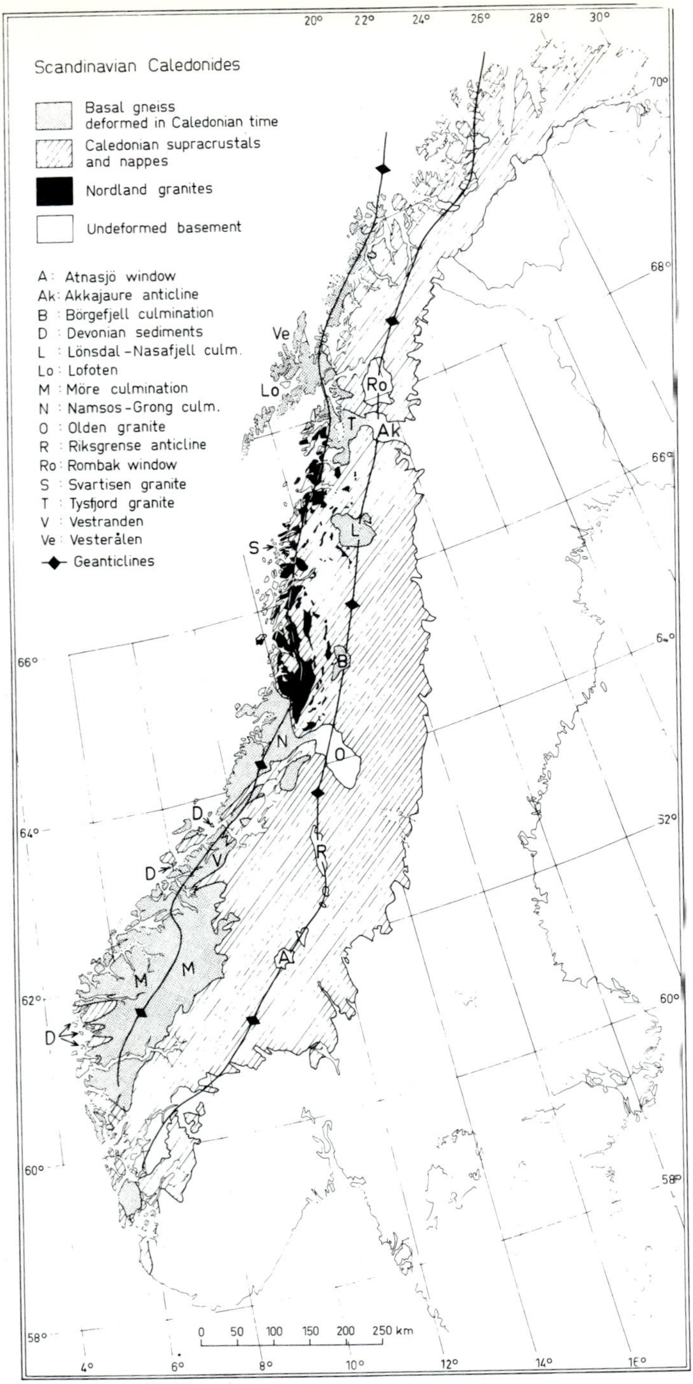

Fig. 15.17. Simplified map of the Scandinavian Caldedonides based on Holtedahl and Dons (1960) and Magnusson *et al.* (1957).

recrystallization, deformation and mobilization during the Caledonian orogenic period.

In two separate areas—the Grong–Olden and the Tysfjord–Akkajaure culminations—the transition from undisturbed basement to strongly mobilized basal gneiss and granite can be followed continuously in the field from the eastern edge of the Caledonides in Sweden to their core zone in Norway.

Isolated basal complexes located closer to the eastern edge of the orogen—e.g. the Riksgrense anticline and the Börgefjell and Lönsdal culminations—were less affected by Caledonian recrystallization and deformation than were the culminations located farther away from the eastern edge, e.g. the Möre–Vestranden–Namsos complexes. The row of basement windows in the sparagmite area east and south-east of the Trondheim synclinorium, was also only mechanically altered in Caledonian time. In the peripheral basement culminations the grade of Caledonian metamorphism is low, and the grade in the cover often increases upward, the higher nappe units exhibiting higher grade than the cover rocks closer to the basement. This is in marked constrast to the conditions in the core of the orogen. Here the grade of metamorphism usually increases downward toward the basement, but where the basement gneiss occurs as anticlinal filling in recumbent folds the grade increases upward in the inverted flank.

The regions of exposed basal gneiss are topographical culminations in the basement, culminations which have risen during Caledonian time, and which only to a lesser degree, perhaps, reflect a primary unevenness of the geosynclinal floor.

The row of peripheral basement windows from Atnasjö in the south to Raipas in the north consists of culminations rising probably from a long ridge or anticline in the basement. This row of culminations marks the boundary between the miogeosyncline to the east–south-east, and the eugeosyncline to the west and north-west (see Fig. 15.17).

The Möre culmination, Vestranden, the Namsos gneiss, the so-called Nordland granites and the granites and gneisses from Tysfjord to Ringvassöy, constitute another row of basement culminations, most likely to be interpreted as more or less remobilized proturberances rising from a continous basement anticline—the internal ridge—lying along the core of the orogen.

It is tempting to compare the external and the internal anticlinal ridges in the basement of the Caledonian syncline with the buoyant ridges that developed in many of our model experiments (see e.g. Figs 7.2, 7.6 and 11.62).

In his work on geosynclines Aubouin (1965) shows that basement ridges similar to those described above are common in orogenic belts. In analogy

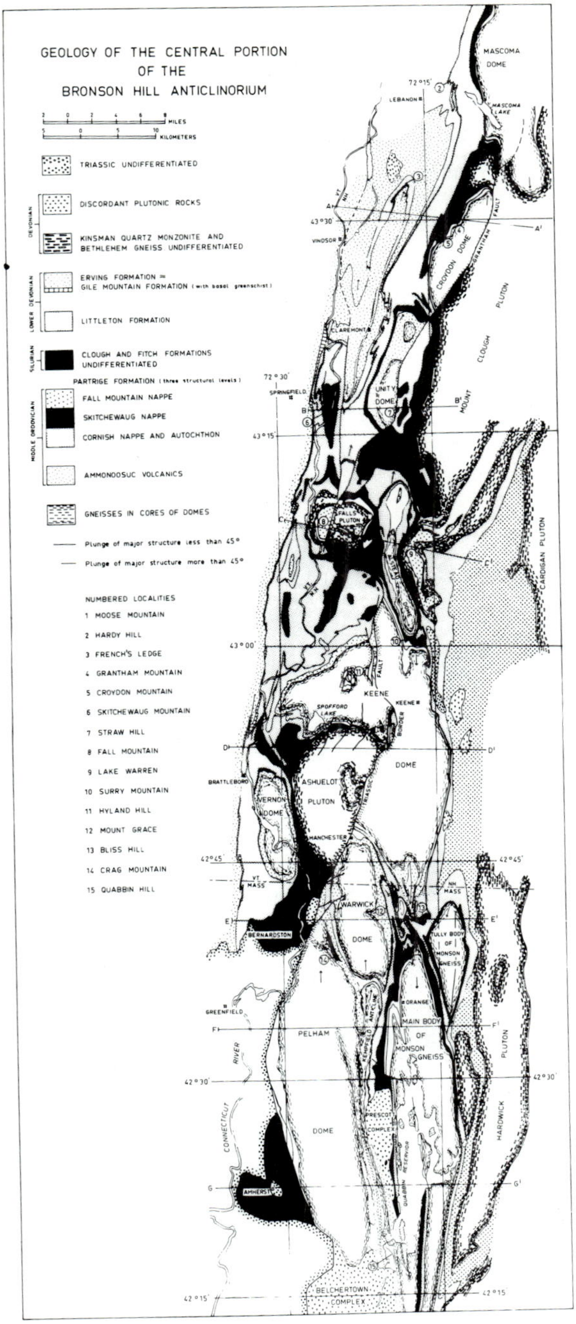

Fig. 15.18. Map of part of the Bronson Hill anticlinorium, Northern Appalachians (from Zen *et al.*, 1968).

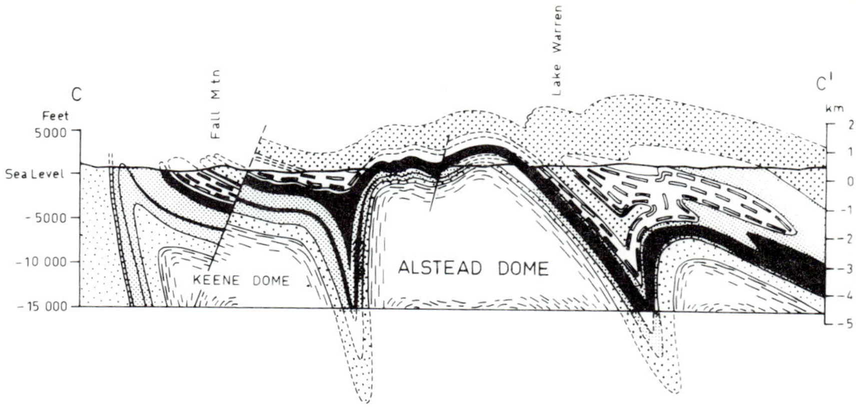

Fig. 15.19. Section along C–C′ showing basement–cover relationship across the Bronson Hill anticlinorium (from Zen *et al.*, 1968).

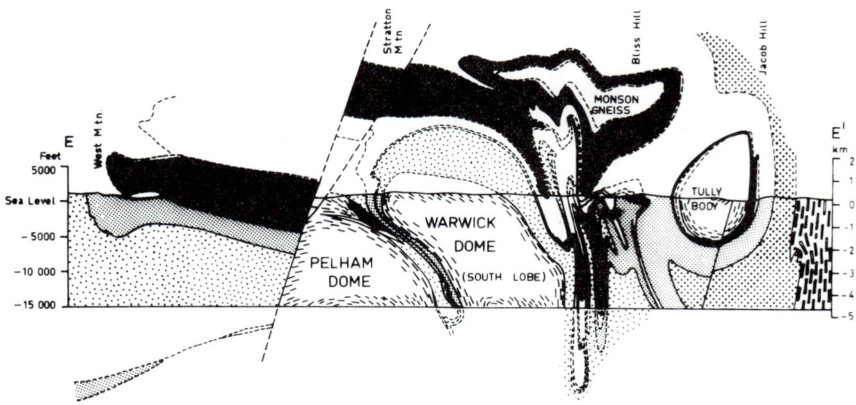

Fig. 15.20. Section along E–E′ showing basement–cover relationship across the Bronson Hill anticlinorium (from Zen *et al.*, 1968).

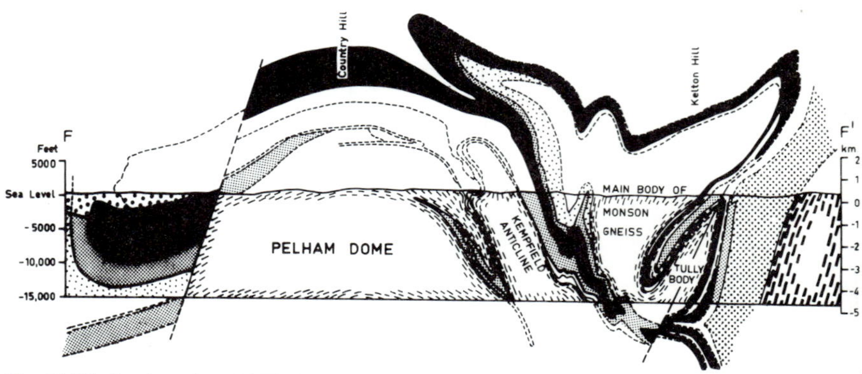

Fig. 15.21. Section along F–F′ showing basement–cover relationship across the Bronson Hill anticlinorium (from Zen *et al.*, 1968).

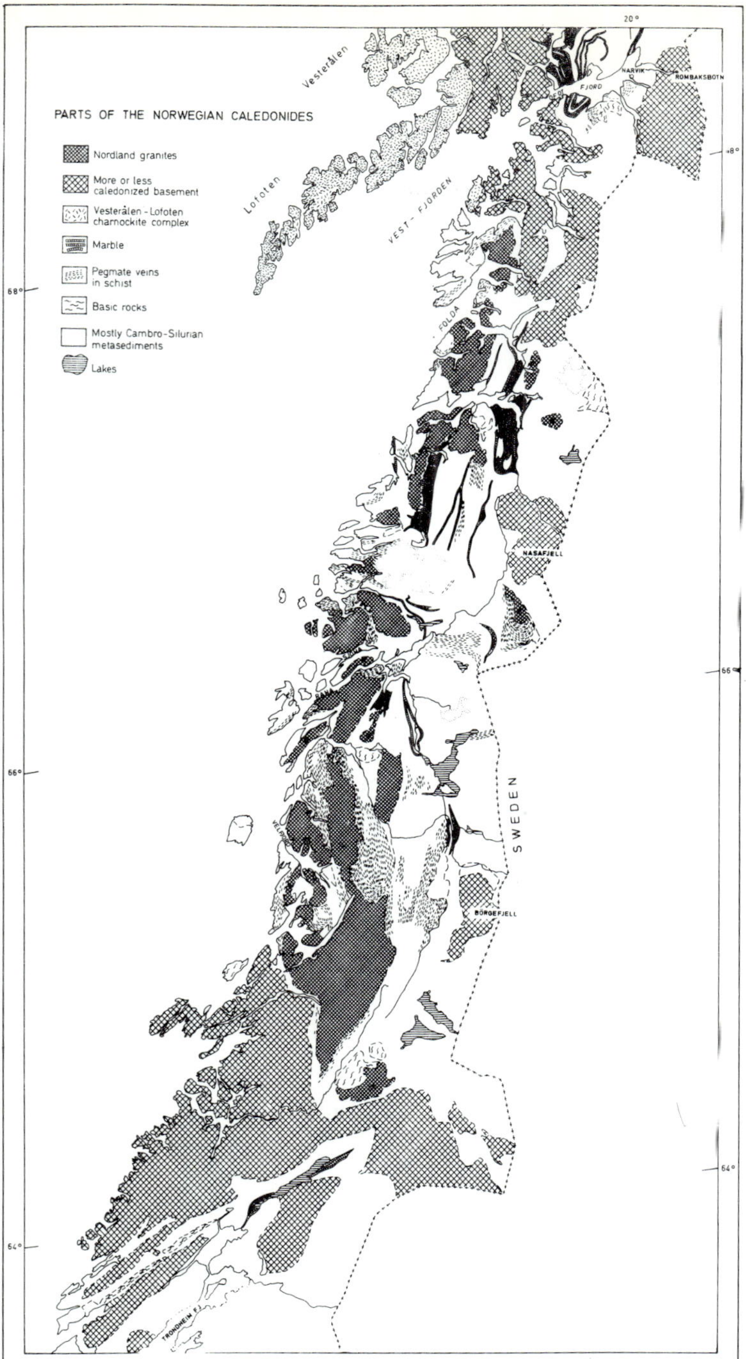

Fig. 15.22. The Norwegian Caledonides between Trondheim fjord and Lofoten-Veterålen. Based on Holtedahl and Dons (1960).

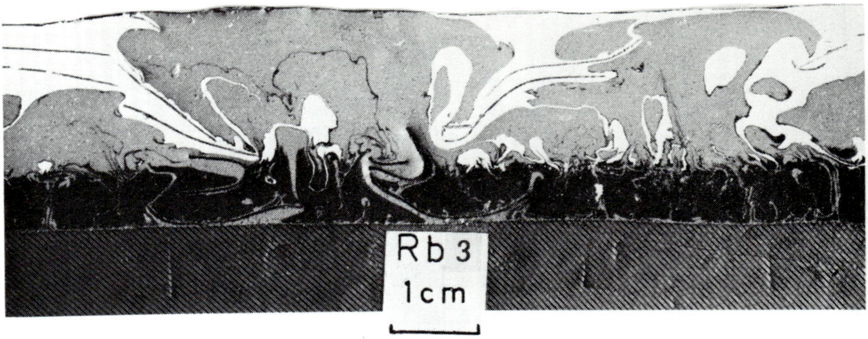

Fig. 15.23. Detail of centrifuged model shown in Fig. 11.51 which should be consulted for explanation. Compare the pattern with the natural gneiss domes in Figs. 15.20 and 15.21.

with Aubouin's model of the Hellenides in Greece, the central ridge from Möre to Ringvassöy may be referred to as the eugeanticline ridge, and the peripheral one from Atnasjö to Raipas as the miogeanticline ridge.

Similar basement uplifts are conspicuous also in the Appalachians and the Alps. In the Appalachians the row of basement culminations extending from the Blue Ridge in the south to the Green Mountains in the north may well be compared with the peripheral row in the Caledonides, and the Bronson Hill–Baltimore gneiss domes (Figs 15.18–15.21) compare excellently with the row of Nordland granites in the Caledonides (Fig. 15.22). Incidentally, model Rb-3 (Fig. 15.23) has developed an irregular structure

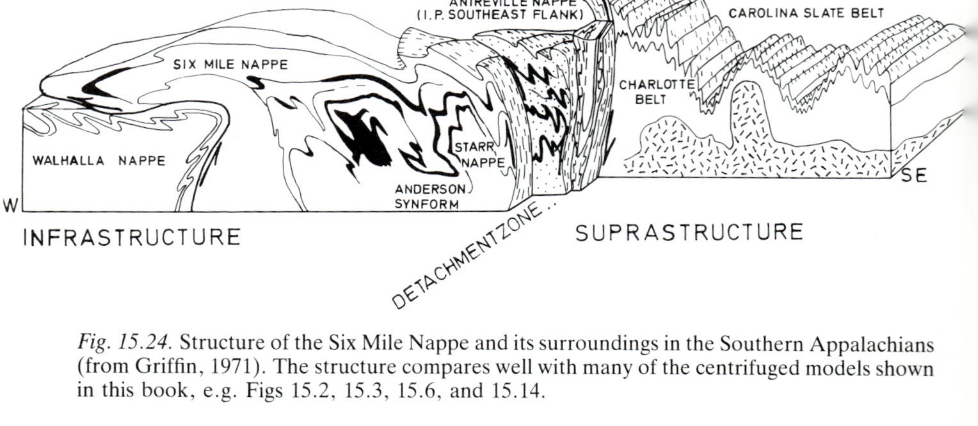

Fig. 15.24. Structure of the Six Mile Nappe and its surroundings in the Southern Appalachians (from Griffin, 1971). The structure compares well with many of the centrifuged models shown in this book, e.g. Figs 15.2, 15.3, 15.6, and 15.14.

which well can be compared with the complicated Bronson Hill domes. It is also easy to find pennine-type nappes in some models (e.g. Figs 15.2, 15.6 and 15.14) which are realistic imitations of, for example, the Six Mile nappe in the southern Appalachians (Fig 15.24).

In the Alps the external massifs of Hercynian gneisses play a similar role as the peripheral culminations in the Caledonides, and the core-gneiss in the pennine nappes may be compared with the internal row of basement domes or culminations in the Scandinavian Caledonides.

The row of Nordland granites—or the "bottom granites" as these bodies of granitoid affinity often are called—extends from the Namsos–Grong–Olden culmination toward the Lofoten-Vesterålen area where a complex of deep-seated gneisses of charnockitic affinity is met with. Many of the Nordland granites are plastic protuberances of the sialic basement into the overburden of metasediments and metalavas. Accordingly, the Nordland granites generally show a gneissosity which is parallel to the boundary and parallel to the schistosity of the adjacent supracrustals. Not uncommonly the Nordlands granites spread on top of the country rocks, constituting the core of lobes extending from the main bodies.

Whole-rock Rb-Sr age determinations show the age of some of the Nordland granite to be the same as of the craton east of the orogen (e.g. Wilson, 1973), viz. around 1700 Ma. There are also examples on Nordland granites which give Caledonian radiometric whole-rock age, e.g. 424 ± 26 Ma for the southernmost granite body, the Bindal granite (Priem *et al.*, 1975). This body shows a more massive and magmatic-looking features than most of the Nordland granites; its young age could therefore be explained by an especially strong mobilization (partly melting) in Caledonian time.

In the Lofoten–Vesterålen complex one finds a wide spectrum of Precambrian radiometric ages, varying between 900 and 2700 Ma (Griffin *et al.*, 1978). The Caledonian influence in this complex is weak and restricted to the boundary toward the cover in the east and to local zones within the complex. It seems that the Lofoten–Vesterålen complex is a part of the basement which has been elevated as a rather rigid block in Caledonian time. The internal structure and mineral facies have retained their Precambrian character, the Caledonian overprinting being limited to the boundary and local internal zones where the relative movement has been strong in Caledonian time. In this respect the Lofoten–Vesterålen complex is unlike the Möre–Vestranden–Namsos gneiss culmination where penetrative deformation and mobility have been pronounced in Caledonian time. It seems that much of the metamorphism and most of the deformation are features produced during the Caledonian orogenesis. Much of the structure is characterized by gneiss-filled anticlines, often spread out as refolded recumbent, isoclinal structures, and domes with complicated overturned lobe pattern. Referring to the Opdal area in the Möre-gneiss complex, for example, Holmsen (in Holmsen and Strand, 1960) writes:

> The large-scale structural picture of the culmination area thus exhibits a number of dome-shaped anticlines, partly overturned in diverging directions. The domes are separated by deeply downfolded, elongated and curved synclines, containing the younger rocks and the overturned and inverted parts of anticlines of the basement gneisses.

These structures of the basement culminations in the core of the orogen as described by Holmsen are remarkably similar to many of the domal structures produced in the experimental models such as recorded in e.g. Figs 15.1–15.14 among many others. The deeply down-folded synclines filled with younger supracrustals found between the gneiss domes, are well developed in our experiments; examples are found in Figs 11.13, 11.16, and, say, 15.2.

The writer has suggested that Vestranden (Fig. 15.17) has achieved its tight homoaxial folding and strong axial stretching because of its critical location in the region between two super-culminations or super-domes, viz. the Möre and the Namsos culminations, each of which is probably a composite polydome structure (Ramberg, 1966). A model simulation is shown in Fig. 11.115.

Though the folding in Vestranden could possibly be explained by a regional Caledonian compressive stress in a N.W.–S.E. direction, the strong stretching parallel to the gently waving fold axis is hard to reconcile with such a model. The difficulty of explaining the commonly encountered lineation and elongation in deep-seated crystalline schists parallel to the main trend of orogenic belts has long been recognized. It is hard to see how the entire mountain chain has been lengthen in the horizontal direction; yet such a result would follow if the local elongation is integrated along the whole belt. It has been suggested

(1) that the mineral lineation, when parallel to the fold axis (*b*-structure) does not correspond to a geometric extension of the rock, or
(2) that the horizontal lengthening is taken up by the arched structure exhibited by some orogenic belts.

Without going into the argument here, the author agrees with those workers who claim that preferred mineral lineation in one direction means local lengthening of the rock in the same direction, quite independent of the angular relation between the elongation and the flow direction (in convergent flow the elongation is likely to parallel the flow lines; in divergent flow the elongation makes an obtuse angle with the flow lines). In other words, mineral lineation is generally parallel to the long axis in the strain ellipsoid. Hence, we consider possibility (1) to be out of the question.

With reference to possibility (2), one notes that the folds in Vestranden are unusually straight, striking E.N.E.–W.S.W. over a stretch of more than 200 km from Kristiansund in the south to Namsos in the north. Bending in the horizontal plane therefore does not explain the horizontal stretching.

The author believes that the following model is realistic. It holds that the deep-seated horizontal lineation and elongation represent a true stretching caused by nearly horizontal flow from a widely distributed granitoid source—the mobile basement—toward two superculminations—the Möre complex and the Namsos–Grong complex—which are close enough for their source area to overlap.

Experimental tests of this kind of structure are discussed on pp. 315–321.

In the Caledonides, Vestranden represents one example of deep-seated stretching between domal structures. If several basal culminations line up along an orogenic chain, as they often do, it is easy to visualize elongation parallel to the main trend as a common feature in deep sections of mountain chains.

The much less elevated peripheral culminations on the miogeanticlinal ridge seem to have been lifted without much plastic flow, the upheaval having occurred along fault planes. Such structures developed in the models when the overburden or the upper part of the buoyant layer was brittle. In model S-69, for example, a low-density buoyant layer of powdered wax soaked in oil was placed above the plastic bouncing putty. The result was a block-shaped upheaval with fractures along the edge (Fig. 11.42). The lifting of rigid blocks by subjacent diapirism is also demonstrated in models (see Figs 11.70–11.79).

The interesting inclusions of eclogite in the Möre gneiss (e.g. Eskola, 1921; Bryhni *et al.*, 1977; Griffin and Råheim, 1973) are consistent with the theory that the gneiss complex rose from great depth. The present erosion surface may well represent a level which was buried below some 20–30 km of overburden rocks in the Calendonian geosyncline. Under such conditions some basic inclusions would develop the characteristic eclogite minerals. When the huge "dome" slowly rose and became eroded from the top, pressure and temperature decreased, but the sluggish chemical reaction of basic silicates was not able to change the assemblage to low-grade minerals; hence, the eclogite inclusions remained as relics while the surrounding quartzo-felspathic rocks adjusted mineralogically to the decreasing temperature and pressure.

The huge Jotun nappes south-east of the Möre basement culmination consist of plutonic rocks of charnockitic affinity (granulite facies) resting at least partly on less metamorphosed Cambro-Silurian sediments. The plutonic masses appear to have moved over the sediments in a direction from north-west. If the Jotun nappes represent a huge recumbent fold—a tremendous lobe of the Möre culmination—it is not difficult to find a parallel in our experimental models. There are, however, facts which make this simple model doubtful:

(1) Granulite-facies rocks which could be likely sources for the Jotun nappes do occur, but are not abundant in the Möre gneiss region;
(2) Much of the rocks in the Jotun nappes are quite basic and thus perhaps too heavy to constitute the buoyant core of a recumbent anticline.

South of the Jotun nappes there are gneiss–schist "klippen" which have been moved laterally over the almost undisturbed Precambrian foreland. In these "klippen" the metamorphism is inverted with schists showing low to intermediate metamorphism below, and high-grade granitoid gneiss above. Such features certainly suggest remnants of overturned folds with gneiss cores of pennine-type nappes.

Along the eastern and south-eastern edge of the huge Möre culmination of basement gneisses there is a deep syncline in the basement surface. Toward the west and north-west the Möre culmination is cut by the steep Norwegian coastline. The syncline may well be compared with the rim syncline which generally forms around domes, partly to compensate for the mass which rises, partly as a result of adjustment to the stresses related to the flow pattern in the source layer (see p. 350). A part of the rim syncline is filled with the folded and metamorphosed strata of the Trondheim synclinorium; another part is filled by the Jotun nappes. At this point it is pertinent to note the consistent south-easterly plunge of the lineation along the north-western boundary of the Jotun nappes, in the region between Vågå and Sognefjord. This is strong indication that the Jotun nappes not only were moved laterally toward the south-east, but indeed also were sagging down in a rim syncline.

The folding of the Trondheim schists may partly arise from the spreading of the huge Möre culmination, partly from sagging down in the depression on the basal floor. In the region of Nordland granites, folds strike in all directions, giving a strong impression of being controlled by the granite bodies much as are the folded strata in experimental dome models (e.g. Ny-9, Fig. 11.36).

Spreading Nappes

A major structure in the Scandinavian Caledonides is of course the famous nappes or thrust sheets which have moved across the belt from the west and north-west to the east and south-east, the maximum displacement probably being several hundred kilometres (Törnebohm, 1888, 1896; Oftedahl, 1949, 1956; Gee, 1975; and others.) In recent years it has become increasingly accepted by geologists that gravity plays a major part in the nappe movement (see e.g. Bucher, 1956; van Bemmelen, 1960; Elliott, 1976; Price, 1973; Ramberg, 1967, 1977, 1980; Ramberg and Sjöström, 1973; Huddleston, 1977). The forward movement of a nappe has been compared with the spreading of a lobe of the hat of a huge mushroom-shaped dome (Ramberg, 1966; Ramberg and Sjöström, 1973), or with the spreading of an ice-cap (Elliot, 1976; Ramberg, 1977; Williams and Zwart in Saxena and Bhattacharji, 1977). The movement pattern of an active ice-cap and of a spreading lobe of a rising dome is in principle the same, the chief difference being that the flow in the ice-cap is fed from above, from the snow precipitated on the surface, while the spreading dome is fed from below, either from buoyant materials or from materials which for other reasons have been forced upward.

The reason for the forward movement of the front of a gravity-driven nappe which moves along a horizontal or an upward-sloping base is in fact the shortening of the vertical dimension of the nappe. This vertical shortening means a lowering of the centre of gravity and hence a decrement of the gravity potential. The shortening in the vertical direction is generally associated with a lengthening in the horizontal plane, the result being a forward movement of the front of the nappe relative to sections farther back.

If the material of the nappe was strongly compressible, the tendency of horizontal movement would be negligible. Instead of strong compressibility one could think of the effect of pressure solution: if the vertical shortening was completely accounted for by solution along horizontal seams or stylolites, then the gravity-potential decrement would balance the energy associated with the dissolution and the transport of the dissolved material away from the sites of dissolution. There would be no energy left to move the whole nappe horizontally. There are, however, no grounds for assuming that pressure solution occurs so widely in nappes that a major part of the vertical compression is not ordinarily compensated for by a horizontal extension. Therefore, the rocks in the dynamic model discussed in Chapter 9 are regarded as incompressible viscous or plastic bodies. Any shortening in one direction is automatically compensated for by a volume-preserving lengthening in directions normal to the shortening, and the decrement of the gravity potential associated with the vertical shortening is completely balanced by the strain energy of the collapsing nappe and the frictional energy at the sole. There is accordingly a noteworthy differences between a gravity-driven nappe which moves horizontally or even uphill along a rigid support and a rock complex which slides downhill because of the force of gravity. In both cases the motion is caused by—or perhaps more correctly associated with—a decrement of the gravity potential, i.e. a lowering of the centre of gravity. The downhill sliding takes place even if the rocks are completely rigid. It is important to realize, however, that the horizontal or uphill movement of the nappe is only possible if the rocks are weak enough to sag and spread under their own weight.

The Scandinavian Caledonian belt is quite straight as it extends from southwest Norway to the northernmost coast of Finmark, and the nappes are shaped as elongate sheets whose length is much larger than their width. During the collapse, therefore, the spreading in the south-west–north-east direction will be negligible. As a good first approximation we may accordingly regard the movement and strain in the nappe as two-dimensional in the plane normal to the belt, the horizontal lengthening being maximum in a direction more or less normal to the trend of the belt.

It is well established by numerous field studies that the strain in several of the nappes show a pronounced extension across the trend of the belt in the west-north-west–east-south-east direction (Kvale, 1953; Strand, 1945; Oftedahl, 1948; Zachrisson, 1964; Ramberg, 1966, 1977 and numerous unpublished field observations; Ghosh *et al.*, 1979). This is demonstrated by mineral lineation, by boudinage, and by the rotation of small-scale fold axis.

It is true that the structural evidence for a stretching in the horizontal direction cannot unconditionally be taken as proof of vertical shortening, because quasi-horizontal extension will also be produced by a large simple-shear angle generated by a large horizontal displacement of the upper sheets relative to the lower sheets. There needs be no vertical shortening, yet the quasi-horizontal lengthening may be considerable.

It is difficult to distinguish between a stretching caused by extensive simple shear alone and a stretching caused by a combination of pure shear with vertical shortening and simple shear parallel to the sole.

The reason for believing that there has been a considerable amount of vertical shortening in the Scandinavian Caledonides is that the lineation together with other evidence of a horizontal lengthening is a penetrative feature observed throughout some of the nappes, rather that being limited to the narrow zones of high shear strain between the various thrust sheets. This is particularly evident in the large Seve nappe. It is difficult to see how the major part of the column of nappes can be exposed to a penetrative process of simple shear unless combined with a simultaneous vertical shortening. A purely simple shearing requires that unknown masses above the column of sheared nappes have been pushed forward in order to transmit shear stress to the nappes below. Of course, this is not impossible, but is seems more acceptable that the horizontal lengthening is caused by a combination of pure and simple shear, involving both vertical shortening and shear along the thrust direction in a quasi-horizontal plane. In this model the primary process is the slow plastico-viscous gravity collapse of the system of nappes.

The dynamics of spreading nappes is discussed in Chapter 9.

16 Submarine Ridges and Continental Rift Systems

In some of the earlier centrifuged models run in 1960 structures had already developed which were believed to simulate oceanic ridges as well as continental rift-valley systems. In these models the top part of rising diapirs spread underneath a crust which has a lower density than the buoyant mass. In models of this nature only a small volume of the rising material finds its way to the surface through fractures in the model crust; by far the larger volume spreads to a progressively thinner pillow underneath the crust. Some models of this type are shown in Figs 16.1–16.5.

With reference to the theory of oceanic-ridge evolution and the evolution of continental rift valleys, the crucial feature of the experimental models is the stretching of the crustal layer due to tensile stress created by friction coupling to the spreading hat of the diapir below. Ramberg (1963a) p. 72 states that:

> As a consequence of the model studies described above, the author suggests an alternative hypothesis [alternative to the Expanding Earth Model and the Mantle Convection Model], viz. that the submarine rises are bulges caused by buoyancy-injected bodies which occur relatively close to the surface of the ocean bottom and are now slowly spreading laterally, thereby producing lateral tension and rift-valley morphology along the crest of the ridges. These buoyancy-injected bodies cannot be of chiefly granitoid composition, because the low density of such rocks or magmas would make the ridges rise above sea level, or force the magmas to pierce the surface and emerge above sea level in the form of volcanic islands. On the other side, the material must be less dense than the average mantle underneath the oceans. It seems reasonable to propose that the ascending material is of gabbroic composition, and rises through the more dense material of the mantle in much the same manner that, at higher levels in the earth, granitic and granodioritic batholiths ascend through an averagely more dense crust. It is possible that the rising masses have their source in the low-velocity subcrustal layer . . .
>
> All morphological features of the submarine rises, the known heat-flow distributions across such ridges, and the measured gravity anomalies and seismic activities discussed by Menard in connection with the convection hypothesis, may equally well be employed in support of the hypothesis proposed by the writer. It seems that more field observations, better theories and more mean-

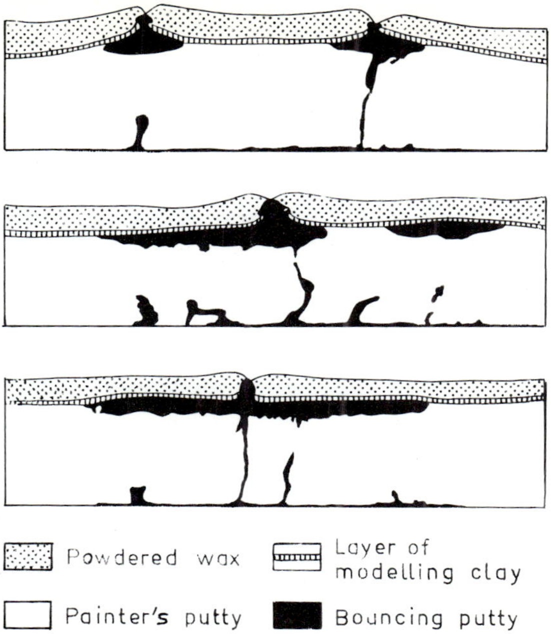

Fig. 16.1 Model showing spreading of buoyant silicone-putty diapir ($\rho = 1{\cdot}12\ \mathrm{g\ cm^{-3}}$) underneath surface layer with less density ($\rho < 1{\cdot}0\ \mathrm{g\ cm^{-3}}$) than the diapir (from Ramberg, 1963a).

Fig. 16.2. Surface of initial stage of model CD-2 (Fig. 16.3).

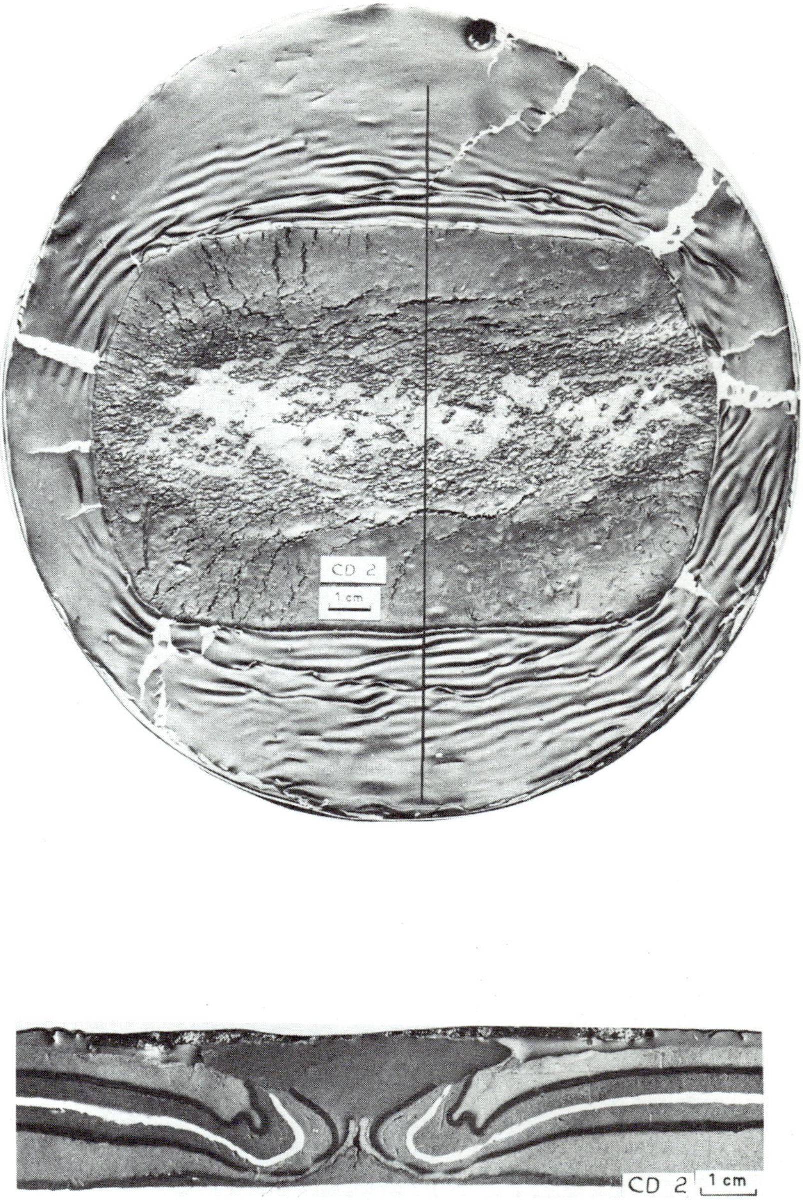

Fig. 16.3. Top view and profile of model CD-2 showing mushroom-shaped diapir which has thinned the overlying crust and torn it to boudins. Run 1 min 15 sec at 1300 *g*, 1 min 47 sec at 2000 *g*, 1 min 3 sec at 3000 *g* and 23 sec at 1500 *g*.

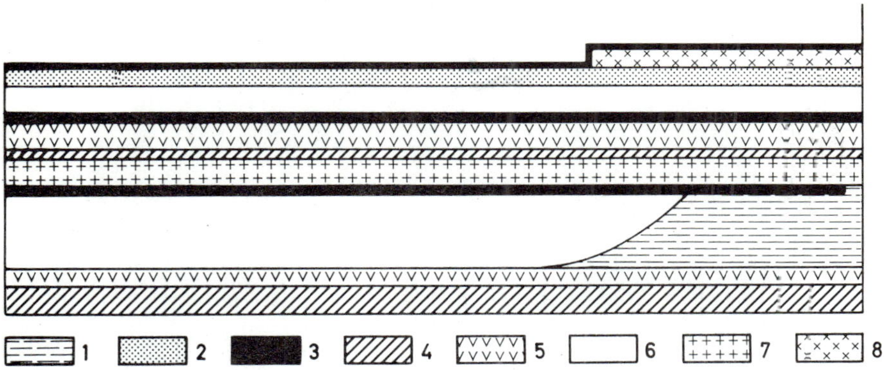

Fig. 16.4. Left-half section of initial structure of model CD-2 (Fig. 16.3). (1) Silicone putty with magnetite powder, $\rho = 1{\cdot}55\ \mathrm{g\ cm^{-3}}$; (2) silicone putty, $\rho = 1{\cdot}28\ \mathrm{g\ cm^{-3}}$; (3) dark grey modelling clay, $\rho = 1{\cdot}68\ \mathrm{g\ cm^{-3}}$; (4) white modelling clay, $\rho = 1{\cdot}78\ \mathrm{g\ cm^{-3}}$; (5) green painter's putty, $\rho = 1{\cdot}85\ \mathrm{g\ cm^{-3}}$; (6) painter's putty, $\rho = 1{\cdot}85\ \mathrm{g\ cm^{-3}}$; (7) yellow painter's putty, $\rho = 1{\cdot}85\ \mathrm{g\ cm^{-3}}$; (8) powdered wax mixed with oil. Length of section 9·7 cm.

ingful model experiments are needed to discriminate among existing hypotheses, or to suggest alternative explanations.

The model by Menard referred to above is from Menard (1962), and since the above quotation was written Menard reconsidered his model (1965).

It was also suggested in Ramberg (1963a) that continental rift-valley systems have a similar origin:

> Continental rift-valley systems may well have an evolutionary history similar to that described above for submarine ridges. Subsurface injection of plastic but crystalline rocks, of a crystal-melt mush, or of chiefly molten materials will first elevate the surface in gentle swells. As spreading takes subsequently place, collapse and stretching of the central part of the gentle swells and levelling of the domed surface give rise to a geomorphology characteristic of rift-valley systems. Cloos (1939) has demonstrated how rift-valley systems generally occur within domed shields of the Earth's crust. Whereas Cloos assumed that the bending of a rigid crust is sufficient to generate tension fractures and rift valleys [*Die Bildung von Gräben und Vulkanspalten ist eine Folge konvexer Durchbiegung der Oberkruste* (p. 481)], the author considers spreading of buoyancy-injected subcrustal bodies to be an essential feature of the mechanism that creates tension and rift-valley collapse of the superincumbent crust.

It speaks well for the applicability of the centrifuge method in tectonic research that later geophysical studies of both continental rift systems and oceanic ridges have verified the existence of a more or less pillow-shaped

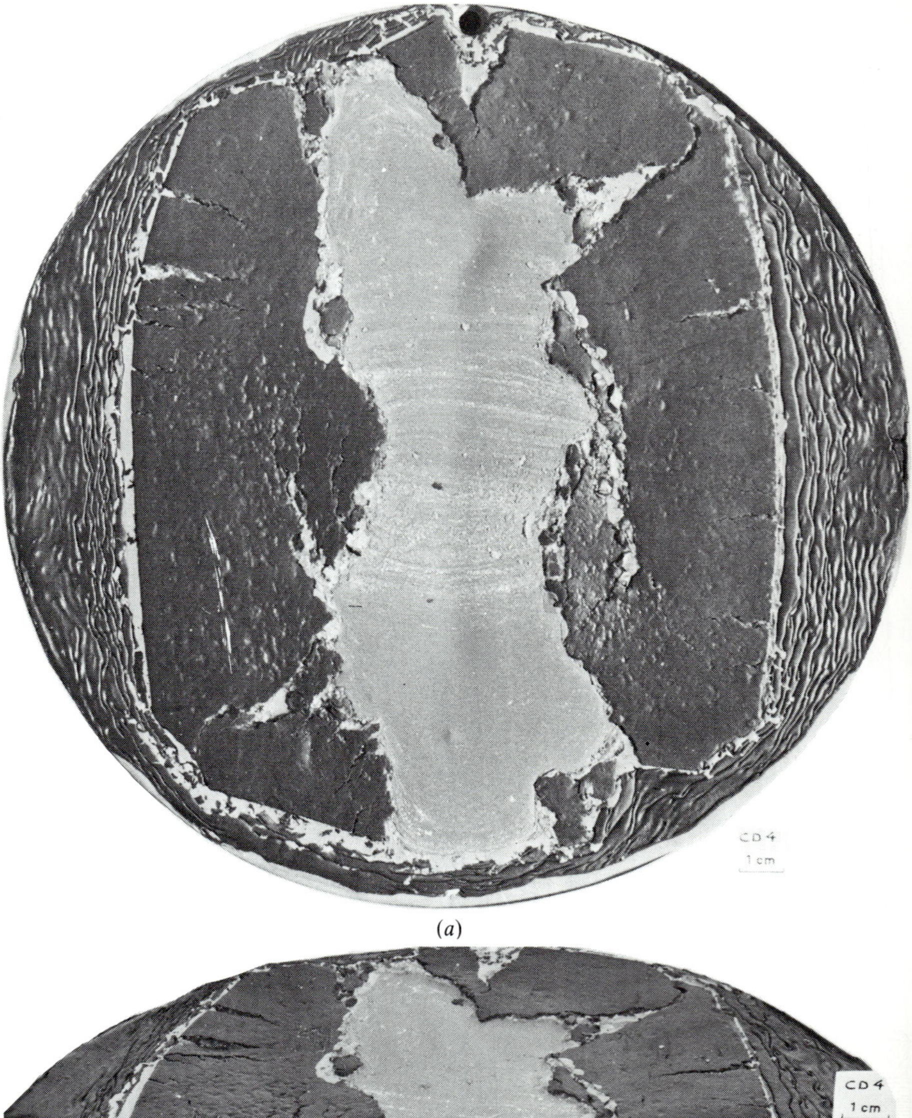

(*a*)

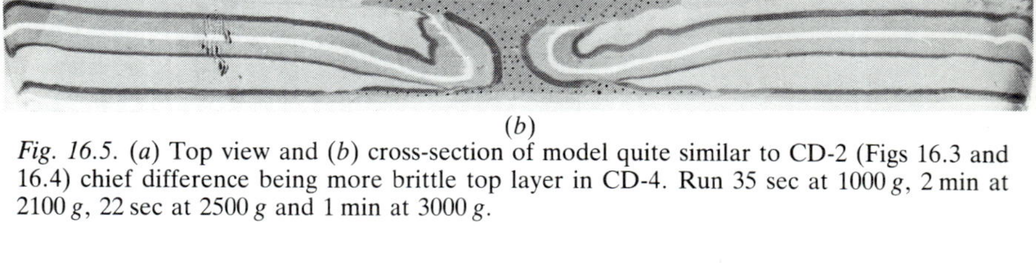

(*b*)

Fig. 16.5. (*a*) Top view and (*b*) cross-section of model quite similar to CD-2 (Figs 16.3 and 16.4) chief difference being more brittle top layer in CD-4. Run 35 sec at 1000 *g*, 2 min at 2100 *g*, 22 sec at 2500 *g* and 1 min at 3000 *g*.

body of abnormally ductile material below the Mid-Atlantic ridge (Talwani *et al.*, 1965), below the Rhine graben (Meissner and Berckhemer, 1967; Mueller *et al.*, 1967; Illies, 1969; Illies and Mueller, 1970), below the Oslo graben (I. B. Ramberg and Neumann, 1978), underneath the Afar rift system (Pilger and Rösler, 1975) and below the Red Sea (Illies and Mueller, 1970). Also the Baikal rift zone (Logatchev *et al.*, in I. B. Ramberg and Neumann, 1978) and the Rio Grande rift (Bridwell, in I. B. Ramberg and Neumann, 1978; Riecker, 1979) are underlaid by abnormal cushions. The same seems to be true for much of the rift system of East Africa according to Baker *et al.* (1972).

The pillow of abnormal mantle material underneath the Mid-Atlantic ridge is less dense than the normal mantle at the same depth according to gravity-anomaly measurement as summarized in Talwani *et al.* (1965) whereas the cushion below the Oslo graben seems to be a little more dense than the surrounding crust according to contributions in I. B. Ramberg and Neumann (1978). In all rift systems studied, however, it seems that the mantle and/or the lower crust underneath the rift are more ductile—show lower seismic velocity—than the adjacent rocks. Whether or not the density is less than normal is not always ascertained.

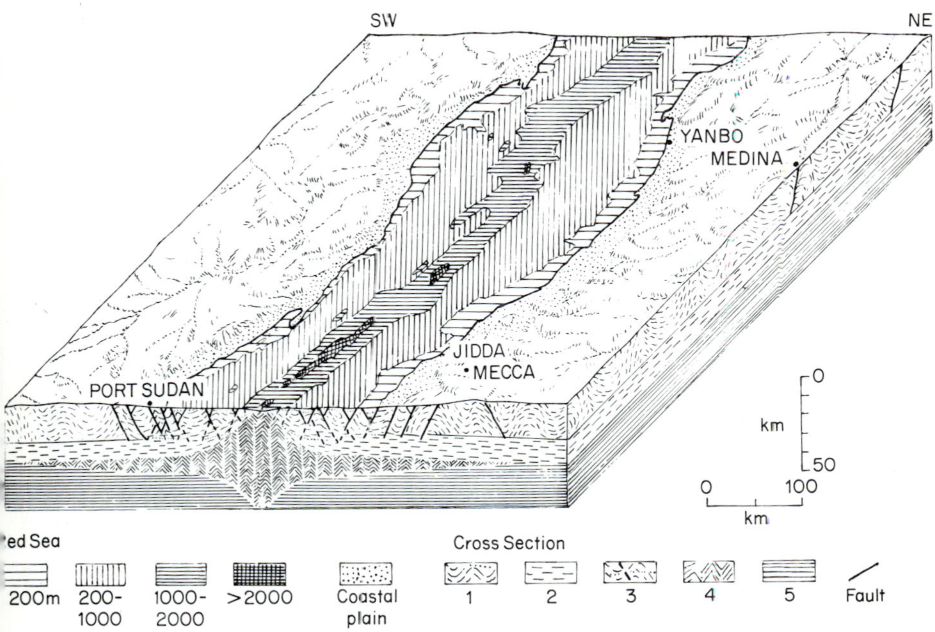

Fig. 16.6. Block diagram of the Red Sea (from Illies, 1969). (1) Basement rock; (2) lower crust; (3) metamorphosed crust; (4) anomalous mantle; (5) mantle.

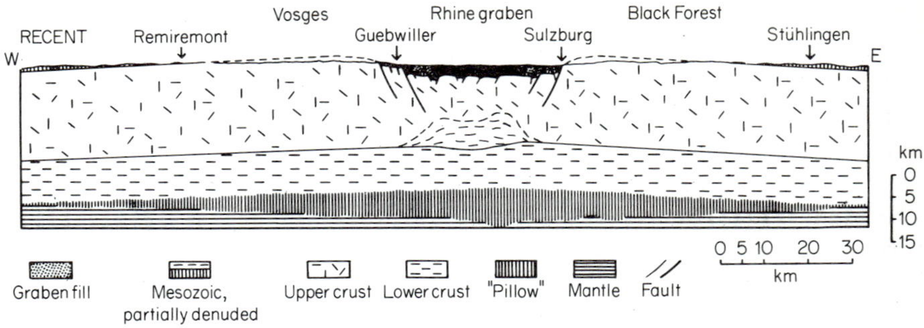

Fig. 16.7. Section through the Rhine graben (from Illies and Mueller, 1970).

For comparison with the experimental models illustrations of the Mid-Atlantic ridge, the Rhine graben and the Red Sea are reproduced in Figs 16.6–16.8.

The Rhine graben has been particularly well studied both from a geophysical and a geological point of view in recent years. It is therefore significant that here the model of a spreading subjacent pillow has been accepted as the driving motor for the rifting of the crust and the differential vertical movement of the crustal fragments. Illies (in Illies and Mueller, 1970, p. 12) is in agreement with the writer's experiment-supported view of 1963: "The separation of the graben's lateral abutments can be explained by frictional bounding following the lateral spreading of the subcrustal pillow."

Since the first centrifuged models were run in 1960 a few additional experimental models supposed to imitate both oceanic ridges and continental rift valleys have been studied. Some models have also been calculated numerically (pp. 100 and 105).

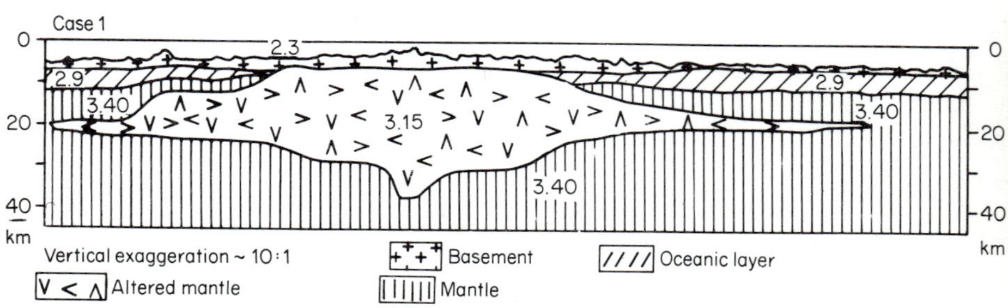

Fig. 16.8. Section across the mid-Atlantic ridge (Talwani *et al.*, 1965).

Figure 16.4 is a cross-section of the initial structure of models CD-2 and CD-4 which were closely similar except for the mechanical property of the top layer which was more brittle in model CD-4. After runs in the centrifuge at an acceleration up to 3000 *g* the structure had developed as shown in Figs 16.3 and 16.5.

In model CD-2 the top plane—the Earth-crust imitation—consisted of compressed oil-wetted collophony powder which partly developed pinch-and-swell structure, partly broke along numerous fractures in the tensile stress field caused by the spreading subjacent mushroom-shaped diapir.

The Earth-crust imitation in model CD-4 was made of compressed oil-wetted paraffin wax which was a little more brittle than the material used in CD-2. Accordingly, the behaviour under tension was different. Instead of pinching and swelling and the development of numerous small fractures, the crustal plate in CD-4 broke along one long zigzag fracture extending from the one end of the model to the other.

Undoubtedly these models show many features in common with the structure of the Mid-Atlantic ridge, the Rhine graben and the Red Sea. It does not stretch the imagination unreasonably to realize how small changes of mechanical properties in the material of the model crust could give near perfect models of both sea-floor spreading and continental rift valleys. Such structures would form if the model material used to simulate the earth's crust would fracture along sharp and reasonably straight planes rather than develop pinch-and-swell structures. Unfortunately, none of the model materials available tends to form clean and planar joints in tension.

Model CD-7 (Figs 16.9 and 16.10) imitates the opening up of two oceans underneath a continental crust which has broken and separated by the spreading of two parallel diapir ridges rising from the low-velocity layer. Note the compression structure in the collision zone between the two continental flakes in the centre of the model.

There is one possibly serious difficulty with the idea that the Mid-Atlantic ridge is produced by a diapiric anticline or a row of individual diapirs rising from the low-velocity layer. According to characteristic wavelength calculation based on reasonable values of densities, effective viscosities and thicknesses of the layers in the crust and upper part of the mantle, the spacing between the anticlines or rows of diapirs should be less than the half-width of the Atlantic ocean (see p. 100). Consequently, there should be at least one additional ridge on either side of the Mid-Atlantic ridge. As discussed in connection with the instability of layered systems in the field of gravity, however, (p. 85) the characteristic wavelength does not always determine the spacing between diapirs rising from a buoyant stratum. The theoretical characteristic wavelength does not control the spacing if the system initially shows sufficiently large geometrical and/or rheological

(*a*)

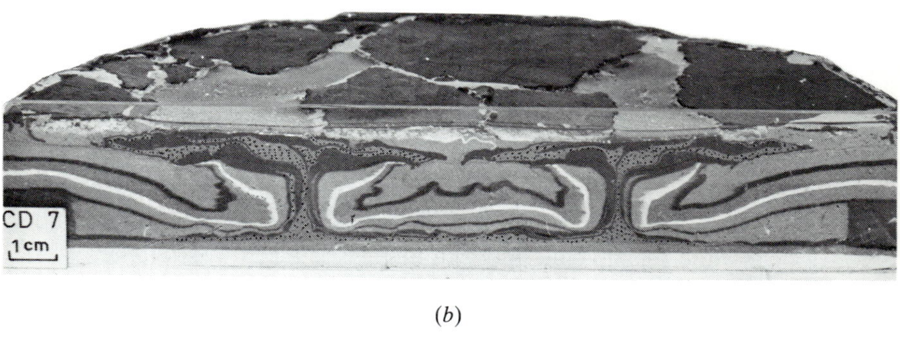

(*b*)

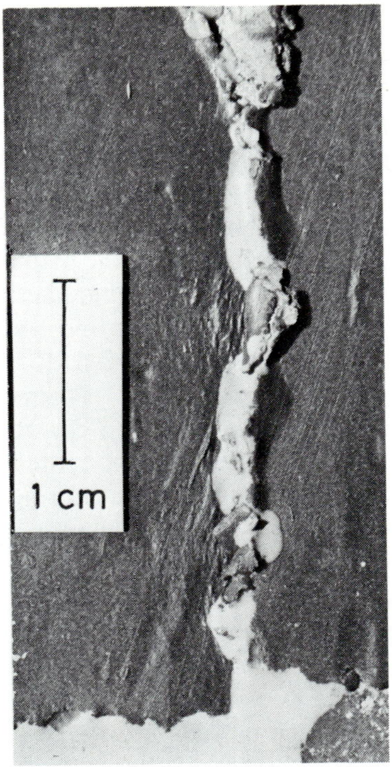

(*c*)

Fig. 16.9. (*a*) Top view and (*b*) cross-section of model whose initial structure is shown in Fig. 16.10. Note the relative powerful collision of the two central fragments (*c*) of the initially unfractured top plate of compressed wax powder "wetted" with oil.

Fig. 16.10. Left half cross-section of initial structures of model CD-7, Fig. 16.9. (1) Modelling clay, $\rho = 1{\cdot}7$ g cm^{-3}; (2) painter's putty, $\rho = 1{\cdot}85$ g cm^{-3}; (3) black modelling clay, $\rho = 1{\cdot}68$ g cm^{-3}; (4) and (6) silicone putty, $\rho = 1{\cdot}28$ g cm^{-3} in lowest layer (4), $\rho = 1{\cdot}62$ g cm^{-3} in highest of the two layers (6); (5) silicone putty with magnetite powder, $\rho = 1{\cdot}47$ g cm^{-3}; (7) compressed wax powder soaked in oil; top cover: 0·6 mm thick sheet of modelling clay. Length of section 9·7 cm.

irregularities such as e.g. thickness variation or density variation along the layering. A change in viscosity because of horizontal temperature variations may also invalidate the characteristic wavelength as a controlling property.

Another possible explanation for the unexpected long distance between the diapirs or ridges protruding from the low-velocity layer is that their spacing is perhaps not controlled by the layering of the crust and the uppermost part of the mantle, but by conditions deeper in the mantle. Diapirs or plumes rising from great depth in the mantle can develop greater spacing than what is determined by the characteristic wavelength of a 100-km-thick low-velocity layer in the uppermost part of the mantle. In the section on the theory of layered systems, p. 105, it was demonstrated numerically that a buoyant layer situated at 1000 km depth and being only a few tens of kilometres thick—would show a characteristic wavelength of several thousand kilometres depending upon the viscosity and density relationships. The theoretically predicted surprisingly long characteristic wavelength of deep-seated strata in a model mantle has been substantiated experimentally as shown in Figs 16.11 and 11.93.

Prior to centrifugation the model shown in Fig. 16.11 consisted of a 0·2-cm-thick buoyant layer of silicone, overlaid by a 3-cm-thick putty overburden and resting on a 3-cm-thick substratum of the same material as that in the overburden. The relevant data are given in the figure legend.

After a run in the centrifuge at 1000 *g* for a few minutes the buoyant layer had risen in the form of a funnel-shaped body, pierced the surface and spread horizontally. (Since in this experiment we were not interested in the velocity of the process the running time was not recorded.)

That the wavelength/thickness ratio is exceptionally large is evident when this model is compared with model S-205 (Fig. 11.1). In the latter model the overburden is not thicker than the buoyant layer itself, and the latter rested on a rigid basement. The photograph demonstrates that the wavelength on an average is only about 50% longer than the thickness of the buoyant layer.

In the model with the deep-seated buoyant layer it is crucial that only *one* dome developed (except the usual rises at the ends of the layer, due to the edge effect not shown in the photograph) on the 10-cm-long layer. It follows then that the wavelength/thickness ratio is at least $5:0{\cdot}2 = 25$. This is much longer than we get in models with thinner overburden relative to the buoyant stratum at moderate viscosity ratios.

A second test (Fig. 11.93) of a similar model gave principally the same results: a thin buoyant layer sandwiched between a thick overburden and a thick substratum of identical material, does show an exceptionally long distance between the diapiric rises. Apparently, the interdiapiric distance is longer than the width of the model.

Another significant feature of the two models is the effective sucking-up of the basal layer. In both models the basal layer consists of the same material as the overburden. Accordingly, the basal layer is not buoyant, yet it is pulled to the surface in the core of the diapir. This is in excellent agreement with the theoretical information presented in Table 7.11. Here it is shown that the amplitude of the deflection is practically the same on

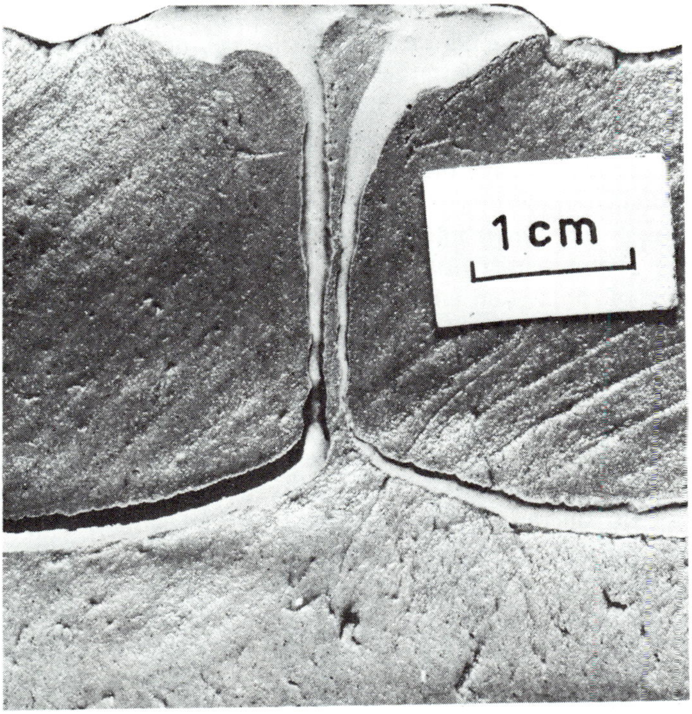

Fig. 16.11. Section through diapiric anticline of thin silicone-putty layer, $\rho = 1.14$ g cm^{-3}, having ascended through an overburden of painter's putty, $\rho = 1{\cdot}85$ g cm^{-3} Note the sucking-up of the heavy substratum ($\rho = 1{\cdot}85$ g cm^{-3}) in the core of the diapir. Model run at 1000 *g*. See also Fig. 11.93.

the top ($\bar{y}_2$) and on the bottom ($\bar{y}_3$) of the buoyant layer (layer 3 in the model), whereas the amplitude on the free surface ($\bar{y}_1$) is much smaller. As we have repeatedly remarked the theory is only valid when the amplitude is small compared with the wavelength. The more interesting it is that the experimental difference between the two amplitudes, $\bar{y}_2$ and $\bar{y}_3$, on the boundaries of the buoyant layer are small even when the diapir has reached the surface. On the other hand, the amplitude of the surface deflection, $\bar{y}_1$, as produced by the rising diapir is much less than $\bar{y}_2$ and $\bar{y}_3$ both in theory and experiment.

The reason why a thin buoyant layer such as that present in the two experimental models is capable of lifting a relatively large volume of the basal material up to the free surface is that the density of the basal material is not larger than the density of the overburden. Hence there is no net increment of the gravity potential associated with the elevation of a volume of the basement into the overburden, provided that the movement at depth does not augment the amplitude of the surface perturbations. Under these conditions the volume of basal material introduced into the region of the overburden equals exactly the volume of overburden material sunk into the region of the basement. Since the density of overburden and basement is the same, the exchange of place of the two materials does not change the potential energy.

In this connection it is informative to note how even quite moderate increments of the density of the substratum affect both the characteristic wavelength of the system and the difference between the amplitudes $\bar{y}_2$ and $\bar{y}_3$ (Table 7.11 and Fig. 7.23). The characteristic wavelength decreases rapidly when the density of the basement increases relative to the density of the overburden, the same density change also causing the amplitude $\bar{y}_3$ to decrease relative to amplitude $\bar{y}_2$. The latter condition means that the heavier the basement the less it will be elevated in the core of the diapirs.

The reason for the shortening of the characteristic wavelength as a consequence of increased density of the basement is that, for any given amplitude of the deflections of the buoyant layer and given thickness of the layer, the induced amplitude at the basal boundary will be smaller the shorter the characteristic wavelength of the buoyant layer. A smaller amplitude of the waves on the basal boundary means less increment of the potential energy.

It is of consequence for geodynamics that buoyant layers situated deep in the Earth's mantle may develop exceptionally long characteristic wavelengths relative to their thickness. When the wavelength reaches several thousand kilometres and hence the ascending diapirs are correspondingly far apart, the dynamics of such deeply buried strata may well constitute a motor for continental drift and sea-floor spreading.

There are several reasonably realistic processes in the solid mantle, which could generate low-density layers at great depths. Beside partial melting, an example would be the change of Mg_2SiO_4 from spinel structure, which is stable at high pressure, to olivine structure, which can occur at the same pressure as the spinal structure but requires higher temperature. The transition spinel→olivine is associated with some 10% decrease of density and an unknown increment in heat content and entropy. That is, the transition requires heat—it is endothermic.

We have little detailed knowledge of the mineralogical and chemical composition of the Earth's mantle, but it seems reasonable that ferro-magnesium silicates constitute a major part of the mantle. It is generally believed that the upper part of the mantle—perhaps down to some 400 km depth—essentially consists of $(Mg,Fe)_2SiO_4$ with olivine structure. Across

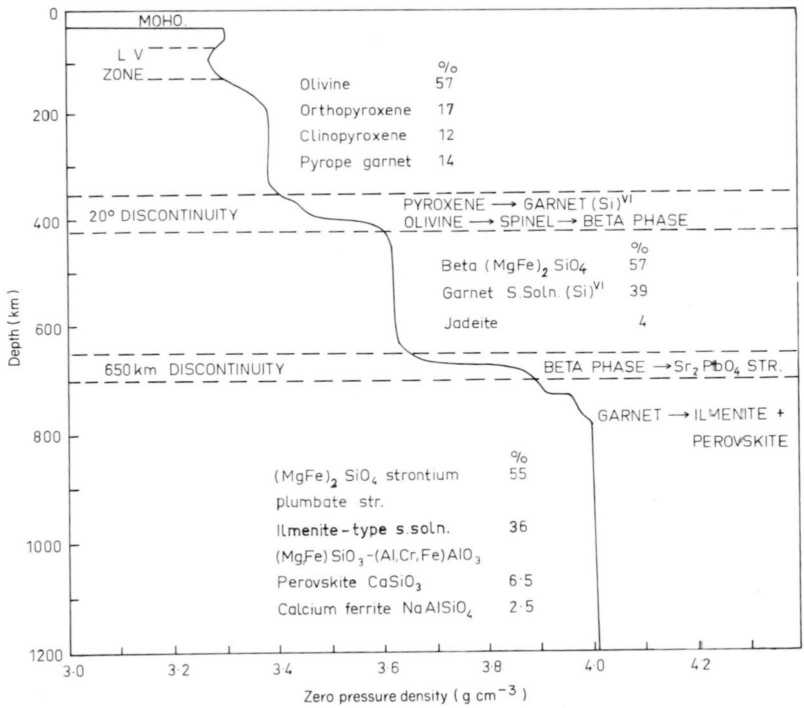

Fig. 16.12. Assumed mineralogical composition of the upper 1200 km of the Earth (from Ringwood in Robertson, 1972).

a rather thin transition zone at about 400 km depth the olivine structure changes to the spinel structure (or possibly a β-Mg_2SiO_4 phase) which continues down to a depth of about 700 km. Here the spinel or β-structure of $(Mg,Fe)_2SiO_4$ changes to a still denser phase, viz. the strontium plumbate structure of the ferromagnesium orthosilicate. This as yet uncertain structure of the Earth's mantle (Fig. 16.12) is simplified after Ringwood (in Robertson, 1972).

In all likelihood the transition strotium plumbate structure→spinel structure is endothermic since the transition is associated with a volume expansion.

Consider now the conditions at say 1000 km down in the mantle. The pressure is about 400 kilobars, but the temperature is not known except that it is below the melting "point" of the material at the pressure in question. For the sake of argument assume that the normal temperature in the Earth at 1000 km is a little below the equilibrium temperature for the transition: strontium plumbate structure ⇌ spinel structure at 400 kilobars. A small local rise in temperature at this level accordingly tends to stabilize the spinel structure. The heat needed for the temperature rise can be created by radioactive decay of the small but not negligible amount of radioactive substances which are present in the surrounding mantle. (The relationship between the geothermal gradient, the phase diagram of Mg_2SiO_4 and the local heating is assumed to be such that the newly formed layer of spinel is separated from the 700 km normal transition level for the reaction strontium plumbate ⇌ spinel.) While the transition to the less dense spinel structure goes on, the released radiogenic heat does not continue to heat the surroundings but instead is spent in the endothermic transition from strontium plumbate to spinel structure. Produced by radioactive decay in the surroundings, heat diffuses into the zone where the spinel structure grows. Consequently, if the mantle is uniform at the 1000 km level, a shell of spinel structure will slowly grow in thickness, extending all around the world and being embedded between two layers of strontium plumbate structure.

Being less dense than the strontium plumbate structure of the immediately overlying mantle, the spinel layer is gravitationally unstable and tends to rise. Experiments and theory demonstrate how the velocity of rise depends upon the rheological properties, the density contrast and not least upon the thickness of the layer. Some relevant data are presented in Table 7.12. It is significant that the velocity of rise of anticlinal perturbations on a buoyant layer is proportional to the square of the thickness when geometrically similar models of same materials are compared. Hence, not until the growing stratum of spinel structure has reached a certain minimum thickness will the tendency to rise be strong enough to overcome the resistance of the crystalline overburden. Diapirs of spinel structure will then protrude from the anticlines on the spinel layer and rise through the strontium plumbate structure up to the 700 km level. Above that level the spinel structure is stable even at the "normal" geothermal conditions.

Coming from greater depth the spinel diapirs are hotter—in spite of some adiabatic cooling—than what corresponds to the geothermal gradient at the 700 km level. They consequently rise across the transition boundary and enter the spinel layer. Here the density contrast is much reduced inasmuch as it is now only determined by the thermal expansion of the spinel structure and the temperature contrast between the diapirs and their surroundings.

As the cooling of the diapirs is chiefly due to the adiabatic expansion—at least in the core of the diapirs—their interior remains appreciably hotter than the surroundings whose temperature is determined by the geothermal gradient. Sooner or later, therefore, the temperature-pressure conditions in the core of the rising diapirs must cross the transition curve: spinel$\rightleftharpoons$olivine. This happens while the diapirs still are within the spinel layer of the mantle, prior to reaching the lower boundary of the olivine shell at the 400 km level.

The transition from spinel to olivine is endothermic with an enthalpy change per gram probably considerably larger than the heat capacity of spinel. There is no heat source available for the endothermic transition other than the heat content of the diapir itself (the radiogenic heat production is too slow to be of significance at this stage). The spinel diapirs must therefore surpass the spinel$\rightleftharpoons$olivine transition curve by a vertical distance which corresponds to Δt degree "super-heating" before the whole diapir has changed to olivine. (The "super-heating" is of course not due to a rise in temperature but to a decrement of pressure.) If ΔH is the specific heat of transition and C_p the specific heat capacity, then the amount of "super-heating" needed to change 100% of a diapir from spinel to olivine is $\Delta t = \Delta H/C_p$. At less degree of "super-heating" (or rather "underpressurizing"), correspondingly smaller portions of the diapir undergoes the change from spinel to olivine structure.

On reaching the 400 km level the diapir consists chiefly of olivine, the interior part of the body being hotter than the normal temperature at the 400 km level. Analogous to what happens at the 700 km level, the hot diapir crosses the transition zone and ascends into the olivine layer, not because its mineral composition or structure is different from the surroundings, but purely on account of the temperature contrast between materials of identical composition and structure. The rise is now in the form of true thermal diapirs—perhaps thermal plumes is now a more appropriate name. Since there is neither compositional nor structural contrast between the rising body and its surroundings the process has become more like an ordinary thermal convection in a homogeneous medium.

The next discontinuity which the rising masses encounter is the bottom of the low-velocity layer. In all likelihood this boundary represents a change in chemical composition and an appreciable drop in density. Hence the diapirs are probably unable to cross the boundary and do not continue far into the low-velocity layer. They nevertheless have a triggering effect upon the movements in the low-velocity layer provided the latter is gravitationally unstable in the sense that its density is less than the overlying part of the mantle. The heat brought into the low-velocity layer by the thermal diapirs helps to locate the protrusions from the layer into the overlying mantle and the Earth's crust. Being triggered by thermal diapirs rising from great

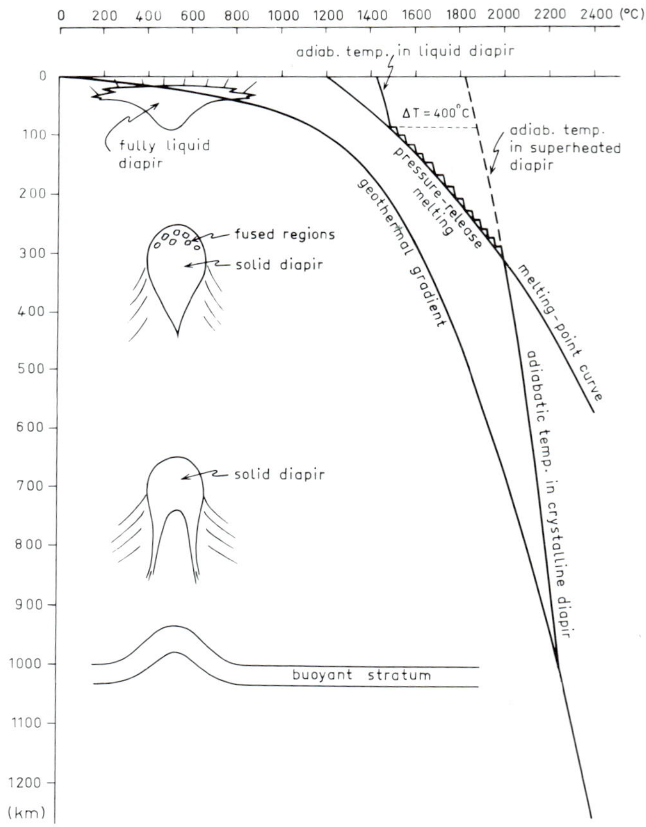

Fig. 16.13. Progressive melting due to diminishing pressure acting on diapir rising from a low-density stratum deep in the mantle. See also text.

depth the spacing of the protuberances from the low-velocity layer may be considerably greater than the characteristic wavelength of the layer.

As the hotter-than-environment diapirs of LV-materials rise through the layer above the low-velocity layer, melting due to depressurization commences provided the melting P, T curve is crossed. A simplified example is shown in Fig. 16.13. The process is quite analogous to the transitions strontium plumbate→spinel, and spinel→olivine which occurred at greater depths. The volume expansion is probably greater for the melting, and the change of effective viscosity is drastic. The low viscosity enables the newly formed melts to progress along narrow fractures and to extend the fractures far into the solid crust due to the wedge-effect of the fluid magma which may pierce the Earth's surface and form volcanoes.

The melting of rising mantle diapirs will of course be more complicated than described above, and probably also much more realistic, if the diapirs are treated as mixtures of various silicates, sulphides and oxides, rather than bodies of pure magnesium silicate. Partial melting of such composite diapirs could well explain much of the diversity of vulcanism and magmatic plutonism encountered on and in the Earth's crust.

17 Emplacement of Basic and Acidic Rocks

The fact that basic and acidic rocks generally occur in contrasted geological settings has led to much discussion in geological literature. The author believes that certain features of the model experiments presented here may contribute to the clarification of the question of emplacement of plutonic bodies of acidic, as opposed to basic, composition.

Field studies have established that basic extrusive rocks are much more abundant than extrusive rocks of acidic composition, whereas the converse is true of plutonic rocks. This important observation has been discussed in detail by Kennedy and Anderson (1938). It is more significant in connection with our model studies that the shape and structure of plutonic basic and acidic bodies are characteristically different. The acidic rocks occur chiefly as dome- or batholith-like bodies with a vertical dimension which usually is considerably relative to the lateral extension, and with a roof which is convex upward. Structural conformity between pluton and surroundings is common. The very shape suggests a rising tendency of the masses. Often closer structural studies support the impression that such bodies have been rising in relation to the surroundings (cf. the "Granit-tektonik" as discussed by Cloos, 1925). The compilation work by Buddington (1959) may be studied for details. See also Grout (1945) and Newall and Rast (1970).

In contrast, the known large plutonic bodies of basic composition are generally shaped as rather thin convex-concave lenses with the convex side facing downward. Their shape suggests a tendency to sink, as does in many instances the structure in the surroundings. Such lopoliths of huge dimensions are among the most well-known plutonic complexes, e.g. the Sudbury and the Duluth complexes in North America and the Bushveld complex in Africa. Disregarding some anorthosite massifs, there is no known large basic pluton that may be interpreted as a dome or batholith.

Examples of basic rocks which have subsided through the crust are naturally less abundant than occurrences of granitoid masses which have risen. In greenstone belts, however, subsided basic rocks occupy considerable areas as pointed out by, e.g. Gorman *et al.* (1978). Good examples of amphibolite bodies which have sunk into a granitoid basement are described from Finland by Ehlers (1976). Petraske *et al.* (1978) have also considered the subsidence of heavy basic magmas or rocks into a sialic substratum.

Basic and acidic rocks differ also in their hypabyssal facies. Diabase and dolerite dykes usually occur as rather thin but extensive sheets that run tens and hundreds of miles across geological formations, whereas acidic dykes are often irregular in shape, and limited in extension. Dykes† of acidic rocks are rare except in regions close to volcanic vents and obvious subvolcanic intrusions. Basic dykes—diabase and dolerite—are found in extensive swarms, often without apparent connection with volcanic vents. Such swarms probably were the feeders to huge flows of plateau basalt. These general impressions from the field support our model experiments in advancing the following theory of emplacement.

Acidic plutons, such as granodiorites and granites, are, in general, less dense than the average rock that makes up the geologically accessible part of the Earth's crust, i.e. the part which may be removed by extreme erosion, or the surface layer which extends down to the level of granulite facies of regional metamorphism. Hence, bodies of granodiorite and granite within the accessible part of the Earth's crust are, in general, enclosed in surroundings that give buoyancy to the embedded plutons. Many of these masses have thus been actively rising relative to their surroundings, until they became "quenched" owing to increased viscosity and strength as they approached the surface of the crust. Structure and shape indicative of upward motion are therefore to be expected of a majority of granitoid plutons, particularly deep-seated ones.

Basic rocks show higher density than the exposed average upper part of the Earth's crust. Basic plutons accessible for direct study accordingly occur in surroundings which generally do not give buoyancy to the enclosed plutons, at least not in their crystalline state. Consequently, basic bodies in general do not display structure and shape that indicate buoyant motion relative to the environment. On the contrary, such bodies exhibit features indicative of lateral spreading (i.e. extensive sheet-shaped plutons), and sinking (i.e. the convex surface facing downward). Such geometric and structural patterns are similar to those developed in the models when ascending masses encountered layers less dense than themselves; see, for example, Figs 16.1, 11.30 and 11.32.

As discussed on p. 350 a flat, uniformly thick sheet that sinks through a fluid material develops a curved shape with the convex side facing upward.

† Pegmatites and aplites are not included in the types of dykes considered here. Pegmatites are related to metamorphism and metasomatism, processes which do not concern us in this work.

This does not hold if the heavy layer is lens shaped and much thicker in the middle than along the edges, as, obviously, most sheet-shaped intrusives are. In this case the high pressure underneath the thick central part of the sheet makes the centre sink more rapidly than the thin edges of the body.

If tension fractures occur in the rigid crust above the body of a basic melt, the melt may of course be forced up through the fractures, irrespective of the relative density of melt and immediate surroundings. The buoyant push from deeper levels is sufficient to elevate the melt through the fractures.

It is tempting to assume that basic rocks, including anorthosites, rise in dome- or batholith-like fashion within the dense surroundings of the deep crust and the mantle in much the same manner that granites and granodiorites rise as a result of buoyant forces within the shallower parts of the crust. The high-melting monomineralic anorthosites are probably wholly crystalline during their ascent, but the gabbroic bodies may well be partly molten, the molten part being able to penetrate into the crust and spread as lopoliths, or to reach the surface through fracture systems.

In conclusion, the author suggests that an important reason for the observed contrasted geometry and structure of acidic and basic intrusive rocks is the condition that the exposed acidic bodies are situated in an environment that exposes the masses to buoyant forces *in situ*, so to speak, whereas the basic bodies that are accessible for direct study are not enclosed in buoyancy-creating surroundings. Basic bodies occurring in the upper part of the crust owe their presence there to an upwardly directed push originated at deeper levels, possibly due to buoyant forces acting on the lower part of the mass which is surrounded by the dense material of the mantle. Besides, most basic bodies encountered in the crust were emplaced as melts which generally are about 10% less dense than corresponding crystalline rocks.

Since plutonic shapes such as domes and batholiths are characteristic of masses which rise as a result of buoyancy *in situ*, such shapes are generally not displayed by the basic rocks. It is conceivable, however, that some anorthosite masses that typically occur within the most deeply eroded regions of the crust, viz. within granulite-facies areas, represent the top portion of the basic batholiths which have roots down in a subcrust more dense than the anorthosite itself.

18 Models of Convection Currents in the Mantle

We know much too little of the physical properties of the mantle to be able to make a reasonably true-to-scale dynamic model of it. Even if the properties were known in sufficient detail, it would hardly be possible to find model materials with the required mechanical and thermal characteristics. For example, the density of the mantle changes from little more than 3 g cm^{-3} at the top to more than 5 g cm^{-3} at the bottom. Clearly, if this is due to compression of a homogeneous mantle, as is often assumed, rather than change in chemical composition and phase changes, it is hard to see how this density change of more than 50% can be matched by a model material across a reasonable thickness of, say, 1 m or so for non-centrifuged models, and some 10 cm for centrifuged models.

As long as our interest is focused on the mechanical aspect of convective movements in the mantle, especially the reflection on the crust of such deep-seated movements, thermal convection in the mantle may to some extent be simulated by purely mechanical models, i.e. models not involving thermal energy. A classical example of this is the model devised by Griggs (1939), who let two drums with horizontal axes rotate in a viscous fluid with a crust floating on the top. The flow pattern around the rotating drums would simulate a part of the flow pattern of a convection cell. However, such experiments, informative as they may be with reference to the structures produced in a buckling crust, are not informative on the geometric pattern of free convection, or on the true forces involved—the torque on the drum was not controlled by a gravitational instability of the system.

In a model devised by the author, Griggs' drums were replaced by a heavy object embedded in a somewhat less dense viscous or plastic material (stitching wax, painter's putty or silicone putty), the crust being simulated by a layer of modelling clay or stiff wax (Ramberg, 1963a). When exposed to a centrifugal force of appropriate magnitude, the heavy object would

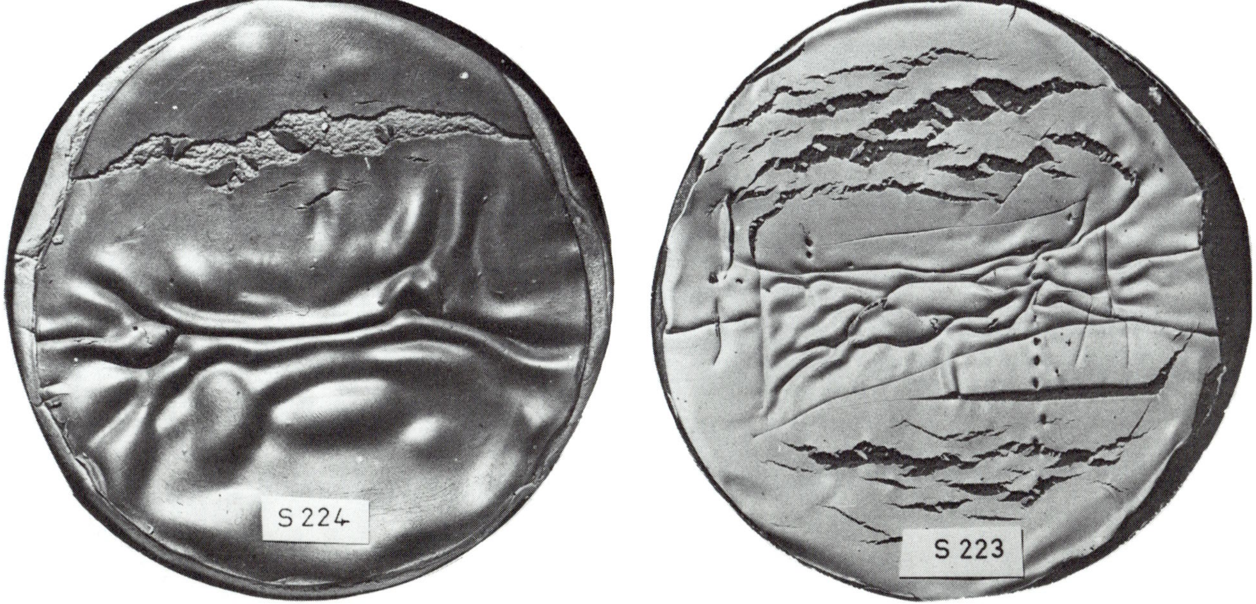

(a) (b)

Fig. 18.1. Buckling and fracturing of surface sheet of modelling clay over painter's putty in which a heavy block has sunk as shown in Fig. 18.2.

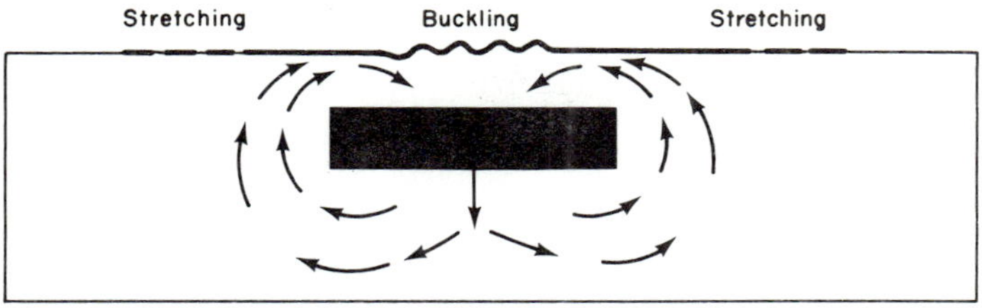

Fig. 18.2. Flow lines around a heavy body sinking through a viscous mass. Deformation in competent crust in indicated. See Figs 18.1 and 14.18.

sink through the viscous surroundings and generate a flow pattern similar to that around the downward-moving limb of a convection cell. Stretching and buckling occurred at predictable regions in the overlying crust. This type of model is, perhaps, a step closer to reality than the rotating-drum type, because in the former the driving force is at least generated by gravity (or corresponding body-force field). Figures 18.1 and 18.2 show cross-section and "bird's eye" views of such models.

However, little information can be gained from these tests as to the cellular pattern developing in free convection, particularly the lateral distribution of the cells, their width relative to their thickness, etc. Some information on this point may be gained by constructing an isothermal model mantle whose density increases upward, containing no lateral heterogeneities that can predetermine the cellular pattern. The spontaneous overturn of this unstable arrangement should correspond rather closely to the pattern of thermal convection during a half-cycle. The theory of this kind of convective systems is presented on pp. 110ff, and numerical examples are found on p. 116 and below.

Instead of making the vertical density variation gradual it was found convenient to make it discontinuous in the experiments. This agrees well with the theory and the numerical tests which are not suited for the handling

Figs. 18.3. Model S-226 prior to run in centrifuge. (1) Silicone putty, $\rho = 1{\cdot}14\ \mathrm{g\ cm^{-3}}$; (2) silicone putty with magnetite, $\rho = 1{\cdot}24\ \mathrm{g\ cm^{-3}}$; (3) silicone putty with tungspar powder, $\rho = 1{\cdot}34\ \mathrm{g\ cm^{-3}}$; (4) silicone putty with magnetite powder, $\rho = 1{\cdot}44\ \mathrm{g\ cm^{-3}}$; (5) silicone putty with tungspar powder, $\rho = 1{\cdot}523\ \mathrm{g\ cm^{-3}}$; (6) silicone putty with magnetite powder, $\rho = 1{\cdot}63\ \mathrm{g\ cm^{-3}}$. Thickness of layers 2 mm. Top layer: 0·5 mm-thick silicone putty with four thin sheets of modelling clay. Length of model 10 cm. See Fig. 18.4.

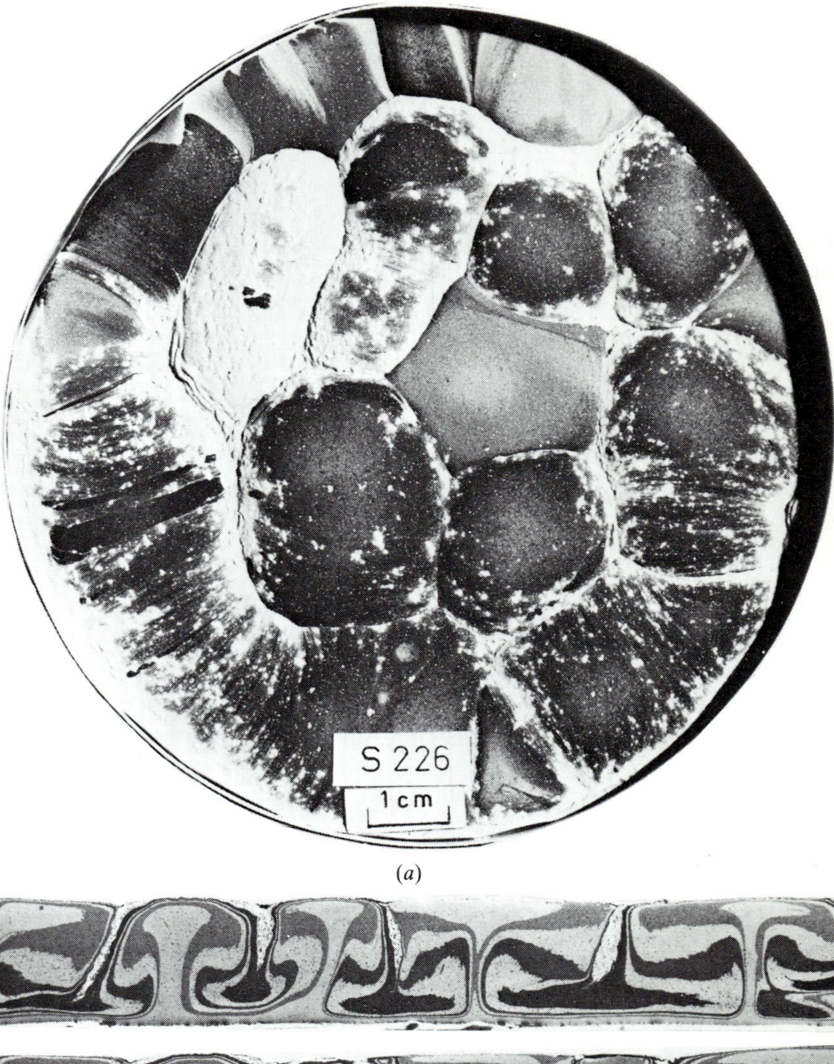

(*a*)

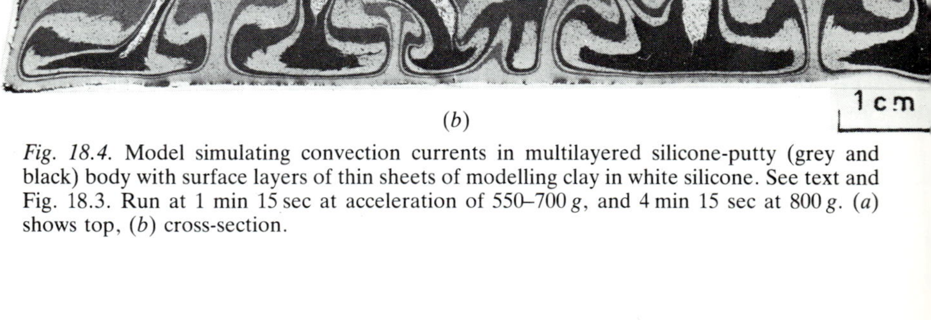

(*b*)

Fig. 18.4. Model simulating convection currents in multilayered silicone-putty (grey and black) body with surface layers of thin sheets of modelling clay in white silicone. See text and Fig. 18.3. Run at 1 min 15 sec at acceleration of 550–700 *g*, and 4 min 15 sec at 800 *g*. (*a*) shows top, (*b*) cross-section.

of continuous variations in rheological properties and densities. In the experiments the variation in density is accomplished by using sheets of silicone mixed with fine powder of magnetite and/or tungspar, each sheet consisting of a uniform mixture.

Model S-226, which also was tested numerically (see below), contained three silicone–magnetite layers alternating with three silicone–tungspar layers, all layers being 2 mm thick. On the top of this pile of strata was a 0·5-mm-thick "crust" of silicone in which four very thin sheets of modelling clay were embedded in order to give a certain stiffness to the crust.

Reckoned from top to bottom (not counting the 0·5-mm-thick "crust") the densities and viscosities of the layers were as follows: $\rho_2 = 1{\cdot}63$,

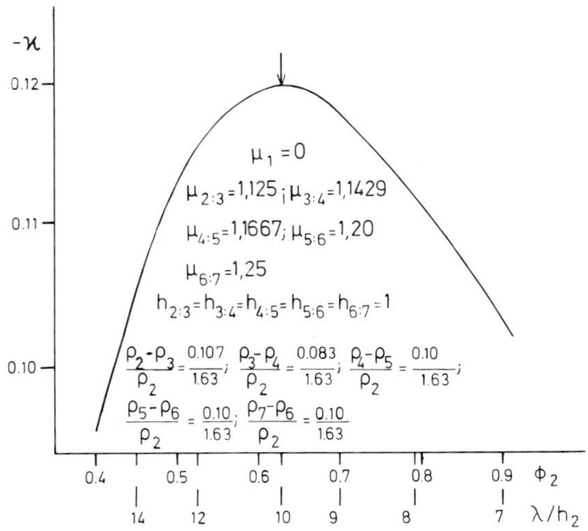

Fig. 18.5. Eigenvalue, κ_1, as a function of ϕ_2 and λ/h_2 of the model shown in Figs 18.3, 18.4 and 18.6.

$\rho_3 = 1{\cdot}523$, $\rho_4 = 1{\cdot}44$, $\rho_5 = 1{\cdot}34$, $\rho_6 = 1{\cdot}24$, $\rho_7 = 1{\cdot}14$, all in g cm^{-3}; and $\mu_2 = 4{\cdot}5\times10^5$, $\mu_3 = 4{\cdot}0\times10^5$, $\mu_4 = 3{\cdot}5\times10^5$, $\mu_5 = 3{\cdot}0\times10^5$, $\mu_6 = 2{\cdot}5\times10^5$, and $\mu_7 = 2{\cdot}0\times10^5$, all in poise. The density and viscosity of the thin and weak surficial sheet were disregarded in the numerical calculation.

The initial model is shown in Fig. 18.3, and the result after centrifugation for 1 min 15 sec at 600 *g* and 4 min 15 sec and 800 *g* is presented in Fig. 18.4.

In the numerical test the following dimensionless ratios:

$$\mu_i\mu_{(i+1)},\ h_i/h_{(i+1)} \text{ and } (\rho_{i-1)} - \rho_i)/(\rho_i - \rho_{(i+1)})$$

are needed as well as the quantities

$$q_i = \frac{(\rho_i - \rho_{(i+1)})\rho a h_{(i+1)}}{2\mu_{(i+1)}}.$$

All those quantities are implied in the measured values of the viscosities, the densities, the thicknesses and the centrifugal acceleration.

That the surficial layer is disregarded in the numerical calculation is probably not of much consequent because the layer is mechanically weak, it is thin relative to the composite thickness of the model and its density is low.

The calculated principal eigenvalue is shown in Fig. 18.5, where the maximum coincides with $\lambda/h_2 \approx 10$ or $\lambda/\Sigma h = 1{\cdot}60$, Σh being the composite thickness of the pile (i.e. $\Sigma h = 6{\cdot}25 \times h_2$).

It is interesting that the average spacing/composite thickness ratio of the experimental diapirs gives exactly the same figure. If the seven central diapirs are considered in the model (the other diapirs are disturbed by the boundary drag since they are in direct contact with the wall of the container) the average distance between the centres of the diapirs is about 2·0 cm. Since the composite thickness is 12·5 mm, the ratio $\lambda/\Sigma h = 1{\cdot}60$. This exact agreement with the theoretical prediction is evidently coincidental because the theory is based on two-dimensional models while the experimental test is three-dimensional.

The steadied proportion between the amplitudes of the deflections at the various boundaries is theoretically given by the proportion between

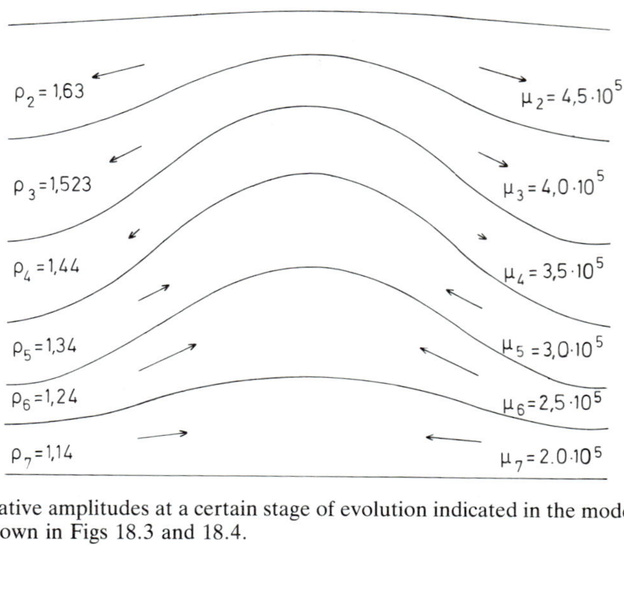

Fig. 18.6. Relative amplitudes at a certain stage of evolution indicated in the model discussed in text and shown in Figs 18.3 and 18.4.

the components of the eigenvector which belongs to the principal eigenvalue. These are:

$$\bar{y}_1/\bar{y}_2/\bar{y}_3/\bar{y}_4/\bar{y}_5/\bar{y}_6 = 0{\cdot}10292/0{\cdot}57258/0{\cdot}90435/1/0{\cdot}78338/0{\cdot}32160$$

as indicated schematically in Fig. 18.6.

For comparison with the experimental time of evolution of the centrifuged model it is interesting to study the theoretically predicted timing.

Assuming that the amplitudes are in the steadied proportions during the diapiric rise, and moreover that none of the amplitudes exceeds 10% of the dominant wavelength, then the amplitude at time t and the initial amplitude are related as follows'

$$\bar{y}_i/\bar{y}_{0i} = \exp(\kappa_1 q_1 t). \tag{18.1}$$

Here

$$q_1 = \frac{-\rho_2 a h_2}{2\mu_2} = \frac{-1{\cdot}63 \times 600 \times 981 \times 0{\cdot}2}{2 \times 4{\cdot}5 \times 10^5} = -0{\cdot}213204 \text{ sec}^{-1}, \tag{18.2}$$

valid for the first period of centrifugation at $a = 600\ g$, and

$$q_1' = \frac{-1{\cdot}63 \times 800 \times 981 \times 0{\cdot}2}{2 \times 4{\cdot}5 \times 10^5} = -0{\cdot}284272 \text{ sec}^{-1}, \tag{18.3}$$

valid for the second period at $a = 800\ g$ (i takes the values from 1 to 6).

The maximum on the principal eigenvalue is $\kappa_1 = -0{\cdot}1199$ as shown in Fig. 18.5. From these data we obtain the amplitudes of the several deflecting interfaces at selected times of centrifugation as related to the corresponding initial amplitudes (Table 18.1).

We see that the rate of amplification is explosive after about 250 sec. This is interesting in view of the observation that the model seemed to develop very rapidly just before the centrifuge was stopped after a total run of 330 sec. It is true that the experimental diapir models in general seem to develop quite suddenly just as one should expect from the exponential nature of the theoretical amplitude-growth equation.

When preparing the initial stages of the tested models care is taken to make each layer uniform as regard thickness, density and crucial rheological

Table 18.1. Theoretical relationship between time and amplitude in model S–226

t (sec)	10	50	75	100	150	200
$\bar{y}_i/\bar{y}_{0i}$	1·29	3·59	6·80	15·95	87·70	482·2
t (sec)	250	300	350	400	500	600
$\bar{y}_i/\bar{y}_{0i}$	2651	14576	80138	440597	$13{\cdot}3\times10^6$	$4{\cdot}125\times10^8$

properties, and to make the contacts between the layers as smooth and parallel to the equipotential surfaces as possible. Yet, the fact that diapirs always do grow in centrifuged models with inverted density stratification must mean that some irregularities, however faint, are present.

Let us assume that in model S-226 the amplitude $\bar{y}_{04}$ of the initial uncontrolled largest deflection (on interface 4, see Fig. 18.4) was 0·01 mm. The analytic theory provides a good description of the evolution of these initial deflections until they have grown to about 10% of the dominant wavelength. For model S-226 this means a ratio $\bar{y}_4/\bar{y}_{04}$ equals to 2 mm/0·01 mm = 200. Here 2 mm is 10% of the dominant wavelength, and 0·01 mm is the initial amplitude. We can now calculate the length of time theoretically needed for the gentle initial deflections to grow from 0·01 mm to the size which corresponds to the validity limit of the theory. By application of eqn (18.1) we find that during the first 75 sec, while the centrifugal acceleration is 600 g, the amplification is 6·8, i.e. the amplitude has grown to 0·068 mm. This value is the initial amplitude to use for the second period in which $a = 800\ g$. To find the time needed in the second period to reach $\bar{y}_4 = 2$ mm we put

$$\ln \frac{2}{0{\cdot}068} = q_1'\kappa_1, t = 0{\cdot}284272 \times 0{\cdot}1199 \times t \qquad (18.4)$$

and obtain $t = 99{\cdot}2$ sec. The total time needed for the initial deflections with amplitude 0·01 mm to reach 2 mm is then 174·2 sec.

When $\bar{y}_4 > 2$ mm, the theory with its characteristic exponential amplification term is no longer valid. The velocity of amplitude growth increases now much less with increasing amplitude. For our purpose we may well assume that the rate of growth at $\bar{y} > 0{\cdot}1\lambda$ is constant and equal to the rate when $\bar{y} = 0{\cdot}1\lambda$. For model S-226 the theoretical rate of amplitude growth at $\bar{y} = 0{\cdot}1\lambda$ is

$$v_4 = y_4 q'_1 \kappa_1 = 0{\cdot}2 \times 0{\cdot}284272 \times 0{\cdot}1199 = 0{\cdot}006817 \text{ cm sec}^{-1}. \qquad (18.5)$$

This we take as the average velocity when $\bar{y}_4 \geqslant 0{\cdot}1\lambda = 2$ mm. On this basis the time needed for the crests of the deflections at interface 4 to reach the top of the pile—i.e. for the amplitude to grow from 2 mm to 6·25 mm—is

$$t = \frac{4{\cdot}25 \text{ mm}}{0{\cdot}06817 \text{ mm sec}^{-1}} = 62.3 \text{ sec}.$$

Theoretically then, the total time needed for initial deflections only 0·01 mm high to grow to diapirs which pierce the surface is 236·5 sec. This is at least the same order of magnitude as the experimental duration of the centrifugation, viz. 330 sec.

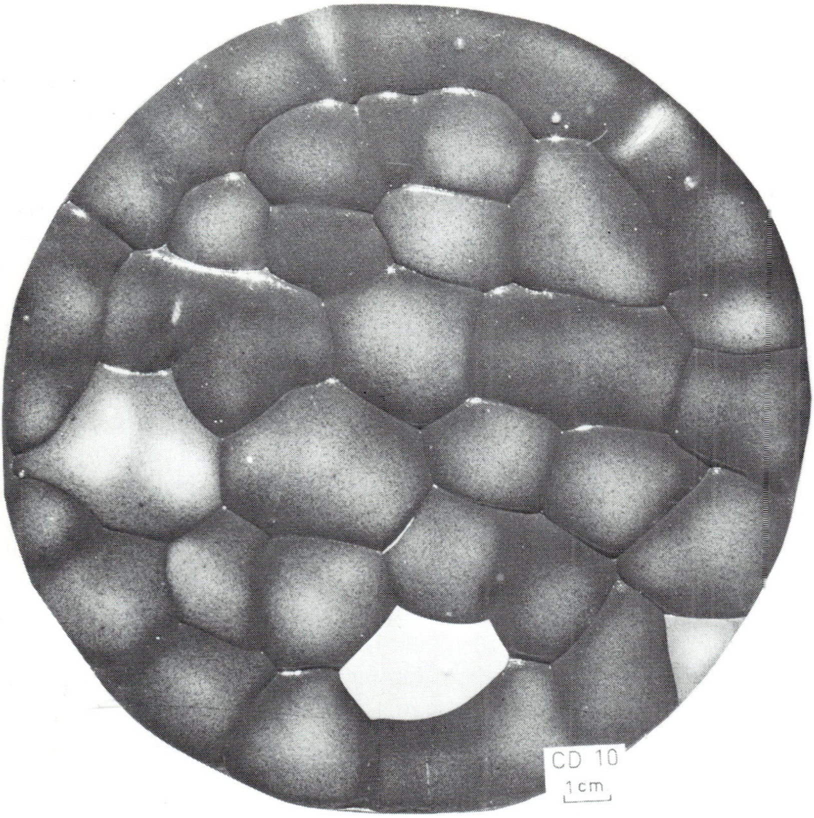

(*a*)

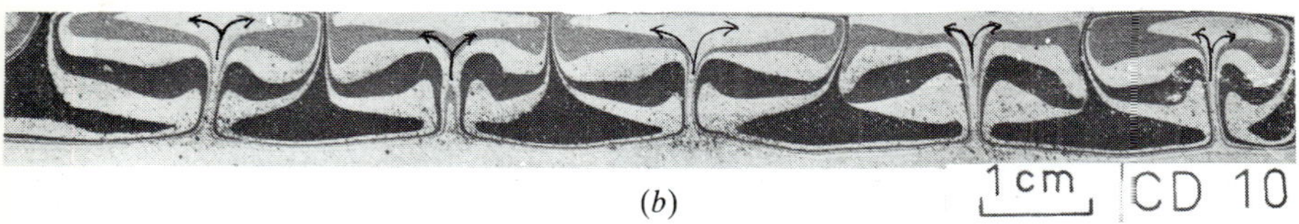

(*b*)

Fig. 18.7. Layered centrifuged model simulating thermal convection. Because of the larger horizontal dimension of this model compared with S-226 (Fig. 18.4) the geometry of the cells is less controlled by the boundary drag at the edge of the model. See also Fig. 18.3 and text. Run 35 sec at acceleration increasing from 0 to 2000 *g*, 3 min 12 sec at 2000 *g* and 1 min 30 sec at between 1500 *g* and 2000 *g*.

Model CD-10 (Figs 18.7 and 18.8) is very similar to S-226, except that CD-10 is built of somewhat more viscous silicone–magnetite and silicone–tungspar mixtures than S-226, and that the diameter of CD-10 is 20 cm, thus making the edge effect less disturbing within the central region of the model. The cell pattern in the middle of the model is probably the

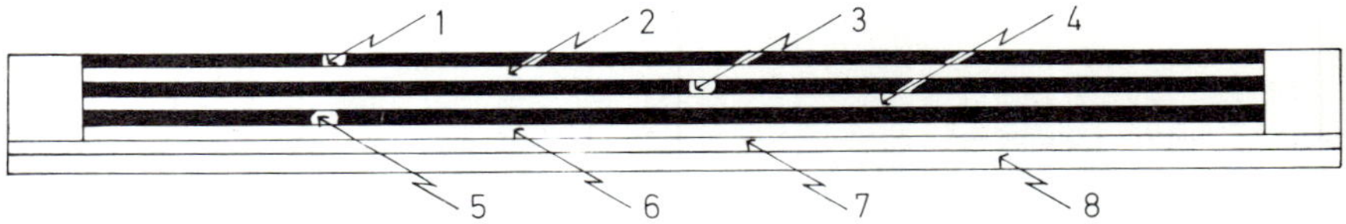

Fig. 18.8. Initial structure of CD-10. (1–6) Layers of silicone putty with $\rho = 1{\cdot}62$, 1·51, 1·41, 1·30, 1·22, and 1·12, all in g cm^{-3}; (7) painter's putty; (8) modelling clay which also constitutes the two square fields at the ends. Length of model 19·5 cm.

same as it would have been if the lateral extent of the model was inifite. Here the average cell-width/thickness ratio is about 2.

In model S-229 (Fig. 18.9) the crust is a thin sheet of modelling clay. In this model the buckled crust (incidentally, note overthrust structures!) is pulled less down into the mantle despite the stronger centrifugal force (7 min at 1800 *g* and 5 min 30 sec at 2900 *g*). This is probably due to the higher density of the mantle materials in S-229 than in S-226.

Model CD-9 (Figs 18.10 and 18.11) is informative in the sense that it shows a fracture system and movement pattern of a rigid low-density crust which does not mimic the pattern of the convection cells below. Yet the latter constitute the driving energy of the surface deformation. Perhaps this model gives experimental support to the claim often occurring in plate-tectonic discussions that plate geometry and kinematics need not mimic the pattern of the driving convection cells in the mantle.

It is clear that the models above are not dynamically similar to a realistic mantle–crust system. Because of the rather large density difference between top and bottom of the model mantle, the driving force of the turnover—the buoyancy—is too high relative to the centrifugal, respecitve gravita-

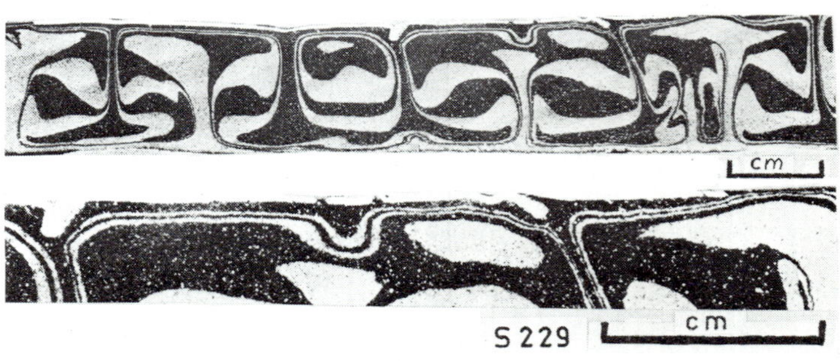

Fig. 18.9. Details of model similar to S-226, Fig. 18.4 except that model S-229 contained a top sheet of relatively stiff modelling clay. Note the deformation structure in the top sheet. Run 7 min at 1800 *g* and 5 min 30 sec at 2900 *g*.

Fig. 18.10. Model simulating convection cells in a silicone-putty multilayer overlaid by a relatively brittle crust. Note lack of geometric correspondence between the width of the cells and the size and shape of the fragments into which the crust has been broken.

tional, pull when compared with the earth. This means, for example, that the lateral compressive stress acting on the model crust gives buckles with longer wavelength than what would correspond to a true-to-scale model. Another effect of excessive buoyant force is that the pushing force of low-angle thrusts—which have developed in model S-229—relative to the friction along the sole is greater in the model than in nature. Accordingly, the development of low-angle thrusts in model S-229 is no decisive indication that convection in the mantle is capable of generating similar thrusts in the Earth's crust.

On the other hand, the equidimensional cellular pattern in the models, as seen on the surface as well as in cross-sections should correspond rather well to convection cells in the mantle, if such exist. For true thermal convection in experimental mantle models the study by Richter and Parsons (1975) is particularly interesting.

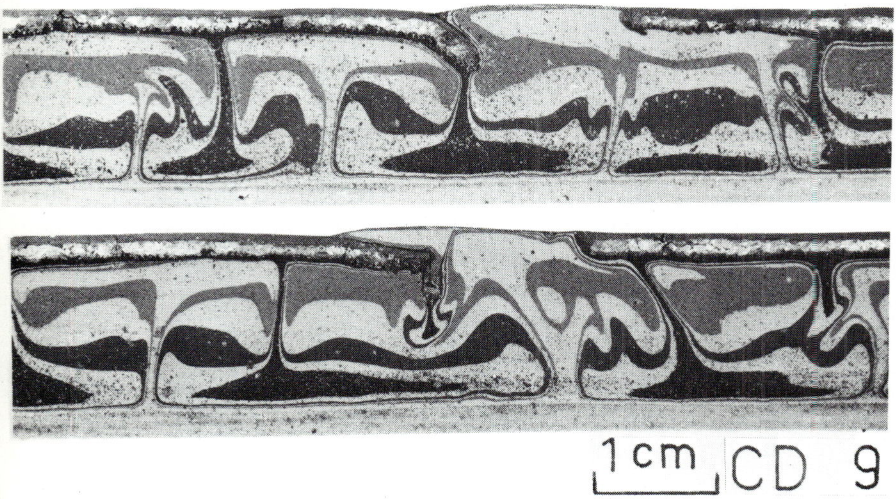

Fig. 18.11. Enlarged view of parts of model CD-9 (Fig. 18.10).

Many field geologists acquainted with the heterogeneous chemical and mineralogical composition of the accessible part of the Earth's crust look with suspicion on the concept of a chemically uniform mantle, a suspicion shared by many geophysicists. Seismic observation points in fact toward considerable differences in mantle properties below oceans on the one hand and below continents on the other. See papers by e.g. Dziewonski *et al.* (1977), Jordan (1979), Green (1975) and Chase (1979) who find that the mantle is heterogeneous horizontally and/or vertically, while Hager and O'Connell (1979), Davis (1977) and Elsasser *et al.* (1979) see no serious evidence against a homogeneous mantle.

If the mantle is heterogeneous and more or less gravitationally differentiated with increasing concentration of heavy chemical components downward, then homogeneous convection is excluded. Heterogeneous convection, however, characterized by phase changes, is a possible process in a gravitationally differentiated multilayered system. In homogeneous thermal convection, the thermal expansion of the convecting medium is the most significant buoyancy-providing parameter. In heterogeneous convection, however, of the kind simulated in many models in this book, buoyancy is provided by the volume change associated with a change of phase. At the high temperature below certain depths buoyant bodies are produced by melting, or by a change from dense to less dense minerals. These bodies will rise until buoyancy is no longer provided because of transformation back to more dense phases and/or because the bodies encounter environments with less density. The convective cycle is completed when the bodies again have sunk down to their original level.

Heterogeneous convection of this kind has many advantages over homogeneous convection. For example, it does not require a uniform mantle, and need not be seriously hampered even if the mantle is differentiated into chemically and mineralogically constrasted layers. Heterogeneous convection may have an irregular flow pattern determined by the particular structure of the convective system. Since the system may be complex and irregular without preventing heterogeneous convection, there is nothing in principle against the irregular tectonic pattern of the Earth being produced by convections of this nature.

According to the author's view there is direct evidence that heterogeneous convection takes place in the Earth. He believes that the extrusion of basaltic magma, for example, and the rise of granitic batholiths constitute parts of heterogenous convection cycles.

From an energy point of view heterogeneous convection is more efficient than the corresponding homogeneous process. The volume changes associated with phase transformations usually provide greater buoyant force than does the thermal expansion of phases. Moreover, the absorption of the heat of melting at depth and the liberation of the same heat closer to the Earth's surface provide a more efficient cooling of the globe than that accomplished by homogeneous convection if compared on basis of same velocity of flow.

Bibliography

Anderson, D. L. (1962). The plastic layer of the earth's mantle. *Sci. Am.* July, 52.

Anderson, D. L. (1965). Recent evidence concerning the structure and composition of the earth's mantle. *In*: "Physics and Chemistry of the Earth", Vol. VI, p. 1. Pergamon, Oxford.

Anderson, D. L. (1965). Conditions for a density minimum in the upper mantle. *J. Geophys. Res.* **70**, 1457.

Anderson, D. L. and Sammis, C. (1970). Partial melting of the upper mantle. *Phys. Earth Planet. Interiors,* **3**, 41.

Anderson, D. L. (1975). Chemical plumes in the mantle. *Geol. Soc. Am. Bull.* **36**, 1593.

Arrhenius, S. (1912). Zur Physik der Salzlagerstätten. *Meddelanden från K. Vetenskapsakademiens Nobelinstitut,* **II**, No. 20.

Arrhenius, S. and Lachmann, R. (1912). Die phys.-chem. Bedingungen bei den Bildungen der Salzlager und ihre Anwendungen auf geologische Probleme. *Geol. Rundschau,* **III**, 139.

Artyushkov, E. V. (1970). Density differentiation of the earth. *Phys. Earth Planet. Interiors,* **2**, 318.

Artyushkov, E. V. (1971). Convective instability in geotectonics. *J. Geophys. Res.* **75**, 1397.

Aubouin, J. (1965). *"Geosynclines"*, Elsevier, New York and Amsterdam.

Avgherinos, P. J. and Schofield, A. N. (1969). Drawdown failure of centrifuged models. *Proc. 7th Int. Conf. on Soil Mech. and Found. Eng.* **2**, 497.

Baker, B. H., Mohr, P. A. and Williams, L. A. J. (1972). Geology of the Eastern Rift System of Africa. *Geol. Soc. Am., Special Paper* **136.**

Barton, D. C. (1926). The American salt-dome problems in the light of the Roumanian and German salt domes. *In*: "Geology of Salt-dome Oil Fields" (R. C. Moore, Ed.). *Am. Ass. Pet. Geol.* Special volume, p. 167.

Bassett, R. H. (1973). Centrifugal model tests of embankments on soft alluvial foundations. *Proc. 8th Int. Conf. on Soil Mech. and Found. Eng.* **2.2**, 23.

Beloussov, V. V. (1960). Tectonophysical investigations. *Geol. Soc. Am. Bull.* **71**, 1255.

Beloussov, V. V. (1961). The origin of folding in the earth's crust. *J. Geophys. Res.* **66**, 2241.

Beloussov, V. V. and Vikhert, A. V. (Eds.) (1975). *"The Earth and Upper Mantle",* Vol. 2, Moscow University Press, Moscow.

van Bemmelen, R. W. (1933a). On the geophysical foundation of the undation theory. *Koninkl, Ned. Akad. Wetenschap., Proc., Ser.* **36**, 686.

van Bemmelen, R. W. (1933b). The unduation theory and the development of the earth's crust. *Congr. Geol. Intern., Compt. Redn.* **16**, (2), 965.

van Bemmelend, R. W. (1960). New view of east Alpine orogenesis. *Intern. Geol. Congr., 21st, Copenhagen, 1960, Rept. Session, Norden, Pt* **18**, 99.

van Bemmelen, R. W. (1973). Geodynamic models of the Alpine type of orogeny. Test case II: The Alps in Central Europe. *Tectonophysics* **18**, 33.

Berner, H., Ramberg, H. and Stephansson, O. (1972). Diapirism in theory and experiment. *Tectonophysics,* **15**, 197.

Berthelsen, A. (1960). Geology of Tovqussap Nunâ. *Medd. Grönland,* **123.**

Biot, M. A. (1959). The influence of gravity on the folding of a layered viscoelastic medium under compression. *J. Franklin Inst.* **267**(3), 211.

Biot, M. A. (1961). Theory of folding of stratified viscoelastic media and its implications in tectonics and orogenesis. *Geol. Soc. Am. Bull* **72**, 1595.

Biot, M. A. (1963). Stability of multilayered continua including the influence of gravity and viscoelasticity. *J. Franklin Inst.* **276**, 231.

Biot, M. A. and Odé, H. (1965). Theory of gravity instability with variable overburden and compaction. *Geophysics,* **30**, 213.

Blackett, P. M. S., Bullard, Sir E. and Runcorn, S. K. (Eds.) (1965). Symposium on Continental drift. *Phil. Trans. Roy. Soc. Lond. No.* **1088.**

Braunstein, J. and O'Brien, G. (Eds) (1968). "Diapirism and Diapirs". *Am. Ass. Pet. Geol. Mem.* **8.**

Bridwell, R. J. (1978). The Rio Grande Rift and a diapiric mechanism for continental rifting. *In*: "Tectonics and Geophysics of Continental Rifts" (I. B. Ramberg and E.-R. Neumann, Eds). Reidel, Dordrecht, Holland.

Brun, J.-P. (1980). The cluster-ridge patterns of mantled gneiss domes in E. Finland: Evidence for large-scale gravitational instability in the proterozoic crust. *Earth Planet Letters* **47**, 441.

Brun, J.-P. (1977). La zonation structurale des dômes gneissiques. Un example: Massif de St. Malo, France. *Canad. J. Earth Sci.* **14**, 1697.

Bryhni, I., Kroghand, E. and Griffin, W. L. (1977). Crustal derivation of Norwegian Eclogites, a review. *Neues Jb. Min. Abh.* **130**, 49.

Bucher, W. H. (1956). Role of gravity in orogenesis. *Geol. Soc. Am. Bull* **67**, 1295.

Bucky, P. B. (1931). The use of models for the study of mining problems. *Tech. Publ. Am. Inst. Min. Engrs.* **425.**

Buddington, A. F. (1959). Granite emplacement with special reference to North America. *Geol. Soc. Am. Bull.* **70**, 671.

Cassen, P. (Ed.) (1979). Solid convection in the terrestrial planets. *Phys. Earth Planet. Interiors,* **19** (No. 2), 106.

Cater, F. W. and Elston, D. P. (1963). Structural development of salt anticlines of Colorado and Utah. *In*: "Backbone of the Americas: a symposium". *Am. Ass. Pet. Geol., Mem.* **2**, 152.

Cathles, L. M. (1975). "The Viscosity of the Earth's Mantle", Princeton University Press, Princeton, New Jersey.

Chandrasekhar, S. (1955). The character of the equilibrium of an incompressible heavy viscous fluid of variable density. *Proc. Cambridge Phil. Soc.* **51**, 162.

Chandrasekhar, S. (1961). "Hydrodynamic and Hydromagnetic Stability", Clarendon Press, Oxford.

Chapman, R. E. (1979). Mechanics of unlubricated sliding. *Geol. Soc. Am. Bull.* **90**, 19.

Charlesworth, J. K. (1957). "The Quaternary Era", Edward Arnold, London.

Chase, Cl. G. (1979). Subduction, the geoid, and lower mantle convection. *Nature,* **282**, 464.

Childs, O. E. and Beebe, W. B. (Eds) (1963). Backbone of the Americas: a symposium. *Am. Ass. Pet. Geol. Mem.* **2**.

Clark, S. P. (Ed.) (1966). "Handbook of Physical Constants." Memoir **97**, Geol. Soc. Amer.

Cloos, H. (1925). "Einführung in die tektonische Behandlung magmatischer Erscheinungen (Granittektonik)". Borntraeger, Berlin.

Cloos, H. (1930). Kunstliche Gebirge. *Natur und Museum, Senckenbergische Naturf. Ges. Frankfurt.* Part 2, Heft **6**, 258.

Cloos, H. (1936). "Einführung in der Geologie". Borntraeger, Berlin.

Cloos, H. (1935). Die Kartierung der Grundgebierges in SW Africa. *Geol. Rundschau,* **26**, 241.

Cloos, H. (1939). Hebung, Spaltung, Vulkanismus. *Geol. Rundschau.* **30**, 403.

Closs, H., Roeder, D. and Schmidt, K. (1978). "Alps, Apennines, Hellenides" (IUCG report 38), Schweizerbart, Stuttgart.

Cole, G. H. A. (1962). "Fluid Dynamics". Methuen, London.

Collet, L. W. (1927). "The Structure of the Alps". Edward Arnold, London.

Courant, R. (1956). "Differential and Integral Calculus, I and II". Interscience Publishers, New York.

Craig, G. Y. (1965). "The Geology of Scotland". Oliver and Boyd, Edinburgh.

Crittenden, M. D., Jr. (1963). Effective viscosity of the earth derived from isostatic loading of Pleistocene lake Bonneville. *J. Geophys. Res.* **68**, 5517.

Daly, A. R. (1947). The Vredefort ring-structure of South Africa. *J. Geol.* **55**, 125.

Daneš, Z. F. (1964). Mathematical formulation of salt-dome dynamics. *Geophysics,* **29**, 414.

Davis, G. F. (1977). *Geophys. J. Roy. Astr. Soc.* **49**, 459.

DeGolyer, E. (1926). Origin of North American salt domes. *In:* "Geology of Salt-dome Oil Fields" (R. C. Moore, Ed.). *Am. Ass. Pet. Geol.* Special volume, p. 1.

DeJong, K. A. and Scholten, R. (Eds) (1973). "Gravity and tectonics". Wiley–Interscience, New York.

Dennis, J. G. (1967). International Tectonic Dictionary. *Am. Ass. Pet. Geol. Mem.* **7**.

Dennis, J. G. and Häll, R. (1978). Jura-type platform folds: A centrifuge experiment. *Tectonophysics,* **45**, T15.

Dixon, J. M. (1975). Finite strain and progressive deformation in models of diapiric structures. *Tectonphysics,* **28**, 89.

Drury, S. A. (1977). Structures induced by granite diapirs in the Archean greenstone belt at Yellowknife, Canada: implications for Archean geotectonics. *J. Geol.* **85**, 345.

Du Toit, A. L. (1954). "The Geology of South Africa". Oliver and Boyd, Edinburgh.

Dyrelius, D., Gee, D. G., Gorbatschev, R., Ramberg, H. and Zachrisson, E., (1980). A profile through the central Scandinavian Caledonides. *Tectonophysics,* **69**, 271.

Dziewonski, A. M., Hager, B. H. and O'Connell, R. J. (1977). Large-scale heterogeneities in the lower mantle. *J. Geophys. Res.* **82**, 239.

Eardley, A. J. (1963). Relation of uplifts to thrusts in Rocky Mountains. *In*: "Backbone of the Americas: a symposium". *Am. Ass. Pet. Geol. Mem.* **2**, 209.

Ehlers, C. (1976). Homogeneous deformation in precambrian supracrustal rocks of Kumlinge area, SW. Finland. *Precamb. Res.* **3**, 481.

Elliott, D. (1976). The motion of thrust sheets. *J. Geophys. Res.* **81**, 949.
Elsasser, W. M., Olson, P. and Marsh, B. D. (1979). The depth of mantle convection. *J. Geophys. Res.* **84**, 147.
Escher, B. G. and Kuenen, P. H. (1929). Experiments in connections with salt domes. *Leidse Geol. Mededel.* **3**, 151.
Eskola, P. (1921). On the eclogites of Norway. *Det. norske videnskapakademis skrifter, Kristiania* **8**.
Eskola, P. (1948). The problem of mantled gneiss domes. *Quart. J. Geol. Soc. London,* **104**, 461.
Fisher, G. W., Pettijohn, F. J., Reed, J. C. and Weaver, K. N. (1970). "Studies of Appalachian Geology: Central and Southern". Interscience, New York, London, Sydney and Toronto.
Fisher, O. (1909). Convection currents in the Earth's interior. *Geol. Mag.* **6**, 8.
Fletcher, R. C. (1967). A Finite Amplitude Method for the Emplacement of Gneiss Domes and Salt Domes. Ph.D. thesis, Brown University, Providence, RI, U.S.A.
Fletcher, R. C. (1977). Folding of a single viscous layer: exact infinitesimal-amplitude solution. *Tectonophysics,* **39**, 593.
Fletcher, R. C. (1979). Folding of a single viscous layer: exact infinitesimal-amplitude solution, reply. *Tectonophysics,* **56**, 324.
Fletcher, R. C. (1979). Folding of a single viscous layer infinitesimal-amplitude solution, reply. *Tectonophysics,* **56**, 324.
Foslie, S. (1941). Tysfjords geologi. *Norg. Geol. Unders.* No. 149.
Fyson, W. K. and Frith, R. A. (1979). Regional deformation and emplacement of granitoid plutons in the Hackett River greenstone belt. *Can. J. Earth Sci.* **16**, 1187.
Gansser, A. (1964). "Geology of the Himalayas". Interscience, London.
Gansser, A. (1973). Facts and theories on the Andes. *J. Geol. Soc. London,* **129**, 93.
Gary, M., McAfee, R., Jr. and Wolf, C. L. (1972). "Glossary of Geology". American Geological Institute, Washington, D.C.
Gee, D. G. (1975a). A tectonic model for the central part of the Scandinavian Calendonides. *Am. J. Sci.* **275A**, 468.
Gee, D. G. (1975b). A Geotraverse through the Scandinavian Calendonides: Östersund to Trondheim. *Sver. Geol. Unders. C* **717**.
Gee, D. G. (1977). A reply to Ramberg's comments. *Am. J. Sci.* **277**, 657.
Ghosh, S. K. and Ramberg, H. (1968). Buckling experiments on intersecting fold patterns. *Tectonophysics,* **5**, 89.
Ghosh, S. K., Ray, N. and Troëng, B. (1979). The Åreskutan "Klippe" in the Scandinavian Caledonides. *Geol. Fören. i Stockholm Förh.* **101**, 85.
Gibbs, J. W. (1948). "The Collected Works of J. W. Gibbs, Ph.D., LL.D.", Vol. I, Yale University, New Haven, Connecticut.
Gillet-Lamont, F. P. W. (1799). Observations géologiques sur le gisement et la forme des replis successifs. *J. Mines,* 59, Ann. **7**, 449.
Girdler, R. W. (1978). Comparison of the E. African rift system and the Permian Oslo rift. *In*: "Tectonics and Geophysics of Continental Rifts" (I. B. Ramberg and E.-R. Neumann, Eds). Reidel, Dordrecht, Holland.
Gorman, B. E., Pearce, T. H. and Birkett, T. C. (1978). On the structure of Archean Greenstone Belts. *Precamb. Res.* **6**, 23.
Green, A. G. (1975). On the postulated Hawaiian plume with emphasis on the limitations of seismic arrays for detecting deep mantle structures. *J. Geophys. Res.* **80**, 4028.

Griffin, V. S., Jr. (1971). Stockwork tectonics in the Appalachian Piedmont of S. Carolina and Georgia. *Geol. Rundschau*, **60**, 868.
Griffin, V. S., Jr. (1974). Analysis of the piedmont in N.W. South Carolina. *Geol. Soc. Am. Bull.* **85**, 1123.
Griffin, W. L. and Råheim, (1973). Convergent metamorphism of eclogites and dolerites, Kristiansund area, Norway. *Lithos,* **6**, 21.
Griffin, W. L. *et al.* (1978). Archaean and Proterozoic crustal evolution in Lofoten-Vesterålen, N. Norway. *J. Geol. Soc. London,* **135**, 629.
Griggs, D. T. (1939). A theory of mountain building. *Am. J. Sci.* **237**, 611.
Grout, F. F. (1945). Scale models of structures related to batholiths. *Am. J. Sci.* **243**, 260.
Gutenberg, B. (1954). Low-velocity layers in the earth's mantle. *Geol. Soc. Am. Bull.* **65**, 337.
Gutenberg, B. (1959). The asthenosphere low velocity layer. *Ann. Geofis.* **12**, 439.
Gustavson, M. (1974a) Harsta. *Norg. Geol. Unders. Nr.* **309**, *Skr. 14.*
Gustavson, M. (1974b). Ofoten. *Norg. Geol. Unders. Nr.* **310**, Skr. *15.*
Guterman, V. G. (1980). Model studies of gravitational gliding tectonics. Tectonophysics (in press).
Gwinner, M. P. (1971). "Geologie der Alpen". Schweizerbart, Stuttgart.
Haarmann, E. (1930). "Die Oszillationstheorie". Ferdinand Enke, Stuttgart.
Hadley, G. (1965). "Linear Algebra". Addison-Wesley, Reading, Palo Alto and London.
Hager, B. H. and O'Connell, R. J. (1979). Kinimatic models of large-scale flow in the earth's mantle. *J. Geophys. Res.* **84**, 1031.
Haller, J. (1955). Der Zentrale Metamorph Komplex von Nordostgrönland. *Medd. Grönland,* **73**, I(3), 1.
Haller, J. (1971). "Geology of the East Greenland Caledonides". Interscience, New York.
Halbouty, M. T. (1967). "Salt Domes, Gulf Region, United States and Mexico", Gulf Publishing Company, Houston.
Hammarling, S. J. (1970). "Latent Roots and Latent Vectors", The University of Toronto Press, Toronto.
Harris, A. L., Holland, C. H. and Leake, B. E. (1979). "The Caledonides of the British Isles—reviewed." Scottish Academic Press, Edinburgh.
den Hartog, J. P. (1948). "Mechanics". Dover, New York.
Haskel, N. A. (1935). The motion of a viscous fluid under a surface load. *Physics,* **6**, 265.
Hayashi, D. and Kizaki, K. (1979). Numerical experiments of migmatite rise based on continuum dynamics. *Tectonophysics,* **60**, 61.
Heard, H. C. (1976). Comparison of the flow properties of rocks at crustal conditions. *Phil. Trans. R. Soc. Lond. A***283**, 173.
Heacock, J. G., Keller, G. V., Oliver, J. E. and Simmons, G. (1977). The earth's crust: its nature and physical properties. *Am. Geophys. Union, Geophys, Monogr.* **20.**
Heezen, B. C. (1960). The rift in the ocean floor. *Sci. Am.* **203**, 98.
Heiskanen, W. A. and Vening Meinesz, F. A. (1958). "The Earth and Its Gravity Field". McGraw-Hill, New York.
Heye, D. (1978). Experimente mit viskosen Flüssigkeiten zur Nachahmung von Salzstrukturen. *Geol. Jb. E* **12**, 31.

Hide, R. (1955). The character of the equilibrium of an incompressible heavy viscous fluid of variable density: an approximate theory. *Proc. Cambridge Fhil. Soc.* **51**, 179.

Hills, E. S. (1963). "Elements of Structural Geology". Methuen, London.

Hoek, E. (1963). Experimental study of rock-stress problems in deep-level mining. *In*: "Experimental Mechanics". Pergamon, Oxford.

Hoek, E. (1965). The design of a centrifuge for the simulation of gravitational force-fields in mine problems. *J. S. African Inst. Min. Met.* **455**.

Holmes, A. (1965). "Principles of Physical Geology". Nelson, London.

Holmsen, P. and Strand, T. (1960). Caledonized basal gneisses in a north-western area. *Norg. Geol. Unders.* **212**, I, (Excursion guide).

Holtedahl, O. (1960). Geology of Norway. *Norg. Geol. Unders.* **208.**

Holtedahl, O. and Dons, J. (1960). Geological Map of Norway. *Norg. Geol. Unders.*

Hsü, K. J. (1969). Role of cohesive strength in the mechanics of overthrust faulting and on landslides. *Geol. Soc. Am. Bull.* **80**, 927.

Hubbert, M. K. (1937). Theory of scale models as applied to the study of geologic structures. *Geol. Soc. Am. Bull.* **48**, 1459.

Hubbert, M. K. (1945). Strength of the earth. *Bull. Am. Ass. Pet. Geol.* **29**, 1630.

Hubbert, M. K. and Rubey, W. W. (1959). Role of fluid pressure in mechanics of overthrust faulting. *Geol. Soc. Amer. Bull.* **70**, 115.

Huddleston, P. J. (1977). Similar folds, recumbent folds, and gravity tectonics in ice and rocks. *J. Geol.* **85**, 113.

Huddleston, P. J. and Stephansson, O. (1973). Layer shortening and fold-shape development in the buckling of single layers. *Tectonophysics* **17**, 299.

Huddleston, P. J. and Hooke, R. Le B. (1978). Origin of foliation in glaciers. *J. Glac.* **20**, 285.

Illies, J. H. (1969). An intercontinental belt of the world rift system. *Tectonophysics* **8**, 5.

Illies, J. H. and St. Mueller (Eds) (1970). "Graben Problems". Schweizerbart, Stuttgart.

Ince, E. L. (1956). "Ordinary Differential Equations", Dover, New York.

Jaeger, J. C. (1956). "Elasticity, Fracture and Flow". Methuen, London.

Jaritz, W. (1973). Zur Entstehung der Salzstrukturen Nordwestdeutschland *Geol. Jb. A* **10.**

Johnson, A. M. (1970). "Physical Processes in Geology." Freeman, Cooper, San Francisco.

Johnson, M. R. W. and Stewart, F. H. (1963). "The British Calendonides". Oliver and Boyd, Edinburgh.

Jordan, T. H. (1979). The deep structure of the continents. *Sci. Amer.*, January, 70.

Kamb, W. B. (1961). The thermodynamic theory of nonhydrostatically stressed solids. *Geophys. Res.* **66**, 259.

Kautsky, G. (1953). Der geologische Bau der Sulitelma-Salojauregebietes in den nordschwedischen Kaledoniden. *Sveriges Geol. Unders. Årsbok, Ser.* **C.** No. 528.

Kennedy, W. Q. and Andersson, E. M. (1938). Crustal layers and the origin of magmas. *Bull. Volcanol.* **3**, 23.

King, P. B., Hadley, J. B., Dietrich, R. U., Cloos, E., Rodgers, J. and Cooper, B. N. (1964). Tectonics of the Southern Appalachians. *Symp. VPI, Dept. Geol. Sci., Roanoke, Virginia, U.S.A.*

Kline, J. S. (1965). "Similtude and Approximation Theory". McGraw-Hill, New York.

Koenigsberger, G. and Morath, O. (1913). Theoretische Grundtlagen der experimentillen Tektonik. *Zeitsch. Dentschen Geologischen Gesellschaft*, Monat. vol. **65**, 65.

Kornberg, W. (Ed.) (1979). What is fail-safe for a dam? Mosaic, National Science Foundation, **10**, 2, Washington.

Kreider, D. L., Kuller, R. G. and Ostberg, D. R. (1968). "Elementary differential equations". Addison–Wesley, Reading, Palo Alto and London.

Kvale, A. (1945). Skyvning og friksjon. *Nor. Geol. Tidsskr.* **24**, 32.

Kvale, A. (1953). Linear structures and their relation to movement in the Caledonides of Scandinavia and Scotland. *J. Geol. Soc. London,* **109**, 51.

Kühn, K. A. (1819). "Handbuch der Geognosie", Vol. 1. Leipzig.

Langhaar, H. L. (1962). "Dimensional analysis and theory of models". Wiley, New York.

Lehman, I. (1967). *In*: "The Earth's Mantle" (T. F. Gaskel, Ed.). Academic Press, London.

Lewis, D. J. (1950). The instability of liquid surfaces when accelerated in a direction perpendicular to their planes, II. *Proc. R. Soc. London, A* **202**, 81.

Lliboutry, L. (1974). Rheological properties of lithosphere. *Tectonophysics,* **24**, 13.

Long, R. G. (1953). Fluid models in geophysics. "Symposium on the Use of Models in Geophysical Fluid Dynamics". Johns Hopkins University, Baltimore.

Logatchev, N. A., Rogozhina, V. A., Solonenko, V. P. and Zorin, Yu. A. (1977). Deep structure and evolution of the Baikal Rift zone. *In*: "Tectonics and Geophysics of Continental Rifts" (I. B. Ramberg and E.-R. Neumann, Eds), p. 49. Reidel, Dordrecht, Holland.

Lundegårdh, P. H., Lundqvist, J. and Lindström, M. (1964). "Berg och jord i Sverige". Almqvist and Wiksell, Stockholm.

MacDonald, G. J. F. (1965). Continental structure and drift. Symposium on Continental Drift. *Phil. Trans. R. Soc. No. 1088,* 109.

Macgregor, A. M. (1951). Some milestones in the precambrian of Southern Rhodesia. *Transactions and Proceedings of the Geol. Soc. of S. Africa.* Johannesburg, **LIV.**

Magnitsky, V. A. (1970). Temperature gradient and the evolution of the earths's mantle. *Phys. Earth Planet. Interiors,* **2**, 363.

Magnusson, N. H., Lundqvist, G. and Regnell, G. (1962). Beskrivning till karta över Sveriges berggrund. *Sveriges Geol. Unders., Ser. Ba,* No. 16.

Malushitsky, Y. N. (1975). "The Stability of Slopes and Embankments" (Centrifugal Model Testing), Budivelnik Publishing House, Kiev.

Marsh, B. D. and Carmichael, I. S. E. (1974). Benioff zone magmatism. *J. Geophys. Res.* **79**, 1196.

Marsh, B. D. (1975). Plume Spacing and Soúrce. *Nature,* **256**, 240.

McDonald, G. F. J. (1957). Thermodynamics of solids under non-hydrostatic stress, with geologic applications. *Amer. J. Sci.* **255**, 266.

McElhinny, M. W. (1979). "The Earth: Its Origin, Structure and Evolution." Academic Press, London and New York.

McKenzie, D. P. (1977). Surface deformation, gravity anomalies and convection. *Geophys. J. R. Astron. Soc.* **48**, 211.

McKenzie, D. P., Roberts, J. M. and Weiss, N. O. (1974). Convection in the Earth's mantle: toward a numerical simulation. *J. Fluid Mech.* **62**, 465.

McLellan, A. G. (1968). The chemical potential in thermodynamic systems under non-hydrostatic stresses. *Proc. R. Soc. London,* **A,307**, 1.

McLellan, A. G. (1970). Non-hydrostatic thermodynamics of chemical systems. *Proc. R. Soc. London, A.* **314**, 443.

Meissner, R. and Berckhemer, H. (1967). Seismic refraction measurements in the northern Rhine graben. *Abh. geol. Landesamt Baden-Würtemberg,* **6**, 105.

Menard, H. W. (1962). The east Pacific rise. *Sci. Am.* **52**.

Menard, H. W. (1965). The world-wide oceanic rise-ridge system. Symposium on Continental Drift. *Phil. Trans. R. Soc. London. No.* **1088**, 109.

Milnes, A. G. and Pfiffner, U. A. (1980). Tectonic evolution of the Central Alps in the cross section St. Gallen–Como. *Eclogae geol. Helv.* **73/2**, 619.

Mohr, P. A. and Wood, C. A. (1976–1977). Volcano spacings and lithospheric attenuation in the east rift of Africa. *Earth Planet. Letters,* **33**, 126.

Mörner, N-A., Ed. (1980). Earth Rheology, Isostasy and Eustacy. J. Wiley, Chichester.

Mueller, St., Peterschmitt, E., Fuchs, K. and Ansorge, J. (1967). The rift structure of the crust and upper mantle beneath the Rhine graben. *Abh. geol. Landesamt Baden-Würtemberg,* **6**, 108.

Naumann, C. F. (1849). "Lehrbuch der Geognosie", 1st Ed. Wilhelm Engelman, Leipzig.

Naumann, C. F. (1858). "Lehrbuch der Geognosie", 2nd Ed. Wilhelm Engelman, Leipzig.

Nettleton, L. L. (1934). Fluid mechanics of salt domes. *Bull. Am. Ass. Pet. Geol.* **18**, 1175.

Nettleton, L. L. (1943). Recent experimental and geophysical evidence of mechanics of salt-dome formation. *Bull. Am. Ass. Pet. Geol.* **27**, 51.

Nettleton, L. L. (1955). History of concepts of Gulf Coast salt-dome formation. *Bull. Am. Ass. Pet. Geol.* **39**, 2373.

Nettleton, L. L. and Elkins, T. A. (1947). Geologic models made from granular materials. *Trans. Am. Geophys. Union,* **28**, 451.

Newall, G. and Rast, N. (Eds) (1970). "Mechanism of Igneous Intrustions". Gallery Press, Liverpool.

Nicholson, R. and Rutland R. W. R. (1969). A section across the Norwegian Caledonides: Bodö to Sulitjelma. *Norg. Geol. Unders.* **269**.

Nuttli, O. (1963). Seismological evidence pertaining to the structure of the earth's upper mantle. *Rev. Geophys.* **1**, 351.

Nye, J. F. (1952). The mechanism of glacial flow. *J. Glac.* **7**, 59.

O'Connell, R. J. (1971). Pleistocene glaciation and the viscosity of the lower mantle. *Geophys. J. R. Astron. Soc.* **23**, 299.

Odé, M. (1966). Gravitational instability of a multilayered system of high viscosity. *Verh. K. Ned. Akad. Wet., Reeks,* **1**, No. *24*.

Oftedahl, C. (1943). Om sparagmiten og dens skyvning innom kartbladet Övre Rendal. *Norg. Geol. Unders.* **161**.

Oftedahl, C. (1945). Om tilliten i det centrale norske sparagmitområde. *Nor. Geol. Tidsskr.* **25**, 295.

Oftedahl, C. (1948). Deformation of quartz conglomerate in central Norway. *J. Geol.* **56**, 476.

Oftedahl, C. (1949). Skyvedekker i det centrale Norges sparagmitformasjon. *Nor. Geol. Tidsskr.* **27**, 164.

Oftedahl, C. (1956). Om Grong kulminasjonen og Grongfeltets skyvdekker. *Norg. Geol. Unders.* **195**, 57.

Osgood, W. F. (1958). "Advanced Calculus". MacMillan, New York.
Panek, L. A. (1952). Centrifugal testing apparatus for mine structure stress analysis. *U.S. Bureau of Mines,* **RI**, 4883.
Parker, T. J. and McDowell, A. N. (1955). Model studies of salt-dome tectonics. *Bull. Am. Ass. Pet. Geol.* **39**, 2384.
Parmentier, E. M. and Turcotte, D. L. (1975). Numerical experiments on the structure of mantle plumes. *J. Geophys. Res.* **80**, 4417.
Parmentier, E. M., Torrance, K. E. and Turcotte, D. L. (1976). Studies of finite amplitude non-newtonian thermal convection with application to convection in the earth's mantle. *J. Geophys. Res.* **81**, 1839.
Paterson, M. S. (1973). Nonhydrostatic thermodynamics and its geological applications. *Rev. of Geophys. and Space Phys.* **11**, (2), 355.
Paterson, W. S. B. (1969). "The Physics of Glaciers". Pergamon Press, Oxford.
Peacy, J. S. (1964). Reconnaissance of the Tömmerås anticline. *Norg. Geol. Unders.* **227**, 13.
Peltier, W. R. (1976). Glacial-isostatic adjustment I. *Geophys. J. R. Astron. Soc.* **46**, 605.
Peltier, W. R. and Andrews, J. T. (1976). Glacial-isostatic adjustment II. *Geophys. J. R. Astron. Soc.* **46**, 609.
Petraske, A. K., Hodge, D. S. and Shaw, R. (1978). Mechanics of emplacement of basic intrusions. *Tectonophysics,* **46**, 41.
Pidgeon, R. T. and Råheim, A. (1972). Geochronological investigation of the geneisses and minor intrusive rocks from Kristiansund, W. Norway. *Nor. Geol. Tidsskr.* **52**, 241.
Pilger, A. and Rösler, A. (Eds) (1975). "Afar depression of Ethiopia". Schweizerbart, Stuttgart.
Pitcher, W. S. and Berger, A. R. (1972). "The Geology of Donegal: A Study of Granite Emplacement and Unroofing". Wiley-Interscience, New York.
Playfair, J. (1964). "Illustration of the Huttonian Theory of the Earth". Dover, New York.
Pokrovsky, G. I. and Fyodorov, I. S. (1936). Studies of soil pressures and soil deformations by means of a centrifuge. *Proc. 1st Int. Conf. on Soil Mech.* **1**, 70.
Porter, A. W. (1958). "The Method of Dimensions". Methuen, London.
Powell, Marjorie (1978). Crystallization conditions of low-pressure cumulate modules from the lesser Antilles Island arc. *Earth Planet. Sci. Letters,* **39**, 162.
Prandtl, O. G. and Tietjens, L. (1934). "Applied Hydro- and Aeromechanics". Dover, New York.
Price, R. A. (1973). Large-scale gravitational flow of supracrustal rocks, Southern Canadian Rockies, *In*: "Gravity and Tectonics" (K. A. De Jong and R. Scholten, Eds). Wiley, New York.
Priem, H. N. A., Boelrijk, N. A. I. M., Hebeda, E. H., Verdurmen, E. A. Th. and Verschure, R. H. (1975). Isotopic dating of the calendonian bindal and svenningdal granitic massifs, Central Norway. *Norges Geol. Unders. Bull.* **32**, No. **319**, 29.
Press, F. (1970). Earth models consistent with geophysical data. *Phys. Earth Planet. Interiors,* **3**, 3.
Rambach, J.-M. and Deramond, J. (1979). Constant thickness overthrust on a visco-plastic sole. *Tectonophysics,* **60**, Tl.
Ramberg, H. (1945). The thermodynamics of the earth's crust II. *Nor. Geol. Tidsskr.* **25**, 307.

Ramberg, H. (1955). Natural and experimental boundinage and pinch-and-swell structures. *J. Geol.* **63**, 512.

Ramberg, H. (1960). The Gibbs free energy of crystals under stress. *Annais da Escola de Minas de Ouro Preto, Brazil.*

Ramberg, H. (1963a). Experimental study of gravity tectonics by means of centrifuged models. *Bull. Geol. Inst. Univ. Upsala,* **42**, 1.

Ramberg, H. (1963b). Fluid dynamics of viscous buckling applicable to folding of layered rocks. *Bull. Am. Assoc. Petrol. Geologists,* **47**, 484.

Ramberg, H. (1964a). Strain distribution and geometry of folds. *Bull. Geol. Inst. Univ. Upsala,* **42**, 1.

Ramberg, H. (1964b). Selective buckling of composite layers with contrasted rheological properties. *Tectonophysics,* **1**, 307.

Ramberg, H. (1964c). A model for the evolution of continents, oceans and orogens. *Tectonophysics,* **2**, 159.

Ramberg, H. (1965a). The effect of gravity on folding and other deformations in the earth's crust. *Bergmekanik, IVA Medd.* **142**, 40.

Ramberg, H. (1965b). Experimental models of fold mountains. *Geol. Foren. Stockholm Forh.* **87**, 484.

Ramberg, H. (1966). The Scandinavian Caledonides as studied by centrifuged dynamic models. *Bull. Geol. Inst. Univ. Upsala,* **43**, 1.

Ramberg, H. (1967a). Model experimentation of the effect of gravity on tectonic processes. *Geophys, J. R. astr. Soc.* **13**.

Ramberg, H. (1967b). "Gravity, Deformation and the Earths Crust". 1st edition, Academic Press, London, New York.

Ramberg, H. (1968a). Fluid dynamics of layered systems in the field of gravity. *Phys. Earth Planet. Interiors*, **1**, 63.

Ramberg, H. (1968b). Instability of layered systems in the field of gravity I and II. *Phys. Earth Planet. Interiors,* **1**, 427.

Ramberg, H. (1970a). Folding of laterally compressed multilayers in the field of gravity, 1, *Phys. Earth Planet. Interiors,* **2**, 20.

Ramberg, H. (1970b). Folding of laterally compressed multilayers in the field of gravity, 2, Numerical examples. *Phys. Earth Planet. Interiors,* **4**, 83.

Ramberg, H. (1972a). Theoretical models of density stratification and diapirism in the Earth. *J. Geophys. Res.*, **77**, 877.

Ramberg, H. (1972b). Mantle diapirism and its tectonic and magmagenetic consequences. *Phys. Earth Planet. Interiors*, **5**, 45.

Ramberg, H. (1973). Model studies of gravity-controlled tectonics by the centrifuge technique: chapter in *Gravity and Tectonics,* 49. J. Wiley–Interscience, New York.

Ramberg, H. (1977a). Comments on D. G. Gee: A tectonic model for the central part of the Scandinavian Caledonides. *Am. J. Sci.* **277**, 647.

Ramberg, H. (1977b). Some remarks on the mechanism of nappe movement. *Geo. För. i Stockholm Förh.* **99**, 110.

Ramberg, H. (1979). Folding of a single viscous layer—Discussion. *Tectonophysics,* **56**, 321.

Ramberg, H. (1980). Diapirism and gravity collapse in the Scandinavian Caledonides. *J. Geol. Soc. London,* **137** (in press).

Ramberg, H. and Ghosh, S. K. (1968). Deformation structures in the Hovin-group schists in the Hommelvik–Hell region. *Tectonophysics*, **6**, 311.

Ramberg, H. and Sjöström, H. (1973). Experimental geodynamical models relating to continental drift and orogenesis. *Tectonophysics,* **19**, 105.

Ramberg, H. and Stephansson, O. (1964). Compression of floating elastic and viscous plates affected by gravity, etc. *Tectonophysics,* **1**, 101.

Ramberg, H. and Stephansson, O. (1965). Note on centrifuged models of excavations in rocks. *Tectonophysics,* **2**, 281.
Ramberg, H. and Strömgard, K.-E. (1971): Experimental tests of modern buckling theory applied on multilayered media. *Tectonophysics,* **11**, 461.
Ramberg, I. B. and Neumann, E.-R. (1978). "Tectonics and Geophysics of Continental Rifts". Reidel, Dordrecht, Holland.
Rayleigh, Lord (1900). Investigation of the character of the equilibrium of an incompressible heavy fluid of variable density. *Scientific Papers, ii,* 200, Cambridge, England.
Reade, T. M. (1908). Mechanics of overthrusts. *Geol. Mag.* **5**, 518.
Reade, T. M. (1909). Mechanics of overthrusts. *Geol. Mag.* **6**, 75.
Rekstad, J. (1929). Salta. *Norg. Geol. Unders.* **134.**
Reyer, E. (1888). "Theoretische Geologie". Stuttgart.
Richter, F. M. and Parsons, B. (1975). On the interaction of two scales of convection in the mantle. *J. Geophys. Res.* **80**, 2529.
Riecker, R. E. (Ed). (1979). "Rio Grande Rift". Am. Geoph. Union, Washington.
Ringwood, A. E. (1970). Phase transformations and the constitution of the mantle. *Phys. Earth Planet. Interiors* **3**, 109.
Ringwood, A. E. (1972). *In*: "The Nature of the Solid Earth" p. 86, (E. C. Robertson, Ed.), McGraw–Hill, New York.
Ringwood, A. E. and Major, A. (1970). The system Mg_2SiO_4 at high pressures and temperatures. *Phys. Earth Planet. Interiors,* **3**, 89.
Robertson, E. C. (Ed.) (1972). "The Nature of the Solid Earth". McGraw–Hill, New York.
Rodgers, J. (1970). "The Tectonics of the Appalachians". Wiley–Interscience, New York.
Roeder, P. L. and Dixon, J. M. (1977). A centrifuge furnace for separating phases at high temperature in experimental petrology. *Can. J. Earth Sci.* **14**, 1077.
Runcorn, S. K. (1965). Changes in the convection pattern in the Earth's mantle and continental drift: evidence for a cold origin of the Earth. *Phil. Trans. R. Soc.* **258**, 228.
Rutland, R. W. R. and Nicholson, R. (1965). Tectonics of the Caledonides of a part of Nordland, Norway. *J. Geol. Soc. London,* **121**, 73.
Råheim, A. (1977). A Rb/Sr study of the rocks of the Surnadal Syncline. *Nor. Geol. Tidsskr.* **51**, 193.
Saklani, P. S. (1978). "Tectonic Geology of the Himalayas." Today and Tomorrow's Printers and Publishers, New Delhi.
Sanneman, D. (1968). *In*: "Diapirism and Diapirs" (J. Braunstein and G. D. O'Brien, Eds). *Am. Ass. Pet. Geol. Mem.* **8**.
Saxena, S. K. and Bhattacharji (1977). "Energetics of Geological Processes". Springer–Verlag, New York.
Schardt, H. (1893). Sur l'origine des Préalpes romandes (Zone du Chablais et du Stockhorn). *Ecologae Geol. Helv.* **4**, 129.
Schmid, S. M. (1975). The Glarus overthrust: field evidence and mechanical model. *Ecologae Geol. Helv.* **68**, 247.
Schrijver, K. (1975). Deformed root of a composite diapir in granulite facies. *Geotektonische Forschungen,* **49**, 1.
Schwerdtner, W. M., Sutcliffe, R. H. and Troëng, B. (1978). Patterns of local strain in the crestal region of immature diapirs. *Canadian J. Earth Sci.* **15**, 1437.

Seidel, E. (1913). Beiträgezur Morphologie und Genesis der permischen salzlagerstätten Mitteldeutschland. *Zeits. Dent. Geol. Ges.,* **124**.

Selig, F. (1965). A theoretical prediction of salt-dome patterns. *Geophysics.* **30**, 633.

Shumskii, P. A. (1964). "Principles of Structural Glaciology". Dover, New York.

Skjeseth, S. and Sörensen, H. (1953). An example of granitization in the central zone of the Caledonides of Northern Norway. *Norg. Geol. Unders.* **184**, 154.

Smith, R. B. (1975). Unified theory of the onset of folding, boudinage and mullion structure. *Geol. Soc. Am. Bull.* **86**, 1601.

Smith, R. B. (1979). The folding of a strongly non-Newtonian layer. *Am. J. Sci.* **279**, 272.

Smithson, S. B. (1971). Densities of Metamorphic Rocks. *Geophysics,* **36**, 690.

Smith, C. H. and Kopp, H. E. (1963). The Moskox intrusion, a newly discovered layered intrusion, etc. *Am. Min. Spec. Paper* **1**, 30.

Smoluchowski, M. (1909). Über ein gewisses Stabilitätsproblem der Elastizitätslehre und dessen Beziehung zur Entstehung von Faltengebirgen. *Abhandl. Akad. Wiss. Krakau, Math. Kl.* 3.

Smoluchowski, M. (1910). Versuche über Faltungsercheinungen schwimmender elastischer Platten. *Abhandl, Akad. Wiss. Krakau, Math. Kl.* 727.

Spencer, A. M. (Ed.) (1974). "Mesozoic–Cenozoic Orogenic Belts". Scottish Academic Press, Edinburgh.

Stille, H. (1940). "Einführung in der Bau Nordamerikas". Borntraeger, Berlin.

Strand, T. (1945). Structural petrology of the Bygdin conglomerate. *Nor. Geol. Tidsskr.* **24**, 14.

Strand, T. (1961). The Scandinavian Caledonides, a review. *Am. J. Sci.* **259**, 161.

Strand, T. and Kulling, O. (1972). "Scandinavian Caledonides". Wiley–Interscience, New York.

Strand, T. and Holmsen, P. (1960). Stratigraphy, petrology and Caledonian nappe tectonics of central southern Norway: Caledonized basal gneisses in a northwestern area. *Norg. Geol. Unders.* **212**, 1.

Strömgård, K.-E. (1973). Stress distribution during formation of boudinage and pressure shadows. *Tectonophysics,* **16**, 215.

Sycheva-Mikhailova, A. M. (1970). Modelling of the formation process of diapiric domes and asthenoliths with the help of a centrifuge. *Geotectonica, I,* 30.

Sycheva-Mikhailova, A. M. (1973). A mechanism of tectonic processes under conditions of density inversion in rocks. Publ. office "Nedra", Moscow

Talbot, C. J. (1977). Inclined and asymmetric upward-moving gravity structures. *Tectonophysics,* **42**, 159.

Talwani, M., Le Pichon, X. and Ewing, M. (1965). Crustal structure of the mid-oceanic ridges. *J. Geophys. Res.* **70**, 341.

Tarakanov, R. Z. and Leviy, N. V. (1968). A model for the upper mantle with several channels of low velocity and strength. *In*: "The Crust and Upper Mantle of the Pacific Area". (L. Knopoff, C. L. Drake and P. J. Hart, Eds). *Am. Geophys. Unim,* Geophys. Monogr., *12.*

Taylor, Sir Geoffrey (1950). The instability of liquid surfaces when accelerated in a direction perpendicular to their planes, I. *Proc. R. Soc. London, A* **201**, 192.

Toksöz, M. N. (1978). Numerical modelling in geodynamics. *Tectonophysics,* **50**.

Tozer, D. C. (1967). Towards a theory of thermal convection in the mantle. *In*: "The Earth's Mantle" (T. F. Gaskell, Ed.). Academic Press, New York and London.

Trusheim, F. (1957). Über Halokinese und ihre Bedeutung für die strukturelle Entwicklung Norddeutschland. *Z. Deut. Geol. Ges.* **109**, 111.

Trusheim, F. (1960). Mechanism of salt migration in northern Germany. *Bull. Am. Ass. Pet. Geol.* **44**, 1579.

Turcotte, D. L. and Oxburgh, E. R. (1972). Mantle convection and the new global tectonics. *Annu. Rev. Fluid Mech.* **4**, 33.

Törnebohm, A. E. (1888). Om fjällproblemet. *Geol. Fören. Stockholm Förh.* **10**, 328.

Törnebohm, A. E. (1896). Grunddragen av det centrala Skandinaviens bergbyggnad. *K. Sven. Vetenskapakad. Handl.* **28**, 5.

Uyeda, S., Murphy, R. W. and Kubayaski, K. (1979). "Geodynamics of the Western Pacific." Center of Academic Publications, Tokyo.

Verhoogen, J. (1951). The chemical potential of a stressed solid. *Trans. Am. Geophys. Union,* **32**, 251.

Verhoogen, J. (1965). Phase changes and convection in the Earth's mantle. *Phil. Trans. R. Soc. London. Ser. A* **258**, 276.

Vinogradov, A. P. and Udintsev, G. B. (Eds). (1975). "Rift Zones of the World Ocean." J. Wiley, Chichester.

Voight, B. (Ed.) (1976). "Mechanics of thrust faults and decollement", *Benchmark Papers in Geology 32.* Dowden, Hutchinson and Ross, Stroudsburg, U.S.A.

Wegmann, C. E. (1930). Über Diapirismus. *Bull Comm. Geol. Finland,* **92**, 58.

Wegmann, C. E. (1935). Deutung der Migmatite. *Geol. Rundschau,* **96**, 305.

Wenk, H. R. and Wenk, E. (1969). Physical constants of alpine rocks. *Beitr. Geol. Schweiz, Kleinere Mitt.* **45**, 343.

Weertman, J. (1957). Deformation of floating ice shelves. *J. Glaciol.* **3**, 38.

Weertman, J. (1962). Mechanism for Continental Drift. *J. Geophys. Res.* **67**, 1133.

Weertman, J. (1979). Height of mountains on Venus and the creep properties of rocks. *Phys. Earth Planet. Interiors,* **19**, 197.

Whitehead, J. A., Jr. and Luther, D. S. (1975). Dynamics of laboratory diapirs and plume models. *J. Geophys. Res.* **80**, 705.

Williams, P. E. and Zwart, H. J. (1977). *In*: "Energetics of Geological Processes" (S. K. Saxena and S. Bhattacharji, Eds). Springer-Verlag, New York.

Willies, B. (1893). Mechanics of Appalachian structure. *13th Ann. Rept. U.S. Geol. Surv.* 211.

Wilson, M. R. (1973). "Some Age and Isotope Studies in the Scandinavian Caledonides." Report from the Z.W.O. Laboratory for Isotopen-Geologie, Amsterdam.

Wiltschko, D. V. (1979). Partitioning of energy in a thrust sheet and implication concerning driving forces. *J. Geophys. Res.* **84**, 6050.

Windley, B. F. (1977). "The Evolving Continents." J. Wiley, Chichester.

Woidt, W. D. (1978). Finite element calculations applied to salt-dome analysis. *Tectonophysics,* **50**, 369.

Zachrisson, E. (1964). The Remdalen Syncline. *Sveriges Geol. Unders. C* **596**.

Zeil, W. (1979). "The Andes." Gebrüder Borntraeger, Berlin.

Zen, E-an, White, W. S., Hadley, J. B. and Thompson, J. B., Jr. (1968). "Studies of Appalachian Geology: Northern and Maritime", Wiley–Interscience, New York.

Zwart, H. J. (1974). Structure and metamorphism in the Seve-Köli nappe complex...*Centenaire de la Société Géologique de Belgique, Géologie des domaines cristallins,* Liège.

Subject Index